W0106566

D. Schomburg · M. Salzmann (Eds.)
GBF – Gesellschaft für Biotechnologische Forschung

Enzyme Handbook
4

Class 3: Hydrolases

Springer-Verlag
Berlin Heidelberg GmbH

Professor Dr. Dietmar Schomburg
Margit Salzmann
GBF – Gesellschaft für Biotechnologische Forschung mbH
Mascheroder Weg 1
W-3300 Braunschweig, FRG

This collection of datasheets was generated from the database „BRENDA"

ISBN 978-3-642-48986-0   ISBN 978-3-642-84437-9 (eBook)
DOI 10.1007/978-3-642-84437-9

This work is subject to copyright. All rights are reserved, whether the whole or part of the material is concerned, specifically the rights of translation, reprinting, reuse of illustrations, recitation, broadcasting, reproduction on microfilms or in other ways, and storage in data banks. Duplication of this publication or parts thereof ist only permitted under the provisions of the German Copyright Law of September 9, 1965, in its current version, and a copyright fee must always be paid. Violations fall under the prosecution act of the German Copyright Law.

© Springer-Verlag Berlin Heidelberg 1991

Originally published by Springer-Verlag Berlin Heidelberg New York in 1991

Softcover reprint of the hardcover 1st edition 1991

The use of registered names, trademarks, etc. in this publication does not imply, even in the absence of a specific statement, that such names are exempt from the relevant protective laws and regulations and therefore free for general use.

The publisher cannot assume any legal responsibility for given data, especially as far as directions for the use and the handling of chemicals and biological materials are concerned. This information can be obtained from the instructions on safe laboratory practice and from the manufacturers of chemicals and laboratory equipment.

Media conversion: Brühlsche Universitätsdruckerei, Giessen
Production of the plasticfiles: Lux-Plastik oHG, Murnau
51/3130-543210 – Printed on acid-free paper

Preface

Recent progress in enzyme immobilisation, enzyme production, coenzyme regeneration and enzyme engineering has opened up fascinating new fields for the potential application of enzymes in a large range of different areas. As more progress in research and application of enzymes has been made the more apparent has become the lack of an up-to-date overview of enzyme molecular properties. The need for such a data bank was also expressed by the EC-task force "Biotechnology and Information". Therefore we started the development of an enzyme data information system as part of protein-design activities at GBF. The present book "Enzyme Handbook" represents the printed version of this data bank. In future it is also planned to make a computer searchable version available.

The enzymes in the Handbook are arranged according to the 1984 Enzyme Commission list of enzymes and later supplements. Some 3000 "different" enzymes are covered. Frequently very different enzymes are included under the same E. C. number. Although we intended to give a representative overview on the molecular variability of each enzyme, the Handbook is not a compendium. The reader will have to go to the primary literature for more detailed information. Naturally it is not possible to cover all numerous, up to 40 000, literature references for each enzyme if data representation is to be concise as is intended.

The authors are grateful to the following bioligists and chemists for invaluable help in the compilation of data, expecially Cornelia Munaretto, Dr. Ida Schomburg, Dr. Sabine Vogel-Ziebolz, Uwe Hirschgänger, Inka Siegmund and Roland Vogt. Mrs. C. Munaretto and Dr. I. Schomburg are also thanked for the correction of the final manuscript.

Braunschweig   Margit Salzmann
June, 1990   Dietmar Schomburg

BRENDA – Enzyme Data for Research and Production

Enzymes are used in all parts of the living world for catalysis of innumerable biochemical reactions. It has been known for some time that they represent potentially highly interesting catalysts for chemical research and production because of their high efficiency, stereospecifity, and their regio- and enantioselectivity. Enormous progress has been made in recent years in enzyme immobilisation, stabilisation, coenzyme regeneration etc., while gene technology has made possible the production of large quantities of otherwise inaccessible enzymes. Enzyme-design methods and recent work on enzyme behaviour in organic solvents have opened further possibilities for their use in new application areas. In addition to the chemical industry their use in the food industry and environmental technology is worthy of mention.

A number of problems still have to be overcome before enzymes take their place besides the other commonly used catalysts in the chemical laboratory, productions and in the awareness of the chemist and the general public. So the current industrial use of enzymes is more or less limited to proteases and carbohydrases, i. e. hydrolytic enzymes. Other synthetically important reactions such as the forming of C-C bonds are rarely achieved enzymatically at present. In addition to the real problems (price, too high selectivity etc.) reservations on the part of the chemists prohibit decisions about their potential use.

This is mainly caused by the undeniable fact that information about enzymes is not as easily obtained as that about other synthetic means and catalysts. Results of enzyme research are often published in journals which are rarely read by most chemists and are often not available in organic-chemistry libraries. The published data on molecular weight, stability etc. can be contradictory, the use of a number of different names for the same enzyme is common, and the systematic classification of the enzyme is usually unsatisfactory.

The apparently simple question:

"Is there an enzyme, that catalyzes the enantioselective replacement of an hydrogen atom in an α-position to an aromatic ring and is stable in a water solution with a certain pH-value and a given temperature?"

can often only be answered after intensive work in a library or the use literature data bases. A rational choice between several potential enzyme catalyzed reaction paths is imposible in a reasonable time. When planning research projects in the area of enzyme design, information about potential enzymes and use of the correct initial enzyme is also essential.

In addition to the above types of chemical problem, systematic biochemical investigations are central to the wide interest in enzymes, therefore a great deal of information can be gained from establishment of a comprehensive collection of enzyme data.

In the GBF research programme, enzyme technology has always played a significant role. This is documented by innovative contributions in the field of cofactor regeneration, enzyme production by genetic engineering and down-stream processing after fermentation.

The area has gained new significance from the activities in protein design where the central theme is protein structure determination and biocomputing.

It soon became evident that the next logical step was the development of an enzyme data bank which could be undertaken by the group for molecular structure research. We hope that publication of this comprehensive, critical literature evaluation will be of wide interest to external users and, in this way, we hope the GBF has provided a further important contribution to the in-formation infrastructure in biotechnology.

Braunschweig
June, 1990

Joachim Klein
GBF, Scientific Director

List of Abbreviations

A	adenosine	*E. coli*	*Escherichia coli*
Ac	acetyl	EDTA	ethylene diaminetetraacetate
ADP	adenosine 5'-diphosphate	EGTA	ethylene glycol bis (β-aminoethyl ether) tetraacetate
Ala	alanine		
All	allose		
Alt	altrose	ER	endoplasmic reticulum
AMP	adenosine 5'-monophosphate	Et	ethyl
		EXAFS	extended x-ray absorption fine structure
Ara	arabinose		
Arg	arginine		
Asn	asparagine	FAD	flavin-adenine dinucleotide
Asp	aspartic acid		
ATP	adenosine 5'-triphosphate	FMN	flavin mononucleotide (riboflavin 5'-monophosphate)
c	cytidine		
cal	calories	Fru	fructose
CDP	cytidine 5'-diphosphate	Fuc	fucose
CDTA	trans-1,2-diaminocyclohexane-N,N,N,N-tetra-acetic acid	G	guanosine
		Gal	galactose
		GDP	guanosine 5'-diphosphate
CMP	cytidine 5'-monophosphate	Glc	glucose
		GlcN	glucosamine
CoA	coenzyme A	GlcNAc	N-acetylglucosamine
CTP	cytidine 5'-triphosphate	Gln	glutamine
Cys	cysteine	Glu	glutamic acid
d	deoxy-	Gly	glycine
D- and L-	prefixes indicating configuration	GMP	guanosine 5'-monophosphate
DFP	diisopropyl fluorophosphate	GSH	glutathione
		GTP	guanosine 5'-triphosphate
DNA	deoxyribonucleic acid		
DPN	diphosphopyridinium nucleotide (now NAD)	Gul	gulose
		H_4	tetrahydro
DTNB	5,5'-dithiobis (2-nitrobenzoate)	His	histidine
		HPLC	high pressure liquid chromatography
DTT	dithiothreitol (i.e. Cleland's reagent)	Hyl	hydroxylysine
EC	number of enzyme in Enzyme Commission's system	Hyp	hydroxyproline
		IAA	iodoacetamide
		Ig	immunoglobulin

Ile	isoleucine	nm	nanometre (10^{-9} metre)
Ido	idose	NMN	nicotinamide mononucleotide
IDP	inosine 5'-diphosphate		
IMP	inosine 5'-monophosphate	NMP	nucleoside 5'-monophosphate
ITP	inosine 5'-triphosphate	NTP	nucleoside 5'-triphosphate
K_m	concentration of substrate giving half maximum velocity (Michaelis constant)	o-	ortho-
		Orn	ornithine
		p-	para-
L-	see D-	PCMB	p-chloro-mercuribenzoate
Leu	leucine		
Lys	lysine	PEP	phosphoenolpyruvate
Lyx	lyxose	pH	$-\log_{10}$ [H]
M	gm molecule (1 mole) per litre	Ph	phenyl
		Phe	phenylalanine
m-	meta-	PIXE	proton-induced X-ray emission
Man	mannose		
MES	2-(N-morpholino)ethane sulfonate	PMSF	phenylmethane-sulfonylfluoride
Met	methionine	Pro	proline
mM	10^{-3} Mole	Q_{10}	temperature coefficient for a reaction
Mur	muramic acid		
MW	molecular weight	Rha	rhamnose
NAD	nicotinamide-adenine dinucleotide (state of oxidation unspecified)	Rib	ribose
		RNA	ribonucleic acid
		mRNA	messenger RNA
		rRNA	ribosomal RNA
NAD^+	nicotinamide-adenine dinucleotide (oxidized form)	tRNA	transfer RNA
		Sar	N-methylglycine (sarcosine)
NADH	reduced NAD		
NADP	NAD phosphate (state of oxidation unspecified)	SDS-PAGE	sodium dodecyl sulphate (= sodium lauryl sulphate)-polyacrylamide gel electrophoresis
$NADP^+$	NADP (oxidized form)		
$NAD(P)^+$	indicates either NAD^+ or $NADP^+$		
NADPH	reduced NADP	Ser	serine
NAD(P)H	indicates either NADH or NADPH	T	ribosyl-thymine
		$t_{1/2}$	time for half-completion of reaction
NDP	nucleoside 5'-diphosphate		
		Tal	talose
NEM	N-ethylmaleimide	TDP	ribosylthymine 5'-diphosphate
Neu	Neuraminic acid		

Thr	threonine	u	uridine
TMP	ribosylthymine 5'-monophosphate	U/mg	µmol/(mg*min)
		UDP	uridine 5'-diphosphate
Tos-	tosyl- (p-toluenesulfonyl-)	UMP	uridine 5'-monophosphate
TPN	triphosphopyridinium nucleotide (now NADP)	UTP	uridine 5'-triphosphate
		Val	valine
		Xaa	symbol for an amino acid of unknown constitution in peptide formula
Tris	tris(hydroxymethyl)- aminomethane		
Trp	tryptophan		
TTP	ribosylthymine 5'-triphosphate	XAS	X-ray absorption spectroscopy
Tyr	tyrosine	Xyl	xylose

Index

1 NOMENCLATURE

EC number
  3.2.1.1

Systematic name
  1, 4-Alpha-D-glucan glucanohydrolase

Recommended name
  Alpha-amylase

Synonymes

Glycogenase	Clarase
Endoamylase	Bactosol TK
Taka-amylase A	Spitase CP 1
Amylase, .alpha.-	Takatherm
Buclamase	Pivozin
Fortizyme	Thermolase
Amylopsin	Kleistase L 1
Maxilase	G 995
Amylase THC 250	Fungamyl 800L [9]
Maxamyl	Termamyl [24]
Thermamyl	Amylotherm [5]
Ptyalin	

CAS Reg. No.
  9000-90-2

2 REACTION AND SPECIFICITY

Catalysed reaction
  Polysaccharide containing alpha-(1–4)-linked glucose units $+ H_2O \rightarrow$
  $\rightarrow$ maltooligosaccharides (mechanism [2, 78], endohydrolysis of 1,4-alpha-
  D-glucosidic linkages in polysaccharides containing three or more 1,4-
  alpha-linked D-glucose units)

Reaction type
  O-Glycosyl bond hydrolysis (endohydrolysis)

Natural substrates
  Polysaccharides with alpha-(1–4)-linkages $+ H_2O$

Enzyme Handbook © Springer-Verlag Berlin Heidelberg 1991
Duplication, reproduction and storage in data banks are only
allowed with the prior permission of the publishers

Substrate spectrum
  1 Starch + H_2O (soluble, raw [15, 19, 21], specific for alpha-(1–4) linkages, exception: Thermoactinomyces vulgaris, hydolysis of alpha-(1–4)-and alpha-(1–6)-linkages [36])
  2 Amylose + H_2O [3, 17, 18, 24, 26, 27, 39, 40, 55, 57, 62, 78]
  3 Amylopectin + H_2O [3, 17, 19, 24, 26, 27, 39, 40, 62]
  4 Glycogen + H_2O [17, 19, 24, 27, 39, 40, 55, 61]
  5 Beta-limit dextrin + H_2O [3]
  6 p-Nitrophenylmaltoheptaoside + H_2O [11]
  7 p-Nitrophenyl-alpha-D-maltoside + H_2O [15, 24]
  8 Phenyl-3'-O-methyl-alpha-maltoside + H_2O (and similar substrates) [2]
  9 Isopanose + H_2O [36]
 10 Pullulan + H_2O [38]
 11 More (cereal amylase: fast hydrolysis of high MW substrates, mammalian amylase: fast hydrolysis of large and small MW substrates [1], classification according to saccharifying and liquifying type [2], preference of maltodextrins with small MW [62], no transferase [1]) [1, 2, 62]

Product spectrum
  1 Maltooligosaccharides (glucose to maltoheptaose, depending on organism and reaction time) [1, 3–5, 11, 15, 17, 18, 24, 26, 30, 39, 45, 47, 48, 50, 55, 57, 62, 75, 77]
  2 Maltooligosaccharides
  3 Maltooligosaccharides
  4 Maltooligosaccharides
  5 ?
  6 ?
  7 p-Nitrophenol + maltose [15]
  8 3'-O-Methyl-maltose + phenol [2]
  9 Glucose + maltose + isomaltose (depending on substrate concentration) [36]
 10 Panose [38]
 11 ?

Inhibitor(s)
  Hg^{2+} [1, 12, 15, 19, 27, 29, 30, 37, 39, 43, 45, 55, 62]; Ag^+ [1, 10, 15 , 27, 39, 40, 62, 72]; Cu^{2+} [1, 10, 15, 19, 24, 27, 37, 39, 40, 43, 45, 55 , 61, 62]; Pb^{2+} [1, 15, 29, 62]; Mg^{2+} [10, 19]; Co^{2+} [10, 39, 40, 45]; Ca^{2+} [10]; Cs^{2+} [11]; Rb^{2+} [11]; Fe^{3+} [15]; Zn^{2+} [15, 19, 24, 27, 30, 39, 40, 45, 55, 61]; Al^{3+} [15, 37]; Cd^{2+} [15, 40, 61 , 62]; Fe^{2+} [24, 27, 39, 40, 45, 55, 61, 62]; Ni^{2+} [39]; Ba^{2+} [55]; Iodoacetate (not: hog pancreas) [1, 2, 40, 72]; Phenylmercuric chloride [1]; Ammonium molybdate [1]; Ascorbic acid [1]; Iodoacetamide [2]; 2, 4-Fluorodinitrobenzene [2]; 2, 4-Dinitrobenzene-1-sulfonate [2]; EGTA [3]; Phenylmercuric acetate [4]; p-Choromercuribenzoate [4, 19, 37,

40]; p-Hydroxymercuribenzoate [4]; N-Bromosuccinimide [10]; EDTA [17, 18, 40, 44, 57, 61, 62, 72]; Maltose (non-competetive) [21, 29, 58]; Trestatin C [21]; Alpha-amylase inhibitor protein (from wheat [25, 41, 58], from Streptomyces sp. No. 280 [92]) [25, 41, 58, 92]; Maltotriitol [37]; Panitol [37]; Isopanitol [37]; F^- [59]; Cl^- [72]; N-Ethylmaleimide [72]

Cofactor(s)/prostethic group(s)
Histidine (activation) [43]; Dithiothreitol (activation) [4]; Mercaptoethanol (activation) [4]

Metal compounds/salts
Ca^{2+} (requirement [1, 2, 18, 22, 57, 59, 72], no requirement [29]) [1, 2, 18, 22, 29, 57, 59, 72]; Cl^- (requirement for maximal activity) [1]; Ba^{2+} (stimulating effect) [10]; Mn^{2+} (activity enhancement) [30]; Ag^+ (activity enhancement) [30]; Fe^{2+} (activity enhancement) [30, 43]; Co^{2+} (activity enhancement) [30]; Phosphate (stimulating effect) [40]; MoO_4^{2-} (stimulating effect) [40]; Anions (activation) [40]; Zn^{2+} (activation) [51]

Turnover number (min^{-1})

Specific activity (U/mg)
11018 (1U: micromol maltose produced per min. per mg protein [22], similar values [47]) [22]; More (14.4–1701 , 1U: reducing sugar equivalent to 1 micromole glucose or maltose per min. per mg [3, 13, 38, 38, 47], assay methods [1, 31–35, 42, 64–68]) [1, 3, 10, 12, 13, 16, 22, 28, 29, 31–35 , 37–46, 50, 52, 55–57, 60–69]

K_m-value (mM)
More (0.3–2.5 mg/ml soluble starch [3, 10, 17, 18, 22, 29, 37, 40 , 42, 43, 47, 50, 56, 72], 1.8 mg/ml amylose [40], 1.9 mg/ml amylopectin [40], 2.3 mg/ml glycogen [40], 14 glucosidic bonds at 50°C, 22 glucosidic bonds at 70°C [60]) [3, 10, 17, 18, 22, 29, 37, 40, 42, 43, 47, 50, 56, 59, 60, 62, 72]

pH-optimum
2.0 [47]; 3.5–8.0 (depending on species) [42]; 3.5 (dependency on strain) [24]; 4.0 [26, 44]; 4.0–5.0 [11]; 4.0–7.0 [12]; 4.2 [21]; 4.5 [4, 36]; 4.5–5.5 [43]; 4.5–10.5 [53]; 4.8–5.0 [29]; 4.8–6.0 (depending on temperature [74]) [16, 74]; 5.0–7.5 [17]; 5.0 [38, 75]; 5.0–9.0 [10]; 5.6 (presence of Ca^{2+} or EDTA [7], depending on temperature [18]) [7, 18]; 5.2–5.8 (depending on presence of Ca^{2+}) [72]; 5.4–6.1 [73]; 5.5–6.5 [3, 50, 56]; 5.5 [62]; 5.8 [59]; 5.9 [77]; 6.0 [15]; 6–7 [27, 41, 60]; 6.2 [55]; 6.3 [57]; 6.9 [22]; 7 (animal [1]) [1, 30, 39]; 7–9 [48]; 8.0 (crab [1]) [1, 14 , 45]; 8.5–9.5 (depending on species) [49]; 9 [40]

pH-range
1.0 (70% of maximal activity) [47]; 2.2–5.5 (less than 50% of maximal activity above pH 5.5 and below pH 2.2) [24]; 2.5–6.0 [11]; 2.5–9.5 (depending on temperature) [12]; 3–10 [40]; 4.0–6. 8 [16]; 4–8 (less than 30% of maximal activity

Enzyme Handbook © Springer-Verlag Berlin Heidelberg 1991
Duplication, reproduction and storage in data banks are only allowed with the prior permission of the publishers

below pH 4 and above pH 8) [55]; 4.8–7.5 (less than 50% of maximal activity below pH 4.8 and above pH 7.5) [56, 57]; 5–10 (less than 74% of maximal activity below pH 5 and above pH 10) [30]; 5.5–6.5 [77]; 5.5–8.2 (less than 50% of maximal activity above pH 8.2 and below pH 5.5) [39]

Temperature optimum (°C)
36–40 [72]; 43 [57]; 45 [44, 55]; 45–92 (Bacilli, depending on species) [42]; 50 [27, 29, 39, 59, 62]; 53–58 [43, 73]; 55 [7]; 55–70 [11]; 60 (dependency on pH, presence of Ca^{2+} [7]) [4, 7, 15, 16, 19, 45, 61]; 55–62 [50]; 62 [21]; 65 (second optimum at 45°C for some strains [77]) [56, 77]; 65–73 [74]; 70 (presence of Ca^{2+} [73]) [24, 26, 47, 60, 73]; 80 [10, 12, 30]; 85 (Bacillus coagulans) [49]; 85–90 [18]; 90 [14, 40, 48]; 91 (Bacillus licheniformis CUMC305) [49]

Temperature range (°C)
35–60 (less than 50% of maximal activity above 60°C and below 35°C) [39, 45]; 45–80 [11]; 55 (50% of maximal activity) [40]; 55–95 (less than 50% of maximal activity above 95°C and below 55°C) [24]; 80 (decrease of activity above) [12]; 90 (Bacillus coagulans: no activity above) [49]; 100 (90% of maximal activity [14, 40], 74% of maximal activity [30], 64% of maximal activity [48]) [14, 30, 40, 48]; 110 (Bacillus licheniformis CUMC 305: 45–50% of maximal activity) [49]

---

3 ENZYME STRUCTURE

Molecular weight
61500 (Streptomyces venezuelae, calculation form nucleotide sequence) [8]
52000 (Streptomyces hygroscopius, calculation from nucleotide sequence) [20]
89000–96000 (Clostridium butyricum, SDS-PAGE [19], Bacillus subtilis, SDS-PAGE [41], sedimentation equilibrum centrifugation [51], Bacteroides amylophilus, SDS-PAGE [57], Pseudomonas MSI, SDS-PAGE [62]) [19, 41, 51, 57, 62]
67000–76000 (Bacillus subtilis, SDS-PAGE [15], Lipomyces starkeyi, gel filtration [26], Paecilomyces sp., gel filtration [44], Bacillus circulans, gel filtration [45], Bacillus subtilis, membrane-bound and extracellular form, gel filtration, SDS-PAGE [52], Tenebrio molitor, SDS-PAGE, gel filtration [59]) [15, 26, 44, 45, 52, 59]
41000–63000 (Pisum sativum, gel chromatography, SDS-PAGE [3], Bacillus brevis, SDS-PAGE [10], Bacillus stearothermophilus, non-denaturing gel electrophoresis, SDS-PAGE [12], Saccharomycopsis fibuligera, gel filtration [16], Lysobacter brunescens, glycerol gradient centrifugation [17], Candida antarctica, SDS-PAGE, gel electrophoresis, 6 M guanidine HCl [21], rat pancreas, SDS-PAGE, gel filtration [22], equilibrum centrifugation [70], Aspergillus awamori, SDS-PAGE [29], Bacillus circulans, gel filtration [39],

Bacillus sp 11–1S, SDS-PAGE [47], Bacillus licheniformis, SDS-PAGE [48],
Dioscorea dumetorum, 3 isozymes, gel filtration [50], hog pancreas,
ultracentrifugation [54], Thermomonospora curvata, SDS-PAGE [56],
thermophilic bacterium V-2, SDS-PAGE [60], human pancreas, parotid
gland, various isozymes, SDS-PAGE [69, 71], Aspergillus niger, acid-stable
and acid-unstable form, sedimentation equilibrum centrifugation [76]) [3,
10, 12, 16, 17, 21, 22, 29, 39, 47, 48, 50, 54, 56, 69, 70, 71, 76]
22500–28000 (Lactobacillus cellobiosus, SDS-PAGE, gel filtration [23],
Bacillus subtilis, gel filtration [27], Bacillus licheniformis, SDS-PAGE [40])
[23, 27, 40]
More (overview bacteria [2, 42], overview molds [2]) [1, 2, 42, 82, 83, 87]

Subunits
Monomer (1 × 22500–67000, Pisum sativum [3], Candida antarctica [21],
rat pancreas [22], rat parotid gland [70], Lactobacillus cellobiosus [23],
Bacillus subtilis [52], human pancreas and parotid gland [69, 71], methods:
SDS-PAGE or equilibrum centrifugation) [3, 21–23, 52, 69–71]
Dimer (2 × 48000, Bacillus subtilis, analysis of cyanogen bromide fragments
[51], 2 × 46000, Bacteroides amylophilus) [52, 57]

Glycoprotein/Lipoprotein
Glycoprotein (Taka-amylase A [2], group A of human parotid gland [71])
[2, 11, 13, 21, 71]; Phospholipoprotein (membrane-bound form) [52]; No
glycoprotein [4, 54]

---

4 ISOLATION/PREPARATION

Source organism
Plants [1]; Animals (e.g. mouse [46], rat [22, 70, 86, 88], man [63, 69, 71, 85,
87], pig [54, 84]) [1, 22, 46, 54, 63, 69, 70, 71 , 78, 84–88]; Bacteria [2, 42, 53];
Molds [2]; Pisum sativum [3]; Bacillus caldovelox [4]; Bacillus licheniformis
(44MB82 [5], TCRDC-B13 [14], CUMC305 [40, 49], NCIB6346 [48], CUMC512
[49]) [5, 14, 40, 48, 49, 79]; Bacillus stearothermophilus [12, 73, 74]; Bacillus
brevis [10]; Bacillus subtilis [15, 27, 41, 51, 52]; Bacillus acidocaldarius [24];
Bacillus circulans [39, 45]; Bacillus coagulans [49]; Bacillus sp. 11–1S [47];
Bacillus amyloliquefaciens [77, 90]; Bacillus megaterium [91]; Streptomyces
limosus [6]; Streptomyces venezuelae [8]; Streptomyces hygroscopius [20];
Aspergillus flavus (var. columnaris) [7]; Aspergillus oryzae [9]; Aspergillus
awamori [29]; Aspergillus niger [75, 76]; Talaromyces emersonii [11]; Saccharomycopsis fibuligera [16]; Lysobacter brunescens [17]; Clostridium
thermohydrosulfuricum [18]; Clostridium butyricum T-7 [19]; Candida
antarctica [21]; Lactobacillus cellobiosus (D-39) [23]; Lipomyces starkeyi
(CBS 1809) [26]; Lipomyces kanonenkoae [82]; Hordeum vulgare (barley)
[28]; Thermoactinomyces sp. [30]; Thermoactinomyces vulgaris [36–38];
Actinomyces sp. (overview) [42]; Calvatia gigantea (edible puffball) [43];

Paecilomyces sp. [44]; Dioscorea dumentorum (yam) [50]; Pichia burtonii [55]; Pichia polymorpha [81]; Thermomonospora curvata [56]; Thermomonospora vulgaris [61]; Bacteroides amylophilus [57]; Tenbrio molitor L. (yellow mealworm) [58, 80]; Pseudomonas MSI [62]; Callosobruchus chinensis [72, 89]; Schwanniomyces castelli [83]

Source tissue

Culture supernatant (bacteria, molds, fungi); Almost all mammalian tissues and fluids (esp. saliva [1, 13, 46], pancreas [1, 22, 46, 54, 69, 70], urine [63], serum [63]) [1]; Shoots [3]; Seedlings [3]; Tuber [50]; Larvae [58, 72]; Cell suspension [52]

Localisation in source

Extracellular (bacteria, molds, fungi, exception: Bacillus subtilis, membrane bound form [52], Pichia burtonii, cell wall bound form [55])

Purification

Bacillus subtilis [2, 15, 27, 41, 52]; Pisum sativum [3]; Bacillus caldovelox [4]; Bacillus brevis [10]; Talaromyces emersonii [11]; Bacillus stearothermophilus (from commercial preparation [12]) [12, 73]; Pig (2 forms [13]) [13, 84]; Saccharomycopsis fibuligera [16]; Lysobacter brunescens [17]; Clostridium butyricum [19]; Candida antarctica [21]; Lactobacillus cellobiosus [23]; Rat [22, 86, 88]; Bacillus acidocaldarius [24]; Lipomyces starkeyi [26]; Aspergillus awamori [29]; Thermoactinomyces sp. [30]; Thermoactinomyces vulgaris [38]; Bacillus circulans [39, 45]; Bacillus licheniformis [40, 48]; Calvatia gigantea (partial) [43]; Paecilomyces sp. [44]; Thermomonospora curvata [56]; Bacteroides amylophilus [57]; Tenebrio molitor [80]; Pseudomonas MSI [62]; Man [69, 85, 87]; Aspergillus niger [75]; Bacillus amyloliquefaciens [77]; Callosobruchus chinensis [72, 89]

Crystallization

[1, 2, 23, 54, 75, 77, 85, 87]

Cloned

[6, 8, 20, 28, 42, 46]

Renaturated

[2]

5 STABILITY

pH

1–11 (depending on Bacillus sp.) [42]; 1.8–7.0 (acid stable form) [75]; 3
(50% inactivation) [39]; 3–7 [7]; 3–9 [4]; 3.5 (10–30% activity, depending on
form of enzyme) [50]; 3.5–6.5 [29]; 4.0–5.5 [24]; 4.0–9.0 [44]; 4.0–9.4 [16];
4.0–10.0 [49]; 4.5–7.6 [62]; 4.5–8.0 (depending on strain) [77]; 4.5–10.5 [18];
5–10 [45]; 5.5–8.0 (higher plants and mammals) [1]; 5.5–10 [14]; 5–10.5
(Taka-amylase A) [2]; 5.8–7.5 [57]; 5.8–8.5 [59]; 6–8 [27]; 6.3–7.9 [23];
6.0–9.0 [15, 39]; 6–11 [74]; 6.5–9.0 [72]; 7–9 [40]; 7–10 [48]; 12 (up to,
Bacillus subtilis) [2]

Temperature (°C)

8 (optimal value for stability) [30]; 40 (unstable above) [39]; 45 (unstable
above) [27, 29, 44, 55]; 40–60 [48]; 25 (optimal value for stability, depending
on strain) [77]; 50 (up to, absence of Ca^{2+} [16], up to [23], 3 h, 100% activity [40], below [45]) [16, 23, 40, 45]; 55–60 (inactivation above, all forms of
enzyme) [50]; 60 (100% stable [4], unstable above [15], stable in presence
of Ca^{2+} [16, 38], 1 h or more [40, 49, 61], inactivation [72]) [4, 15, 16, 38, 40,
49, 61, 72]; 65 (at least 2 h [18], 15 min., 100% activity [47]) [18, 47]; 70 (1 h,
60% activity [4], 1 h, 70% activity [40], Bacillus licheniformis: 1 h, 100% activity [49]) [4, 40, 49]; 80 (30 min., 50% activity [10, 12], 45 min. stable [40],
Bacillus coagulans: 20 min., 16% activity [49], 2 h, no loss of activity [60])
[10, 12, 40, 49 , 60]; 90 (20 min., 50% activity [10], 1 h, pH 4.5, 80% activity
[24], 30 min., 20% activity [74]) [10, 24, 74]; 100 (addition of soluble starch,
4 h stable) [40]; More (increase of thermostability by Ca^{2+} [1–3, 7, 12,
14–19, 27, 38, 39, 42, 44, 73], by polyols [9], by substrate [40, 49], by bovine
serum albumin [17, 61, 73], by Na^+ [73], no protection by Ca^{2+} [24, 45, 61,
72]) [3, 7, 9, 16–19, 24, 27, 38–40 , 44, 45, 61, 72, 73]

Oxidation

Organic solvent

General stability information

Ca^{2+} (protection against proteolyis [1], protection against thermal inactivation) [1–3, 7, 12, 14–19, 38, 39, 42, 44, 73]; Urea (2–8 M, no denaturation) [2,
61, 73]; Guanidine hydrochloride (denaturation) [73]; Enzyme-substrate
complex (or enzyme-product complex: stabilization of molecular conformation, prevention of formation of aggregates induced by heat denaturation)
[2]; Polyols (e. g. sorbitol, stabilization) [9]; Carbohydrates (stabilization)
[12]; Na^+ (stabilization) [14]; Resistant to proteolysis [3]

Storage

–5°C [7]; 0°C or –20°C, 25 mM Tris-HCl, pH 7.5, 100 mM KCl, at least 9 days
[17]; 4°C or –20°C, parotid gland: several months, pancreas: several days
[69]; –15°C, 1 mM phosphate buffer, pH 7.3, 30 mM $CaCl_2$ [73]

Enzyme Handbook © Springer-Verlag Berlin Heidelberg 1991
Duplication, reproduction and storage in data banks are only
allowed with the prior permission of the publishers

6 CROSSREFERENCES TO STRUCTURE DATABANKS

PIR/MIPS code

ALHUS (precursor, salivary, human); ALHUP (precursor, pancreatic, human); ALMSS (precursor, salivary and hepatic, mouse); ALMSP (A2, precursor, pancreatic, mouse); ALMSP1 (A1, precursor, pancreatic, mouse); ALMSPC (BC, precursor, pancreatic, mouse); ALMSPA (BA, precursor, pancreatic, mouse); ALRTP (precursor, pancreatic rat, fragment); ALPGP (pancreatic, pig); ALBS (precursor, Bacillus subtilis); ALBSNA (precursor, Bacillus subtilis strain NA64); ALBSN7 (precursor, Bacillus subtilis strain N7); ALBSN (precursor, Bacillus amyloliquefaciens); ALBSL (precursor, Bacillus licheniformis); ALBSF (precursor, Bacillus stearothermophilus); ALBSK (precursor, Bacillus circulans); ALBH (precursor, barley); ALBHB (B, precursor, barley); A21663 (Bacillus licheniformis, fragment); A34648 (precursor, Bacillus sp. strain B1018); A35282 (Aspergillus niger); JS0101 (Streptomyces violaceus); S00064 (precursor, yeast, Saccharomycopsis fibuligera); S05486 (2, precursor, wheat, fragment); S05487 (2, precursor, wheat, fragment); S05488 (2, precursor, wheat, fragment); S05489 (2, precursor, wheat, fragment); S05490 (2, precursor, wheat, fragment); S06115 (Schwanniomyces occidentalis); S06357 (precursor, wheat); S10013 (1, precursor, clone lambda-OSg2, rice); S01312 (B, Dictyoglomus thermophilum); S01313 (C, Dictyoglomus thermophilum); A19506 (Bacillus amyloliquefaciens, fragments); B24549 (Bacillus licheniformis); A26151 (Bacillus licheniformis, fragment); S01031 (precursor, Bacillus megaterium); A24549 (Bacillus stearothermophilus); A24436 (precursor, version 2, Bacillus stearothermophilus); S02098 (Bacillus subtilis); B28391 (precursor, Streptomyces albidoflavus); A21826 (clone 103, barley, fragment); C21826 (clone 168, barley, fragment); B21826 (clone 96, barley, fragment); S06275 (1, precursor, clone p141.117, barley); A24457 (2, barley, fragments); S07040 (2, precursor, clone p155.3, barley); A26267 (B, precursor, barley, fragment); A30759 (B, precursor, 1, barley); B30759 (B,precursor 2, barley); B31960 (precursor, 46, barley, fragment); A31960 (precursor, 6–4, barley); A10627 (A, Aspergillus oryzae, fragments); A29347 (red flour, beetle, fragment); A25529 (precursor, fruit fly); A28293 (2A, precursor, pancreatic, human, fragment); B28293 (2B, precursor, pancreatic, human, fragment); JS0165 (precursor, carcinoid, human); A29614 (pancreatic, human); A17230 (pancreatic, version 2 pig); A18952 (precursor, pancreatic, mouse, fragments); A14059 (precursor, salivary, rat, fragment)

Brookhaven code

2TAA (Aspergillus oryzae)

7 LITERATURE REFERENCES

[1] Thoma, J.A., Spradlin, J.E., Dygert, S. in "The Enzymes", 3rd. Ed. (Boyer, P.D., Ed.) 5, 115–189 (1971) (Review)
[2] Takagi, T., Toda, H., Isemura, T. in "The Enzymes", 3 rd. Ed. (Boyer, P.D., Ed.) 5, 235–271 (1971) (Review)
[3] Beers, E.P., Duke, S.H.: Plant Physiol., 92, 1154–1163 (1990)
[4] Bealin-Kelly, F.J., Kelly, C.T., Fogarty, W.M.: Biochem. Soc. Trans., 18, 310–311 (1990)
[5] Dobreva, E., Ivanova, V.: Acta Biotechnol., 9, 549–554 (1989)
[6] Virolle, M.-J., Bibb, M.J.: Mol. Microbiol., 2, 197–208 (1988)
[7] Ali, F.S., Abdel-Moneim, A.A.: Zentralbl. Mikrobiol., 144, 615–621 (1989)
[8] Virolle, M.-J., Long, C.M., Chang, S., Bibb, M.J.: Gene, 74, 321–334 (1988)
[9] Graber, M., Combes, D.: Enzyme Microb. Technol., 11, 673–677 (1989)
[10] Tsvetkov, V.T., Emanuilova, E.I.: Appl. Microbiol. Biotechnol., 31, 246–248 (1989)
[11] Bunni, L., McHale, L., McHale, A.P.: Enzyme Microb. Technol., 11, 370–375 (1989)
[12] Brumm, P.J., Hebeda, R.E., Teague, W.M.: Food Biotechnol., 2, 67–80 (1988)
[13] Furuichi, Y., Takahashi, T.: Agric. Biol. Chem., 53, 293–294 (1989)
[14] Bajpai, P., Bajpai, P.K.: Biotechnol. Bioeng., 33, 72–78 (1989)
[15] Hayashida, S., Teramoto, Y., Inoue, T.: Appl. Environ. Microbiol., 54, 1516–1522 (1988)
[16] Gogoi, B.K., Bezbaruah, R.L., Pillai, K.R., Baruah, J.N.: J. Appl. Bacteriol., 63, 373–379 (1987)
[17] Von Tigerstrom, R.G, Stelmaschuk, S.: J. Gen. Microbiol., 133, 3437–3443 (1987)
[18] Melasniemi, H.: Biochem. J., 246, 193–197 (1987)
[19] Tanaka, T., Ishimoto, E., Shimomura, Y., Taniguchi, M., Oi, S.: Agric. Biol. Chem., 51, 399–405 (1987)
[20] Hoshiko, S., Makabe, O., Nojiri, C., Katsumata, K., Satoh, E., Nagaoka, K.: J. Bacteriol., 169, 1029–1036 (1987)
[21] De Mot, R., Verachtert, H.: Eur. J. Biochem., 164, 643–654 (1987)
[22] Reddy, M.K., Heda, G.D., Reddy, J.K.: Biochem. J., 242, 681–687 (1987)
[23] Sen, S., Chakrabarty, S.L.: J. Appl. Bacteriol., 60, 419–423 (1986)
[24] Kanno, M.: Agric. Biol. Chem., 50, 23–31 (1986)
[25] Buonocore, V., De Biasi, M.-G., Giardina, P., Poerio, E., Silano, V.: Biochim. Biophys. Acta, 831, 40–48 (1985)
[26] Kelly, C.T., Moriarty, M.E., Fogarty, W.M.: Appl. Microbiol. Biotechnol., 22, 352–358 (1985)
[27] Takasaki, Y.: Agric. Biol. Chem., 49, 1091–1097 (1985)
[28] Rogers, J.C.: J. Biol. Chem., 260, 3731–3738 (1985)
[29] Bhella, R.S., Altosaar, I.: Can. J. Microbiol., 31, 149–153 (1985)
[30] Obi, S.K.C., Odibo, F.J.C.: Can. J. Microbiol., 30, 780–785 (1984)
[31] Pierre, K.J., Tung, K.-K. in "Methods Enzym. Anal.", 3rd. Ed. (Bergmeyer, H.U., Ed.) 4, 146–151 (1984)
[32] Rauscher, E. in "Methods Enzym. Anal.", 3rd. Ed. (Bergmeyer, H.U., Ed.) 4, 152–157 (1984)
[33] Rauscher, E. in "Methods Enzym. Anal.", 3rd. Ed. (Bergmeyer, H.U., Ed.) 4, 157–161 (1984)
[34] Wahlefeld, A.W. in "Methods Enzym. Anal.", 3rd. Ed. (Bergmeyer, H.U., Ed.) 4, 161–167 (1984)

Enzyme Handbook © Springer-Verlag Berlin Heidelberg 1991
Duplication, reproduction and storage in data banks are only
allowed with the prior permission of the publishers

[35] Foo, Y., Rosalki, S.B. in "Methods Enzym. Anal.", 3rd. Ed. (Bergmeyer, H.U., Ed.) 4, 167–177 (1984)

[36] Sakano, Y., Fukushima, J., Kobayashi, T.: Agric. Biol. Chem., 47, 2211–2216 (1983)

[37] Sakano, Y., Hiraiwa, S.-i., Fukushima, J., Kobayashi, T.: Agric. Biol. Chem., 46, 1121–1129 (1982)

[38] Shimizu, M., Kanno, M., Tamura, M., Suekane, M.: Agric. Biol. Chem., 42, 1681–1688 (1978)

[39] Takasaki, Y.: Agric. Biol. Chem., 47, 2193–2199 (1983)

[40] Krishnan, T., Chandra, A.K.: Appl. Environ. Microbiol., 46, 430–437 (1983)

[41] Orlando, A.R., Ade, P., Di Maggio, D., Fanelli, C., Vittozzi, L.: Biochem. J., 209, 561–564 (1983)

[42] Kindle, K.L.: Appl. Biochem. Biotechnol., 8, 153–170 (1983) (Review)

[43] Kekos, D., Macris, B.J.: Appl. Environ. Microbiol., 45, 935–941 (1983)

[44] Zenin, C.T., Park, Y.K.: J. Ferment. Technol., 61, 109–112 (1983)

[45] Takasaki, Y.: Agric. Biol. Chem., 46, 1539–1547 (1982)

[46] Schibler, U., Pittet, A.-C., Young, R.A., Hagenbüchle, O., Tosi, M., Gellman, S., Wellauer, P.K.: J. Mol. Biol., 155, 247–266 (1982)

[47] Uchino, F.: Agric. Biol. Chem., 46, 7–13 (1982)

[48] Morgan, F.J., Priest, F.G.: J. Appl. Bacteriol., 50, 107–114 (1981)

[49] Medda, S., Chandra, A.K.: J. Appl. Bacteriol., 48, 47–58 (1980)

[50] Emiola, L.O.: J. Biochem., 87, 289–295 (1980)

[51] Detera, S.D., Friedberg, F.: Int. J. Pept. Protein Res., 14, 364–372 (1979)

[52] Mäntsälä, P., Zalkin, H.: J. Biol. Chem., 254, 8540–8547 (1979)

[53] Ingle, M.B., Erikson, R.J.: Adv. Appl. Microbiol., 24, 257–278 (1978) (Review)

[54] Kluh, I.: Collect. Czech. Chem. Commun., 44, 288–293 (1979)

[55] Moulin, G., Glazy, P.: Z. Allg. Mikrobiol., 18, 269–274 (1978)

[56] Glymph, J.L., Stutzenberger, F.J.: Appl. Environ. Microbiol., 34, 391–397 (1977)

[57] McWethy, S.J., Hartman, P.A.: J. Bacteriol., 129, 1537–1544 (1977)

[58] Buonocore, V., Poerio, E., Pace, W., Petrucci, T., Silano, V., Tomasi, M.: FEBS Lett., 67, 202–206 (1976)

[59] Buonocore, V., Poerio, E., Silano, V., Tomasi, M.: Biochem. J., 153, 621–625 (1976)

[60] Hasegawa, A., Miwa, N., Oshima, T., Imahori, K.: J. Biochem., 79, 35–42 (1976)

[61] Allam, A.M., Hussein, A.M., Ragab, A.M.: Z. Allg. Mikrobiol., 15, 393–398 (1975)

[62] Kato, K., Sugimoto, T., Amemura, A., Harada, T.: Biochim. Biophys. Acta, 391, 96–108 (1975)

[63] Rick, W., Stegbauer in "Methoden Enzym. Anal.", 3rd. Ed. (Bergmeyer, H.U., Ed.) 1, 919–923 (1974)

[64] Rauscher, E. in "Methoden Enzym. Anal.", 3rd. Ed. (Bergmeyer, H.U., Ed.) 1, 923–927 (1974)

[65] Wahlefeld, A.W. in "Methoden Enzym. Anal.", 3rd. Ed. (Bergmeyer, H.U., Ed.) 1, 927–931 (1974)

[66] Street, H.V. in "Methoden Enzym. Anal.", 3rd. Ed. (Bergmeyer, H.U., Ed.) 1, 931–936 (1974)

[67] Hillmann, G. in "Methoden Enzym. Anal.", 3rd. Ed. (Bergmeyer, H.U., Ed.) 1, 936–943 (1974)

[68] Hobbs, J.R., Aw, S.E. in "Methoden Enzym. Anal.", 3rd. Ed. (Bergmeyer, H.U., Ed.) 1, 943–949 (1974)

[69] Stiefel, D.J., Keller, P.J.: Biochim. Biophys. Acta, 302, 345–361 (1973)

[70] Sanders, T.G., Rutter, W.J.: Biochemistry, 11, 130–136 (1972)

[71] Keller, P.J., Kauffman, D.L., Allan, B.A., Williams, B.L.: Biochemistry, 10, 4867–4874 (1971)
[72] Podoler, H., Applebaum, S.W.: Biochem. J., 121, 321–325 (1971)
[73] Pfueller, S.L., Elliott, W.H.: J. Biol. Chem., 244, 48–54 (1969)
[74] Ogasahara, K., Imanishi, A., Isemura, T.: J. Biochem., 67, 65–75 (1970)
[75] Minoda, Y., Koyano, T., Arai, M., Yamada, K.: Agric. Biol. Chem., 32, 104–109 (1968)
[76] Arai, M., Koyano, T., Ozawa, H., Minoda, Y., Yamada, K.: Agric. Biol. Chem., 32, 507–513 (1968)
[77] Welker, N.E., Campbell, L.L.: Biochemistry, 6, 3681–3689 (1967)
[78] Fischer, E.H., Stein, E.A. in "The Enzymes", 2nd. Ed. (Boyer, P.D., Lardy, H., Myrbäck, K., Eds.) 4, 313–343 (1966) (Review)
[79] Madsen, G.B., Norman, B.E., Slott, S.: Starch/Staerke, 25, 304–308 (1973)
[80] Buonocore, V., Poerio, E., Gramenzi, F., Silano, V.: J. Chromatogr., 114, 109–114 (1975)
[81] Moulin, G., Boze, H., Galzy, P.: Folia Microbiol., 27, 377–381 (1982)
[82] Spencer-Martins, I., Van Uden, N.: Eur. J. Appl. Microbiol. Biotechnol., 6, 241–250 (1979)
[83] Oteng-Gyang, K., Moulin, G., Glazy, P.: Z. Allg. Mikrobiol., 21, 537–544 (1981)
[84] Fischer, E.H., Stein, E.A.: Arch. Sci., 7, 9 (1954)
[85] Kauffmann, D.L., Zager, N.I., Cohen, E., Kelly, P.J.: Arch. Biochem. Biophys., 137, 325–339 (1970)
[86] Loyter, A., Schram, A.: Biochim. Biophys. Acta, 65, 200–206 (1962)
[87] Mutzbauer, H., Schultz, G.V.: Biochim. Biophys. Acta, 102, 526–532 (1965)
[88] Vandermeers, A., Christophe, J.: Biochim. Biophys. Acta, 154, 110 (1968)
[89] Podoler, H., Applebaum, S.W.: Biochem. J., 121, 317 (1971)
[90] Takkinen, K., Petterson, R.F., Kalkkinen, N., Palva, I., Söderlund, H., Kaariainen, L.: J. Biol. Chem., 258, 1007–1013 (1983)
[91] Stark, J.R., Stewart, T.B., Priest, F.G: FEMS Microbiol. Lett., 15, 295–298 (1982)
[92] Tajiri, T., Koba, Y., Ueda, S.: Agric. Biol. Chem., 47, 671–679 (1983)

Enzyme Handbook © Springer-Verlag Berlin Heidelberg 1991
Duplication, reproduction and storage in data banks are only
allowed with the prior permission of the publishers

1 NOMENCLATURE

EC number
   3.2.1.2

Systematic name
   1, 4-Alpha-D-glucan maltohydrolase

Recommended name
   Beta-amylase

Synonymes
   Saccharogen amylase
   Saccharogenamylase
   Glycogenase
   Amylase, .beta.-
   Beta amylase

CAS Reg. No.
   9000-91-3

---

2 REACTION AND SPECIFICITY

Catalysed reaction
   Polysaccharides containing alpha-(1–4)-linked glucose units $+ H_2O \rightarrow$
   $\rightarrow$ maltose (mechanism [2, 4, 34], hydrolysis of 1,4-alpha-D-glucosidic
   linkages in polysaccharides so as to remove successive maltose units from
   the non-reducing ends of the chains)

Reaction type
   O-Glycosyl bond hydrolysis (exohydrolysis)

Natural substrates
   Polysaccharides containing alpha-(1–4)-linked D-glucose units $+ H_2O$

Substrate spectrum
   1 Polysaccharides containing alpha-(1–4)-linked D-glucose units $+ H_2O$
      (specific for alpha-(1–4)-linkages [4], requirement for non-reducing
      chain end [31])
   2 Starch $+ H_2O$
   3 Amylopectin $+ H_2O$ [4, 8, 20, 24, 28, 30]
   4 Glycogen $+ H_2O$ [8, 30]
   5 Amylose $+ H_2O$ [20, 24, 30, 37]
   6 Alpha-D-glucopyranosyl-(1–4)-2-deoxy-D-glucal (maltal) [14]

Enzyme Handbook © Springer-Verlag Berlin Heidelberg 1991
Duplication, reproduction and storage in data banks are only
allowed with the prior permission of the publishers   1

Product spectrum
  1 Maltose (+ beta-limit dextrin [8, 20]) [4, 8, 15, 20, 24, 25, 27, 28, 30, 31]
  2 Maltose + ?
  3 Maltose + ?
  4 Maltose + ?
  5 Maltose
  6 Alpha-D-glucopyranosyl-(1–4)-2-deoxy-D-glucose (2-deoxymaltose)

Inhibitor(s)
  Glucose [1]; Maltose (competetive) [1, 4, 20]; Alpha-methyl-glucoside [1];
  Cyclodextrins (e.g. cyclohexaamylose [1]) [1, 4, 8, 31]; o-Iodosobenzoate
  [2]; Ag^+ [2, 11, 29, 30]; Hg^{2+} [2 , 4, 11, 15, 16, 23, 29, 30, 32]; Cu^{2+} [2, 4, 15,
  16, 30]; N-Ethylmaleimide [4, 6, 29]; p-Chloromercuribenzoate (reversed by
  DTT or mercaptoethanol [33], reversed by glutathione [11]) [4, 6, 8, 9, 11,
  15, 17, 23, 29, 30, 33]; Pb^{2+} [11, 16]; Ni^{2+} [11]; Cd^{2+} [11]; Fe^{3+} [11]; Zn^{2+}
  [11, 16]; Mn^{2+} [11]; Mg^{2+} [11]; EDTA [11, 16]; Beta-amylase inhibitor (from
  Streptomyces sp. No. 54) [18]; K_2PtCl_6 [23]; K_2PtCl_4 [23]; K_2IrCl_6 [23];
  Na_2PdCl_6 [23]; CO_3^{2-} [29]; Iodoacetamide [29]

Cofactor(s)/prosthetic group(s)
  No cofactors [2, 37]

Metal compounds/salts
  No metal ion requirement [2, 8, 9]; K^+ (reactivity enhancement) [29]; Na^+
  (reactivity enhancement) [29]; Ca^{2+} (reactivity enhancement) [26]

Turnover number (min⁻¹)

Specific activity (U/mg)
  More (assay methods [1], 1077 [6], 122.5 [8], 1 U: formation of 1 micromol
  maltose per min. per mg protein [6, 8, 20, 25, 29]) [1, 3, 6, 8, 9, 11, 17, 20, 21,
  25, 28, 29, 30 , 33–35, 37]

K_m-value (mM)
  More (1.65–6.8 mg/ml soluble starch [4, 8, 20, 29, 32], 1.65–2.25 mg/ml am-
  lyopectin [21]) [4, 8, 17, 20, 21, 29, 30, 32, 35]

pH-optimum
  4–6 [1]; 4.5–5.5 (substrate raw starch) [9]; 5.4 [21]; 5.5 (native form of enzyme [26]) [8, 26, 32]; 5.5–6.0 (substrate soluble starch [9]) [2, 9, 15]; 6.0
  (crystalline and immobilized enzyme [26]) [4, 20, 26, 34]; 6.5–7.0 [24]; 6.9
  [11]; 7.0 [6, 30]; 7.0–7.5 [10]; 7.0–8.0 [24]

pH-range
2.5–9.0 (less than 43% of maximal activity below pH 2.5 and above pH 9.0) [15]; 3.0 (70% of maximal activity) [34]; 3.4–7.8 (less than 80% of maximal activity below pH 3.4 and above pH 7.8) [32]; 4.0–7.0 [9]; 4.5–7.5 (less than 80% of maximal activity below pH 4.5 and above pH 7.5) [20]; 5.0–6.0 (more than 80% of maximal activity) [8]; 5.5 (38% of maximal activity) [24]; 6.9–7.5 [11]

Temperature optimum (°C)
40 [17]; 45 [25, 32]; 50 (native and crystalline form [26]) [6, 24, 26, 30]; 50–55 [2]; 55 (immobilized form) [26]; 60 [10]; 75 [15]; 75–80 (substrate raw starch) [9]; 80 [11]

Temperature range (°C)
70 (70% of maximal activity) [10]; 80 (85% of maximal activity) [15]; 85 (83% of maximal activity) [11]; 90 (32% of maximal activity) [11]

---

3 ENZYME STRUCTURE

Molecular weight
62830 (Bacillus circulans, calculation from nucleotide sequence) [5]
280000 (barley zymogen, activation by proteolysis and mercaptoethanol, gel filtration) [35]
193000–215000 (sweet potato, gel filtration [3], measurement of crystal density at different solvent densities [36], Clostridium thermosulfurogenes, gel filtration [8], wheat, calculation from sedimentation, diffusion, and specific volume data [37]) [3, 8, 36, 37]
160000 (barley zymogen, activation by mercaptoethanol, gel filtration) [35]
127314 (Bacillus polymyxa, nucleotide sequence of a gene encoding for alpha-and beta-amylase) [7]
107000 (Vicia faba, sucrose density centrifugation, gel filtration) [34]
53000–68000 (Pisum sativum, gel filtration [4], Medicago sativa, gel filtration [6, 20], Bacillus stearothermophillus, gel filtration [11], extracellular and intracellular form, gel filtration [16], Bacillus cereus, SDS-PAGE [12, 17], gel filtration, 5 M guanidine-HCl, sedimentation equilibrum centrifugation [17], Bacillus polymyxa, SDS-PAGE [13], soybean, sedimentation equilibrum centrifugation [23], rice, SDS-PAGE [28], gel filtration, disc gel electrophoresis [29], barley, sedimentation equilibrum centrifugation [33], gel filtration [35]) [4, 6, 11–13, 16, 17, 20, 23, 28, 29, 33, 35]
41700–44000 (Medicago sativa, gel filtration [20], Bacillus polymyxa, gel filtration [25]) [20, 25]
35000 (Bacillus cereus var. mycoides, gel filtration) [30]

Enzyme Handbook © Springer-Verlag Berlin Heidelberg 1991
Duplication, reproduction and storage in data banks are only
allowed with the prior permission of the publishers

Subunits

Tetramer (4 × 47500, sweet potato, gel filtration, 6 M guanidine HCl [1],
4 × 55500, sweet potato, SDS-PAGE [3], 4 × 51000, identical, Clostridium
thermosulfurogenes, SDS-PAGE [8], 4 × 26000, Vicia faba, SDS-PAGE +
mercaptoethanol [34]) [1, 2, 8, 34]
Monomer (1 × 55000, Pisum sativum, SDS-PAGE, 1 × 60000, Bacillus
cereus, SDS-PAGE, gel filtration, 6M guanidine HCl [17], 1 × 56000, barley,
active enzyme, gel filtration [35]) [4, 17, 35]

Glycoprotein/Lipoprotein

Glycoprotein [8, 43]; No glycoprotein [4, 11, 16, 23]

---

4 ISOLATION/PREPARATION

Source organism

Higher plants [1, 2]; Sweet potato (Iponemia batata) [1–3, 14, 26, 36]; Oats
[1]; Corn [1]; Oryza sativa (rice) [1, 28, 29]; Sorghum [1, 38]; Pisum sativum
[4]; Medicago sativa (alfalfa) [6, 20]; Barley (Hordeum sp.) [33, 35, 39];
Wheat (Triticum sp.) [37]; Glycine max (soybean) [21–23, 41]; Glycine
gracilis [21]; Glycine ussuriens [21]; Pear [32]; Vicia faba [34]; Bacillus circulans [5]; Bacillus polymyxa [7, 13, 25, 31]; Bacillus stearothermophilus
[11, 16]; Bacillus cereus (hyperproducing strain [12], var. mycoides [30])
[12, 17, 30]; Bacillus megaterium [19, 24]; Clostridium thermosulfurogenes
[8, 9, 15, 42]; Eimeria nidulans [10]; Mustard [44]; Entamoeba histolytica
[43]

Source tissue

Seeds [1, 2, 21–23, 27–29, 33, 35, 37]; Malt [1, 27]; Tuber [1, 26]; Epicocytes
[4]; Roots [20]; Leaves [34]; Culture supernatant (bacteria)

Localisation in source

Extracelluar (bacteria, exception: Bacillus stearothermophhilus, extracelluar
and intracellular form [16]); Aleurone layer of seeds (chemically attached to
glutenin [40]) [1, 40]; Cytoplasm (of epidermal cells of leaf) [34]

Purification

Plants (freeing from alpha-amylase) [1]; Sweet potato [3, 26]; Pisum
sativum [4]; Medicago sativa [6, 20]; Clostridium thermosulforogenes [8];
Bacillus stearothermophilus [11]; Glycine max (1 of 7 components [21]) [21,
41]; Bacillus megaterium (partial) [24]; Bacillus polymyxa (form I and II)
[25]; Barley (4 major forms [39]) [33, 39]; Wheat [27, 37]; Rice [28, 29];
Bacillus cereus var. mycoides [30]; Pear (partial, 2 forms) [32]; Vicia faba
[34]

Crystallization

[1–3, 23, 33, 36, 41]

Cloned
[5, 7, 13, 42]

Renaturated
—

5 STABILITY

pH
3.5–6.5 [15]; 3.5–7.0 [8]; 4.0–8.0 [10]; 4.0–9.0 [25]; 5.0–9.0 [24]; 6.0–8.0 [29]; 6–9 [30]; 7.0–11.5 [11]

Temperature (°C)
40 (denaturation above) [4]; 50 (1 hour stable [10], inactivation above [25, 30], inactivation of native and crystalline enzyme [26]) [10, 25, 26, 30]; 55 (15 min., 50% activity [29], inactivation of immobilized enzyme [26]) [26, 29]; 60 (120 min., 96% inactivation [6], 10 min., 95% inctivation [32]) [6, 32]; 70 (presence of Ca^{2+}: 120 min. [8], up to [15]) [8, 15]; More (effect of immobilization on stability) [24]

Oxidation
Oxidation with $K_3Fe(CN)_6$ with resulting dimerization [29]; o-Iodosobenzoate (oxidation) [2]

Organic solvent
Ethanol (3% v/v stable) [15]

General stability information
Ca^{2+} (stabilization) [8]; Surface denaturation [22, 33]; Freezing (partial denaturation [22, 34], denaturation of pure enzyme [33]) [22, 33, 34]; Mercaptoethanol (catalysis of polymerization, no reduction of activity [33], stabilization [34]) [33, 34]; Unstable in highly diluted solutions [22]; Denaturing agents: change of conformation [22]

Storage
4°C 20% v/v ethanol, at least 4 days [8]; 4°C, presence of Ca^{2+}, more than 6 months [5]; 4°C, interconversion of form E1 to form E2 [32]; 4°C, suspension in ammonium sulfate, prevention of polymerization by reducing agent (e.g. dithiothreitol) [33]; 3°C, 0.05 M mercaptoethanol, 1 week, 50% activity [34]

Enzyme Handbook © Springer-Verlag Berlin Heidelberg 1991
Duplication, reproduction and storage in data banks are only
allowed with the prior permission of the publishers

6 CROSSREFERENCES TO STRUCTURE DATABANKS

PIR/MIPS code

A32251 (alpha-amylase EC 3.2.1.1 precursor, Bacillus polymyxa, fragments); S03745 (precursor, Bacillus circulans); A29108 (precursor, Bacillus polymyxa, fragment); A29130 (precursor, Bacillus polymyxa, fragment); A31389 (precursor, thermophilic Clostridium thermosulfurigenes); A29291 (soybean); A24416 (soybean, fragment); A29292 (sweet potato); S00222 (barley)

Brookhaven code

7 LITERATURE REFERENCES

[1] Thoma, J.A., Spradlin, J.E., Dygert, S. in "The Enzymes", 3rd. Ed. (Boyer, P.D., Ed.) 5, 115–189 (1971) (Review)

[2] French, D. in "The Enzymes", 2nd. Ed. (Boyer, P.D., Lardy, H., Myrbäck, K., Eds.) 4, 345–368 (1960) (Review)

[3] Ann, Y.-G., Iizuka, M., Yamamoto, T., Minamiura, N.: Agric. Biol. Chem., 54, 769–774 (1990)

[4] Lizotte, P.A., Henson, C.A., Duke, S.H.: Plant Physiol., 92, 615–621 (1989)

[5] Siggens, K.W.: Mol. Microbiol., 1, 86–91 (1987)

[6] Kohno, A., Nanmori, T., Shinke, R.: J. Biochem., 105, 231–233 (1989)

[7] Uozumi, N., Sakurai, K., Sasaki, T., Takekawa, S., Yamagata, H., Tsukagoshi, N., Udaka, S.: J. Bacteriol., 171, 375–382 (1989)

[8] Shen, G.-J., Saha, B.C., Lee, Y.-E., Bhatnagar, L., Zeikus, J.G.: Biochem. J., 254, 835–840 (1988)

[9] Saha, B.C., Shen, G.-J., Zeikus, J.G.: Enzyme Microb. Technol., 9, 598–601 (1987)

[10] Chatterjee, B.S., Das, A.: Biotechnol. Lett., 10, 143–147 (1988)

[11] Srivastava, R.A.K.: Enzyme Microb. Technol., 9, 749–754 (1987)

[12] Nanmori, T., Numata, Y., Shinke, R.: Appl. Environ. Microbiol., 53, 768–771 (1987)

[13] Friedberg, F., Rhodes, C.: J. Bacteriol., 165, 819–824 (1986)

[14] Hehre, E.J., Kitahata, S., Brewer, C.F.: J. Biol. Chem., 261, 2147–2153 (1986)

[15] Hyun, H.H., Zeikus, J.G.: Appl. Environ. Microbiol., 49, 1162–1167 (1985)

[16] Srivastava, R.A.K.: Enzyme Microb. Technol., 6, 422–426 (1984)

[17] Nanmori, T., Shinke, R., Aoki, K., Nishira, H.: Agric. Biol. Chem., 47, 941–947 (1983)

[18] Arai, M., Sumida, M., Nakatani, S., Murao, S.: Agric. Biol. Chem., 47, 183–185 (1983)

[19] Stark, J.R., Stewart, T.B., Priest, F.G.: FEMS Microbiol. Lett., 15, 295–298 (1982)

[20] Doehlert, D.C., Duke, S.H., Anderson, L.: Plant Physiol., 69, 1096–1102 (1982)

[21] Mikami, B., Aibara, S., Morita, Y.: Agric. Biol. Chem., 46, 943–953 (1982)

[22] Mori, E., Mikami, B., Morita, Y., Jirgensons, B.: Arch. Biochem. Biophys., 211, 382–389 (1981)

[23] Morita, Y., Yagi, F., Aibara, S., Yamashita, H.: J. Biochem., 79, 591–603 (1976)

[24] Thomas, M., Priest, F.G., Stark, J.R.: J. Gen. Microbiol., 118, 67–72 (1980)

[25] Murao, S., Ohyama, K., Arai, M.: Agric. Biol. Chem., 43, 719–726 (1979)

[26] Ohba, R., Shibata, T., Ueda, S.: J. Ferment. Technol., 57, 146–150 (1979)

[27] Hon, C.C, Reilly, P.J.: Biotechnol. Bioeng., 21, 505–512 (1979)

[28] Okamoto, K., Akazawa, T.: Agric. Biol. Chem., 42, 1379–1384 (1978)

[29] Matsui, H., Chiba, S., Shimomura, T.: Agric. Biol. Chem., 41, 841–847 (1977)
[30] Takasaki, Y.: Agric. Biol. Chem., 40, 1523–1530 (1976)
[31] Marshall, J.J.: FEBS Lett., 46, 1–4 (1974)
[32] Pech, J.C., Bonneau, G., Fallot, J.: Phytochemistry, 12, 299–305 (1973)
[33] Visuri, K., Nummi, M.: Eur. J. Biochem., 28, 555–565 (1972)
[34] Chapman, G.W., Pallas, J.E., Mendicino, J.: Biochim. Biophys. Acta, 276, 491–507 (1972)
[35] Shinke, R., Mugibayashi, N.: Agric. Biol. Chem., 35, 1381–1390 (1971)
[36] Colman, P.M., Matthews, B.W.: J. Mol. Biol., 60, 163–168 (1971)
[37] Rexova, L., Kopec, Z., Keil, B.: Collect. Czech. Chem. Commun., 32, 678–684 (1967)
[38] Botes, D.P., Joubert, F.J., Novellie, L.: J. Sci. Food Agric., 18, 409–414 (1967)
[39] Lundgard, R., Svensson, B.: Carlsberg Res. Commun., 52, 313–326 (1987)
[40] Rowsell, E.V., Goad, L.J.: Biochem. J., 84, 73P (1962)
[41] Morita, Y., Aibaras, S., Yamashita, H., Yagi, F., Suganuma, T., Hiromi, K.: J. Biochem., 77, 343–351 (1975)
[42] Kitamoto, N., Yamagata, H., Kato, N., Tsakagoshi, N., Udaka, S.: J. Bacteriol., 170, 5848–5854 (1988)
[43] Nebinger, P.: Biol. Chem. Hoppe-Seyler, 367, 161–167 (1986)
[44] Subbaramaiah, K., Sharma, R.: J. Biochem. Biophys. Methods, 10, 315–320 (1985)

Enzyme Handbook © Springer-Verlag Berlin Heidelberg 1991
Duplication, reproduction and storage in data banks are only
allowed with the prior permission of the publishers

1 NOMENCLATURE

EC number
3.2.1.3

Systematic name
1, 4-Alpha-D-glucan glucohydrolase

Recommended name
Glucan 1, 4-alpha-glucosidase

Synonymes
Glucoamylase
Amyloglucosidase
Gamma-amylase
Lysosomal alpha-glucosidase
Exo-1, 4-alpha-glucosidase
Glucose amylase
Alpha-1, 4-glucan glucohydrolase
Acid maltase
Agidex
Diazyme
AMG 50L
AMG 200L

CAS Reg. No.
9032-08-0

2 REACTION AND SPECIFICITY

Catalysed reaction
Alpha-D-glucan + H_2O →
→ beta-D-glucose (hydrolysis of alpha-(1–4) and alpha-(1–6)-glucan with cleavage of glucose from non-reducing ends of chain)

Reaction type
O-Glycosyl-bond hydrolysis (exohydrolysis)

Natural substrates
Alpha-D-glucans + H_2O

Enzyme Handbook © Springer-Verlag Berlin Heidelberg 1991
Duplication, reproduction and storage in data banks are only
allowed with the prior permission of the publishers

Substrate spectrum
1 Starch (soluble, hydrolysis of raw starch [31, 33]) + H_2O
2 Glycogen + H_2O [2, 4, 13, 22, 23, 30, 31, 33–35]
3 Amylose + H_2O [4, 13, 19, 30, 31, 33–35]
4 Amylopectin + H_2O [13, 23, 24, 30, 31, 33–35]
5 Dextrin + H_2O [13, 24, 28, 34]
6 Maltotriose (or maltotetraose, maltopentaose, maltohexaose, maltohep-taose) + H_2O [2, 4, 13, 19, 20, 30, 31, 33, 35]
7 Panose (alpha-glucopyranosyl-(1–6)-maltose) + H_2O [20]
8 Phenyl-alpha-maltoside + H_2O [33]
9 More (preference for high MW molecules with alpha-(1–4)-linkages [2, 9, 24], slow hydrolysis of alpha-(1–6)-linked oligo-or polysaccharides [9, 20], no hydrolysis of alpha-(1–6)-linkages [24])

Product spectrum
1 Beta-D-glucose [4, 16, 22, 33]
2 Beta-D-glucose
3 Beta-D-glucose
4 Beta-D-glucose
5 Beta-D-glucose
6 Beta-D-glucose
7 ?
8 ?
9 ?

Inhibitor(s)
$AgNO_3$ [1]; $KMnO_4$ [1]; Cd^{2+} [1]; Ca^{2+} [1]; Acarbose [4, 9, 13]; Cyclic dex-trins (competetive) [9, 13]; Alpha-D-glucosides (competetive) [9]; Glucose [9]; Tris [9, 33]; 2-Amino-2-ethyl-1, 3-propanediol [9]; Maltitol [9]; 1-Deoxynojirimycin [9]; Trestaine [9]; EDTA [13]; Alpha-methyl glucoside (hydrolysis of starch and maltose [20]) [13, 20]; Hg^{2+} [22, 23, 26, 30, 31, 33]; Mn^{2+} [22, 23, 26]; Pb^{2+} [22, 23, 26, 30, 31]; Cu^{2+} [20]; Ag^{2+} [20]; Co^{2+} [20]; D-Glucono-delta-lactone [24]; Al^{3+} [30]; More [9]

Cofactor(s)/prostethic group(s)
No cofactors [19]; Polyvalent anions (activation) [20]

Metal compounds/salts
No metal ions [19, 27]; Divalent cations (activation) [20]

Turnover number (min^{-1})

Specific activity (U/mg)
5507 [13]; 1125 [15]; 80 [24]; 68 [14]; More [1–4, 6, 7, 16, 17, 19–27, 30, 32, 34–36]

K_m-value (mM)
0.91–18 (maltose) [4, 7, 14, 16, 17, 24]; 0.26–3.79 (maltotriose) [4, 7, 14, 16, 17, 24]; 0.14–4.12 (maltotetraose) [4, 7, 14, 16]; 0.12–4.33 (maltopentaose) [4, 7, 14, 16]; 0.096–3.65 (maltohexaose) [4, 7, 14, 16]; 0.097–4.15 (maltoheptaose) [4, 7, 14, 16]; 36.47–59 (isomaltose) [7, 14, 16]; 12–14 (panose) [14, 16]; 2.75 (p-nitrophenyl-alpha-D-glucoside) [24]; More (values for high MW glucans in mg/ml or %w/v) [7, 21, 22, 25, 28, 29, 33]

pH-optimum
2.4–4.8 [13]; 4.0–4.5 [4]; 4.0–5.0 [6, 19] 4.1–4.7 [29]; 4. 2 [9, 27]; 4.4 [3]; 4.5 (isoenzyme GA-I [1], substrate raw starch [21], enzyme II [32]) [1, 16, 20, 21–23, 25, 30, 32]; 4.6 (enzyme I) [31]; 4.8 (isoenzyme GA-II) [1]; 4.5–5.0 [26]; 5.0 (enzyme I) [32]; 5.4 [35]; 5.5 (substrate boiled soluble starch [21]) [2, 17, 21, 34]; 6.6 [47]

pH-range
2–6 [20]; 2.4–4.8 [13]; 3.0–7.0 [27]; 3.0–8.0 [2]; 3.8–6 [9]; 4.0–5.5 [6]; 4.0–6.0 [17]

Temperature optimum (°C)
40 (Rhizopus sp.) [19]; 50–60 (Aspergillus sp.) [19]; 50 (enzymes II and III) [25]; 55 [1, 13]; 55–60 (enzyme I) [32]; 56 [4]; 57 [9]; 60 (enzyme I [25], enzyme II [32]) [6, 17, 25, 30, 34, 35]; 65 [28]; 70 (presence of 5% starch [6]) [3, 6, 20, 24]

Temperature range (°C)
30–65 [17]; 45–65 [9]

---

3 ENZYME STRUCTURE

Molecular weight
306000 (Saccharomyces diastaticus, high pressure size exclusion chromatography) [17]
225000–250000 (Trichosporon adeninovorans, gel filtration [6], Saccharomyces cerevisiae, polyacrylamide-gradient-gel electrophoresis, equilibrum centrifugation, gel filtration, velocity centrifugation [7]) [6, 7]
190000 (Saccharomyces cerevisiae, sucrose gradient centrifugation of crude enzyme) [1]
48000–91000 (Monascus [1], Aspergillus oryzae, gel filtration [4, 25], SDS-PAGE, ultracentrifugation [4], Thermomyces lanuginosus, SDS-PAGE [5, 24], gel filtration [24], Candida antarctica, gel filtration, SDS-PAGE [9], Rhizopus niveus, SDS-PAGE [10], Candida tsukubaensis, SDS-PAGE [13], Rhizopus delemar, SDS-PAGE [14], Rhizopus sp., sedimentation equilibrum centrifugation [26], Aspergillus candidus Link var. aureus, gel permeation chromatography, SDS-PAGE [15], Paecilomyces varioti, gel filtration, SDS-PAGE [16], Aspergillus awamori, calculation from DNA sequence [18],

Enzyme Handbook © Springer-Verlag Berlin Heidelberg 1991
Duplication, reproduction and storage in data banks are only
allowed with the prior permission of the publishers

calculation from sedimentaiton and diffusion data [30], fungi, values depending on organism and analytical method, secondary and tertiary structure [19], Aspergillus niger, gel filtration [20], Aspergillus saitoi, SDS-PAGE, with and without reducing agents, amino acid and sugar composition [22, 23], Beta vulgaris, SDS-PAGE [29], Penicillium oxalicum, calculation from sedimentation and diffusion data [32], Cephalosporium charticola, gel filtration [35], Mucor rouxianus, SDS-PAGE [36], Endomycopsis fibuligera [44]) [1, 4, 5, 9, 10, 13–16, 18–20, 22–26, 30, 32, 35, 36, 44]

Subunits

Tetramer (4 × 70000, Saccharomyces cerevisiae, polyacrylamide-gradient-gel electrophoreis with SDS and urea) [7] Dimer (2 × 186000, Saccharomyces diastaticus, SDS-PAGE) [17] Monomer (1 × 69000, Paecilomyces varioti, SDS-PAGE [16], fungi [19], multiple forms (Gluc 1, Gluc 2, Gluc 3) differing in both amino acid composition and carbohydrate content: 1 × 74000 (Gluc 1), 1 × 58000 (Gluc 2), 1 × 61400 (Gluc 3), Rhizopus sp., SDS-PAGE [26]) [16, 19, 26]

Glycoprotein/Lipoprotein

Glycoprotein (carbohydrate composition [8])

---

4 ISOLATION/PREPARATION

Source organism

Monascus (sp. No. 3403) [1]; Saccharomyces cerevisiae [2, 7]; Beta vulgaris [3, 29]; Aspergillus oryzae [4, 8, 25, 28]; Aspergillus sp. [37, 38]; Thermomyces lanuginosus (formerly Humicola lanuginosa) [5, 24, 47]; Trichosporon adeninovorans [6]; Candida antarctica [9]; Rhizopus niveus [10, 41]; Rhizopus nodosus [43]; Saccharomyces diastaticus [11, 17, 39, 40]; Aspergillus niger [12, 20]; Candida tsukubaensis [13]; Rhizopus delemar [14]; Aspergillus candidus (Link var. aureus) [15]; Paecilomyces varioti [16]; Aspergillus awamori [18, 30]; Endomycopsis fibuligera [21, 34, 44]; Aspergillus saitoi [22]; Rhizopus sp. [26]; Cephalosporium eichorniae [27]; Cephalosporium charticola [35]; Mucor rouxianus [31, 36]; Penicillium oxalicum [32, 33]; Clostridium acetobutylicum [45]; Flavobacterium [46]; Coniphora cerebella [48]; Fungi (overview [9]) [9, 42]

Source tissue

Cell [1, 4]; Sporulating cells [2]; Suspension culture [3]; Callus tissue [3]; Culture medium [5, –7, 9, 13, 15–17, 19, 20, 21, 24, 25, 27, 28, 34, 35]; Seed [29]; Mycelia [31, 32, 36]

Localisation in source

Vacuole [2]; Extracellular (from suspension culture [3]) [3, 5–7, 9, 13, 15, 16, 17, 19–21, 24, 25, 27, 28, 34, 35]; Intracellular (from callus) [3]; Cell wall (as a soluble form in free space of cell wall) [3]

Purification

Monascus sp. (2 isoenzymes) [1]; Saccharomyces cerevisiae (partial) [2]; Beta vulgaris [3, 29]; Aspergillus oryzae (3 isoenzymes [25, 28]) [4, 25, 28]; Thermomyces lanuginosus [5, 24]; Candida antarctica [9]; Candida tsukubaensis [13]; Rhizopus delemar (3 isoenzymes, from commercial preparation) [14]; Aspergillus candidus [15]; Paecilomyces varioti [16]; Saccharomyces diastaticus [17]; Aspergillus niger [20]; Endomycopsis fibuligera [21, 34]; Aspergillus saitoi (from Molsin, commercial digest of Asp. saitoi) [22, 23]; Rhizopus sp. (from Gluczyme, commercial product) [26]; Cephalosporium eichhorniae [27]; Aspergillus awamori [30]; Penicillium oxalicum [32]; Cephalosporium charticola [35]; Mucor rouxianus (2 isoenzymes) [36]

Crystallization

[36]

Cloned

[11, 18, 37–40]

Renaturated

–

---

5 STABILITY

pH

2.0–11.5 [20]; 2–7 [13]; 2.5–7.5 [22, 23]; 3.0–6.5 [32]; 3. 0–8.0 [19, 29]; 3.6–6.5 (isoenzyme GA-I) [1]; 4.0–7.0 (isoenzyme GA-II) [1]; 4.0–7.5 (isoenzyme II) [31]; 4.0–8.5 [14]; 4.0–8.0 (isoenzyme I [31]) [9, 26, 31]; 4.5–8.5 [4]; 4.0–9.0 [34]; 4.5–10 [16]; 5.0–6.0 [25]; 6.5–8.0 [24]

Temperature (°C)

30 (isoenzyme GA-II, up to [1], 30 minutes [13]) [1, 13]; 37 (2 days) [10]; 40 (isoenzyme GA-I, up to [1], up to [4]) [1, 4]; 45 (up to) [14, 25]; 50 (30 minutes, 15% activity [13], up to [16, 22, 23, 24, 26, 30, 31, 34], stable at high temperature in presence of starch or glycerol [20]) [13, 16, 20, 22–24, 26, 30, 31, 34]; 55 (up to) [32]; 58 (up to, with 1.25% starch) [6]; 60 (inactivation above) [19]; 62 (up to) [27]; 65 (up to) [3]; 75 (inactivation) [17]; 100 (10 minutes, 0.5% starch solution, 70% activity) [47]; More [24]

Oxidation

Organic solvent

Enzyme Handbook © Springer-Verlag Berlin Heidelberg 1991
Duplication, reproduction and storage in data banks are only
allowed with the prior permission of the publishers

General stability information

Starch (stabilization) [6]; Glycerol (stabilization, protection against heat inactivation) [13]; Urea (8M, 20% denaturation [22], 8 M, no denaturation [24]) [22, 24]; Guanidium chloride, (4 M, 50% denaturation) [24]

Storage

−20°C or −56°C, 30% glycerol [2]; −20°C, 10 mM sodium phosphate buffer, pH 7, 50% glycerol, several months [6]; −20°C [22–24, 26]; 4°C, 1 month [10]

---

6 CROSSREFERENCES TO STRUCTURE DATABANKS

PIR/MIPS code

ALBYG (precursor, yeast, Saccharomyces cerevisiae); ALASGR (Aspergillus niger); JP0001 (precursor, Rhizopus oryzae); B26877 (S1, yeast, Saccharomyces cerevisiae, fragment); A26877 (S2, yeast, Saccharomyces cerevisiae, fragment); C26877 (SGA, yeast, Saccharomyces cerevisiae); JT0479 (Aspergillus awamori var. kawachi); A29776 (G2, precursor, Aspergillus awamori); A29166 (precursor, Aspergillus awamori); A27897 (Aspergillus phoenicis, fragment); A32609 (human)

Brookhaven code

---

7 LITERATURE REFERENCES

[1] Yasuda, M., Kuwae, M., Matsushita, H.: Agric. Biol. Chem., 53, 247–249 (1989)
[2] Pugh, T.A., Shah, J.C., Magee, P.T., Calncy, M.J.: Biochim. Biophys. Acta, 994, 200–209 (1989)
[3] Masuda, H., Murata, M., Takahashi, T., Sugawara, S.: Plant Physiol., 88, 172–177 (1988)
[4] Ono, K., Shigeta, S., Oka, S.: Agric. Biol. Chem., 52, 1707–1714 (1988)
[5] Jensen, B., Olsen, J., Allermann, K.: Can. J. Microbiol., 34, 218–223 (1988)
[6] Büttner, R., Bode, R., Birnbaum, , D.: J. Basic Microbiol., 27, 299–308 (1987)
[7] Kleinman, M.J., Wilkinson, A.E., Wright, I.P., Evans, I.H.: Biochem. J., 249, 163–170 (1988)
[8] Carter, R.D., Hu, S., Dill, K.: Int. J. Biol. Macromol., 9, 269–272 (1987)
[9] De Mot, R., Verachtert, H.: Eur. J. Biochem., 164, 643–654 (1987)
[10] Tanaka, Y., Ashikari, T., Nakamura, N., Kiuchi, N., Shibano, Y., Amachi, T., Yoshuzumi, H.: Agric. Biol. Chem. , 50, 1737–1742 (1986)
[11] Pretorius, I.S., Chow, T., Modena, D., Marmur, J.: Mol. Gen. Genet., 203, 29–35 (1986)
[12] Svensson, B., Larsen, K., Gunnarsson, A.: Eur. J. Biochem., 154, 497–502 (1986)
[13] De Mot, R., Van Oudendijck, E., Verachtert, H.: Antonie Leeuwenhoek, 51, 275–287 (1985)
[14] Abe, J.-I., Nagano, H., Hizukuri, S.: J. Appl. Biochem., 7, 235–247 (1985)
[15] Kolhekar, S.R., Mahajan, P.B., Ambedkar, S.S., Borkar, P.S.: Appl. Microbiol. Biotechnol., 22, 181–186 (1985)

[16] Takeda, Y., Matsui, H., Tanida, M., Takao, S., Chiba, S.: Agric. Biol. Chem., 49, 1633–1641 (1985)
[17] Tucker, M., Grohmann, K., Himmel, M. in "Biotechnol. Bioeng. Symp. No.14", 279–293, John Wiley & Sons (1984)
[18] Nunberg, J.H., Meade, J.H., Cole, G., Lawyer, F.C., McCabe, P., Schweickart, V., Tal, R., Wittman, V.P., Flatgaard, J.E., Innis, M.A.: Mol. Cell. Biol., 4, 2306–2315 (1984)
[19] Manjunath, P., Shenoy, B.C., Raghavendra Roa, M.R.: J. Appl. Biochem., 5, 235–260 (1983) (Review)
[20] Fogarty, W.M., Benson, C.P.: Eur. J. Appl. Microbiol. Biotechnol., 18, 271–278 (1983)
[21] Ueda, S., Saha, B.C.: Enzyme Microb. Technol., 5, 196–198 (1983)
[22] Inokuchi, N., Takahashi, T., Irie, M.: J. Biochem., 90, 1055–1067 (1981)
[23] Takahashi, T., Inokuchi, N., Irie, M.: J. Biochem., 89, 125–134 (1981)
[24] Basaveswara Rao, V., Sastri, N.V.S., Subba Rao, P.V.: Biochem. J., 193, 379–387 (1981)
[25] Mitsue, T., Saha, B.C., Ueda, S.: J. Appl. Biochem., 1, 410–422 (1979)
[26] Takahashi, T., Tsuchida, Y., Irie, M.: J. Biochem., 84, 1183–1194 (1978)
[27] Day, D.F.: Curr. Microbiol., 1, 181–184 (1978)
[28] Razzaque, A., Ueda, S.: J. Ferment. Technol., 56, 296–302 (1978)
[29] Chiba, S., Inomata, S., Matsui, H., Shimomura, T.: Agric. Biol. Chem., 42, 241–245 (1978)
[30] Yamasaki, Y., Suzuki, Y., Ozawa, J.: Agric. Biol. Chem., 41, 2149–2161 (1977)
[31] Yamasaki, Y., Tsuboi, A., Suzuki, Y.: Agric. Biol. Chem., 41, 2139–2148 (1977)
[32] Yamasaki, Y., Suzuki, Y., Ozawa, J.: Agric. Biol. Chem., 41, 755–762 (1977)
[33] Yamasaki, Y., Suzuki, Y., Ozawa, J.: Agric. Biol. Chem., 41, 1443–1449 (1977)
[34] Sukhumavasi, J., Kato, K., Harada, T.: J. Ferment. Technol., 53, 559–565 (1975)
[35] Krzechowska, M., Urbanek, H.: Appl. Microbiol., 30, 163–166 (1975)
[36] Tsuboi, A., Yamasaki, Y., Suzuki, Y.: Agric. Biol. Chem., 38, 543–550 (1974)
[37] Boel, E., Hjart, I., Svensson, B., Norris, F., Norris, K.E., Fiil, N.P.: EMBO J., 3, 1097 (1984)
[38] Innis, N.A., Holland, M.J., McCabe, P.C., Cole, G.E., Wittman, V.P., Tal, R., Watt, K.W.K., Gelfand, D.H., Holland, J.P., Meade, J.H.: Science, 228, 21 (1985)
[39] Yamashita, I., Fukui, S.: Agric. Biol. Chem., 47, 2289 (1983)
[40] Yamashita, I., Suzuki, K., Fukui, S.: J. Bacteriol., 161, 567 (1985)
[41] Marshall, J.J., Whelan, W.J.: FEBS Lett., 9, 85–88 (1970)
[42] Manjunath, P., Rao, M.R.R.: J. Biosci., 1, 409–425 (1979)
[43] Muthukumaran, N., Dhar, S.C.: Ital. J. Biochem., 32, 239–253 (1983)
[44] Fukui, T., Nikuni, Z.: Agric. Biol. Chem., 33, 884–891 (1969)
[45] French, D., Knapp, D.W.: J. Biol. Chem., 187, 463 (1950)
[46] Bender, H.: Eur. J. Biochem., 115, 278 (1981)
[47] Taylor, P.M., Napier, E.J., Fleming, I.D.: Carbohydr. Res., 61, 301 (1978)
[48] King, N.J.: Biochem. J., 105, 577–583 (1967)

1 NOMENCLATURE

EC number
3.2.1.4

Systematic name
1, 4-(1, 3;1, 4)-beta-D-Glucan 4-glucanohydrolase

Recommended name
Cellulase

Synonymes

Endo-1, 4-beta-Glucanase
.beta.-1 , 4-Glucanase
Endo-.beta.-(1.fwdarw.4)-
Glucanase
.beta.-1, 4-Endoglucan
hydrolase
Endo-1, 4-.beta.-D-Glucanase
Celluase A
Cellulosin AP
Endoglucanase D
Alkali cellulase
Cellulase A 3
Celludextrinase
9.5 Cellulase [29]
Avicelase
Pancellase SS
More (the name cellulase
frequently is applied to a
complex group of
synergetically acting enzymes,
some of these are: EC 3.2.1.4,
EC 3.2.1.6, EC 3.2.1.21, EC

3.2.1.91)
Cellzyme
Celluclast 2.0L
Meicelase CEP
Meicelase [4, 65]
Onozuka R 10
Enzylon CA 40
Celluclast 250L
Cooper cellulase
Maxazyme [1]
Celluclast 1.5L
Driserase
Driselase [61]
SP 227
Macerozyme R 10
Meicelase CM 8B
Biotex
Zimet 43/803
Zimet 43/802
Zimet 43/804
Genencor 150 L

CAS Reg. No.
9012-54-8

2 REACTION AND SPECIFICITY

Catalysed reaction
Endohydrolysis of 1,4-beta-D-glucosidic linkages in cellulose, lichenin and
cereal beta-D-glucans, also hydrolyses 1,4-linkages in beta-D-glucans also
containing 1,3-linkages

Enzyme Handbook © Springer-Verlag Berlin Heidelberg 1991
Duplication, reproduction and storage in data banks are only
allowed with the prior permission of the publishers

Reaction type

O-Glycosyl bond hydrolysis (endohydrolysis)

Natural substrates

Cellulose + H_2O (synernergism of cellulolytic enzymes [6, 32, 55, 60, 63, 72], action in complementary manner to degrade less crystalline regions of celluloytic material [5])
Xyloglucan + H_2O [12]
Cotton + H_2O [67]

Substrate spectrum

1 Cellulose + H_2O (specific for beta-(1–4)-glucosidic bonds, exeption: Penicillium funiculosum, hydrolysis of beta-(1–4)-, beta-(1–3)-, beta-(1–6)-, alpha-(1–6)-linkages [42] amorphous and crystalline cellulose [6, 42], acid-swollen cellulose [18, 26, 28, 38, 42, 63])
2 Carboxymethylcellulose + H_2O [5, 9, 11, 18, 23, 26, 28, 33, 38, 40, 44, 54, 61, 66, 69]
3 Cellooligosaccharides + H_2O (not: cellobiose [13, 44], rate of hydrolysis increases with chain length [22], mode of action [9, 18, 26, 28, 32, 40, 42, 49, 54, 61])
4 Avicel + H_2O [18, 67, 69]
5 Lichenan + H_2O [22]
6 Cellobial + H_2O [33]
7 p-Nitrophenyl-beta-D-cellobioside + H_2O [66, 67]
8 More (transglycosylation [66], condensation [67], preparation of substrates [37])

Product spectrum

1 Beta-(1–4)-glucans (chain length depending on organism and experimental conditions)
2 Beta-(1–4)-glucans (main products: cellobiose, cellotriose, glucose [43, 80, 82])
3 Cellooligosaccharides (main products: cellobiose [5, 13, 13, 26, 32, 38, 40, 42, 44, 49, 50, 54, 61, 68, 79], glucose [12, 26, 32, 40, 49, 50, 54, 79], cellotriose [44, 49, 54, 61, 79])
4 ?
5 ?
6 2-Deoxycellobiose [33]
7 Glucose + p-nitrophenyl-D-glucoside + p-nitrophenol + cellobiose (molar ratio depending on reaction time and isozyme) [66]
8 ?

Inhibitor(s)

Hg^{2+} (restored by cysteine or NaCl [75]) [13, 21, 33, 40, 43, 49, 50, 54, 57, 59, 65, 67, 68, 70, 75, 79]; Mn^{2+} [13, 54]; Glucose (not: Bacillus subtilis [6]) [6, 63, 80]; Cellobiose (not: Bacillus subtilis [6]) [6, 63, 80]; Glucosylamine [63]; Gluconolactone [63, 83]; Fe^{3+} [40, 59, 68]; N-Bromosuccinimide [40, 54]; N-Bromosuccinic acid [69]; Cu^{2+} [43, 54, 59, 67, 68]; Pb^{2+} [43, 68, 69]; Sn^{2+} [43]; Ag^+ [43, 59, 67, 68, 70]; Fe^{2+} [43]; EDTA [43, 56]; Arsenite [43]; Sulphite [43]; Cd^{2+} [50]; SDS (250 micromol) [54]; NaCN [56]; Iodoacetamide [56]; p-Chloromercuribenzoate [56, 57, 59, 70]; 2, 4-Dinitrophenol [56]; p-Hydroxymercuribenzoate [57]; Alpha, alpha'-dipyridyl [59]; Beta-hydroxyquinoline [59]; $Na_2S_2O_3$ [59]; $KMnO_4$ [65]; Dithiothreitol (cellulase 4.5) [70]; Diethylpyrocarbonate [59]; Ca^{2+} [79]; Dextran (slightly) [83]; Heavy metal ions [85]; Sulfhydryl reagents [85]; Oxidizing and reducing agents [85]

Cofactor(s)/prostethic group(s)

No cofactors [29]; SDS (activation) [13]; Amines (activation) [43]; Polyols (activation) [43]; Dithiothreitol (activation) [43]; Cysteine (activation) [56]

Metal compounds/salts

No metal requirement [7, 40]; Ca^{2+} (activation: Clostridium thermocellum, Cellulomonas [6], Thermomonospora fuscata [79]) [6, 56, 79]; $BaCl_2$ (activation) [13]; NaF (activation) [13]; Na_2MnO_4 (activation) [13]; $Na_2P_2O_7$ (activation) [13]; EDTA (activation) [13]; Co^{2+} (increase of activity) [43]; Na^+ (activation) [50]; K^+ (activation) [50]; Mg^{2+} (activation) [56]; Mn^{2+} (stimulation) [59]

Turnover number (min^{-1})

5–6 (cellobiose) [77]; 50–200 (cellotriose) [77]; 400 (cellotetraose) [77]; 450 (cellohexaose) [77]

Specific activity (U/mg)

60 (1 U: amount of enzyme which degrades carboxymethylcellulose to reducing carbohydrate corresponding to 1 micromol glucose per min.) [15]; 428 [27]; 116.8 [49]; More (values for immobilized enzyme [4], as increase of fluidity of medium [9], staining techniques [19], substrates for assay [25], assay methods [17, 25, 34, 35, 36, 52], differentiation between different cellulolytic enzymes [36]) [3, 4, 9, 10, 16–23, 27, 28, 31, 33–36, 38, 40, 42–44, 50, 52–54, 56, 61, 62, 64, 65, 68–70, 73, 79]

K_m-value (mM)

15.4 (cellotetraose) [9]; 6.7 (cellopentaose) [9]; 0.561 (cellopentaose) [61]; 0.381 (cellohexaose) [61]; More (0.48–80 mg/ml carboxymethylcellulose, dependency on pH [64], comparison of values for fungi [72]) [12, 13, 21, 32, 40, 44, 49, 50, 51, 58, 59, 62, 64, 67, 72]

Enzyme Handbook © Springer-Verlag Berlin Heidelberg 1991
Duplication, reproduction and storage in data banks are only
allowed with the prior permission of the publishers

pH-optimum

3.6–6.0 (depending on isozyme) [3]; 3.8–4.0 [64]; 4.0 (enzyme F V [23]) [23, 49, 61]; 4.0–4.5 [32]; 4–5 [44]; 4.2 (Coniophora cerebella, enzyme B) [60]; 4.4 [62]; 4.5 (enzyme F I [23]) [23, 59, 69]; 4.5–5.0 [22, 65, 67, 68]; 4.5–5. 5 (depending on isoelectric point) [13]; 4.7 (Coniophora cerebella, enzyme A) [60]; 4.8–5.1 [4]; 4.8–5.6 (cellulase 4.8) [70]; 5.0 (enzyme F II [23]) [23, 31, 40, 56, 71]; 5.0–6.0 (Bacillus) [6]; 5.2 [28]; 5.5 [43, 45, 81]; 5.5–5.8 [26]; 5.5–7.5 (cellulase II [30], Sporocytophaga sp. [60]) [7, 30, 60]; 5.6–6.0 [78]; 5.7–6.2 (cellulase 4.5) [70]; 5.9 (enzyme Endo 1) [5]; 6.0 [27, 46, 51, 53, 79]; 6.0–6.5 [12]; 6.2–7.5 [80]; 6.3 [54]; 6.5 [21]; 6.8 [9]; 7.0 (cell-bound form [18, 73], cellulase I [30]) [18, 30, 47, 73]; 8.0 (extracellular forms) [18, 73]; 8.1 [29]; 9.0 (alkalophilic Bacilli [6]) [6, 50]

pH-range

2–6 (more than 50% of maximal activity) [61]; 2.1 (40% of maximal activity) [62]; 3.0–6.0 [51]; 4.0–5.0 [58]; 4.0–6. 5 (more than 50% of maximal activity) [40]; 4–7.5 [53]; 4. 8–5.2 (more than 50% of maximal activity) [27]; 4.8–8.2 (more than 50% of maximal activity) [46]; 6.0–8.5 [29]; 7.5 (15% of maximal activity) [62]; 8.0–10.5 [50]

Temperature optimum (°C)

35–43 (cellulase 4.8) [70]; 40 [71]; 45 [54, 56, 64]; 45–50 [7, 49]; 45–55 (depending on isozyme) [3]; 45–60 (cellulase 4.5) [70]; 48–60 [48]; 50 (enzyme F I [23], form B and C [32], form II-B [65]) [23, 31, 32, 51, 61, 65, 67, 68]; 50–70 (depending on isozyme) [13]; 52 [80]; 55 [44, 47]; 58 (from E2) [79]; 60 (Bacillus [6], form II-A [65]) [6, 9, 40, 45, 59, 65, 69]; 62 [28]; 65 (enzyme F V [23]) [22, 23]; 70 [58]; 74 (enzyme form A [32], form E1 [79]) [32, 79]; 75–80 [26]

Temperature range (°C)

35–60 [61]; 80 (no activity) [58]

---

3 ENZYME STRUCTURE

Molecular weight

210000 (Clostridium thermocellum cellulosome, at least 14 different polypeptides) [39]

110000 (Clostridium thermocellum, enzyme Endo-2) [5]

108000 (Thermomonospora fusca, enzyme E1) [21]

106000 (Thermomonospora fusca, enzyme E4) [21]

94000 (Thermomonospora fusca, enzyme E1) [79]

92000 (Bacillus sp. No. 1139) [50]

83000–94000 (Clostridium thermocellum) [28]

78000 (Sclerotium rolfsii, cellulase C) [32]

71000 (Thermomonospora fusca, enzyme E3 [21], Humicola lanuginosa [56]) [21, 56]

70000 (Pisum sativum [12], Phaseolus vulgaris enzymes 4.5 and 4.8 [70]) [12, 70]

66000 (Aspergillus aculeatus, enzyme FII) [23]

65000 (Clostridium thermocellum) [27, 46]

63000 (Myrothecium verrucaria) [84]

60000 (Clostridium thermocellum, enzyme Endo-1) [5]

58000 (Trichoderma viride, enzyme III) [48]

57000 (Trichoderma viride, enzyme V [48], Humicola insolens [31]) [31, 48]

56000 (Robillarda sp. Y-20 , enzyme CMCase I [40], CMCase II [44] , Penicillium funiculosum [42]) [40, 42, 44]

52000 (Trichoderma reesei, enzyme Endo 1 [15], Sclerotium rolfsii, enzyme A [32], Trichoderma viride, enzyme VI [48] , Sporocytophaga myxococcoides, cellulase II [30, 60], Irpex lacteus [68]) [15, 30, 32, 48, 60, 68]

51000 (Phaseolus vulgaris) [29]

50000 (Trichoderma viride, enzyme I) [48]

49000 (Lathyrus odoratus) [1]

48000 (Streptomyces reticuli, enzyme CMCase 2 [7], Trichoderma reesei, enzyme Endo 2 [15], Trichoderma konigii, enzymes Endo 3a and 3b [16, 63]) [7 , 15, 16, 63]

46500 (Neurospora crassa, enzyme Endo I) [3]

46000 (Sporocytophaga myxococcoides, cellulase I [30, 60], Thermomonospora fusca [79]) [30, 60, 79]

45000 (Clostridium josui [9], Thermomonnospora fusca, enzyme E5 [21], Trichoderma viride, enzyme II [48], Trichoderma viride enzyme II [67], Erwinia chrysanthemi [80]) [9, 21, 48 , 67, 80]

44000 (Dolabella auricularia) [44]

43000 (Trichoderma viride, enzyme II-B) [65]

42000 (Thermomonospora fusca, enzyme E2) [21]

40000 (Streptomyces reticuli, enzyme CMCase 1) [7]

38000 (Aspergillus aculeatus, enzyme FV) [23]

37300 (Neurospora crassa, enzyme Endo III) [3]

35000 (Penicillium notatum [74], Erwinia chrysanthemi [81]) [74, 81]

34000 (Thermoascus aurantiacus) [22]

33000 (Acetivibrio cellulolyticus, enzyme C2) [82]

32000 (Thermoascus aurantiacus) [41]

31000 (Trichoderma koningii, enzyme Endo 4 [16, 63] , Aspergillus niger [33, 49]) [16, 33, 49, 63]

30000 (Neurospora crassa, enzyme Endo IV [3], Ruminococcus albus [38], Trichoderma viride enzyme II-A [65]) [3, 38, 65]

29000 (Lenzites trabea) [62]

28000 (Sclerotium rolfsii, enzyme B) [32]

28300–37500 (Sporotrichum pulverulentum) [20]

26000 (Aspergillus niger) [64]

25000 (Aspergillus aculeatus, enzyme FI [23], Fusarium monoliforme [69]) [23, 69]

Enzyme Handbook © Springer-Verlag Berlin Heidelberg 1991
Duplication, reproduction and storage in data banks are only
allowed with the prior permission of the publishers

23500 (Trichoderma viride, enzyme IV) [48]
23000 (Bacillus subtilis) [43]
21000 (Neurospora crassa [3], Phytophora infestans [53]) [3 , 53]
15500 (Irpex lacteus) [61]
13000 (Trichoderma koningii, enzyme Endo 1) [16, 63]
10400 (Acetivibrio cellulolyticus, enzyme C3) [82]
More (methods: gel filtration, non-denaturing gel electrophoresis, SDS-
PAGE, sedimentation equilibrum centrifugation)

Subunits

Monomer (1 × 65000, Clostridium thermocellum [27, 46], 1 × 51000,
Phaseolus vulgaris [29], 1 × 56500–59000, Penicillium funiculosum [44],
1 × 26000, Aspergillus niger [64], SDS-PAGE) [27, 29, 44, 46, 64]
Dimer (1 × 32000 + 1 × 22000, Sclerotium rolfsii, enzyme A) [32]

Glycoprotein/Lipoprotein

Glycoprotein (enzymes E2 and E3 [21], Endo I-II, V-VI [48] , E2 [79], comparison of fungal cellulases [72]) [2, 3, 21, 22, 27, 31, 32, 40, 42, 48, 54, 61,
65, 68, 72, 73, 79]; No glycoprotein (enzyme E1, E4, E5 [21], Endo IV [48], E1
[79]) [12, 21, 25, 29, 30, 33, 44, 48, 60, 64, 69, 79]

---

4 ISOLATION/PREPARATION

Source organism

Fungi (unidentified strain [2]) [2, 13, 36, 86–88]; Bacteria (overview) [6];
Plant [83]; Lathyrus odoratus (sweet pea) [1]; Phaseolus vulgaris cv. Red
Kidney [1, 11, 29, 70]; Neurospora crassa [3]; Clostridium thermocellum
(overproduced in E. coli [27, 46]) [4, 8, 27, 28, 39, 46]; Streptomyces reticuli
[7]; Clostridium josui [9]; Caldocellum saccharolyticum (gene expressed in
E. coli) [10]; Pisum sativum [12]; Trichoderma viride (mutant strain QM9414
[14]) [14, 48, 65–67]; Trichoderma reesei (formerly Trichderma viride QM6a)
[15, 55, 60]; Trichoderma koningii [16, 60, 63]; Pseudomonas fluorescens
var. cellulosa [18, 73]; Sporotrichum pulverulentum [20, 60];
Thermononospora fusca [21, 79, 89, 90]; Thermoascus aurantiacus [22,
41]; Aspergillus aculeatus [23]; Aspergillus fumigatus Fresenius [24];
Talaromyces emersonii [26]; Sporocytophaga myxococcoides [30, 60];
Humicola insolens [31]; Humicola grisea [31]; Sclerotium rolfsii [32];
Aspergillus niger [33, 49, 64]; Ruminococcus albus [38, 91]; Robillarda sp.
Y-20 [40, 44]; Penicillium funiculsosum [42]; Bacillus subtilis [43, 45];
Bacillus sp No. 1139 (alkalophilic) [50]; Aspergillus candidus [51];
Phytophora infestans [53]; Dolabella auricularia [54]; Humicola lanuginosa
[56]; Acetivibrio cellolyticus [57, 82]; Myceliophthora thermophila D-14 [58];
Aspergillus terreus [59]; Coniophora cerebella [60]; Ascomycetes [60];

Deuteromycetes [60]; Chaetomium thermophile [60]; Irpex lacteus
(Polyporus tulipiferae) [61, 68]; Lenzites trabea [62]; Fusarium monoliforme
[69]; Erwinia chrysanthemi [80, 81]; Bacteroides succinogenes S85 [92];
Pyricularia oryzae [71]; Myrothecium verrucaria [72, 76–78, 84, 85]; Penicillium notatum [74, 75]

Source tissue
Culture filtrate (fungi and bacteria); Cell lysate [27]; Leaf (abscission zone)
[1, 11, 29]; Anther (of flower buds) [1]; Plumules [12]; Hooks [12]; Swollen
apical epicotyl tissue [12]; Germinating cotyledons [70]

Localisation in source
Extracellular (fungi and bacteria); Cellulosome (high MW aggregate of cellulolytic enzymes) [6]; Periplasmic space or cell wall [18]; Cytoplasmic
granules [46]; Plasma membrane (plants) [83]

Purification
Lathyrus odoratus [1]; Phaseolus vulgaris (forms 4.5 and 4. 8 [70]) [1, 29,
70]; Fungi [2]; Neurospora crassa (4 enzymes) [3]; Clostridium thermocellum (endoglucanase 1 and 2 [5], cellulosome [6, 39], 1 of 5 enzymes [28],
overproduced in E. coli [27, 46]) [3, 5, 6, 27, 28, 39, 46]; Streptomyces reticuli
(2 enzymes) [7]; Clostridium josui [9]; Caldocellum saccharolyticum (expressed in E. coli) [10]; Pisum sativum [12]; Trichoderma viride (from commercial preparation [14], Endo I-VI from commercial preparation [48], 2 enzymes from commercial preparation [65], enzyme II from commercial
preparation [67]) [14, 48, 65, 67]; Trichoderma koningii (4 enzymes) [16];
Pseudomonas fluorescens var. cellulosa (2 extracellular, 1 cell wall bound)
[18, 73]; Sporotrichum pulverulentum (5 enzymes) [20]; Thermomonospora
fusca (5 enzymes [21], enzyme E1 and E2 [79]) [21, 79]; Thermoascus
aurantiacus [22, 41]; Aspergillus aculeatus (3 enzymes) [23]; Aspergillus
fumigatus Fresenius [24]; Talaromyces emersonii (4 enzymes) [26];
Sporocytophaga myxococcoides (2 enzymes) [30]; Humicola sp. [31];
Sclerotium rolfsii [32]; Aspergillus niger (from commercial preparation) [33,
49, 64]; Ruminococcus albus [38]; Robillarda sp. Y-20 (CMCase I [40],
CMCase II [44]) [40, 44]; Penicillium funiculosum [42]; Bacillus subtilis [43];
Bacillus sp No. 1139 [50]; Phytophora infestans [52]; Dolabella auricularia
[54]; Humicola lanuginosa [56]; Myceliophthora thermophila D-14 (multi-
enzyme complex) [58]; Aspergillus terreus [59]; Irpex lacteus (from commercial preparation [61], enzyme S-1 [68]) [61, 68]; Trichoderma koningii (4
components) [63]; Lenzites trabea [62]; Fusarium monoliforme [69]; Erwinia
chrysanthemi [80]; Acetivibrio cellulolyticus (components C2 and C3) [82]

Crystallization
[27, 46]

Enzyme Handbook © Springer-Verlag Berlin Heidelberg 1991
Duplication, reproduction and storage in data banks are only
allowed with the prior permission of the publishers

Cloned
[6, 11, 45, 46, 80, 81, 89–91]

Renaturated
[82]

---

5 STABILITY

pH
1–9 [64]; 2.0–9.0 (enzyme FI) [23]; 3.0–6.0 [61]; 3.0–8.0 (depending on isoelectric point [13], enzyme FII [23]) [13, 23, 69, 73]; 3.0–9.0 (enzyme FV) [23]; 3.0–11.0 [31]; 4.0–5.0 [68]; 4.0–7.0 [40, 44]; 4.5 [32]; 4.5–7.0 (enzyme II-B) [65]; 4.5–7.5 [67]; 5.0–7.0 (enzyme II-A) [65]; 5.0–8.0 [33, 49]; 5.5–10.0 [54]; 6–11 [50]; 7 [78]; 7.0–8.0 [18]

Temperature (°C)
0–20 (more than 72 h) [47]; 30 (up to, enzyme FII) [23]; 40 (up to) [54, 71]; 43 (up to) [53]; 45 (up to, enzyme FI) [23]; 50 (up to [6], 1 h, 80% activity [9], below [40, 44, 68], 10 min., 90% activity [50]) [6, 9, 40, 44, 50, 68]; 50–80 (depending on isoelectric point) [13]; 60 (10 min., complete inactivation [18], 30 min., 50% activity [47], 36 h, 5% activity [51], inactivation above [64, 69]) [18, 47, 51, 64, 69]; 65 (up to [22, 27], 5 min. stable [31]) [22, 27, 31]; 70 (7.5 h, 50% activity [10], 10 min., 80% activity [32]) [10, 32]; 75 (1–2 h, 50% activity) [26]; 95 (5 min., 45 % activity) [31]

Oxidation

Organic solvent

General stability information
Insensitive to freezing/thawing [7]; Cellulase 4.5 very unstable [70]; Native enzyme susceptible to endopeptidases, not to exopeptidases [75]

Storage
–20°C, at least 1 year [22]; –15°C, indefinitely [64]; –4 °C, 0.05 M acetate buffer, pH 5.0, at least 2 weeks [31]; Freeze-dried [26]; 4°C, diluted solution, several days [22, 64]; 4°C, several days [7]

---

6 CROSSREFERENCES TO STRUCTURE DATABANKS

PIR/MIPS code
CZCLAM (A, precursor, Clostridium thermocellum); CZCLBM (B, precursor, Clostridium thermocellum); CZCLDM (D, precursor, Clostridium thermocellum); CZCLCA (CCA, precursor, Clostridium cellulolyticum); A32884 (precursor, Pseudomonas solanacearum, fragment); A35136 (Bacillus polymyxa); JA0174 (kidney bean, fragment); JT0268 (C, Clostridium

3.2.1.4

thermocellum); JT0308 (alkalophilic Streptomyces); JT0347 (Clostridium thermocellum); S03767 (Z, precursor, Erwinia chrysanthemi); S02711 (precursor, Caldocellum saccharolyticum, contains: cellulase EC 3.2.1.4., Cellulose 1, 4-beta-cellobiosidase EC 3.2.1.91); JQ0356 (Butyrivibrio fibrisolvens); A25156 (1, Bacillus sp.); B25156 (2, Bacillus sp.); JS0174 (C, precursor, Bacillus sp.); A29003 (alkaline Bacillus sp.); A26874 (precursor, Bacillus subtilis); A27198 (precursor, Bacillus subtilis); A27631 (precursor, Clostridium acetobutylicum); A24993 (Cellulomonas fimi); S06328 (precursor, Avocado); JQ0458 (Aspergillus aculeatus); A25928 (A, precursor, Trichoderma reesei); A25565 (precursor, Trichoderma reesei)

Brookhaven code

---

7 LITERATURE REFERENCES

[1] Sexton, R., Del Campillo, E., Duncan, D., Lewis, L.N.: Plant Sci., 67, 169–176 (1990)
[2] Yamanobe, T., Mitsuishi, Y.: Agric. Biol. Chem., 54, 301–307 (1990)
[3] Yazdi, M.T., Radford, A., Keen, J.N., Woodward, J.R.: Enzyme Microb. Technol., 12, 120–123 (1990)
[4] Taniguchi, M., Kobayashi, M., Fujii, M.: Biotechnol. Bioeng., 34, 1092–1097 (1989)
[5] Halliwell, G., Halliwell, N.: Biochim. Biophys. Acta, 992, 223–229 (1989)
[6] Robson, L.M., Chambliss, G.H.: Enzyme Microb. Technol., 11, 626–644 (1989) (Review)
[7] Wachinger, G., Bronnenmeier, K., Staudenbauer, W.L., Schrempf, H.: Appl. Environ. Microbiol., 55, 2653–2657 (1989)
[8] Shima, S., Kato, J., Igarashi, Y., Kodama, T.: J. Ferment. Bioeng., 68, 75–78 (1989)
[9] Fujino, T., Sukhumavase, J., Sasaki, T., Ohmiya, K., Shimizu, S.: J. Bacteriol., 171, 4076–4079 (1989)
[10] Patchett, M.L., Neal, T.L., Schofield, L.R., Strange, R.C., Daniel, R.M., Morgan, H.W.: Enzyme Microb. Technol., 11, 113–115 (1989)
[11] Tucker, M.L., Sexton, R., Del Campillo, E., Lewis, L. N.: Plant Physiol., 88, 1257–1262 (1988)
[12] Maclachlan, G.: Methods Enzymol., 160, 382–391 (1988)
[13] Yamanobe, T., Mitsuishi, Y.: Agric. Biol. Chem., 54, 309–317 (1990)
[14] Voragen, A.G.J., Beldman, G., Rombouts, F.M.: Methods Enzymol., 160, 243–251 (1988)
[15] Schülein, M.: Methods Enzymol., 160, 234–242 (1988)
[16] Wood, T.M.: Methods Enzymol., 160, 221–233 (1988)
[17] Enari, T.-M., Niku-Paavola, M.-L.: Methods Enzymol., 160, 117–126 (1988)
[18] Yamane, K., Suzuki, H.: Methods Enzymol., 160, 200–210 (1988)
[19] Coughlan, M.P.: Methods Enzymol., 160, 135–144 (1988)
[20] Eriksson, K.-E., Pettersson, B.: Methods Enzymol., 160, 368–376 (1988)
[21] Wilson, D.B.: Methods Enzymol., 160, 314–323 (1988)
[22] Shepherd, M.G., Cole, A.L., Tong, C.-C.: Methods Enzymol., 160, 300–307 (1988)
[23] Murao, S., Sakamoto, R., Arai, M.: Methods Enzymol., 160, 274–299 (1988)
[24] Stewart, J.C., Heptinstall, J.: Methods Enzymol., 160 , 264–274 (1988)
[25] McCleary, B.V.: Methods Enzymol., 160, 74–86 (1988)
[26] Coughlan, M.P., Moloney, A.P.: Methods Enzymol., 160, 363–368 (1988)

Enzyme Handbook © Springer-Verlag Berlin Heidelberg 1991
Duplication, reproduction and storage in data banks are only
allowed with the prior permission of the publishers

[27] Beguin, P., Joliff, G., Juy, M., Amit, A.G, Millet, J., Poljak, R.J., Aubert, J.-P.: Methods Enzymol., 160, 355–362 (1988)
[28] Ng, T.K., Zeikus, J.G.: Methods Enzymol., 160, 351–355 (1988)
[29] Durbin, M.L., Lewis, L.N.: Methods Enzymol., 160, 342–351 (1988)
[30] Goksor, J.: Methods Enzymol., 160, 338–342 (1988)
[31] Hayashida, S., Ohta, K., Mo, K.: Methods Enzymol., 160, 323–332 (1988)
[32] Lachke, A.H., Despande, M.V.: FEMS Microbiol. Rev., 54, 177–194 (1988) (Review)
[33] Okada, G.: Methods Enzymol., 160, 259–264 (1988)
[34] Canevascini, G.: Methods Enzymol., 160, 112–116 (1988)
[35] Wood, T.M., Bhat, K.M.: Methods Enzymol., 160, 87–112 (1988)
[36] Van Tilbeurgh, H., Loontiens, F.G., De Bruyne, C.K., Claeyssens, M.: Methods Enzymol., 160, 45–58 (1988)
[37] Wood, T.M.: Methods Enzymol., 160, 19–25 (1988)
[38] Wood, T.M.: Methods Enzymol., 160, 216–221 (1988)
[39] Lamed, R., Bayer, E.A.: Methods Enzymol., 160, 472–482 (1988)
[40] Yoshigi, N., Taniguchi, H., Sasaki, T.: Agric. Biol. Chem., 52, 1389–1396 (1988)
[41] Feldman, K.A., Lovett, J.S., Tsao, G.T.: Enzyme Microb. Technol., 10, 262–272 (1988)
[42] Sahasrabudhe, N.A., Lachke, A.H., Ranjekar, P.K.: Biotechnol. Lett., 9, 881–886 (1987)
[43] Au, K.-S., Chan, K.-Y.: J. Gen. Microbiol., 133, 2155–2162 (1987)
[44] Uziie, M., Sasaki, T.: Enzyme Microb. Technol., 9, 459–465 (1987)
[45] Park, S.H., Pack, M.Y.: Enzyme Microb. Technol., 8, 725–728 (1986)
[46] Joliff, G., Beguin, P., Juy, M., Millet, J., Ryter, A., Poljak, R., Aubert, J.-P.: Bio/Technology, 4, 896–900 (1986)
[47] Kluepfel, D., Shareck, F., Mondou, F., Morosoli, R.: Appl. Microbiol. Biotechnol., 24, 230–234 (1986)
[48] Beldman, G., Searle-van Leeuwen, M.F., Rombouts, F.M. , Voragen, F.G.J.: Eur. J. Biochem., 146, 301–308 (1985)
[49] Okada, G.: Agric. Biol. Chem., 49, 1257–1265 (1985)
[50] Fukumori, F., Kudo, T., Horikoshi, K.: J. Gen. Microbiol., 131, 3339–3345 (1985)
[51] Ortega, J.: Biotechnol. Lett., 7, 109–112 (1985)
[52] Buchholz, K., Rapp, P., Zadrazil, F. in " Methods Enzym. Anal.", 3rd. Ed. (Bergmeyer, H.U., Ed.) 4, 178–188 (1984)
[53] Bodenmann, J., Heiniger, U., Hohl, H.R.: Can. J. Microbiol., 31, 75–82 (1984)
[54] Anzai, H., Nisizawa, K., Matsuda, K.: J. Biochem., 96 , 1381–1390 (1984)
[55] Montenecourt, B.S.: Trends Biotechnol., 1, 156–161 (1983)
[56] Olutiola, P.O.: Experientia, 38, 1332–1333 (1982)
[57] MacKenzie, C.R., Bilous, D.: Can. J. Microbiol., 28, 1158–1164 (1982)
[58] Sen, S., Abraham, T.K., Chakrabarty, S.L.: Can. J. Microbiol., 28, 271–277 (1982)
[59] Garg, S.K., Neelakantan, S.: Biotechnol. Bioeng., 24, 737–742 (1982)
[60] Goksoyr, J., Eriksen, J.: Econ. Microbiol., 5, 283–330 (1980) (Review)
[61] Kanda, T., Wakabayashi, K., Nisizawa, K.: J. Biochem., 87, 1625–1634 (1980)
[62] Herr, D., Baumer, F., Dellweg, H.: Arch. Microbiol., 117, 287–292 (1978)
[63] Wood, T.M., McCrae, S.I.: Biochem. J., 171, 61–72 (1978)
[64] Hurst, P.L., Nielsen, J., Sullivan, P.A., Shepherd, M.G.: Biochem. J., 165, 33–41 (1977)
[65] Okada, G.: J. Biochem., 77, 33–42 (1975)
[66] Okada, G., Nisizawa, K.: J. Biochem., 78, 297–306 (1975)
[67] Okada, G.: J. Biochem., 80, 913–922 (1976)
[68] Kanda, T., Wakabayashi, K., Nisizawa, K.: J. Biochem. , 79, 977–988 (1976)

[69] Matsumoto, K., Endo, Y., Tamiya, N., Kano, M., Miyauchi, K., Abe, J.: J. Biochem., 76, 563–572 (1974)

[70] Lew, F.T., Lewis, L.N.: Phytochemistry, 13, 1359–1366 (1974)

[71] Sudo, T., Nagayama, H., Tamari, K.: Agric. Biol. Chem., 37, 1651–1659 (1973)

[72] Whitaker, D.R. in "The Enzymes", 3rd. Ed. (Boyer, P. D., Ed.) 5, 273–298 (1971) (Review)

[73] Yamane, K., Suzuki, H., Nisizawa, K.: J. Biochem., 67, 19–35 (1970)

[74] Pettersson, G., Eaker, D.L.: Arch. Biochem. Biophys., 124, 154–159 (1968)

[75] Eriksson, K.E., Pettersson, G.: Arch. Biochem. Biophys., 124, 160–166 (1968)

[76] Larner, J. in "The Enzymes", 2nd. Ed. (Boyer, P.D., Lardy, H., Myrbäck, K., Eds.) 4, 369–378 (1960) (Review)

[77] Whitaker, D.R.: Arch. Biochem. Biophys., 53, 439 (1954)

[78] Kooima, P.: Enzymologia, 18, 371 (1957)

[79] Calza, R.E., Irwin, D.C., Wilson, D.B.: Biochemistry, 24, 7797–7804 (1985)

[80] Boyer, M.-H., Chambost, J.-P., Magnan, M., Cattaneo, J.: J. Biotechnol., 1, 241–252 (1984)

[81] Boyer, M.-H., Cami, B., Chambost, J.-P., Magnan, M., Cattaneo, J.: Eur. J. Biochem., 162, 311–316 (1987)

[82] Saddler, J.N., Khan, A.W.: Can. J. Microbiol., 27, 288–294 (1981)

[83] Koehler, D.E., Leonard, R.T., Vanderwoude, W., Linkins, A.E., Lewis, L.N.: Plant Physiol., 58, 324 (1976)

[84] Whitaker, D.R., Colvin, J.R., Cook, W.H.: Arch. Biochem. Biophys., 49, 257 (1954)

[85] Basu, S.N., Whitaker, D.R.: Arch. Biochem. Biophys., 42, 12 (1953)

[86] Ryu, D.D., Mandels, M.: Enzyme Microb. Technol., 2, 91–102 (1980) (Review)

[87] Mandels, M. in "Annual Reports On Fermentation" (Tsao, G.T., Ed.) 5, 37–78, N.Y. Academic Press (1982) (Review)

[88] Couglan, M.P., Ljurgdahl, L.G. in "Biochemistry And Genetics Of Cellulose Degration" (Aubert, J.-P., Beguin, P., Millet, J., Eds.) pp.11–30, N.Y. Academic Press (1988) (Review)

[89] Ghangas, G.G., Wilson, D.B.: Appl. Environ. Microbiol., 53, 1470–1475 (1987)

[90] Collmer, A., Wilson, D.B.: Bio/Technology, 1, 594–601 (1983)

[91] Kawai, S., Honda, H., Tanase, T., Taya, M., Iijima, S., Kobayashi, T.: Agric. Biol. Chem., 51, 59–63 (1987)

[92] Schellhorn, H.E., Forsberg, C.W.: Can. J. Microbiol., 30, 446–448 (1984)

Enzyme Handbook © Springer-Verlag Berlin Heidelberg 1991
Duplication, reproduction and storage in data banks are only
allowed with the prior permission of the publishers

1 NOMENCLATURE

EC number
3.2.1.6

Systematic name
1, 3-(1, 3;1, 4)-Beta-D-glucan 3(4)glucanohydrolase

Recommended name
Endo-1, 3(4)-beta-glucanase

Synonymes
Beta-1, 3-glucanase [6]
Glucanase, endo-1, 3(4)-.beta.-
.beta.-1, 3–1, 4-Glucanase
Laminarinase
Laminaranase
Endo-1, 3-beta-glucanase
Endo-.beta., 1, 3(4)-glucanase
Endo-. beta.-1, 3–1, 4-glucanase
Endo-beta-(1 --> 3)-D-glucanase [3, 14]
Endo-beta-1, 3-glucanase IV [9]
Endo-1, 3-beta-D-glucanase [6]
Endo-beta-(1–3)-D-glucanase [7]
Endo-1, 3–1, 4-.beta.-D-glucanase

CAS Reg. No.
62213-14-3

---

2 REACTION AND SPECIFICITY

Catalysed reaction
Endohydrolysis of 1, 3- or 1, 4-linkages in beta-D-glucans when the glucose
residue whose reducing group is involved in the linkage to be hydrolysed is
itself substituted at C-3

Reaction type
O-Glycosyl bond hydrolysis

Natural substrates
Beta-D-glucans + H_2O

Enzyme Handbook © Springer-Verlag Berlin Heidelberg 1991
Duplication, reproduction and storage in data banks are only
allowed with the prior permission of the publishers

Substrate spectrum
1. Laminarin + H_2O [2, 3, 17]
2. Laminarin + H_2O [7, 18]
3. Lichenin + H_2O
4. Beta-D-glucans + H_2O
5. Pachyman (short-chain) + H_2O [3, 9]
6. Glucan + H_2O (beta-(1 --> 3)-beta-(1 --> 4)mixed-linked [3], yeast glucan [9]) [3, 9]
7. Laminarin + H_2O [9]
8. Curdlan (paramylon) + H_2O [9]
9. Lichenan + H_2O [9]
10. Paramylon + H_2O [9]
11. Laminarin + H_2O [11]
12. Laminarin + H_2O [15]
13. Laminaribiose + H_2O (0.03% the rate of laminarin [18]) [15, 17]
14. Laminaritriose + H_2O [17]
15. Pustulan + H_2O (slight) [17]
16. Gentiotriose + H_2O (slight) [17]
17. Amygdalin + H_2O (slight) [17]
18. More (transglycosylase activity) [3, 9]

Product spectrum
1. Laminaridextrins + glucose (minor product) [2]
2. Laminaribiose + D-glucose [7, 18]
3. ?
4. ?
5. ?
6. Laminaripentaose (main product) [9]
7. Laminaripentaose (main product) [9]
8. Laminaripentaose (main product) [9]
9. Laminaripentaose (main product) [9]
10. Laminaripentaose (main product) [9]
11. Glucose (70–90%) + laminaribiose (small amount) + glucose oligomers (tetrasaccharide) (Bacillus thetaiotaomicron) [11]
12. Glucose [15]
13. ?
14. ?
15. ?
16. ?
17. ?
18. More [11]

Inhibitor(s)
Mn^{2+} (no effect [9]) [2]; Hg^{2+} [2, 3, 9, 15, 20]; N-Bromosuccinimide [2]; $KMnO_4$ [2]; Ag^+ [9]; Heavy metal ions [20]

Cofactor(s)/prostethic group(s)

Metal compounds/salts

Turnover number (min⁻¹)

Specific activity (U/mg)
66.0 [3]; 366.7 [15]; 0.4 [7]; 350 [18]; More [19, 20]

K_m-value (mM)
0.004 (short-chain pachyman) [3]; 0.0032 (laminarin) [15]; 8.3 (laminaribiose) [15]; More [2]

pH-optimum
5.0 (Trichoderma longibrachiatum, laminarin) [2]; 5.5 (Rizoctonia solani, short-chain pachyman) [3]; 6.5 (Flavobacterium dorminator, laminarin) [9]; 6.0 (yeast, glucan, Flavobacterium dorminator) [9]; 5.5 (Rhizopus niveus) [15]; 5.8 (Spisula sacchalinensis) [18]

pH-range
3.5–6.5 (3: 20% of maximum activity, 6.5: 85% of maximum activity, Spisula sacchalinensis) [18]; 2.0–8.0 (5–10% of maximal activity at 2.0 and 8.0, Rhizopus niveus) [15]

Temperature optimum (°C)
50 (Trichoderma longibrachiatum) [2]; 60 (Rhizoctonia solani) [3]; 40 (Flavobacterium dorminator, yeast glucan) [9]; 55 (Flavobacterium dorminator, laminarin) [9]; 60 (Rhizopus niveus) [15]; 45 (Spisula sacchalinensis) [18]

Temperature range (°C)
30–80 (30: 90% of maximal activity, 80: 40% of maximal activity, Trichoderma longibrachiatum) [2]; 20–80 (20: 25% of maximal activity, 80: 15–20% of maximal activity, Rhizopus niveus) [15]

---

3 ENZYME STRUCTURE

Molecular weight
29000 (SDS-PAGE, Rhizoctonia solani) [3]
160000 (gel exclusion chromatography, Saccharomyces cerevisiae) [7]
33000 (gel filtration, SDS-PAGE, ultracentrifugation, Flavobacterium dorminator) [9]
166000 (molecular-sieve chromatography, Flavobacterium) [14]
22000 (gel filtration, Spisula sacchalinensis) [18]
21000 (SDS-PAGE, gel filtration, Candida utilis) [20]

Subunits

Enzyme Handbook © Springer-Verlag Berlin Heidelberg 1991
Duplication, reproduction and storage in data banks are only
allowed with the prior permission of the publishers

Glycoprotein/Lipoprotein
Glycoprotein (10.3% carbohydrate, 17 hexose units [9], acidic glycoprotein [20]) [9, 20]

4 ISOLATION/PREPARATION

Source organism
Flavobacterium dorminator (var. glucanolyticae) [9]; Saccharomyces cerevisiae [7]; Irpex lacteus [10]; Bacteroides thetaiotaomicron [11]; Bacteroides distasonis [11]; Bacteroides fragilis [11]; Penicillium italicum [12]; Bacillus polymxa [13]; Flavobacterium [14]; Rhizopus niveus [15]; Spisula sacchalinesis (bivalvia) [16, 18]; Saccharomyces lactis [17]; Bacillus macerans [1]; Trichoderma longibrachiatum [2]; Rhizoctonia solani [3]; Lentinus edodes [5]; Clostridium thermocellum [4]; Pisum sativum [6]; Botrytis cinerea [8]; Basidiomycetes [19]; Candida utilis [20]; Rye [21]

Source tissue
Mycelium [8]; Culture medium [6, 8, 15, 20]; Seedlings [6]; Crystalline style [18]

Localisation in source
Extracellular [2, 20]; Cell-bound [11]

Purification
Rhizoctonia solani [3]; Saccharomyces cerevisiae [7]; Rhizopus niveus [15]; Saccharomyces lactis (partial) [17]; Spisula sacchalinensis [18]; Candida utilis [20]

Crystallization
–

Cloned
–

Renaturated
–

5 STABILITY

pH
3.0–9.0 (4°C, 48 hours, Trichoderma longibrachiatum, stable) [2]; 6–9 (10 minutes, 55°C, stable, unstable below 6 and above 9, Flavobacerium dorminator) [9]; 5.0–8. 5 (Rhizoctonia solani) [3]

Temperature (°C)
30 (absence of substrate, 1 hour, stable, Trichoderma longibrachiatum) [2];
60 (absence of substrate, 10 minutes, complete loss of activity, Trichoderma
longibrachiatum) [2]; 50 (stable below, Rhizoctonia solani) [3]; 40 (stable
below, 10 minutes, Flavobacterium dorminator) [9]; 20 (7 days, 8% loss of
activity, Spisula sacchalinensis) [18]; 60 (3 minutes, 80% loss of activity,
Spisula sacchalinensis) [18]

Oxidation

Organic solvent

General stability information

Storage
0°C, 7 days [18]

---

6 CROSSREFERENCES TO STRUCTURE DATABANKS

PIR/MIPS code
JA0088 (leadwort-leaved tobacco)

Brookhaven code

---

7 LITERATURE REFERENCES

[1] Borriss, R., Schröder, K.-L.: Zentralbl. Bakteriol. Parasitenkd. Infektionskr. Hyg., Abt.
    II, 136, 324–329 (1981)
[2] Sharma, A., Nakas, J.P.: Enzyme Microb. Technol., 9, 89–93 (1987)
[3] Totsuka, A., Usui, T.: Agric. Biol. Chem., 50, 543–550 (1986)
[4] Schwarz, W., Bronnenmeier, K., Staudenbauer, W.L.: Biotechnol. Lett., 7, 859–864
    (1985)
[5] Leatham, G.F.: Appl. Environ. Microbiol., 50, 859–867 (1985)
[6] Wong, Y.-S., MacClachlan, G.A.: Plant Physiol., 65, 222–228 (1980)
[7] Villa, T.G., Phaff, H.J.: Eur. J. Appl. Microbiol. Biotechnol., 9, 9–14 (1980)
[8] Martinez, M.J., Reyes, F., Lahoz, R., Perez-Leblic, M. I.: FEMS Microbiol. Lett., 19,
    157–160 (1983)
[9] Nagasaki, S., Mori, H., Yamamoto, S.: Agric. Biol. Chem., 45, 2689–2694 (1981)
[10] Kawai, M., Noguchi, M., Shimura, G., Suga, Y., Samejima, H.: Agric. Biol. Chem., 42,
    333–337 (1978)
[11] Salyers, A.A., Palmer, J.K., Wilkins, T.D.: Appl. Environ. Microbiol., 33, 1118–1124
    (1977)
[12] Santos, T., Villanueva, J.R., Nombela, C.: J. Bacteriol., 129, 52–58 (1977)
[13] Eka, Q.U., Fogarty, W.M.: Eur. J. Appl. Microbiol. Biotechnol., 1, 13–23 (1975)
[14] Manners, D.J., Wilson, G.: Biochem. J., 135, 11–18 (1973)
[15], Horitsu, H., Satake, T., Tomoyeda, M.: Agric. Biol. Chem., 37, 1007–1012 (1973)
[16] Isakov, V.V., Sova, V.V., Denisenko, V.A., Sakharovsky, V.G., Alyakova, L.A.,
    Dzizenko, A.K.: Biochim. Biophys. Acta, 268, 184–186 (1972)

Enzyme Handbook © Springer-Verlag Berlin Heidelberg 1991
Duplication, reproduction and storage in data banks are only
allowed with the prior permission of the publishers

[17] Tingle, M.A., Halvorson, H.O.: Biochim. Biophys. Acta, 250, 165–171 (1971)
[18] Sova, V.V., Elyakova, L.A., Vaskovsky, V.E.: Biochim. Biophys. Acta, 212, 111–115 (1970)
[19] Totani, K., Harumiya, S., Nanjo, F., Usui, T.: Agric. Biol. Chem., 47, 1159–1162 (1983)
[20] Villa, T.G., Notario, V., Villanueva, J.R.: Biochem. J., 177, 107–114 (1979)
[21] Manners, D.J., Marshall, J.J.: Phytochemistry, 12, 547–553 (1973)

1 NOMENCLATURE

EC number
3.2.1.7

Systematic name
2, 1-Beta-D-fructan fructanohydrolase

Recommended name
Inulinase

Synonymes
Inulase
Endoinulinase [1, 3, 5, 6, 7, 12]
Endo-inulinase [5]
Exoinulinase (literature doesn't permit clear classification of exoinulinase to
EC 3.2.1.7 or 3.2.1.26) [3, 4, 6, 8, 12, 13, 17, 18]

CAS Reg. No.
9025-67-6

---

2 REACTION AND SPECIFICITY

Catalysed reaction
Endohydrolysis of 2, 1-beta-D-fructosidic linkages in inulin; More (different
definitions of reaction catalyzed by EC 3.2.1.7: splitting-off fructose moieties
from the nonreducing end of inulin molecule or from sugars displaying
fructose unit at beta-2, 1-position [9, 11, 14, 15, 16, 19, 20, 25],
exo-inulinase: liberates single fructose molecules from polysaccharide
chain of inulin [12, 13, 19]) [9, 11–16, 19, 20, 25]

Reaction type
O-Glycosyl bond hydrolysis

Natural substrates
Inulin + H_2O

Enzyme Handbook © Springer-Verlag Berlin Heidelberg 1991
Duplication, reproduction and storage in data banks are only
allowed with the prior permission of the publishers

Substrate spectrum

1 Inulin + H_2O (endoinulinase) [1, 2, 3, 7, 11, 12, 16, 23, 25]
2 Inulin + H_2O (exoinulinase) [4, 5, 8, 12, 13, 14, 16, 17, 19]
3 Sucrose + H_2O (low activity [3], no activity [7]) [3, 5, 8, 11, 13, 14, 16, 17, 19, 23, 24, 25]
4 Raffinose + H_2O (exo-acting enzyme [8]) [8, 9, 14, 17, 19, 24]
5 Stachyose + H_2O [9, 14]
6 Nystose + H_2O [7]
7 1-Kestose + H_2O [7]
8 More (endo-acting inulinases : no hydrolysis of sucrose and raffinose [20], not: melizitose [16, 24], exo-acting emzyme: no effect on melizitose and maltose [8], plant inulinase: no invertase activity [11], invertase activity [3, 5], not: levan, sucrose, raffinose, melizitose) [3, 5, 8, 11, 16, 24]

Product spectrum

1 Fructose oligomers (of various chain-lengths [5, 12], inulotriose + inulotetraose + inulopentaose are main products [1, 2], tri-, tetra-, penta- and hexasaccharides [7], fructose + oligofructosides [1]) [1, 2, 5, 7, 12]
2 Fructose (main product [4, 5, 8, 9, 11, 12, 13, 17, 19]) + glucose (traces [5]) [4, 5, 8, 9, 11, 12, 13, 17, 19]
3 Glucose + fructose
4 Fructose [14]
5 Fructose [14]
6 ?
7 ?
8 ?

Inhibitor(s)

Ba^{2+} [11]; Cu^{2+} [11]; Zn^{2+} [11, 17]; I_2 (endoinulinase, not exoinulinase) [12]; $AgNO_3$ [16]; $HgCl_2$ [16]; Phenylhydrazine [16]; Urea [16]; Hydrochloride [16]; SDS [16]; N-Bromosuccinimide [17]; EDTA (no effect [12, 14], exo-and endo-inulinase [5]) [5, 11]; Mg^{2+} (exo-and endo-inulinase [5]) [5, 11]; Mn^{2+} (exo-and endo-inulinase [5], slight effect [17], no effect [12]) [5, 11, 17]; Fe^{3+} (slight effect [17], no effect [12], exo-and endo-inulinase [5]) [11, 17]; Pb^{2+} (slight effect [17]) [11, 17]; Ag^+ (slight effect [17]) [11, 14, 17]; Hg^{2+} [11, 14, 17]; p-Chloromercuribenzoic acid [11, 16]

Cofactor(s)/prostethic group(s)

More (no low-weight molecular cofactor) [17]

Metal compounds/salts

Fe^{3+} (strong activator) [5]

Turnover number (min^{-1})

Specific activity (U/mg)
106.2 [1]; 101 [2]; 20.8 [5]; 61.6 (endoinulinase I) [3]; 328.1 (endoinulinase II)
[3]; 50.3 (endoinulinase III) [3]; More [3, 11, 14, 16, 17, 24]

K$_m$-value (mM)
570 (endoinulinase, inulin) [5]; 60 (exoinulinase, inulin) [5]; 0.21 (inulin) [7];
0.24 (fructo-oligosaccharide, 1^F(1-beta-D-fructofuranosyl)$_8$sucrose) [7]; 0.33
(fructo-oligosaccharide, 1^F(1-beta-D-fructofuranosyl)$_6$sucrose) [7]; 0.85
(fructo-oligosaccharide, 1^F(1-beta-D-fructofuranosyl)$_5$sucrose) [7]; 3.8
(fructo-oligosaccharide, 1^F(1-beta-D-fructofuranosyl)$_4$sucrose) [7]; 2.8
(fructo-oligosaccharide, 1^F(1-beta-D-fructofuranosyl)$_3$sucrose) [7]; 16
(nystose) [7]; 8.4 (1-kestose) [7]; 15.7 (sucrose) [9]; 8.2 (raffinose) [9]; 9.7
(stachyose) [9]; 6.7 (inulin, soluble) [10]; 10 (inulin, immobilized) [10]; 225
(sucrose, exoinulinase) [13]; 3. 65 (sucrose) [14]; 7.5 (stachyose) [14]; 4.79
(raffinose) [14]; More [8, 11, 13, 14, 17, 18, 24]

pH-optimum
4.5 (Candida kefyr) [11]; 3.5 (Candida salmeticensis) [11]; 4.0
(Debaromyces phaffii) [11]; 6.0 (Bacillus sp., thermophilic, exoinulinase)
[13]; 6.0 (Panaeolus papillonaceus) [14]; 4.5 (Kluyveromyces fragilis) [16];
4–5 (Talaromyces flavus) [11]; 4 (Debaromyces phaffii) [17]; 4
(Debaromyces cantarelli) [18]; 4.3–4.4 (Aspergillus niger) [19]; 4.6
(Clostridium acetobutylicum) [24]; 6.0–7. 0 (Chrysosporium pannorum) [1];
4.5 (Aspergillus ficuum, immobilized) [4]; 5.1 (Penicillium purpurogenum)
[7]; 5.8 (Fusarium oxyspoum, mycelial) [8]; 6.2 (Fusarium oxysporum, extracellular) [8]; 4.5–5.0 (inulin, Kluyveromyces marxianus) [9]; 3.5 (sucrose,
Kluyveromyces marxianus) [9]; 6.0 (soluble, Kluyveromyces marxianus) [10];
5.0 (immobilized, Kluyveromyces marxianus) [10]

pH-range
4.2–7.0 (Fusarium oxysporum , extracellular) [8]; 3.0–8.5 (50% of maximum
activity at 8.5, Panaeolus papillonaceus) [14]; 4.0–9.0 (4.0: not active at and
below, 9.0: 45% of maximal activity, exoinulinase, Bacillus sp.) [13]; 2–8 (2:
75–90% of maximal activity, 8: 10–20% of maximal activity, inulin, sucrose,
raffinose, Debaromyces phaffii) [17]; 3–5.5 (3: 40% of maximal activity, 5.5:
60% of maximal activity, exo-acting enzyme, Aspergillus niger) [19]; More
[11, 15]

Enzyme Handbook © Springer-Verlag Berlin Heidelberg 1991
Duplication, reproduction and storage in data banks are only
allowed with the prior permission of the publishers

Temperature optimum (°C)

50 (Chrysosporium pannorum) [1]; 30 (mycelial, Fusarium oxysporum) [8]; 60 (immobilized, Aspergillus ficuum) [4]; 37 (Fusarium oxysporum, extracellular) [8]; 70 (sucrose, Klyveromyces marxianus) [9]; 50 (inulin, Kluyveromyces marxianus) [9]; 50 (soluble, Kluyveromyces marxianus) [10]; 55 (immobilized, Kluyveromyces marxianus) [10]; 45–50 (Penicillium sp.) [11]; 40–50 (Talaromyces flavus) [11]; 45–60 (Aspergillus niger) [11]; 50 (Candida kefyr) [11]; 46 (Candida salmeticensis) [11]; 55 (Kluyveromyces fragilis) [11]; 49 (Kluyveromyces fragilis) [11]; 30 (Debaromyces cantarelli) [11]; 50 (Debaromyces phaffii) [11]; 45 (Bacillus sp., thermophilic, ex-oinulinase) [13]; 60–65 (Panaeolus papillonaceus) [14]; 50 (Debaromyces phaffii) [17]; 50 (Debaromyces cantarelli) [18]; 55–56 (Aspergillus niger) [19]

Temperature range (°C)

30–70 (Aspergillus ficuum) [4]; 10–60 (Bacillus sp., thermophilic, 60°C: 70% of maximal activity) [13]; 49–64 (49°C: 85% of maximal activity, 64°C: 15% of maximal activity, Aspergillus niger) [19]; 10–60 (Debaromyces phaffii, 10°C: 10% of maximal activity, 60°C: 50% of maximal activity) [17]; More [14]

---

3 ENZYME STRUCTURE

Molecular weight

53000 (Aspergillus, HPLC, endoinulinase) [5]
64000 (SDS-disc electrophoresis, gel filtration, Penicillium purpurogenum) [7]
300000 (gel filtration, Fusarium oxysporum) [8]
81000 (HPLC, Aspergillus, exoinulinase) [5]
116000 (gel permeation chromatography, Panaeolus papillonaceus) [14]
58000 (SDS-PAGE, Chrysosporium pannorum) [1]
56000 (gel filtration, Chrysosporium pannorum) [1]
54000 (Aspergillus niger) [2]
64000 (Aspergillus ficuum, gel filtration, endoinulinase) [3]
74000 (Aspergillus ficuum, gel filtration, exoinulinase) [3]

Subunits

Dimer (1 × 60000, 1 × 56000, SDS-PAGE, Panaeolus papillonaceus) [14]

Glycoprotein/Lipoprotein

Glycoprotein (Chrysosporium pannorum: 5% carbohydrate [1], Aspergillus ficuum: endoinulinase I (25%), endoinulinase II (22%), endoinulinase III (33%), exoinulinase I (41%), exoinulinase II (39%), exoinulinase III (24%), exoinulinase IV (23%), exoinulinase V (24%) [3]) [1, 3]

4 ISOLATION/PREPARATION

Source organism
   Chrysosporium pannorum [1, 6]; Aspergillus ficuum (exoinulinase: 5 forms, endoinulinase: 3 forms [3], immobilized [4]) [3, 4]; Aspergillus (endo- and exo-inulinase [5], one exoinulinase and 3 endoinulinases [12]) [5, 12]; Penicillium purpurogenum [7]; Fusarium oxysporum [8]; Kluyveromyces marxianus (immobilized [10]) [9, 10, 22]; Plants [11]; Bacillus sp. (thermophilic strains) [13, 26]; Panaeolus papillonaceus [14]; Debaromyces cantarelli [15, 18]; Kluyveromyces fragilis [16, 21]; Debaromyces phaffii [17]; Aspergillus niger (exo-acting enzyme [19]) [2, 19]; Flavobacterium multivorum [23]; Clostridium acetobutylicum [24, 25]; Bacteria [23, 26]; Talaromyces flavus [11]; Candida kefyr [11]; Candida salmeticensis [11]

Source tissue
   Cell [1]; Culture medium [7, 8, 9]; Mycelium [8]

Localisation in source
   Intracellular [11, 16, 19, 23, 25]; Extracellular [8, 9, 11, 16, 17, 19, 24, 25]; Cell wall (associated with) [9, 15, 17, 21]; Soluble [21]; More (Kluyveromyces marxianus: distribution of inulinase in supernatant, cell-wall and cell bound fraction is dependent on the nature of carbon-limiting substrate, dilution rate, medium composition and growth temperature) [9]

Purification
   Fusarium oxysporum [8]; Panaeolus papillonaceus [14]; Kluyveromyces fragilis [16]; Chrysosporium pannorum [1]; Aspergillus ficuum (5 exoinulinases, 3 endoinulinases) [3]; Aspergillus [5]; Penicillium purpurogenum [7]

Crystallization
   –

Cloned
   –

Renaturated
   –

5 STABILITY

pH
   4.5–8.5 (24 hours, 30°C, Chrysosporium pannorum) [1]; 5.0–7.5 (Penicillium purpurogenum) [7]; 5.0–8.0 (exoinulinase, Bacillus sp.) [13]; 4.0–10.0 (Panaeolus papillonaceus) [14]

Enzyme Handbook © Springer-Verlag Berlin Heidelberg 1991
Duplication, reproduction and storage in data banks are only
allowed with the prior permission of the publishers

Temperature (°C)

45 (stable up to, 10 minutes, Chrysosporium pannorum) [1]; 40 (Aspergillus ficuum, stable up to) [4]; 60 (Chrysosporium, complete loss of activity after 30 minutes) [1]; 55 (Penicillium purpurogenum, stable up to) [7]; 60 (Kluyveromyces marxianus, 8 hours) [9]; 60 (Kluyveromyces marxianus, half-life: 30 minutes) [9]; 30 (Kluyveromyces marxianus, 40 days, immobilized, 24% loss of activity) [10]; 50 (Debaromyces cantarelli, 5% loss of activity after 5 minutes) [11]; 60 (60% loss of activity after 10 minutes, Debaromyces phaffii) [11]; 50 (29% loss of activity after 30 minutes, Candida sameticensis) [11]; 50 (pH 5, 10 minutes, exoinulinase, Bacillus sp.) [13]; 70 (pH 5, 10 minutes, complete loss of activity, Bacillus sp., exoinulinase) [13]; 63 (pH 5, 10 minutes, 50 % loss of activity, Bacillus sp., exoinulinase) [13]; 30 (7 days, Panaeolus papillonaceus) [14]; 40 (4 hours, Panaeolus papillonaceus) [14]; 70 (denatured immediately, Debaromyces phaffii) [17]; 50 (Debaromyces phaffii) [17]; 47 (10 minutes, Clostridium acetobutylicum) [24]; 50.5 (10 minutes, 50% loss of activity, Clostridium acetobutylicum) [24]; 55 (10 minutes, complete loss of activity, Clostridium acetobutylicum) [24]; More (heat stability improved by immobilization [4, 10]) [4, 10, 11, 13, 14, 17, 18]

Oxidation

Organic solvent

General stability information

Operational stability of immobilized enzyme [10]; Freezing, thawing, lyophilization (no effect) [14]; Dialysis (no effect) [17]; Calcium alginate immobilized cells (higher stability) [22]

Storage

4°C, 60 days [13]

6 CROSSREFERENCES TO STRUCTURE DATABANKS

PIR/MIPS code

PT0030 (Exol, Aspergillus ficuum, fragment)

Brookhaven code

7 LITERATURE REFERENCES

[1] Xiao, R., Tanida, M., Takao, S.: J. Ferment. Bioeng., 67, 244–248 (1989)
[2] Nakamura, T., Kurokawa, T., Nakatsu, T., Ueda, S.: Nippon Nogeikagaku Kaishi, 52, 159–166 (1978)
[3] Ettalibi, M., Baratti, J.C.: Appl. Microbiol. Biotechnol., 26, 13–20 (1987)
[4] Kim, C.H., Rhee, S.K.: Biotechnol. Lett., 11, 201–206 (1989)

[5] Azhari, R., Szlak, A.M., Ilan, E., Sideman, S., Lotan, N.: Biotechnol. Appl. Biochem., 11, 105–117 (1989)

[6] Xiao, R., Tanida, M., Takao, S.: J. Ferment. Technol., 66, 553–558 (1988)

[7] Onodera, S., Shiomi, N.: Agric. Biol. Chem., 52, 2569–2576 (1988)

[8] Gupta, A.K., Nagpal, B., Kaur, N., Singh, R.: J. Chem. Technol. Biotechnol., 42, 69–76 (1988)

[9] Rouwenhorst, R.J., Visser, L.E., Van Der Baan, A.A., Scheffers, W.A., Van Dijken, J.P.: Appl. Environ. Microbiol., 54, 1131–1137 (1988)

[10] Bajpai, P., Margaritis, A.: J. Ferment. Technol., 65, 239–242 (1987)

[11] Vandamme, E.J., Derycke, D.G.: Adv. Appl. Microbiol., 29, 139–176 (1983) (Review)

[12] Beck, R.H.F., Praznik, W.: J. Chromatogr., 369, 240–243 (1986)

[13] Allais, J.-J., Hoyos-Lopez, G., Baratti, J.: Carbohydr. Polym., 4, 277–290 (1987)

[14] Murkherjee, K., Sengupta, S.: Can. J. Microbiol., 33, 520–524 (1987)

[15] Beluche, I., Guiraud, J.P., Galzy, P.: Folia Microbiol., 25, 32–39 (1980)

[16] Negoro, H.: J. Ferment. Technol., 56, 102–107 (1978)

[17] Demeulle, S., Guiraud, J.P., Galzy, P.: Z. Allg. Mikrobiol., 21, 181–189 (1981)

[18] Guiraud, J.P., Bernit, C., Galzy, P.: Folia Microbiol., 27, 19–24 (1982)

[19] Derycke, D.G., Vandamme, E.J.: J. Chem. Technol. Biotechnol., 34B, 45–51 (1984)

[20] Nakamura, T., Hoashi, S., Nakatsu, S.: Nippon Nogeikagaku Kaishi, 52, 105–110 (1978)

[21] Workman, W.E., Day, D.F.: Antonie Leeuwenhoek, 50, 349–353 (1984)

[22] Bajpai, P., Margaritis, P.: Enzyme Microb. Technol., 7, 34–36 (1985)

[23] Allais, J.-J., Kammoun, S., Blanc, P., Girard, C., Barattti, J.C.: Appl. Environ. Microbiol., 52, 1086–1090 (1986)

[24] Efstathiou, I., Reysset, G., Truffaut, N.: Appl. Microbiol. Biotechnol., 25, 143–149 (1986)

[25] Looten, P., Blanchet, D., Vandecasteele, J.P.: Appl. Microbiol. Biotechnol., 25, 419–425 (1987)

[26] Allais, J.-J., Hoyos-Lopez, G., Kammoun, S., Baratti, J.C.: Appl. Environ. Microbiol., 53, 942–945 (1987)

Enzyme Handbook © Springer-Verlag Berlin Heidelberg 1991
Duplication, reproduction and storage in data banks are only
allowed with the prior permission of the publishers

1 NOMENCLATURE

EC number
3.2.1.8

Systematic name
1, 4-Beta-D-xylan-xylanohydrolase

Recommended name
Endo-1, 4-beta-xylanase

Synonymes
Endo-(1 -- > 4)-beta-xylanase(1 -- > 4)-beta-xylan 4-xylanohydrolase [39]
Xylanase, endo-1, 4-
Xylanase
Endo-1, 4-.beta.-xylanase
Beta-1, 4-xylanase [4, 44]
Endo-1, 4-xylanase
Endo-.beta.-1, 4-xylanase
Endo-1, 4-.beta.-D-xylanase
.beta.-1, 4-Xylan xylanohydrolase
.beta.-Xylanase
.beta.-1, 4-D-Xylanase
1, 4-Beta-xylan xylanohydrolase [8]
Beta-D-xylanase [32]

CAS Reg. No.
9025-57-4

2 REACTION AND SPECIFICITY

Catalysed reaction
Endohydrolysis of 1, 4-beta-D-xylosidic linkages in xylans (mechanism [6])

Reaction type
O-Glycosyl bond hydrolysis

Natural substrates
Xylan + H_2O (sole carbon source [6], decomposition of lignocellulosic material [11]) [6, 11]

Enzyme Handbook © Springer-Verlag Berlin Heidelberg 1991
Duplication, reproduction and storage in data banks are only
allowed with the prior permission of the publishers

Substrate spectrum

1 1, 4-Beta-D-xylan + H_2O (hydrolyzed by all enzymes mentioned here) [1–56]
2 4-Nitrophenyl-beta-D-xylanopyranoside + H_2O [6]
3 Rhodymenan + H_2O [6, 30]
4 Xylotriose + H_2O [15, 40]
5 Xylotetraose + H_2O [15]
6 Xylopentaose + H_2O [15]
7 Carboxymethyl cellulose + H_2O (slight [19, 20]) [19, 20, 41, 49, 53]
8 p-Nitrophenyl-beta-D-glucoside + H_2O (xylanase I) [41]
9 Cellulose + H_2O (slight [19]) [19, 53, 54]
10 Pectin + H_2O (slight) [19]
11 Starch + H_2O [53]
12 Amylopectin + H_2O [51]
13 Arabinogalactan + H_2O (slight [20]) [20, 51]
14 p-Nitrophenyl glucopyranoside + H_2O (slight) [20]
15 Glucomannan + H_2O (slight) [20]
16 More (no activity on carboxymethyl-cellulose and p-nitrophenyl-beta-D-xyloside [27], transxylosidation activity [6, 8, 9, 12], not: xylobiose [6, 15, 53],) [6, 8, 9, 12, 15, 27, 53]

Product spectrum

1 More (products depend on organism, different combinations and proportion of following products: xylose, xylobiose, arabinose, glucose, xylo-oligosaccharides, arabino-oligosaccharides) [1–56]
2 ?
3 Xylooligosaccharides (containing a 1, 3-beta-linkage) + 1, 4-beta-linked products [6]
4 ?
5 ?
6 ?
7 ?
8 ?
9 ?
10 ?
11 ?
12 ?
13 ?
14 ?
15 ?
16 ?

Inhibitor(s)
Cu^{2+} (no effect [22]) [9, 12, 16, 18, 19, 24, 26]; Hg^{2+} [9, 12, 13, 15, 16, 18, 19, 22, 24, 26, 33, 36, 40, 50, 51, 52, 53]; Ag^+ [12, 33, 50, 51, 52]; EDTA (not [40]) [12]; Fe^{3+} [13, 15, 18]; SDS [13, 15, 40]; N-Bromosuccinimide (5 mM [13]) [13, 15, 50]; Cu^+ [18]; Al^{3+} [18]; Co^{2+} (no effect [22]) [18]; Cr^{3+} [18]; Cr^{2+} [19]; Mn^{2+} [19, 33, 40]; Lead acetate [22]; Pb^{2+} [24]; Zn^{2+} [26]; BrO_3^- [26]; Cl^- [26]; EGTA [26]; Xylose [31]; $AgNO_3$ [33]; $HgCl_2$ [33, 36, 48]; $MnSO_4$ [33]; Xylan [34]; p-Chloromercuribenzoate (not [46]) [36]; 5,5'-Dithiobis(2-nitrobenzoic acid) (slight) [36]; N-Ethylmaleimide [36]; $FeCl_3$ [36]; $FeCl_2$ [36]; $CuSO_4$ [36]; p-Hydroxymercuribenzoate [36]; Xylobiose [38]; Methanol [40]; Ethanol [40]; $KMnO_4$ [48]; 2-Hydroxy-5-nitrobenzyl bromide [50]; N-Acetylimidazole [50]; Fe^{2+} [51]; Iodoacetate [51, 53]; PO_4^{3-} [51]; p-Substituted mercuribenzoate (component I) [12]

Cofactor(s)/prostethic group(s)

Metal compounds/salts
Ca^{2+} (strong activation [20], activation [50]) [20, 50]; Al^{3+} (activates) [53]; Fe^{3+} (activates) [53]; Pb^{2+} (activates) [53]; More (metal ion not required for activity) [46]

Turnover number (min⁻¹)

Specific activity (U/mg)
55.1 [6]; 1780 [7]; 1582 [9]; 2400 [22]; 6.6 [24]; 1582 [9]; More [8, 12, 17, 28, 32, 33, 34, 40, 41, 44, 45, 46, 49, 54]

K_m-value (mM)
More (xylan: 8.27 mg/ml [11], 4.5 mg/ml [12, 16], 0.86 mg/ml [26], 2.9–3.7 mg/ml [36], 0.20 mg/ml [40], xylopentaose: 0.95 mg/ml [40], xylohexaose: 0.66 mg/ml [40], xylooctaose: 0.58 mg/ml [40]) [11, 12, 16, 26, 36, 40]

pH-optimum
5.4 (xylan or 4-nitrophenyl-beta-D-xylopyranoside, Cryptococcus albidus) [6]; 6.5 (Bacillus pumilus) [7]; 5.5–6.0 (Streptomyces sp.) [13, 15]; 6.0–6.5 (Malbranchea pulchella) [8, 17]; 3.5 (xylanase A, Trichoderma lignorum) [9]; 6.5 (xylanase B, Trichoderma lignorum) [9]; 5.0 (Schizophyllum commune) [11]; 6.0 (Gliocladium virens) [53]; 6.0 (acidophilic Bacillus, form I) [12, 16]; 5.5–6.0 (Aspergillus niger, Ia) [54]; 4.0–4.5 (Aspergillus niger, Ib) [54]; 4.0 (Aspergillus niger, IIa) [54]; 6.0 (Aspergillus niger, I) [18]; 5.5 (Aspergillus niger, II) [18]; 5.0 (Aspergillus niger) [19, 20]; 5.1 (Thermoascus auranticus) [22]; 5.5 (Hordeum vulgare) [26]; 6.0 (Streptomyces lividans) [27]; 6–7 (Streptomyces lividans) [29]; 4–5 (Trichoderma reesei) [28]; 5.8 (Cellulomonas uda) [31]; 6.0 (Irpex lacteus) [33]; 5.0 (Trichoderma harzianum) [34]; 5.5–7 (Clostridium stercorarium) [36]; 6.0 (Humicola lanuginosa) [38]; 5.5 (Bacillus subtilis) [39]; 5.5 (Streptomyces sp.) [40]; 6.0–7.0 (Bacillus sp., thermophilic alkalophilic) [42]; 5.0 (Bacillus subtilis)

Enzyme Handbook © Springer-Verlag Berlin Heidelberg 1991
Duplication, reproduction and storage in data banks are only
allowed with the prior permission of the publishers

[43]; 5.0 (Trichoderma viride) [44]; 6.5 (Bacillus pumilus) [45]; 5.0 (Trichosporon cutaneum) [46]; 5.5–7.0 (Bacillus circulans, xylanase A/B) [47]; 4.5–5.5 (Talaromyces byssochlamydoides) [48]; 5.5 (Termitomyces clypeatus) [51]; 5.0 (Cryptococcus albidus) [5]; More [4]

pH-range

3–9 [19, 33]; 4.5–8 [29, 36]; 4–10 (Streptomyces sp.) [40]; 2–8 (Talaromyces) [48]; 3–8 (Aspergillus niger) [54, 55]

Temperature optimum (°C)

45–50 (Bacillus pumilus) [7]; 50 (Schizophyllum commune) [11]; 70 (Malbranchea pulchella) [8, 17]; 60–65 (Streptomyces sp.) [13, 15]; 45 (Trichoderma lignorum) [9]; 65 (Bacillus sp., I) [16]; 70–75 (Talaromyces byssochlamydoides) [48]; 70 (Bacillus sp., II) [16]; 80 (Bacillus sp., thermophilic) [49]; 45 (Aspergillus niger) [18, 19, 20]; 50 (Trametes hirsuta) [50]; 80 (Thermoascus auranticus) [22]; 55 (Termitomyces clypeatus) [51]; 35 (Hordeum vulgare) [26]; 60 (Streptomyces lividans) [27]; 40 (Cryptococcus albidus) [52]; 55–60 (Streptomyces lividans) [29]; 45 (Gliocladium virens) [53]; 60 (Irpex lacteus) [33]; 70 (Irpex lacteus, xylanase I) [33]; 70 (Irpex lacteus, xylanase II) [33]; 50 (Trichoderma harzianum, xylanase I) [34]; 60 (Trichoderma harzianum, xylanase II) [34]; 65 (Humicola lanuginosa) [36]; 55 (Streptomyces sp.) [40]; 65–70 (Bacillus sp., alkalophilic, thermophilic) [42]; 50 (Bacillus subtilis) [43]; 40 (Bacillus pumilus) [45]; 50 (Trichosporon cutaneum) [46]

Temperature range (°C)

50–70 (Streptomyces lividans) [27]; 25–100 (Irpex lacteus) [33]; 25–70 (Streptomyces sp.) [40]; 20–90 (Talaromyces) [48]

---

3 ENZYME STRUCTURE

Molecular weight

26000–28000 (gel filtration, Cryptococcus albidus) [6]
48000 (SDS-PAGE, Cryptococcus albidus) [6]
24000 (SDS-PAGE, Bacillus pumilus) [7]
20000 (equilibrium sedimentation, Bacillus pumilus) [7]
22384 (amino acid analysis, Bacillus pumilus) [7]
56000 (SDS-PAGE, Bacillus sp., acidophilic /thermophilic) [49]
21000 (SDS-PAGE, Trichoderma lignorum, xylanase A) [9]
20000 (SDS-PAGE, Trichoderma lignorum, xylanase B) [9]
21000 (SDS-PAGE, Schizophyllum commune) [11]
22000 (SDS-PAGE, alkalophilic Bacillus, component I) [12, 16]
50000 (SDS-PAGE, alkalophilic Baillus, component II) [12, 16]
50000 (SDS-PAGE, gel filtration, Streptomyces sp., X-I) [13, 15]
25000 (SDS-PAGE, sedimentation analysis, Streptomyces sp. , X-II, X-III) [13, 15]

13000 (gel chromatography, Aspergillus niger) [18]
14000 (gel chromatography, Aspergillus niger) [19]
28000 (SDS-PAGE, Aspergillus niger) [20]
32000 (SDS-PAGE, Thermoascus auranticus) [22]
20000 (SDS-PAGE, Trichoderma harzianum) [23]
29000 (SDS-PAGE, Clostridium acetobutylicum, xylanase A) [24]
65000 (SDS-PAGE, Clostridium acetobutylicum, xylanase B) [24]
43000 (slab-gel electrophoresis, Streptomyces lividans) [27]
32000 (SDS-PAGE, Trichoderma reesei, xylanase I) [28]
23000 (SDS-PAGE, Trichoderma reesei, xylanase II) [33]
55000 (Acacia verek) [32]
38000 (gel filtration, SDS-PAGE, Irpex lacteus, xylanase I) [33]
62000 (gel filtration, SDS-PAGE, Irpex lacteus, xylanase II) [33]
20000 (SDS-PAGE, Trichoderma harzianum, xylanase I) [34]
29000 (SDS-PAGE, Trichoderma harzianum, xylanase II) [34]
44000 (SDS-PAGE, Clostridium stercorarium, xylanase A) [36]
72000 (SDS-PAGE, Clostridium stercorarium, xylanase B) [36]
62000 (SDS-PAGE, Clostridium stercorarium, xylanase C) [36]
21000 (SDS-PAGE, gel permeation chromatography, Humicola lanuginosa) [36]
43000 (SDS-PAGE, gel filtration, Streptomyces sp.) [40]
33000 (SDS-PAGE, Neurospora crassa, xylanase I) [41]
30000 (SDS-PAGE, Neurospora crassa, xylanase II) [41]
32000 (Bacillus subtilis) [43]
13000 (Trichoderma viride) [44]
20000 (SDS-PAGE, Bacillus pumilus) [45]
45000 (SDS-PAGE, Trichosporon cutaneum) [46]
85000 (gel filtration, Bacillus circulans, xylanase A) [47] 15000 (gel filtration, Bacillus circulans, xylanase B) [47]
22000–24000 (gel chromatography, Trametes hirsuta) [50]
90000 (SDS-PAGE, Termitomyces clypeatus) [51]
31000 (gel chromatography, Aspergillus niger, IA, IB) [54]
50000 (gel chromatography, Aspergillus niger, II) [54]

Subunits
Monomer (Schizophyllum commune) [11]
Monomer (SDS-PAGE, equilibrium sedimentation, Bacillus pumilus) [7]
Monomer (alkalophilic Bacillus) [12]
Monomer (SDS-PAGE, gel filtration, Streptomyces sp.) [13]

Enzyme Handbook © Springer-Verlag Berlin Heidelberg 1991
Duplication, reproduction and storage in data banks are only
allowed with the prior permission of the publishers

Glycoprotein/Lipoprotein

Glycoprotein (Irpex lacteus, xylanase I (2.5%), xylanase II (8.0%) [33], Clostridium stercorarium: xylanase A (19%), xylanase B (3%), xylanase C (4%) [36], Talaromyces: X-a (36.6%), X-b-I (31.5%), X-b-II (14.2%), glucose, mannose, fucose [48], Aspergillus niger: 20%, glucose, galactose, glucosamine [55], Cryptococcus albidus: 27% [6], Trichoderma reesei: xylanase I (14%), xylanase II (8%) [28]) [4, 6, 28, 33, 36, 48, 55]; More (no carbohydrate [7, 13], not glycosylated [11, 46]) [7, 11, 13, 46]

4 ISOLATION/PREPARATION

Source organism

Caldocellum saccharolyticum (expressed in E. coli) [2]; Bacillus polymyxa (expressed in E. coli) [1]; Clostridium stercorarium [4]; Streptomyces sp. [4]; Streptomyces exfoliatus [4]; Trichoderma harzianum (2 forms [34, 35]) [4, 23, 24, 25]; Trichoderma reesei (3 forms [28]) [4, 28]; Aeromonas sp. [4]; Penicillium janthinellum [4]; Talaromyces byssochlamydoides [4]; Cryptococcus albidus [6]; Bacillus pumilus [7]; Malbranchea pulchella [8, 17]; Trichoderma lignorum (xylanase A/B) [9]; Aspergillus fumigatus [10, 37]; Schizophyllum commune [11]; Bacillus (alkalophilic, 2 forms I/II [12], alkalophilic /thermophilic [16, 42], thermophilic /acidophilic [49]) [12, 16, 42, 49]; Streptomyces sp. (3 forms: X-I, X-II, X-III [13]) [13, 15, 21, 30, 40]; Termitomyces clypeatus [14]; Streptomyces flavogriseus [21]; Streptomyces olivochromogenes [21]; Thermoascus auranticus [22]; Clostridium acetobutylicum (xylanase A/B) [24]; Neurospora crassa [25, 41]; Hordeum vulgare [26]; Streptomyces lividans [27, 29]; Cellulomonas uda [31]; Acacia verek [32]; Irpex lacteus (Polyporus tulipiferae, 2 forms) [33]; Clostridium stercorarium (3 forms A/B/C) [36]; Humicola lanuginosa [38]; Bacillus subtilis [39, 43]; Trichoderma viride [44]; Bacillus pumilus [45]; Trichosporon cutaneum [46]; Bacillus circulans [47]; Trametes hirsuta [50]; Termitomyces clypeatus [51]; Cryptococcus albidus [52]; Gliocladium virens [53]; Aspergillus niger (5 forms [54], 2 isoenzymes I/II [8]) [4, 18, 19, 20, 54, 55]; Aspergillus foetidus [50]; Talaromyces byssochlamydoides (3 forms: X-a, X-b-I, X-b-II [48]) [4, 48]; More (multiple forms) [4]

Source tissue

Culture medium [6, 40, 41, 49]; Culture filtrate [22, 23, 25, 28, 34]; Aleuron layer [26]; More (commercial preperation of Bacillus subtis alpha-amylase) [39]

Localisation in source

Extracellular [27, 29, 31, 40, 41, 43, 45, 47, 49, 52]; Intracellular (small amount) [41]; Cell wall (associated) [52]

Purification

Caldocellum saccharolyticum (expressed in E. coli, partial) [2]; Cryptococcus albidus [6]; Bacillus pumilus [7]; Malbranchea pulchella (var. sulfurea) [8, 17]; Trichoderma lignorum [9]; Aspergillus fumigatus (partial)
[19]; Schizophyllum commune [11]; Bacillus (alkalophilic /thermophilic [42,
16], alkalophilic, 2 forms I/II [12]) [12 , 16, 42]; Streptomyces sp. (3 forms X-I,
X-II, X-II [13]) [13, 15, 40]; Aspergillus niger [18, 29, 20, 54, 55]; Thermoascus
auranticus [22]; Trichoderma harzianum (2 forms [34]) [23, 34]; Clostridium
acetobutylicum (xylanase A/B) [24]; Neurospora crassa [25, 41]; Hordeum
vulgare [26]; Streptomyces lividans [27]; Trichoderma reesei (3 forms) [28];
Acacia verek [32]; Irpex lacteus (Polyporus tulipiferae, 2 forms) [33];
Clostridium stercorarium (3 forms A/B/C) [36]; Humicola lanuginosa [38];
Bacillus subtilis [43]; Trichoderma viride [44]; Bacillus pumilus [45];
Trichosporon cutaneum [46]; Talaromyces byssochlamydoides (3 forms:
X-a, X-b-I, X-b-II) [48]; Trametes hirsuta [50]; Termitomyces clypeatus [51];
Gliocladium virens [53]

Crystallization

–

Cloned

(Bacillus polymyxa gene in E. coli [1], Caldocellum saccharolyticum gene in
E. coli [3], Aeromonas gene in E. coli [5]) [1, 3, 5, 7]

Renaturated

–

---

5 STABILITY

pH

4.5–10 (1 hour, 40°C, alkalophilic Bacillus) [12, 16]; 3.0–10.5 (Streptomyces
sp.) [13, 15]; 5.0 (highest stability, half-life: 20 minutes, Aspergillus niger,
form I) [18]; 6. 0 (highest stability, half-life: 40 hours, Aspergillus niger, form
II) [18]; 5.5–5.6 (highest stability, half-life: 40 hours, Aspergillus niger) [19];
5.0–6.0 (highest stability, Hordeum vulgare) [26]; 7–12 (Clostridium stercorarium) [36]; 3.0 (50% loss of activity, Clostridium stercorarium, xylanase
B/C) [36]; 4.0 (no activity left at, Clostridium stercorarium, xylanase A) [36];
4–10 (2 hour, Streptomyces sp.) [40]; 4.5–10.5 (45°C, 1 hour, Bacillus sp.,
alkalophilic/thermophilic) [42]; 2.0–6.0 (Bacillus sp.) [49]; 4.0–8.0 (Trametes
hirsuta) [50]; 3–10 (Termitomyces clypeatus) [51]; 4.5–7.0 (Gliocladium
virens) [53]; 6.5 (optimal stability, Bacillus pumilus) [7]; 2.5–6.0 (xylanase A,
Trichoderma lignorum) [9]; 3.5–8.0 (xylanase B, Trichoderma lignorum) [9];
6–8 (Schizophyllum commune) [11]; More [40]

Enzyme Handbook © Springer-Verlag Berlin Heidelberg 1991
Duplication, reproduction and storage in data banks are only
allowed with the prior permission of the publishers

Temperature (°C)
40 (pH 4.5–10, 1 hour, alkalophilic Bacillus) [12]; 55 (pH 5.5, 30 minutes, Streptomyces sp.) [13, 15]; 80 (half-life: 54 minutes, Thermoascus aurantiacus) [22]; 40 (10% loss of activity after 20 hours, xylanase I, 20 hours, xylanase II, Trichoderma reesei) [28]; 60 (80% loss of activity after 20 hours, Trichoderma reesei) [28]; 65 (15 days, Clostridium stercorarium) [36]; 81 (half-life: xylanase A (2 minutes), xylanase B (0.5 minutes), xylanase C (8 minutes), Clostridium stercorarium) [36]; 55 (10 minutes, stable up to, Strep-tomyces sp.) [40]; 45 (loss of activity above, Trichoderma viride) [44]; 95 (5 minutes, X-a (35%), X-b-I (46%), X-b-II (70%) loss of activity after 15 minutes, Bacillus sp.) [49]; 45 (up to 30 minutes stable, Trametes hirsuta) [50]; 60 (stable up to, Termitomyces clypeatus) [51]; 50 (half-life: 7 minutes, Cryptococcus albidus) [6]; 50 (half-life: 30 minutes, Bacillus pumilus) [7]; 60 (complete loss of activity after 15 minutes, Bacillus pumilus) [7]; 50 (pH 6.7, without substrate, Malbranchea pulchella) [8]; 45 (stable up to, Gliocladium virens) [53]; More [19, 22, 46, 54]

Oxidation

Organic solvent

General stability information

Storage
4°C, several months (Cryptococcus albidus) [6]; Frozen state (stable in-definitely, Cryptococcus albidus) [6]; 4°C, 30 days [8]; 4°C, acetate buffer, pH 5.5 months [40]

---

6 CROSSREFERENCES TO STRUCTURE DATABANKS

PIR/MIPS code
WWBSXP (Bacillus pumilus); S06850 (barley, fragment); S06047 (precursor, Pseudomonas fluorescens subsp. cellulosa); S01734 (precursor, Bacillus circulans); S01690 (basidiomycete, Filobasidium floriforme); A24540 (Filobasidium floriforme, fragment); A05147 (Schizophyllum commune, frag-ment)

Brookhaven code

7 LITERATURE REFERENCES

[1] Sandhu, J.S.: Enzyme Microb. Technol., 6, 271–274 (1984)
[2] Patchett, M.L., Neal, T.L., Schofield, L.R., Strange, R.M., Daniel, R.M., Morgan, H.W.: Enzyme Microb. Technol., 11, 113–115 (1989)
[3] Bergquist, P.L.: Biotechnol. Genet. Eng. Rev., 5, 199–244 (1987)
[4] Wong, K.K.Y., Tan, L.U.L., Saddler, J.N.: Microbiol. Rev., 52, 305–317 (1988) (Review)
[5] Kudo, T., Ohkoshi, A., Horikoshi, K.: J. Gen. Microbiol., 131, 2825–2830 (1985)
[6] Biely, P., Vrsanska, M.: Methods Enzymol., 160, 638–648 (1988)
[7] Okada, H., Shinmyo, A.: Methods Enzymol., 160, 632–637 (1988)
[8] Matsuo, M., Yasui, T.: Methods Enzymol., 160, 671–674 (1988)
[9] John, M., Schmidt, J.: Methods Enzymol., 160, 662–671 (1988)
[10] Stewart, J.C., Heptinstall, J.: Methods Enzymol., 160, 264–274 (1988)
[11] Jurasek, L., Paice, M.G.: Methods Enzymol., 160, 659–662 (1988)
[12] Akiba, T., Horikoshi, K.: Methods Enzymol., 160, 655–659 (1988)
[13] Yasui, T., Marui, M., Kusakabe, I., Nakanishi, K.: Methods Enzymol., 160, 648–654 (1988)
[14] Khowala, S., Mukherjee, M., Sengupta, S.: Enzyme Microb. Technol., 10, 563–567 (1988)
[15] Marui, M., Nakanishi, K., Yasui, T.: Agric. Biol. Chem., 49, 3399–3407 (1985)
[16] Okazaki, W., Akiba, T., Horikoshi, K., Akahoshi, R.: Agric. Biol. Chem., 49, 2033–2039 (1985)
[17] Matsuo, M., Yasui, T.: Agric. Biol. Chem., 49, 839–841 (1985)
[18] Frederick, M.M., Kiang, C.-H., Frederick, J.R., Reilly, P.J.: Biotechnol. Bioeng., 27, 525–532 (1985)
[19] Shei, J.C., Fratzke, A.R., Frederick, M.M., Frederick, J.R., Reilly, P.J.: Biotechnol. Bioeng., 27, 533–538 (1985)
[20] Fournier, R., Frederick, M.M., Frederick, J.R., Reilly, P.J.: Biotechnol. Bioeng., 27, 539–546 (1985)
[21] Johnson, K.G., Harrison, B.A., Schneider, H., MacKenzie, C.R., Fontana, J.D.: Enzyme Microb. Technol., 10, 403–409 (1988)
[22] Tan, L.U.L., Mayers, P., Saddler, J.N.: Can. J. Microbiol., 33, 689–692 (1987)
[23] Tan, L.U.L., Yu, E.K.C., Louis-Seize, G.W., Saddler, J.N.: Biotechnol. Bioeng., 30, 96–100 (1987)
[24] Lee, S.F., Forsberg, C.W., Rattray, J.B.: Appl. Environ. Microbiol., 53, 644–650 (1987)
[25] Deshpande, V., Lachke, A., Mishra, C., Keskar, S., Rao, M.: Biotechnol. Bioeng., 27, 1832–1837 (1986)
[26] Benjavongkulchai, E., Spencer, M.S.: Planta, 169, 415–419 (1986)
[27] Morosoli, R., Bertrand, J.-L., Mondou, F., Shareck, F., Kluepfel, D.: Biochem. J., 239, 587–592 (1986)
[28] Lappalainen, A.: Biotechnol. Appl. Biochem., 8, 437–448 (1986)
[29] Kluepfel, D., Shareck, F., Mondou, F., Morosoli, R.: Appl. Microbiol. Biotechnol., 24, 230–234 (1986)
[30] Chen, W.P., Matsuo, M., Yasui, T.: Agric. Biol. Chem., 50, 1195–1200 (1986)
[31] Rapp, P., Wagner, F.: Appl. Environ. Microbiol., 51, 746–752 (1986)
[32] Lienart, Y., Comtat, J., Barnoud, F.: Plant Sci., 41, 91–96 (1985)
[33] Kanda, T., Amano, Y., Nisizawa, K.: J. Biochem., 98, 1545–1554 (1985)
[34] Tan, L.U.L., Wong, K.K.Y., Yu, E.K.C., Saddler, J.N.: Enzyme Microb. Technol., 7, 425–430 (1985)
[35] Tan, L.U.L., Wong, K.K.Y., Saddler, J.N.: Enzyme Microb. Technol., 7, 431–437 (1985)

Enzyme Handbook © Springer-Verlag Berlin Heidelberg 1991
Duplication, reproduction and storage in data banks are only
allowed with the prior permission of the publishers

[36] Berenger, J.-F., Frixon, C., Bigliardi, J., Creuzet, N.: Can. J. Microbiol., 31, 635–643 (1985)

[37] Wase, D.A.J., Raymashasay, S., Wang, C.W.: Enzyme Microb. Technol., 7, 225–229 (1985)

[38] Kitpreechavanich, V., Hayashi, M., Nagai, S.: J. Ferment. Technol., 62, 415–420 (1984)

[39] Kato, Y., Nevins, D.J.: Plant Physiol., 75, 753–758 (1984)

[40] Nakajima, T., Tsukamoto, K.-I., Watanabe, T., Kainuma, K., Matsuda, K.: J. Ferment. Technol., 62, 269–276 (1984)

[41] Mishra, C., Keskar, S., Rao, M.: Appl. Environ. Microbiol., 48, 224–228 (1984)

[42] Okazaki, W., Akiba, T., Horikoshi, K., Akahoshi, R.: Appl. Microbiol. Biotechnol., 19, 335–340 (1984)

[43] Bernier, R., Desrochers, M., Jurasek, L., Paice, M.G. : Appl. Environ. Microbiol., 46, 511–514 (1983)

[44] Labavitch, J.M., Greve, L.C.: Plant Physiol., 72, 668–673 (1983)

[45] Panbangred, W., Shinmyo, A., Kinoshita, S., Okada, H. : Agric. Biol. Chem., 47, 957–963 (1983)

[46] Stüttgen, E., Sahm, H.: Eur. J. Appl. Microbiol. Biotechnol., 15, 93–99 (1982)

[47] Esteban, R., Villanueva, J.R., Villa, T.G.: Can. J. Microbiol., 28, 733–739 (1982)

[48] Yoshioka, H., Nagato, N., Chavanich, S., Nilubol, N., Hayashida, S.: Agric. Biol. Chem., 45, 2425–2432 (1981)

[49] Uchino, F., Nakane, T.: Agric. Biol. Chem., 45, 1121–1127 (1981)

[50] Kubackova, M., Karacsonyi, S., Bilisics, L., Toman, R.: Folia Microbiol., 23, 202–209 (1978)

[51] Ghosh, A.K., Banerjee, P.C., Sengupta, S.: Biochim. Biophys. Acta, 612, 143–152 (1980)

[52] Notario, V., Villa, T.G., Villanueva, J.R.: J. Gen. Microbiol., 114, 415–422 (1979)

[53] Takahashi, M., Kutsumi, S.: J. Ferment. Technol., 57, 434–439 (1979)

[54] John, M., Schmidt, B., Schmidt, J.: Can. J. Biochem., 57, 125–134 (1979)

[55] Gorbacheva, I.V., Rodionova, N.A.: Biochim. Biophys. Acta, 484, 79–93 (1977)

[56] Whistler, R.L., Masak, E.: J. Am. Chem. Soc., 77, 1241–1243 (1955)

1 NOMENCLATURE

EC number
3.2.1.10

Systematic name
Dextrin 6-alpha-D-glucanohydrolase

Recommended name
Oligo-1, 6-glucosidase

Synonymes
Exo-oligo-1, 6-Glucosidase [11]
Limit dextrinase
Isomaltase
Sucrase-isomaltase
Glucosidase, oligo-1, 6-
Oligo-1, 6-glucosidase
Dextrin 6.alpha.-glucanohydrolase
Isomaltase
Limit-dextrinase [3]
Dextrinase, limit
Alpha-limit dextrinase
Dextrin 6-glucanohydrolase

CAS Reg. No.
9032-15-9

---

2 REACTION AND SPECIFICITY

Catalysed reaction
Hydrolysis of 1, 6-alpha-D-glucosidic linkages in isomaltose and dextrins
produced from starch and glycogen by alpha-amylase; More (enzyme from
intestinal mucosa also catalyses the reaction of EC 3.2.1.48) [9, 13, 16, 17, 19]

Reaction type
O-Glycosyl bond hydrolysis

Natural substrates
Isomaltose + H_2O
Dextrins + H_2O (produced from starch and glycogen)

Enzyme Handbook © Springer-Verlag Berlin Heidelberg 1991
Duplication, reproduction and storage in data banks are only
allowed with the prior permission of the publishers

Substrate spectrum
1 Isomaltose + H_2O [2, 9, 20]
2 Dextrins + H_2O
3 Palatinose + H_2O [9]
4 Nigerose + H_2O [2]
5 Phenyl-alpha-D-glucoside + H_2O [6]
6 p-Nitrophenyl-alpha-D-glucose + H_2O [2, 4, 6]
7 Isomaltotriose + H_2O [2, 6]
8 Panose + H_2O [2, 6, 11, 14, 20]
9 Isomaltosaccharides (n between 2 and 6) + H_2O [6, 11, 14]
10 Sucrose + H_2O (not [1]) [6]
11 Turanose + H_2O [6]
12 Alpha-limit dextrins + H_2O [11, 20]
13 Amylopectin + H_2O [19]
14 Starch + H_2O
15 Maltose + H_2O
16 More (not: malto-oligosaccharides, melizitose, melibiose, raffinose, cellobiose, sophorose, gentiobiose, lactose, pullulan, dextran, amylose)

Product spectrum
1 Glucose
2 ?
3 ?
4 Glucose
5 Glucose [6]
6 Glucose [6]
7 Glucose [6]
8 Glucose + maltose [6]
9 Glucose [6, 14]
10 Glucose [6]
11 Glucose + fructose [6]
12 ?
13 ?
14 ?
15 Glucose
16 ?

Inhibitor(s)

N-Ethylmaleimide [2]; Glucose (slight [11]) [2, 6, 11];
1-O-Methyl-alpha-D-glucose [2]; Urea [4]; Ethanol [4]; Cd^{2+} [6]; Pb^{2+} [6,
11]; Zn^{2+} [6, 11]; Co^{2+} [6, 11]; Sn^{2+} [6]; Cu^{2+} [6, 11]; Ni^{2+} [6]; Mb^{2+} [6];
Fe^{3+} [6]; Ba^{2+} (slight [11]) [6, 11]; Mg^{2+} (slight) [6]; Ca^{2+} (slight) [6];
p-Chloromercuribenzoate [6]; Tris [6, 11]; Phenyl alpha-maltoside [6];
Phenyl alpha-D-glucoside [6, 11]; Glucono-delta-lactone [6, 11];
p-Nitrophenol [6, 11]; Tris phenol [6]; Fe^{2+} [11]; Hg^{2+} [6, 11]; Mn^{2+} [11];
D-Glucosamine (slight) [11]; Beta-D-galactopyranoside [11];
Alpha-D-mannopyranoside; Conduritol-B-epoxide [17]; More [11]

Cofactor(s)/prostethic group(s)

Metal compounds/salts

Turnover number (min^{-1})

Specific activity (U/mg)
5.19 [2]; More [9, 10]

K$_m$-value (mM)
6.51 (p-nitrophenyl-alpha-D-glucose) [2]; 0.17 (p-nitrophenyl
alpha-D-glucopyranoside) [6]; 3.0 (isomaltose) [6]; 4.4 (isomaltotriose) [6];
7.4 (isomaltotetraose) [6]; 13 (isomaltopentaose) [6]; 15 (isomaltohexaose)
[6]; 3.5 (palatinose) [9]; 2.5 (isomaltose) [9]; 0.81
(p-nitrophenyl-alpha-D-glucopyranoside) [11]; 2.2
(phenyl-alpha-D-glucopyranoside) [11]; 6.2 (isomaltose) [11]; 3.0
(isomaltotriose) [11]; 2.8 (isomaltotetraose) [11]; 4.6 (isomaltopentaose)
[11]; 11 (isomaltohexaose) [11]; More [14]

pH-optimum
5.6–7.0 [2]; 5.9 [4]; 6.7 [11]

pH-range
3–8.2 [2]; 4.9–9.5 (half-optima: 5.5, 8.2) [11]; 4.3–9.8 (half-optima: 4.9, 7.6)
[4]; 4.1–10.2 (half-optima: 4.8, 7. 3) [6]

Temperature optimum (°C)
70–75 [1]; 74 [4]; 62 [6]; 41 [11]

Temperature range (°C)
58–84 (half-optima) [4]; 47–75 (half-optima) [6]; 27–53 (half-optima) [11]

Enzyme Handbook © Springer-Verlag Berlin Heidelberg 1991
Duplication, reproduction and storage in data banks are only
allowed with the prior permission of the publishers

3 ENZYME STRUCTURE

Molecular weight

60000 (SDS-PAGE, Bacillus thermoglucosidasius, gene expression in E. coli) [1]

30000 (SDS-PAGE, thermoanaerobium) [2]

33000 (gel filtration, thermoanaerobium) [2]

102000 (SDS-PAGE, Hordeum vulgare) [3]

62000 (SDS-PAGE, Bacillus sp.) [4]

60000 (Bacillus coagulans, SDS-PAGE) [6]

245000 (SDS-PAGE, Zalophus californianus) [9]

60000 (gel filtration, SDS-PAGE, Bacillus cereus) [11]

140000 (SDS-PAGE, intact pancreatic duct: 2 polypeptide chains, 140000 + 150000, pig) [16]

150000 (SDS-PAGE, intact pancreatic duct, 2 polypeptide chains, 140000 + 150000, pig) [16]

260000 (SDS-PAGE, pancreas disconnected from duodenum, pig) [16]

Subunits

Monomer (SDS-PAGE, 1 × 245000, Zalophus californianus) [9]

More (enzyme from intestinal mucosa is a single polypeptide chain also catalysing the reaction of EC 3.2.1.48 [9, 13, 16, 17, 19], intact pancreatic ducts: 2 polypeptide chains, pancreas disconnected from duodenum: 1 polypeptide chain [9, 16]) [9, 13, 16–19]

Glycoprotein/Lipoprotein

Glycoprotein [14]; More (no carbohydrate) [11]

---

4 ISOLATION/PREPARATION

Source organism

Bacillus coagulans [4, 6]; Bacillus cereus [4, 7, 10, 11]; Bacillus flavocaldarius [4]; Pisum sativum [5]; Bacillus stearothermophilus [8]; Pig [15, 16]; Zalophus californianus (also maltase activity) [9]; Collarinus ursinus [9]; Human [13, 18]; Rabbit [17]; Oat [19]; Rat [19]; Bacillus thermoglucosidasius (expression in E. coli [1]) [1, 4, 8, 12, 14]; E. coli (Bacillus thermoglucosidasius gene) [1]; Thermoanaerobium [2]; Hordeum vulgare [3]; Bacillus sp. [4]; More [20]

Source tissue

Seeds [3]; Small intestine [9, 17, 18]; Intestine [13, 15, 16, 19]; Cell [2, 4, 5, 6, 8]

Localisation in source

Cytoplasm [1]; Periplasmic space (low activity) [1]; Extracellular (low activity) [1]; Cell-bound [2]; Chloroplast [5]; Membrane (bound) [15]

Purification
 Bacillus thermoglucosidasius (expression in E. coli) [1, 14]; E. coli (Bacillus thermoglucosidasius gene) [1]; Themoanaerobium [2]; Hordeum vulgare [3]; Bacillus sp. [4]; Bacillus coagulans [6]; Bacillus cereus [10, 11]; Pig [10]

Crystallization
 –

Cloned
 (Bacillus thermoglucosidasius gene in E. coli) [1]

Renaturated
 –

---

5 STABILITY

pH
 5.5–10.3 (15 hours, 25°C, stable) [4]; More [4, 6]

Temperature (°C)
 71 (10 minutes, 50% inactivation) [1]; 76 (half-life: 20 minutes) [2]; 70 (half-life: 10 minutes) [2, 4]; 13–40 (pH 6.8 , 10 minutes, no loss of activity) [11]; 50 (pH 6.8, 10 minutes, 75% loss of activity) [11]; 53 (pH 6.8, 10 minutes, stable) [6]; More (thermostability rises with proline content) [4]

Oxidation

Organic solvent
 Ethanol (inactivation [6], 50% inactivation in 29% ethanol [11]) [6, 11]

General stability information
 Substrate (stabilizes) [2]; Mg^{2+} (stabilizes) [2]; Mn^{2+} (stabilizes) [2]; Ca^{2+} (stabilizes) [2]; Ca^{2+} (stabilizes) [2]; Zn^{2+} (destabilizes) [2]; EDTA (destabilizes) [2]; Urea (inactivation [6], 50% inactivation, 3.1 M urea [11]) [6, 11]; SDS (0.0048% SDS, 50% inactivation) [11]; More (high resistance against proteolysis) [12]

Storage

---

6 CROSSREFERENCES TO STRUCTURE DATABANKS

PIR/MIPS code
 A26967 (pig, fragment); A29163 (intestinal, rabbit, fragment); B26967 (pig, fragment); C26967 (rabbit, fragment)

Brookhaven code

Enzyme Handbook © Springer-Verlag Berlin Heidelberg 1991
Duplication, reproduction and storage in data banks are only
allowed with the prior permission of the publishers

7 LITERATURE REFERENCES

[1] Watanabe, K., Iha, H., Ohashi, A., Suzuki, Y.: J. Bacteriol., 171, 1219–1222 (1989)
[2] Plant, A.R., Parratt, S., Daniel, R.M., Morgan, H.W.: Biochem. J., 255, 865–868 (1988)
[3] Lecommandeur, D., MacGregor, A.W., Daussant, J.: J. Chromatogr., 441, 436–442 (1988)
[4] Suzuki, Y., Oishi, K., Nakano, H., Nagayama, T.: Appl. Microbiol. Biotechnol., 26, 546–551 (1987)
[5] Kakefuda, G., Duke, S.H., Hostak, M.S.: Planta, 168, 175–182 (1986)
[6] Suzuki, Y., Tomura, Y.: Eur. J. Biochem., 158, 77–83 (1986)
[7] Yoshigi, N., Chikano, T., Kamimura, M.: Agric. Biol. Chem., 50, 1335–1337 (1986)
[8] Suzuki, Y., Tamagawa, S.: Appl. Microbiol. Biotechnol., 20, 218–220 (1984)
[9] Wacker, H., Aggeler, R., Kretchmer, N., O'Neill, B., Takesue, Y., Semenza, G.: J. Biol. Chem., 259, 4878–4884 (1984)
[10] Suzuki, Y., Takii, Y., Taguchi, H.: Eur. J. Appl. Microbiol. Biotechnol., 18, 254–256 (1983)
[11] Suzuki, Y., Aoki, R., Hayashi, H.: Biochim. Biophys. Acta, 704, 476–483 (1982)
[12] Suzuki, Y., Imai, T.: Biochim. Biophys. Acta, 705, 124–126 (1982)
[13] Skovbjerg, H., Sjöström, H., Noren, O.: FEBS Lett., 108, 399–402 (1979)
[14] Suzuki, Y., Ueda, Y., Nakamura, N., Abe, S.: Biochim. Biophys. Acta, 566, 62–66 (1979)
[15] Marshall, J.J., Sturgeon, C.M., Whelan, W.J.: Anal. Biochem., 82, 435–444 (1977)
[16] Sjöström, H., Noren, O., Christiansen, L., Wacker, H., Samenza, G.: J. Biol. Chem., 255, 11332–11338 (1980)
[17] Quaroni, A., Gershon, E., Semenza, G.: J. Biol. Chem., 249, 6424–6433 (1974)
[18] Asp, N.-G., Dahlqvist, A.: FEBS Lett., 35, 303–305 (1973)
[19] Hauri, H.-P., Quaroni, A., Isselbacher, K.J.: Proc. Natl. Acad. Sci. USA, 76, 5183–5186 (1979)
[20] Manners, D.J.: Nature (New Biol.), 234, 150–151 (1971)

1 NOMENCLATURE

EC number
3.2.1.11

Systematic name
1, 6-Alpha-D-glucan-6-glucanohydrolase

Recommended name
Dextranase

Synonymes
Dextran hydrolase
Endodextranase
Dextranase DL 2
DL 2
Endo-dextranase
Alpha-D-1, 6-glucan-6-glucanohydrolase

CAS Reg. No.
9025-70-1

---

2 REACTION AND SPECIFICITY

Catalysed reaction
Dextran + $H_2O \rightarrow$
$\rightarrow$ n oligosaccharides (endohydrolysis of 1,6-α-D-glucosidic linkages in dextran)

Reaction type
O-Glycosyl bond hydrolysis (endohydrolysis, method for distinction between endo-and exo-dextranases [4])

Natural substrates
Dextran + H_2O

Substrate spectrum
1 Dextran + H_2O (e.g. Dextran T-2000 [16], dextran-gels, Sephadex G-100, G-150, G-200 [17], specific for alpha-1, 6-linkages remote from reducing ends [5], highly specific [7 , 8], only dextran synthesized by Leuconostoc mesenteroides [17], Aspergillus carneus preferentially attacks third alpha-1, 6-glycosidic linkage from the reducing end [13])
2 Isomaltodextrins + H_2O [13]
3 Isomaltose + H_2O (Penicillium luteum) [13]

Enzyme Handbook © Springer-Verlag Berlin Heidelberg 1991
Duplication, reproduction and storage in data banks are only
allowed with the prior permission of the publishers

Product spectrum
1 Isomaltotriose (64.4% [8], 34.9% [1]) + isolamtose (41.5% [1]) + isomaltose + isomaltotetraose (7% [8]) [8, 16] + isomaltopentaose [16] + higher isomaltooligosaccharides (20% [1], 23.7% [8]) + glucose (0.8% [1]) [1, 7–10, 16]
2 ?
3 D-Glucose [13]

Inhibitor(s)
Hg^{2+} [1, 7, 8, 10, 12, 14–16]; SDS [1, 12]; Sodium m-periodate [1]; p-Chloromercuribenzoate [6, 7]; EDTA (immobilized form [6]) [6, 14]; Mercaptoethanol (immobilized form) [6]; Sodium ascorbate (immobilized form) [6]; Sodium azide (immobilized form) [6]; Potassium ferricyanide (immobilized form) [6]; Pb^{2+} [7]; Cu^{2+} [7, 8, 10, 15]; Zn^{2+} [7, 8]; Fe^{3+} [10, 15]; Ag^+ [12, 14, 15]; H_2O_2 [14]; Iodoacetic acid [14]

Cofactor(s)/prostethic group(s)
Auxin (activation) [18]

Metal compounds/salts
Mg^{2+} [8]; Mn^{2+} [8, 14]; Ca^{2+} [8]; Co^{2+} [14]; Cu^{2+} [14]

Turnover number (min^{-1})

Specific activity (U/mg)
1500 [7]; 2330 [15]; 146 [1]; More [2 , 3, 5, 8, 10, 11, 14, 16, 17]

K_m-value (mM)
1.1–1.2 (dextran T-200) [3]; 0.003–0.456 (dextran, value depending on dextran type) [9]; 0.11 (dextran-10 (Pharmacia)) [11]; More [6]

pH-optimum
4–6 [16]; 4.0–5.5 (molds) [18]; 4.5 [5]; 5.0 [1, 9]; 5–6.5 (bacteria) [18]; 5.0–5.5 [15]; 5.0–6.5 [17]; 5.2–5.4 [3]; 5.5 [10, 11]

pH-range
3.5–5.5 [9]; 4.0–8.8 [17]; 4.5–6.5 [5]; 5.3 (Cellovibrio fulva) [18]; 5.5–6.5 (Lactobacillus bifidus) [18]; 5.6–7.0 [12]; 7.5 (below, Penicillium lilacinum) [18]

Temperature optimum (°C)
35 [8]; 36 [3]; 50 [1, 6, 7, 9, 16, 17]; 55 [5, 11]; 60 [15]; 65 [10]

Temperature range (°C)
30–55 [17]; 40–65 [9]

3 ENZYME STRUCTURE

Molecular weight

175000 (Streptococcus sobrinus, dextranase C, SDS-PAGE) [3]
160000 (Streptococcus sobrinus, dextranase D, SDS-PAGE) [3]
114000 (Penicillium aculeatum, gel filtration, SDS-PAGE, gel electrophoresis) [8]
71000 (Aspergillus carneus, ultracentrifugal analysis) [13, 15]
68000 (Lipomyces starkeyi, gel filtration) [1]
70000–77000 (Chaetomium gracile, ultracentrifugal analysis, SDS-PAGE)
[10]
60000 (Cytophaga sp., sedimentation equilibrum centrifugation) [17]
46000 (Penicillium luteum, ultracentrifugal analysis) [13]
44000 (Penicillium funiculosum, gel filtration) [14]
39000 (Fusarium moniliforme, gel filtration) [11]
23000 (Lipomyces starkeyi, gel filtration) [9]

Subunits

Monomer [8]
More (due to interaction with SDS no determination possible) [1]

Glycoprotein/Lipoprotein

–

---

4 ISOLATION/PREPARATION

Source organism

Lipomyces starkeyi [1, 9]; Penicillium aculeatum [2, 6, 7]; Penicillium purpurogenum [2]; Penicillium funiculosum [2, 14]; Streptococcus sobrinus [3];
Paecilomyces lilacinus [5]; Flavobacterium sp. M-73 [8]; Chaetomium
gracile [10]; Fusarium moniliforme [11]; Actinomyces israelii [12];
Aspergillus carneus [13, 15]; Penicillium luteum [13, 16]; Crytophyga sp.
(aerobic bacterium) [17]; Avena sativa [19, 20]; Verticillium sp. [18]; Spicaria
sp. [18]; Lactobacillus bifidus [18]; Cellovibrio fulva [18]

Source tissue

Culture medium [11–3, 5–16]; Cell [17]; Coleoptiles [19, 20]

Localisation in source

Extracellular [1, 3, 5–16]; Cell wall (surface [17]) [17, 19]

Enzyme Handbook © Springer-Verlag Berlin Heidelberg 1991
Duplication, reproduction and storage in data banks are only
allowed with the prior permission of the publishers

Purification

Lipomyces starkeyi [1, 9]; Streptococcus sobrinus (dextranases C + D) [3]; Paecilomyces lilacinus (2 isoenzymes) [5]; Flavobacterium sp. M-73 (dextranase II) [8]; Chaetomium gracile (2 isoenzymes) [10]; Fusarium moniliforme [11]; Penicillium funiculosum (2 isoenzymes) [14]; Aspergillus carneus (dextranases I + II) [15]; Penicillium luteum [16]; Cytophaga sp. [17]

Crystallization

(Chaetomium dextranase I) [10]

Cloned

[3]

Renaturated

[3]

---

5 STABILITY

pH

2.5–6.5 [1]; 3.5–7.0 [16]; 3.5–7.5 [9]; 4.5 (unstable below) [5]; 4.0–8.0 [18]; 4.5–6.5 (2 months) [7]; 4.5–9.0 [15]; 5.0–7.5 [14]; 5.5–7.5 [11]; 5.5–11.0 [10]; 6.0 (solubilized cell wall enzyme) [17]; 6.5–12.0 [8]

Temperature (°C)

30–40 [1]; 35 (below) [8]; 40 (up to) [14]; 44 (up to) [3]; 45 (immobilized form) [6]; 50 (soluble form [6], 60 min. [8], 2h, pH 5.0, 70% activity [9], below [15, 16]) [6, 8, 9, 15, 16]; 55 (below) [10]; 65 (up to) [5]

Oxidation

Organic solvent

General stability information

Lyophilization, freezing, thawing (stable in cell wall preparations) [17]

Storage

4°C, immobilized on bentonite, dry form, or buffer suspension, or sucrose suspension, up to 3 weeks [6]; 4°C, soluble form, acetate buffer, pH 5.6, 90 days, 100% activity [5]; 4°C, in cell wall preparations, pH 5, several months, solubilized enzyme less stable [17]

---

6 CROSSREFERENCES TO STRUCTURE DATABANKS

PIR/MIPS code

Brookhaven code

7 LITERATURE REFERENCES

[1] Koenig, D., Day, D.: Eur. J. Biochem., 183, 161–167 (1989)
[2] Shukla, G.L., Prabhu, K.A.: Enzyme Microb. Technol., 11, 533–536 (1989)
[3], Barrett, J.F., Barrett, T.A., Curtiss, R.: Infect. Immun., 55, 792–802 (1987)
[4] Kobayashi, M., Utsugi, H., Matsuida, K.: Agric. Biol. Chem., 51, 2073–2079 (1987)
[5] Lee, J.M., Fox, P.F.: Enzyme Microb. Technol., 7, 573–577 (1985)
[6] Mahdu, Prablu, K.A.: Enzyme Microb. Technol., 7, 279–282 (1985)
[7] Mahdu, Prabllu, K.A.: Enzyme Microb. Technol., 6, 217–220 (1984)
[8] Kobayashi, M., Tagaki, S., Shiota, M., Mitsuishi, Y., Matsuda, K.: Agric. Biol. Chem., 47, 2585–2593 (1983)
[9] Webb, E., Spencer-Martins, I.: Can. J. Microbiol., 29, 1092–1095 (1983)
[10] Hattori, A., Ishibashi, K., Minato, S.: Agric. Biol. Chem., 45, 2409–2416 (1981)
[11] Simonson, L.G., Liberta, A.E., Richardson, A.: Appl. Microbiol., 30, 855–861 (1975)
[12] Staat, R.H., Schachtele, C.F.: Infect. Immun., 12, 556–563 (1975)
[13] Hiraoka, N., Tsuji, H., Fukumoto, J., Yamamoto, T., Tsuru, D.: Int. J. Pept. Protein Res., 5, 161–169 (1973)
[14] Sugiura, M., Ito, A., Ogiso, T., Kato, K., Asano, H.: Biochim. Biophys. Acta, 309, 357–362 (1973)
[15] Hiraoka, N., Fukumoto, J., Tsuru, D.: J. Biochem., 71, 57–64 (1972)
[16] Fukumoto, J., Tsuji, H., Tsuru, D.: J. Biochem., 69, 1113–1121 (1971)
[17] Janson, J.-C., Porath, J.: Methods Enzymol., 8, 615–621 (1966)
[18] Fischer, E.H., Stein, E.A. in "The Enzymes", 2nd. Ed. (Boyer, P.D., Eds.) 4, 301–312 (1960) (Review)
[19] Heyn, A.N.J.: Proc. Natl. Acad. Sci. USA, 78, 6608–6612 (1981)
[20] Heyn, A.N.J.: Biochem. Biophys. Res. Commun., 38, 831–837 (1970)

1 NOMENCLATURE

EC number
3.2.1.14

Systematic name
Poly(1, 4-(N-acetyl-beta-D-glucosaminide)) glycanohydrolase

Recommended name
Chitinase

Synonymes
Chitodextrinase
1, 4-Beta-poly-N-acetylglucosaminidase
Poly-beta-glucosaminidase
Beta-1, 4-poly-N-acetyl glucosamidinase
Poly-beta-glucosaminidase
Endochitinase

CAS Reg. No.
9001-06-3

---

2 REACTION AND SPECIFICITY

Catalysed reaction
Chitin + H_2O →
→ oligomers of N-acetylglucosamine (random hydrolysis of N-acetyl-beta-D-glucosaminide 1,4-beta-linkages in chitin and chitodextrins)

Reaction type
O-Glycosyl bond hydrolysis (endohydrolysis)

Natural substrates
Chitin (in fungal cell walls [25, 26], lysis of Rhizopus cell walls [34])

Substrate spectrum
1 Chitin (reacylated chitosan [3], colloidal chitin [6, 13, 31], chitin produced in reaction medium by chitin synthetase [3, 24, 28], preformed chitin [24], glycolchitin [6, 9, 13, 23, 31]) + H_2O [3, 6, 9, 13, 23, 24, 28, 31]
2 Chitosan (with different degree of deacylation) + H_2O [10, 17]
3 Chitooligosaccharides (tetramer or larger) + H_2O [8 , 17]
4 N, N'-Diacetylchitobiose + H_2O [10]
5 3, 4-Dinitrophenyltetra-N-acetyl-beta-D-chitotetraoside + H_2O (no hydrolysis of phenyl-beta-chitooligosaccharides [13]) [10, 33]
6 More (some chitinases also display the activity defined in EC 3.2.1.17)

Enzyme Handbook © Springer-Verlag Berlin Heidelberg 1991
Duplication, reproduction and storage in data banks are only
allowed with the prior permission of the publishers

Product spectrum

1 Chitooligosaccharides (N, N'-diacetylchitobiose (main product [1, 6, 11, 12, 25, 28, 31], 67% [24], 5.2% [29]) + chitotriose (87.9% [29]) + chitotetraose [3, 10, 24, 31] + N-acetylglucosamine (long incubation time [26], small amounts [6], sole product [10]))

2 ?

3 ?

4 N-Acetylglucosamine

5 ?

6 ?

Inhibitor(s)

Allosamidin [1, 38, 39]; Histamine [9, 25]; Cu^{2+} [10, 22, 24, 27, 31, 37]; Hg^{2+} [10, 20, 24, 27, 31, 34, 37]; KCN [10]; Mn^{2+} [10]; Zn^{2+} [10, 36]; 2-Acetamido-2-deoxy-D-gluconolactone [10]; Iodoacetate [20]; Phosphate [20]; Arsenate [20]; N-Ethylmaleimide [20]; p-Chloromercuribenzoate [20]; Trypsin [24]; Fe^{3+} (slightly) [24]; N-Acetylglucosamine [25]; EDTA (slightly) [30, 31]; Fe^{2+} [27, 34]; Sn^{2+} [34]; Mg^{2+} [36]; Co^{2+} [36]; Na^+ [36]; Ca^{2+} [36]; Al^{3+} [36]; Colloidal chitin (hydrolysis of 3, 4-dinitrophenyl-tetra-N-acetylchitotetraoside) [33]; Alpha-chitin (hydrolysis of 3, 4-dinitrophenyl-tetra-N-acetylchitotetraoside) [33]

Cofactor(s)/prostethic group(s)

Trypsin (activation by limited proteolysis) [22, 24, 30]; More (phospholipid-enzyme interaction required for microsomal enzyme) [22]

Metal compounds/salts

Ca^{2+} (4 moles per mole of enzyme) [36]

Turnover number (min^{-1})

Specific activity (U/mg)

14.6 [3, 32]; 4.4–6.2 [34]; 1.33 [26]; More (assay method [2, 5, 14, 15]) [1, 4, 6, 8, 9–13, 18, 20, 23–26, 28–35]

K_m-value (mM)

4.4 (chitin, expressed in N-acetylglucosamine equivalents, similar values [3, 28, 32]) [26]; 33.0 (chitin, expressed in N-acetylglucosamine equivalents) [27]; 4.4 (reacylated chitosan) [12]; 0.0075 (3, 4-dinitrophenyl-tetra-N-acetylchitotetraoside) [33]; More (values expressed in mg chitin per ml) [3, 8, 10, 20, 22]

pH-optimum
1.5–2.5 [28]; 1.65 [35]; 1.7 (colloidal chitin) [30]; 2.5–3.0 (glycol chitin) [30];
3.0 (gastric chitinases of vertebrates) [37]; 3.5 (optimum 1) [23]; 3.7 [10];
4.2–4.8 [36]; 4–7 [26]; 4.5 [6]; 4.7 [18]; 5.0 [13, 27, 37]; 5.1 [8]; 5.55 (microsomal form) [22]; 5.65 (cytosolic form) [22]; 5.5–6.5 [34]; 6 (substrate
chitotetraoside [23]) [3, 23]; 6.0–8.0 (substrate colloidal chitin) [31]; 6.3
(Streptomyces griseus) [37]; 6.5 [1, 9, 16, 25]; 6.7 [11, 24]; 7.0 [20]; 8.0–8.5
(optimum 2) [23]; 9–10 (substrate glycol chitin) [23]; 10.5 (substrate glycol
chitin) [31]

pH-range
1.3–3.1 [35]; 3–9 [3, 9, 25]; 3.0–11.0 [13]; 4.4–7.0 [27]; 7.7 (inactive above)
[10]

Temperature optimum (°C)
30 [16]; 40 (at pH 1.5 [30]) [30, 37]; 44 [8]; 45 [1, 20]; 51 (at pH 3.5 [30]) [10,
31]; 55 (chitinase I) [34]; 60 (chitinase II) [34]

Temperature range (°C)
65 (40% of maximum activity) [10]; 70 (up to) [13]

---

3 ENZYME STRUCTURE

Molecular weight
105000–110000 (Aeromonas hydrophila subsp. anaerogenes, SDS-PAGE,
gel filtration) [20]
60000–63000 (Vibrio sp., gel filtration, SDS-PAGE [31], goat, gel filtration
[35]) [31, 35]
50000–58000 (Serratia liquefaciens, SDS-PAGE [4], Serratia marcescens,
SDS-PAGE [12, 26]) [4, 12, 26]
47000–48000 (Stomoxys calcitrans, SDS-PAGE [27], calf, gel filtration [30])
[27, 30]
30000–38000 (Phaseolus vulgaris, SDS-PAGE [9], gel filtration [9, 25],
Pycnoporus cinnabarinus, gel filtration [6] , Streptomyces erythraeus,
SDS-PAGE [13], Mucor rouxii, chitinase I, gel filtration [16], Pisum sativum,
chitinases I -III, SDS-PAGE [18], Dioscorea opposita, SDS-PAGE [23], wheat,
SDS-PAGE, sedimentation equilibrium centrifugation [32], goat, chitinase I,
gel filtration [35], Streptomyces antibioticus, sedimentation equilibrium
centrifugation [37]) [6, 9, 13, 16, 18, 23, 25, 32, 35, 37]
20600–29000 (wheat, gel filtration, SDS-PAGE, sedimentation equilibrium
centrifugation [3], Lycopersicon esculentum, gel filtration [8], Neurospora
crassa, gel filtration [11], Mucor rouxii, chitinase II, gel filtration [16], Streptomyces orientalis, chitinase II, gel filtration [34] , Streptomyces sp.,
sedimentation equilibrium centrifugation [36]) [3, 8, 9, 11, 16, 34, 36]

Enzyme Handbook © Springer-Verlag Berlin Heidelberg 1991
Duplication, reproduction and storage in data banks are only
allowed with the prior permission of the publishers

Subunits
Monomer [3]

Glycoprotein/Lipoprotein
Glycoprotein [20, 23, 28]

---

4 ISOLATION/PREPARATION

Source organism
Candida albicans [1]; Wheat [3, 32]; Serratia liquefaciens [5]; Pycnoporus cinnabarinus [6]; Soybean [7]; Calvatia cyathiformis (puffball) [7]; Lycoperdon candidum (puffball) [7]; Lycopersicon esculentum (tomato) [8]; Phaseolus vulgaris [9, 25]; Verticillium albo-atrum [10]; Neurospora crassa [11, 24]; Serratia marcescens [12, 17, 21, 26]; Streptomyces erythraeus [13]; Mucor rouxii [16]; Pisum sativum [18]; Trichoderma harziana [19]; Streptomyces antibioticus [17, 37]; Streptomyces sp. [17, 36]; Streptomyces griseus [17]; Aeromonas hydrophila subsp. anaerogenes [20]; Mucor mucedo [22]; Dioscorea opposita (yam) [23]; Stomoxys calcitrans (stable fly) [27]; Saccharomyces cerevisiae [28]; Cupiennius salei (hunting spider) [29]; Calf [30]; Vibrio sp. [31]; Vibrio alginolyticus [33]; Goat [35]; Streptomyces orientalis [34]; Vertebrates (insectivorous fishes, amphibians, reptiles, birds, mammals [37], comparison of chitinase levels in body fluids and organs of animals [35]) [35, 37, 40]; Bacteria [37]; Fungi [37]; Plants [37]; Onchocera gibsoni (female) [38]; Bombyx mori [39]; Coprinus lagopus [41]

Source tissue
Culture medium [4, 6, 10, 12, 13, 19, 20, 26, 31, 33, 34, 36, 37]; Germ [2, 32]; Cell [1, 28]; Seeds (soybean) [7]; Leaves [9, 25]; Mycelium [11, 16, 22, 24]; Glebas (spore mass chambers) [7]; Pods [18]; Pupae [27]; Digestive fluid [29]; Serum [30, 35]; Glandular tissues [37]; Pancreas [40]; Gastric mucosa [40]

Localisation in source
Extracellular [4, 6, 10, 12, 13, 19, 20, 26, 31, 33, 34, 36, 37]; Central vacuole [9]; Cytoplasm [11, 16, 22, 24]; Microsomes (zymogen) [22]; Cell wall [24]; Lysosomal vacuoles [41]

Purification
Wheat germ [3, 32]; Serratia liquefaciens (partial) [4]; Pycnoporus cinnabarinus [6]; Lycopersicon esculentum [8]; Phaseolus vulgaris [9, 25]; Verticillium albo-atrum [10], Neurospora crassa [11]; Serratia marcescens [12, 26]; Streptomyces erythraeus [13]; Mucor rouxii (2 isoenzymes) [16]; Pisum sativum (3 isoenzymes) [18]; Aeromonas hydrophila subsp. anaerogenes [20]; Dioscorea opposita (3 isoenzymes) [23]; Stomoxys calcitrans [27]; Saccharomyces cerevisiae [28]; Cupiennius salei [29]; Calf [30]; Vibrio sp. [31]; Streptomyces orientalis (2 isoenzymes) [34]; Goat [35]; Streptomyces sp. [36, 37]

Crystallization
[36]

Cloned
[21]

Renaturated
–

---

5 STABILITY

pH
1.7–8.9 (3–6 optimum stability) [30]; 3.5–6.5 [35]; 4–10 [13]; 4.8–8.0 [6]; 4.5–7 [26]; 5–11 [23]; 6.0–9.0 [20]; 3.0 (stable at) [28]

Temperature (°C)
40 (up to) [9, 16, 25, 31]; 50 (15 min., 15% activity [16], 4 h, 65% activity [25], below [34]) [16, 25, 34]; 60 (up to [13, 23, 30], 1h, 50% activity [35]) [13, 23, 30, 35]; 65 (10 min., 55% activity) [36]; More (unstable at elevated temperatures) [6]

Oxidation

Organic solvent

General stability information
Bovine serum albumin (stabilization) [10]; Freezing/thawing (no inactivation) [1, 27]; Urea (5 M, stable) [28]; Ca^{2+} (stabilization) [36]; Chitin (stabilization) [37]

Storage
–20°C, or 4°C, 3 months [1]; –20°C, or 2°C, several weeks [3]; –80°C, Tris buffer, pH 8.5, 0.5 M NaOH, 0.02% azide, 7 years [3]; Frozen, crude or purified enzyme, several months [8]; –20°C, lyophilized, at least 2 years [9]; 2°C, several months [10]; 0°C or below, 50 mM KH_2PO_4-NaOH buffer, pH 6.2, 10 mM $MgCl_2$, 1 mM EDTA, 3 mM NaN_3, several weeks [11]; 0.05 M phosphate buffer, pH 6.3, at least 4 years [12]; –20°C, microsomal enzyme increase in activity over 7 days, cytosolic enzyme stable 3 days [22]; 0°C, or below, several weeks [24]; –20°C [27]; 0–5°C, sodium citrate buffer, pH 3, 0.02% NaN_3, at least 2 months [28]; Ammonium buffer, pH 6.5 [30]; –20°C, or 2°C, pH 8.5, 0.02% azide, several weeks [32]; –20°C, 3 weeks, 14% activity, 25°C, 3 weeks, 100% activity [35]; 2°C, lyophilized [36]

Enzyme Handbook © Springer-Verlag Berlin Heidelberg 1991
Duplication, reproduction and storage in data banks are only
allowed with the prior permission of the publishers

6 CROSSREFERENCES TO STRUCTURE DATABANKS

PIR/MIPS code
A25090 (Serratia marcescens); A25898 (kidney bean); A29074 (tobacco, fragment); A31455 (cucumber, TNV-infected leaves); S04856 (B, precursor, Serratia marcescens); JX0076 (Saccharopolyspora erythraea); A29912 (Streptomyces plicatus, fragment); A24215 (kidney bean, fragment); S05426 (precursor, potato); S04131 (barley, fragment); A29104 (barley, fragments)

Brookhaven code

7 LITERATURE REFERENCES

[1] Dickinson, K., Keer, V., Hitchcock, C.A., Adams, D.J.: J. Gen. Microbiol., 135, 1417–1421 (1989)
[2] Cabib, E., Sburlati, A.: Methods Enzymol., 161, 457–459 (1988)
[3] Cabib, E.: Methods Enzymol., 161, 498–501 (1988)
[4] Joshi, S., Kozlowski, M., Richens, S., Comberbach, D. M.: Enzyme Microb. Technol., 11, 289–296 (1989)
[5] Cabib, E.: Methods Enzymol., 161, 424–426 (1988)
[6] Ohtakara, A.: Methods Enzymol., 161, 462–470 (1988)
[7] Zikakis, J.P., Castle, J.P.: Methods Enzymol., 161, 490–497 (1988)
[8] Pegg, G.F.: Methods Enzymol., 161, 484–489 (1988)
[9] Boller, T., Gehri, A., Mauch, F., Vögeli, U.: Methods Enzymol., 161, 479–484 (1988)
[10] Pegg, G.F.: Methods Enzymol., 161, 474–479 (1988)
[11] Arroyo-Begovich, , A.: Methods Enzymol., 161, 471–474 (1988)
[12] Cabib, E.: Methods Enzymol., 161, 460–462 (1988)
[13] Hara, S., Yamamura, Y., Fujii, Y., Mega, T., Ikenaka, T.: J. Biochem., 105, 484–489 (1989)
[14] Boller, T., Mauch, F.: Methods Enzymol., 161, 430–435 (1988)
[15] Ohtakara, A.: Methods Enzymol., 161, 426–430 (1988)
[16] Pedraza-Reyes, M., Lopez-Romero, E.: J. Gen. Microbiol., 135, 211–218 (1989)
[17] Ohtakara, A., Izume, M., Mitsutomi, , M.: Agric. Biol. Chem., 52, 3181–3182 (1988)
[18] Mauch, F., Hadwiger, L.A., Boller, T.: Plant Physiol. , 87, 325–333 (1988)
[19] Ridout, C.J., Coley-Smith, J.R., Lynch, J.M.: Enzyme Microb. Technol., 10, 180–187 (1988)
[20] Yabuki, M., Mizushina, K., Amatatsu, T., Ando, A., Fujii, T., Shimada, M., Yamashita, M.: J. Gen. Appl. Microbiol., 32, 25–38 (1986)
[21] Jones, J.D.G., Grady, K.L., Suslow, T.V., Bedbrook, J.R.: EMBO J., 5, 467–473 (1986)
[22] Humphreys, A.M., Gooday, G.W.: J. Gen. Microbiol., 130, 1359–1366 (1984)
[23] Tsukamoto, T., Koga, D., Ide, A., Ishibashi, T., Horino-Matsushige, M., Yagishita, K., Imoto, T.: Agric. Biol. Chem., 48, 931–939 (1984)
[24] Zarain-Herzberg, A., Arroyo-Begovich, A.: J. Gen. Microbiol., 129, 3319–3326 (1983)
[25] Boller, T., Gehri, A., Mauch, F., Vögeli, U.: Planta, 157, 22–31 (1983)
[26] Roberts, R.L., Cabib, E.: Anal. Biochem., 127, 402–412 (1982)
[27] Chen, A.C., Mayer, R.T., DeLoach, J.R.: Arch. Biochem. Biophys., 216, 314–321 (1982)

[28] Correa, J.U., Elango, N., Polacheck, I., Cabib, E.: J. Biol. Chem., 257, 1392–1397 (1982)
[29] Mommsen, T.P.: Biochim. Biophys. Acta, 612, 361–372 (1980)
[30] Lundblad, G., Elander, M., Lind, J., Slettengren, K.: Eur. J. Biochem., 100, 455–460 (1979)
[31] Ohtakara, A., Mitsutomi, M., Uchida, Y.: J. Ferment. Technol., 5, 169–177 (1979)
[32] Molano, J., Polacheck, I., Duran, A., Cabib, E.: J. Biol. Chem., 254, 4901–4907 (1979)
[33] Aribisala, O.A., Gooday, G.W.: Biochem. Soc. Trans., 6, 568–569 (1978)
[34] Tominaga, Y., Tsujisaka, Y.: Agric. Biol. Chem., 40, 2325–2333 (1976)
[35] Lundblad, G., Herderstedt, , B., Lind, J., Steby, M.: Eur. J. Biochem., 46, 367–376 (1974)
[36] Skujins, J., Pukite, A., McLaren, A.D.: Enzymologia, 39, 353–370 (1970)
[37] Jeuniaux, C.: Methods Enzymol., 8, 644–650 (1966) (Review)
[38] Gooday, G.W., Brydon, L.J., Chappell, , L.H.: Mol. Biochem. Parasitol., 29, 223–225 (1988)
[39], Koga, D., Isogai, A., Sakuda, S., Matsumoto, S., Suzuki, A., Kimura, A., Ide, A.: Agric. Biol. Chem., 51, 471–476 (1987)
[40] Jeuniaux, C.: Nature, 192, 135–136 (1961)
[41] Iten, W., Matile, P.: J. Gen. Microbiol., 61, 301–309 (1970)

Enzyme Handbook © Springer-Verlag Berlin Heidelberg 1991
Duplication, reproduction and storage in data banks are only
allowed with the prior permission of the publishers

1 NOMENCLATURE

EC number
3.2.1.15

Systematic name
Poly(1, 4-alpha-D-galacturonide)glycanohydrolase

Recommended name
Polygalacturonase

Synonymes
Pectin depolymerase
Pectinase
Endopolygalacturonase
Pectolase
Pectin hydrolase
Pectin polygalacturonase
Endo-polygalacturonase
Poly-alpha-1, 4-galacturonide glycanohydrolase
Pectic depolymerase
Endogalacturonase
Endo-D-galacturonase
Endopolygalacturonate lyase
Endopectinase
Endo-D-galacturonanase
Phylendonase
D-Galacturonase
Pectinase SS
Remanase
Liquifying polygalacturonase [44]
Pectozyme
Pectic hydrolase
Rapidase [3]
Pectolyase [7]
Rohament P [30]
Pectinex Ultra [28]
Klerzyme
Ultrazym 100
Pektopol PT

CAS Reg. No.
9032-75-1

Enzyme Handbook © Springer-Verlag Berlin Heidelberg 1991
Duplication, reproduction and storage in data banks are only
allowed with the prior permission of the publishers

2 REACTION AND SPECIFICITY

Catalysed reaction
Pectic acid + n H_2O →
→ tetra-, tri-, di-, and monogalacturonic acid (random hydrolysis of 1,4-alpha-D-galactosiduronic linkages in pectate and other galacturonans)

Reaction type
O-Glycosyl-bond hydrolysis

Natural substrates
Polygalacturonic acid + H_2O (e.g. in cell walls, maceration of plant tissues [41])

Substrate spectrum
1 Polygalacturonate (pectic acid) + H_2O (dependency of rate of hydrolysis on viscosity of medium [9], effect of size of polygalacturonic acid on rate of hydrolysis [40], requirement of at least 3 galacturonic acid residues at the non-reducing end of substrate [42])
2 Heptagalacturonic acid + H_2O [28]
3 Hexagalacturonic acid + H_2O [28, 32]
4 Pentagalacturonic acid + H_2O [28, 32, 42]
5 Tetragalacturonic acid + H_2O [28, 32, 42]
6 Trigalacturonic acid + H_2O [28, 32]
7 Pectin (low reactivity) + H_2O [1, 4, 33, 41]

Product spectrum
1 Galacturonic acid oligomers (2–6 residues [31], monogalacturonic acid Botrytis cinerea enzyme only [37])
2 Hexagalacturonic acid + pentagalacturonic acid + tetragalacturonic acid + trigalacturonic acid + digalacturonic acid + galacturonic acid [28]
3 Pentagalacturonic acid [32] + tetragalacturonic acid [28, 32] + trigalacturonic acid [28, 32] + digalacturonic acid [28, 23] + galacturonic acid [28]
4 Tetragalacturonic acid [28] + trigalacturonic acid [28, 32, 42] + digalacturonic acid [28, 32, 24] + galacturonic acid [28]
5 Trigalacturonic acid [28, 32, 42] + galacturonic acid [28, 32, 42]
6 Digalacturonic acid [28, 32] + galacturonic acid [28, 32]
7 Oligomers of methyl-D-galacturonic acid

Inhibitor(s)

Polygalacturonate (substrate inhibition) [1, 2]; Polygalacturonase inhibitory protein (from Phaseolus vulgaris) [6]; Pectin [2]; Mn^{2+} [9, 29, 38]; Zn^{2+} [9]; Hg^{2+} [13, 14, 21, 26, 29, 38]; Ba^{2+} [13, 21, 38]; Cd^{2+} [13, 14, 21]; Cu^{2+} [13, 21, 26, 38]; Sn^{2+} [13]; Sr^{2+} [13]; Pb^{2+} [14, 38]; Ag^{2+} [21, 26]; Ca^{2+} [21, 26, 38]; Hg^{+} [21, 26]; Co^{2+} [26]; p-Chloromercuribenzoate [29, 37]; Tris [29]; Diethylpyrocarbonate [35, 38]; Triethanolamine [37]; Urea [37, 38]; NaCl (more than 0.15 mM) [39]; EDTA [38, 43]; Mg^{2+} [38, 45]; 2, 4-Dinitrophenol [38]; Iodoacetic acid [38]; Cysteine [38]; Sodium azide [38]; Potassium cyanide [38]; 8-Hydroxyquinoline [38]

Cofactor(s)/prostethic group(s)

Metal compounds/salts

Na^{+} (activation) [13, 34, 40]; Fe^{3+} (activation) [9]; Co^{2+} (activation) [9]; Mg^{2+} (activation) [9]; Ca^{2+} (activation) [43]

Turnover number (min⁻¹)

Specific activity (U/mg)

34900 [12]; 126.2 [4]; 963.0 [9]; More (assay method [5]) [1, 7, 13, 16, 21, 22, 27–29, 32, 33, 34, 36–38, 41, 42]

K_m-value (mM)

3.98 (trigalacturonic acid, similar values [26, 28]) [21]; 2.77 (tetragalacturonic acid, similar values [26, 28]) [21]; 8.4 (tetragalacturonic acid) [42]; 0.71 (pentagalacturonic acid, similar value [26]) [21]; 4.8 (pentagalacturonic acid) [42]; 0.05 (polygalacturonic acid, similar value [26]) [21]; 0.7 (hexagalacturonic acid) [28]; 0.89 (heptagalacturonic acid) [28]; 0.34 (polygalacturonic acid) [28]; 0.014 (sodium pectate MW 33000) [34]; More [2, 9, 14, 17, 29, 30, 37, 38]

pH-optimum

2.5 [4]; 3.5 (substrate trigalacturonic acid) [13]; 3.75 [14]; 3–4.5 (Aspergillus aculeatus) [18]; 4.0 (PG-B [7], PG-II [2], substrate tetragalacturonic acid [13]) [2, 7, 13, 25, 37, 45]; 4.0–5.0 (protopectinase activity [21]) [21, 29]; 4.2 [42]; 4.5 (PG-I [2], Mucor pusillus [18], substrate pectic acid [13]) [2, 13, 18, 34, 39, 40, 41]; 4.68 [1]; 4.6–4.8 [38]; 4.7 (PG-A) [7]; 4.8 [16]; 4.9 [28]; 5.0 (substrate pectic acid [21]) [9, 21, 26, 32, 33]; 5.1 [16]; 5.4 [17]; 5.5 [24]; 5.5–6.0 [22]; 10.0 [43]

pH-range

3.5–5 [39]; 3.5–5.5 (PG-I) [40]; 5–5.5 (PG-II) [40]; 4.8–5.2 [32]

Temperature optimum (°C)

37 [14]; 40 [25, 28]; 40–45 [32]; 45 [9]; 50 (Mucor pusillus [18]) [18, 21, 26, 29]; 50–60 [12]; 60 (Aspergillus aculeatus) [18]; 65 [43]; 70 [13]; 75 [24]

Enzyme Handbook © Springer-Verlag Berlin Heidelberg 1991
Duplication, reproduction and storage in data banks are only
allowed with the prior permission of the publishers

Temperature range (°C)
50–75 [43]

---

3 ENZYME STRUCTURE

Molecular weight

199500 (Lycopersicon esculentum, PG-I, gel filtration) [23]
84000 (Lycopersicon esculentum, PG-I, gel filtration) [40]
75900 (Kluyveromyces marxianus, PG-I, gel filtration) [25]
69000 (Botrytis cinerea, gel filtration) [37]
50000–57000 (Penicillium capsulatum [1], Botrytis cinerea, PG-II, SDS-
PAGE, gel filtration [2], Saccharomyces fragilis, PG-II, SDS-PAGE [29],
Lycopersicon esculentum, sedimentation equilibrium centrifugation [34]) [1,
2, 29, 34]
40000–48000 (Aureobasidium pullulans, SDS-PAGE [12], Stereum purpureum, SDS-PAGE, gel filtration, sedimentation equilibrium centrifugation
[13] Cryptococcus albidus, gel filtration [14], Lycopersicon esculentum, PG-
II, gel filtration [23, 34, 40], Saccharomyces fragilis, PG-I, SDS-PAGE [29],
Aspergillus niger, gel filtration [32], Aspergillus niger, gel filtration, value
depending on buffer [35], Prunus persica, gel filtration [39]) [12, 13, 14, 23,
29, 32, 34, 35, 40]
30000–39000 (Botrytis cinerea, gel filtration, SDS-PAGE [2], Aspergillus
japonicus, PG-B, gel filtration [7, 41], Cryptococcus albidus, SDS-PAGE
[14], Kluyveromyces fragilis, enzyme II, gel filtration, sedimentation equilibrium centrifugation [20], Galactomyces reesei, gel filtration, sedimentation equilibrium centrifugation [21], Kluyveromyces marxianus, PG-II, gel
filtration, SDS-PAGE [25], Trichosporon penicillatum, gel filtration,
sedimentation equilibrium centrifugation [27], Saccharomyces fragilis, PG-
II, SDS-PAGE [29], Rhizoctonia fragariae, both isoenzymes, gel filtration
[33], Aspergillus niger, PG-I, gel filtration [35]) [2, 7, 20, 21, 25, 27, 29, 33, 35,
41]
25000 (Aspergillus japonicus, PG-A, gel filtration) [7]

Subunits

Tetramer (Lycopersicon esculentum, PG-I) [23]
Monomer (Fusarium oxysporum [36], Kluyveromyces marxianus, enzyme II,
SDS-PAGE [25]) [25, 36]
Oligomer (Kluyveromyces marxianus, enzyme I, 47900 + 28100, SDS-PAGE)
[25]

Glycoprotein/Lipoprotein

Glycoprotein [20, 21, 23, 27, 36]

4 ISOLATION/PREPARATION

Source organism
Penicillium capsulatum [1]; Botrytis cinerea [2, 37]; Corticum rolfsii [4];
Aspergillus niger [3, 15, 28, 30, 32, 35]; Aspergillus japonicus [7, 41];
Lycopersicon esculentum (tomato, 2 isoenzymes, extract from tomato fruit
converts PG-II to PG-I [19]) [8, 19, 23, 34, 40, 48]; Rhizopus stolonifer [9, 38];
Erwinia carotovora subs. atroseptica and carotvora [10]; Aspergillus al-
leacus [11]; Aureobasidium pullulans [12]; Stereum pupureum [13]; Cryptococcus albidus var. albidus [14]; Trichoderma reesei [16]; Bacteroides
thetaiotaomicron [17]; Aspergillus aculeatus [18]; Mucor pusillus [18];
Kluyveromyces fragilis [20]; Galactomyces reesei [21]; Pseudomonas
solanacearum [22]; Clostridium thermosulfurogenes [24]; Kluyveromyces
marxianus [25]; Trichosporon penicillatum [26, 27]; Saccharomyces fragilis
[29]; Trichoderma koningii [31]; Rhizoctonia fragariae [33]; Fusarium
oxysporum f. sp. lycopersici [36]; Prunus persica (peach) [39];
Acrocylindrium sp. [42]; Bacillus No. P-4-N [43]; Verticillium albo-atrum [46];
Pseudomonas cepacia [47]; Fungi [44, 45]

Source tissue
Culture medium; Fruit [8, 19, 23, 34, 39, 40, 48]; Cell [17]

Localisation in source
Extracellular; Membrane (inner, associated) [17]

Purification
Penicillium capsulatum [1]; Botrytis cinerea (2 isoenzymes [2]) [2, 37];
Aspergillus niger (from commercial enzyme preparations: Rapidase C-80
[3], Pectinex Ultra [29], Rohament P [30], Pectinol [35], product from Koch-
Light, England [32]) [3, 15, 29, 30, 32, 35]; Corticum rolfsii [4]; Aspergillus
japonicus (from commercial preparation Pectolyase [7]) [7, 41]; Rhizopus
stolonifer [9]; Aureobasidium pullulans [12]; Stereum purpureum [13]; Cryptococcus albidus [14]; Trichoderma reesei [16]; Bacteroides
thetaiotaomicron (partial) [17]; Kluyveromyces fragilis (4 isoenzymes) [20];
Galactomyces reesei [21]; Lycopersicon esculentum (2 isoenzymes [23, 40])
[23, 34, 40]; Trichosporon penicillatum [27]; Saccharomyces fragilis (3
isoenzymes) [29]; Trichoderma koningii [31]; Rhizoctonia fragariae (2
isoenzymes) [33]; Fusarium oxysporum (2 isoenzymes) [36]; Prunus persica
[39]; Acrocylindrium [42]; Verticillium albo-atrum [46]; Pseudomonas
cepacia [47]

Crystallization
[12, 20, 21, 27]

Cloned
[8, 48]

Enzyme Handbook © Springer-Verlag Berlin Heidelberg 1991
Duplication, reproduction and storage in data banks are only
allowed with the prior permission of the publishers

Renaturated

—

5 STABILITY

pH

0.6 (crude enzyme, 20 minutes, 30–50% activity) [45]; 2.0–8.0 [4]; 2.5–7.5 [42]; 3.0–5.0 [9, 28]; 3–6 [26]; 3.0–7.0 [21, 32]; 3.5–6.0 [29]; 4–5 [41]; 4.0–6.0 [12, 38]; 4.0–8.0 [14]; 4.0–9.0 [13]; 4.3 (PG-I, best stability) [40]; 5.6 (PG-II, best stability) [40]; 6.5 (best stability) [43]; 10 (75% activity) [13]

Temperature (°C)

30 (up to) [9]; 30–50 [38]; 40 (below [12, 14, 42], 5 minutes [7]) [7, 12, 14, 42]; 42 (below) [28]; 50 (below) [26, 29, 36, 41]; 57 (PG-II, 50% activity) [40]; 60 (crude enzyme, 3 hours, 50% activity, purified enzyme, 3.8 minutes, 50% activity) [1]; 70 (crude enzyme, 15 minutes, 50% activity, purified enzyme, 1 minute, 50% activity) [1]; 75 (up to) [43]; 78 (PG-I, 50% activity) [40]; More (polygalacturonic acid protects against heat inactivation) [42]

Oxidation

Organic solvent

General stability information

Storage

–20°C [22]; –20°C, 20 mM sodium acetate buffer, pH 5.0 [31]; 4°C, crude enzyme: acetone powder preparation or dissolved in citrate-phosphate buffer, pH 4.8 or distilled water, partially purified enzyme: freeze-dried [38]; Lyophilized [2, 35]; Below 5°C, several months [4]; Frozen, concentrated solution, several weeks [7]

6 CROSSREFERENCES TO STRUCTURE DATABANKS

PIR/MIPS code

JA0156 (precursor, tomato, fragment); A25534 (precursor, tomato); S06340 (precursor, tomato)

Brookhaven code

7 LITERATURE REFERENCES

[1] Gillespie, A.-M., Coughlan, M.P.: Biochem. Soc. Trans. , 17, 384–385 (1989)

[2] Schejter, A., Marcus, L.: Methods Enzymol., 161, 366–373 (1988)

[3] Rozie, H., Somers, W., Bonte, A., Visser, J., Van't Riet, K., Rombouts, F.M.: Biotechnol. Appl. Biochem., 10, 346–358 (1988)

[4] Tagawa, K., Kaji, A.: Methods Enzymol., 161, 361–365 (1988)

[5] Collmer, A., Ried, J.L., Mount, M.S.: Methods Enzymol., 161, 329–335 (1988)

[6] Cervone, F., De Lorenzo, G., Degra, L., Salvi, G., Bergami, M.: Plant Physiol., 85, 631–637 (1987)

[7] Baldwin, E.A., Pressey, R.: Plant Physiol., 90, 191–196 (1989)

[8] Sheehy, R.E., Pearson, J., Brady, C.J., Hiatt, W.R.: Mol. Gen. Genet., 208, 30–36 (1987)

[9] Manachini, P.L., Fortina, M.G., Parini, C.: Biotechnol. Lett., 9, 219–224 (1987)

[10] Ried, J.L., Collmer, A.: Appl. Environ. Microbiol., 52, 305–310 (1986)

[11] Sreenath, H.K., Kogel, F., Radola, B.J.: J. Ferment. Technol., 64, 37–44 (1986)

[12] Sakai, T., Takaoka, A.: Agric. Biol. Chem., 49, 449–458 (1985)

[13] Miyairi, K., Okuno, T., Sawai, K.: Agric. Biol. Chem. , 49, 1111–1118 (1985)

[14] Federici, F.: Antonie Leeuwenhoek, 51, 139–150 (1985)

[15] Lobarzewski, J., Fiedurek, J., Ginalska, G., Wolski, T.: Biochem. Biophys. Res. Commun., 131, 666–674 (1985)

[16] Markovic, O., Slezarik, A., Labudova, I.: FEMS Microbiol. Lett., 27, 267–271 (1985)

[17] McCarthy, R.E., Kotarski, S.F., Salyers, A.A.: J. Bacteriol., 161, 493–499 (1985)

[18] Foda, M.S., Rizk, I.R., Gibriel, A.Y., Basha, S.I.: Zentralbl. Mikrobiol., 139, 463–469 (1984)

[19] Pressey, R.: Eur. J. Biochem., 144, 217–221 (1984)

[20] Sakai, T., Okushima, M., Yoshitake, S.: Agric. Biol. Chem., 48, 1951–1961 (1984)

[21] Sakai, T., Yoshitake, S.: Agric. Biol. Chem., 48, 1941–1950 (1984)

[22] Ofuya, C.O.: Curr. Microbiol., 10, 141–146 (1984)

[23] Moshrefi, M., Luh, B.S.: Eur. J. Biochem., 135, 511–514 (1983)

[24] Schink, B., Zeikus, J.G.: FEMS Microbiol. Lett., 17, 295–298 (1983)

[25] Call, H.-P., Emeis, C.-C. in "Util. Enzymes Technol. Aliment.", Symp. Int. (Dupuy, P., Ed.) 513–517 (1982)

[26] Sakai, T., Okushima, M., Sawada, M.: Agric. Biol. Chem., 46, 2223–2231 (1982)

[27] Sakai, T., Okushima, M.: Agric. Biol. Chem., 46, 667–676 (1982)

[28] Heinrichova, K., Dzurova, M.: Collect. Czech. Chem. Commun., 46, 3145–3156 (1981)

[29] Lim, J., Yamasaki, Y., Suzuki, Y., Ozawa, J.: Agric. Biol. Chem., 44, 473–480 (1980)

[30] Rexova-Benkova, L., Mrackova, M., Babor, K.: Collect. Czech. Chem. Commun., 45, 163–168 (1980)

[31] Fanelli, C., Cacace, M.G., Cervone, F.: J. Gen. Microbiol., 104, 305–309 (1978)

[32] Heinrichova, K., Rexova-Benkova, , L.: Collect. Czech. Chem. Commun., 42, 2569–2576 (1977)

[33] Cervone, F., Scala, A., Foresti, M., Cacace, M.G., Noviello, C.: Biochim. Biophys. Acta, 482, 379–385 (1977)

[34] Markovic, O., Slezarik, A.: Collect. Czech. Chem. Commun., 42, 173–179 (1977)

[35] Cooke, R.D., Ferber, C.E.M., Kanagasabapathy, L.: Biochim. Biophys. Acta, 452, 440–451 (1976)

[36] Strand, L.L., Cordon, M.E., MacDonald, D.L.: Biochim. Biophys. Acta, 429, 870–883 (1976)

Enzyme Handbook © Springer-Verlag Berlin Heidelberg 1991
Duplication, reproduction and storage in data banks are only
allowed with the prior permission of the publishers

[37] Urbanek, H., Zalewska-Sobczak, J.A.: Biochim. Biophys. Acta, 377, 402–409 (1975)
[38] Trescott, A.S., Tampion, J.: J. Gen. Microbiol., 80, 401–409 (1974)
[39] Pressey, R., Avants, J.K.: Plant Physiol., 52, 252–256 (1973)
[40] Pressey, R., Avants, J.K.: Biochim. Biophys. Acta, 309, 363–369 (1973)
[41] Ishii, S., Yokotsuka, T.: Agric. Biol. Chem., 36, 1885–1893 (1972)
[42] Kimura, H., Uchino, F., Mizushima, S.: J. Gen. Microbiol., 74, 127–137 (1973)
[43] Horikoshi, K.: Agric. Biol. Chem., 36, 285–293 (1972)
[44] Deuel, H., Stutz, E.: Adv. Enzymol. Relat. Subj. Biochem., 20, 341–382 (1958) (Review)
[45] Lineweaver, H., Jansen, E.J.: Adv. Enzymol. Relat. Subj. Biochem., 11, 267–295 (1951) (Review)
[46] Wang, M.C., Keen, N.T.: Arch. Biochem. Biophys., 141, 749 (1970)
[47] Ulrich, J.M.: Physiol. Plant Pathol., 5, 37 (1975)
[48] Grierson, D., Tucher, G.A., Keen, J., Ray, J., Bird, C.R., Schuch, W.: Nucleic Acids Res., 14, 8595–8603 (1986)

1 NOMENCLATURE

EC number
3.2.1.17

Systematic name
Peptidoglycan N-acetylmuramoyl-hydrolase

Recommended name
Lysozyme

Synonymes
Muramidase
Globulin G
Mucopeptide glucohydrolase
Globulin G1
N, O-Diacetylmuramidase
Lysozyme g
L-7001
1, 4-N-Acetylmuramidase [3]
Mucopeptide N-acetylmuramoylhydrolase
PR1-Lysozyme [26]

CAS Reg. No.
9001-63-2

---

2 REACTION AND SPECIFICITY

Catalysed reaction
Hydrolysis of 1, 4-beta-linkage between N-acetylmuraminic acid and N-acetyl-D-glucosamine residues in peptidoglycan and between N-acetyl-D-glucosamine residues in chitodextrin; More (glycosyl transfer to saccharides, other alcohols and H_2O)

Reaction type
O-Glycosyl bond hydrolysis

Natural substrates
Peptidoglycan + H_2O (unspecific defence mechanism against bacterial infections associated with monocyte-macrophage system [1], involvement in host defence [4], anti-tumor activity [4], anti-metastatic activity [4], ruminants: digestive enzyme [13], phage T4: cell lysis from within, at the end of latent period, cell lysis from without, at the begin of infection [20]) [1, 4, 13] More [14]

Enzyme Handbook © Springer-Verlag Berlin Heidelberg 1991
Duplication, reproduction and storage in data banks are only
allowed with the prior permission of the publishers

Substrate spectrum

1 Peptidoglycan + H_2O (of: Micrococcus luteus [8, 9], Pseudomonas aeruginosa [26], Micrococcus lysodeikticus [29], Staphylococcus epidermis: Staphylococcus aureus enzyme, not chicken egg-white enzyme [49]) [8, 9, 14, 19, 49]

2 Chitin + H_2O (not: Streptomyces erythraeus enzyme [27]) [8, 9, 14, 19]

3 Glycol chitin + H_2O (not: Streptomyces erythraeus enzyme) [8]

4 Ethylene glycol chitin + H_2O (not: Streptomyces erythraeus enzyme [27]) [8]

5 Chitotetraose + H_2O (not: Nephthys hombergii enzyme [40]) [13, 18, 19]

6 Chitopentaose + H_2O [40]

7 More (not: p-nitrophenyl-N-acetylglucosaminide [9], phage T4: e lysozyme more specific than hen egg-white lysozyme, e lysozyme /only hydrolysis of murein chains in which N-acetylmuraminic acid is substituted by peptide side chains L-Ala-D-Glu-meso-diaminopimelic acid-D-Ala [20], Pseudomonas aeruginosa, pyocinogenic: no activity towards intact cells of gram-negative and gram-positive bacteria, lysis of chloroform-killed gram-negative and gram-positive bacteria [26], gram-negative bacteria better substrate than gram-positive bacteria [45], Papaya lysozyme: high chitinase activity [48], lysozyme often acts as chitinase [8, 14, 17]) [8, 9, 14, 17, 20, 26, 45, 48]

Product spectrum

1 N-Acetylamino sugars (disaccharide units attached to the muropeptides [19], C_3 and C_6 muropeptides [20]) [19, 20, 14]

2 Sugars (reducing) [8]

3 Sugars (reducing) [8]

4 Sugars (reducing) [8]

5 Chitotriose + N-acetylglucosamine [13]

6 ?

7 ?

Inhibitor(s)

Poly-L-lysine [8]; Glycol chitosan [8]; $ZnCl_2$ [8]; $HgCl_2$ [8]; $AgNO_3$ [8]; Histamine [8, 40, 54]; Chitotetraose [13]; Tri-N-acetylchitotriose [14]; N-Acetylglucosamine (dimer: Nephthys hombergii, slight [40], trimer [40], alpha and beta anomeric forms [14], fig: poor inhibitor [54]) [40, 14, 54, 17, 29, 47]; Phenyl-beta-D-chitobioside [14]; N-Acetylglucosaminono-(1, 5)-lactone [14]; N-Acetylmuraminic acid [17]; Nuclear lysozyme inhibitor (other subcellular lysozymes except nuclear are unaffected [22]) [22]; $MgCl_2$ [45]; NaCl [45]; KCl [45]; N-Bromosuccinimide (pH 4) [48]; ICl [48]; EDTA (above 0.1 mM) [49]; Sodium citrate (above 0.1 M) [49]; $CuSO_4$ [49]; SDS [49]; I^- [57]; Sodium dodecyl sulfonate [57]; Sodium dodecanate [57]; More (high ionic strength) [13]

Cofactor(s)/prostethic group(s)
More (no metal cofactor required [8], practically inactive in absence of Triton X-100, hydrolysis of murein catalysed only when in contact with lipophilic components [16]) [8, 16]

Metal compounds/salts
Mg^{2+} (increases activity in phage T4 e lysozyme [21], no effect: phage T4 ghost lysozyme [20]) [20]; NaCl (activates) [49]; Ag^+ (activates) [49]; $CaCl_2$ (activates) [49]; More (ionic strength optimum: 0.05–0.1 M [20], 0.04–0. 07 [29]) [20, 29]

Turnover number (min⁻¹)
30 (GlcNAc-MurNAc) [14]; 19.8 ((GlcNAc)$_5$) [14]; 9–15 ((GlcNAc)$_6$) [14]; More [14]

Specific activity (U/mg)
More (139.81/min × mg: decrease in absorbance [9]) [8, 9, 13, 17, 18, 22, 25, 27, 29, 33, 34, 41, 42, 45, 46, 53, 55, 56, 49]

K_m-value (mM)
More (166 mg/ml purified Micrococcus cell walls [8]) [8, 14]; 0.01 ((GlcNAc)$_5$, (GlcNAc)$_6$) [14]; 0.8 ((GlcNAc)$_3$-p-nitrophenol) [14]; 1.7 ((GlcNAc)$_2$-Ph2, 4-diNO$_2$) [14]; 4 ((GlcNAc)$_2$-p-nitrophenol) [14]

pH-optimum
6.0 (wheat, substrate: Micrococcus cells) [8]; 5.5 (trout) [9]; 6.5 (rabbit) [23]; 5.0 (narrow range, ruminants) [13]; 5–8 (broad, non-ruminants, chicken, pig, monkey) [13]; 6.0 (phage T4) [16]; 5.0 (calf) [17]; 7.6 (phage T4: e lysozyme) [20]; 5.5–6.0 (phage T4: ghost lysozyme) [20]; 6.4 (pyocinogenic Pseudomonas aeruginosa) [26]; 4.0 (Streptomyces erythraeus) [27]; 7.5 (tortoise) [29]; 6.5 (Ceratitis capitata) [33]; 6–7 (phage T2) [41]; 7.2 (phage F1, F5) [41]; 7.5–8.5 (phage A-22) [41]; 7–8 (phage P22) [45]; 4.5–5.5 (turnip) [47]; 6.35 (human) [56]; 6.0–9.4 (Staphylococcus aureus) [49]; More [14]

pH-range
3–9 (wheat) [8]; 4–8 (trout) [9]; 4–8.5 (phage T4) [16]; 3–7 (calf) [17]; 2.5–8.5 (20% of maximal activity at 2.5 and 8.5, phage T4, ghost) [20]; 4.5–8.5 (rabbit) [23]; 5–9 (pyocinogenic Pseudomonas aeruginosa) [26]; 2.5–8 (Streptomyces erythraeus) [27]; 4.5–9.5 (tortoise) [29]; 3–8 (Ceratitis capitata) [33]; 5–10 (phage P22) [45]; 3.5–7. 0 (turnip) [47]; 4.5–9.5 (Staphylococcus aureus) [49]; More [14]

Temperature optimum (°C)
60 (wheat) [8]; 45 (trout) [9]; 50 (turnip) [47]; 37 (Staphylococcus aureus) [49]; More [14]

Temperature range (°C)
10–90 (wheat, inactive above) [8]; 10–65 (trout) [9]; 15–70 (turnip) [47]

Enzyme Handbook © Springer-Verlag Berlin Heidelberg 1991
Duplication, reproduction and storage in data banks are only
allowed with the prior permission of the publishers

3 ENZYME STRUCTURE

Molecular weight

25400 (SDS-PAGE, wheat, W1A) [8]
14400 (SDS-PAGE, trout, I/II) [9]
15000 (SDS-PAGE, human, milk and pancreatic juice) [15]
15000 (gel filtration, SDS-PAGE, phage T4) [16]
15000 (gel filtration, cattle) [17]
14300 (gel filtration, rabbit) [23]
24000 (gel filtration, slab gel electrophoresis in SDS, Pseudomonas aeruginosa, pyocinogenic) [26]
18500 (ultracentrifugation, Streptomyces erythraeus) [27]
15400 (sedimentation and diffusion coefficient values, low speed sedimentation (without reaching equilibrium), amino acid composition, Trionyx gangeticus Cuvier) [29]
23200 (SDS-PAGE, sedimentation velocity centrifugation in sucrose gradient, gel filtration, Ceratitis capitata) [33]
15000 (amino acid analysis, Papio cynocephalus) [46]
14400–14800 (human leukemia lysozyme) [34]
18600 (amino acid composition, phage T2) [41]
11000–13000 low speed sedimentation (without reaching equilibrium)
17600 (sedimentation analysis, sequential analysis, phage lambda) [41]
19000 (low speed sedimentation (without reaching equilibrium),
14400 (Sephadex G-100 chromatography, horse) [42]
200000 (gel filtration, phage P22) [45]
250000 (ultracentrifugal studies, papaya) [48]
25000 (gel filtration, turnip) [47]
15000–17000 (analytical ultracentrifugation, human) [53]
29000 (ultracentrifugal studies, fig) [54]
17900 (sedimentation equilibrium data, sedimentation and diffusion coefficient, bacteriophage lambda synthetized in E. coli) [55]
14000–16000 (sedimentation velocity, short-column sedimentation equilibrium, amino acid analysis, human) [56]
More (primary structure [32, 51, 52]) [14, 32, 41, 51, 52, 57]

Subunits

Dimer (reversible dimerization, dimer: predominant species above pH 9, between pH 5 and 9 higher polymers) [14]
Monomer (Pseudomonas aeruginosa, SDS slab gel electrophoresis) [15]
Monomer (SDS-PAGE, tortoise) [29]
Monomer (SDS-PAGE, Ceratitis capitata) [33]
Monomer [55]
Monomer (papaya) [48]
More (glucose-induced polymerization [12], polymerization favored by increased pH [14]) [12, 14]

Glycoprotein/Lipoprotein

–

---

4 ISOLATION/PREPARATION

Source organism

Chicken (gene expression in: Lactobacillus lactis subsp. lactis [1], E. coli [3], Saccharomyces cerevisiae [5], human cell lines HeLa and MCF-7 [6]) [1, 3, 5, 6, 7, 11, 12, 14, 24, 25, 30, 57]; Human (gen expression in: Saccharomyces cerevisiae [2, 4], primary structure [52]) [2, 4, 14, 15, 34, 43, 53, 56]; E. coli (phage-induced [5], phage T4 induced [10, 16]) [5, 10, 16]; Wheat (multiple electrophoretic forms [8]) [8]; Salmo gairdneri (trout, c-type) [9]; Phage T4 (enzyme is product of e gene (lysis from within), or from gene 5 (lysis from without) [20], wilde-type and mutant [28]) [5, 10, 16, 20, 28, 38]; Ruminants (type c) [13]; Cow (gastric mucosa: 3 lysozymes c, other tissues: no lysozyme c, low levels of another lysozyme, g class?) [13]; Sheep [13]; Calf [17]; Deer [13]; Duck (Kaki and wilde-type [44], Peking-duck [18, 44], multiple forms [44], primary structure of lysozyme II [51]) [14, 18, 44, 51]; Turkey [14]; Japanese quail [14]; Goose (primary structure [52]) [14, 52]; Rat (primary structure [32]) [22, 32, 35]; Fig [54]; Rabbit [23]; Bacteriophage lambda (synthesized in E. coli) [55]; Pseudomonas aeruginosa (pyocinogenic) [26]; Streptomyces erythraeus [27]; Trionyx gangeticus Cuvier (tortoise) [29]; Ceratitis capitata (dipterous) [33]; Bovine [39]; Dog [39]; Nephthys hombergii (annelid) [40]; Phages (T2, T4, N20F, lambda, F1, F5, Kp, 2, Pf15, A-22, F12, 13, 14, P1, P14, N1 and other ones) [41]; Horse (mare) [42]; Phage P22 (enzyme induced in Salmonella typhimurium) [45]; Papio cynocephalus (baboon) [46]; Turnip [47]; Papaya [48]; Turkey [50]; Staphylococcus aureus [49]; Colitis bacteriophage [58]; More [14, 19, 57]

Source tissue

Egg white (turkey [14, 50], Japanese quail [14], goose [14], duck [14, 18, 51], hen [1, 5, 6, 7, 14, 24, 25, 29, 30, 57]) [1, 5, 6, 7, 14, 25, 29, 30, 57, 51]; Histolytic lymphoma cell line U-937 (human) [2]; Milk (horse [42], baboon [46], human [4, 14, 15, 43, 56]) [4, 14, 15, 43, 56, 42, 46]; Kidney (trout [9], rat [35]) [9, 35]; Germ (wheat) [8]; Stomach (ruminants, fundic region highest activity) [13]; Gastric mucosa (cow, 3 lysozymes c) [13]; Tears (human) [14, 34]; Cell [26]; Saliva (human [14], human parotid [53]) [14, 53]; Placenta (human) [14]; Spleen serum (human) [14]; Leucocytes (human) [14]; Serum (human) [34]; Blood (human, leukemia patients) [14]; Eggs [33]; Urine (human, leukemia patients) [14, 34]; Cartilage [39]; Pancreatic juice (human) [15]; Serum [39]; Rennet (cattle) [17]; Abomasal secretions [17]; Ghosts (phage T4) [20]; Liver (rat) [22]; Macrophages (alveolar, rabbit) [23]; Latex [48, 54]

Enzyme Handbook © Springer-Verlag Berlin Heidelberg 1991
Duplication, reproduction and storage in data banks are only
allowed with the prior permission of the publishers

Localisation in source

Nucleus [22, 35]; Mitochondria [22]; Microsomes [22]; Soluble [22]

Purification

Wheat (multiple electrophoretic forms) [8]; Salmo gairdneri (trout, 2 forms: I/II, c-type) [9]; Cow [13]; Duck (3 electrophoretic forms, Peking duck) [18]; Human (leukemia lysozyme [34]) [15, 34, 43, 53, 56]; Phage T4 [16, 41]; Calf [17]; Rat [22, 35]; Rabbit [23]; Chicken [25, 57]; Pseudomonas aeruginosa (pyocinogenic) [26]; Streptomyces erythraeus [27]; Trionyx gangeticus Cuvier [29]; Ceratitis capitata [33]; Bovine [39]; Dog [39]; Phage T2 [41]; Equus caballus (horse) [42]; Phage P22 (host: Salmonella typhimurium) [45]; Papio cynocephalus (baboon) [46]; Turnip [47]; Turkey [50]; Fig [54]; Bacteriophage lambda (host: E coli) [55]; Papaya [48]; Stapylococcus areus [49]; More (from several sources, affinity chromatography [36], with tri-(N-acetylglucosamine)-agarose [37]) [36, 37, 14, 41]

Crystallization

(chicken: crystallographic studies of denaturation and renaturation [31, 30], atomic and molecular displacements [11], crystallization conditions [7], E. coli, phage T4 induced: study of structural basis of thermal stability [10], human [14], human leukemia lysozyme [34], Streptomyces erythraeus [27], Trionyx gangeticus Cuvier [29], bovine [39], Phage T4 [41], turkey [50]) [7, 11, 10, 14, 27, 29, 34, 39, 41, 50]

Cloned

(chicken, hen egg-white gene, expression in Lactobacillus lactis subsp. lactis [1], E. coli [3], Saccharomyces cerevisiae [5], human cell lines HeLa and MCF-7 [6], human, histolytic lymphoma cell line U-937 [2] gene, expression in Saccharomyces cerevisiae [2, 4], E. coli, phage induced, expression in E.coli [5], colitis bacteriophage gen, expression in E. coli [58]) [1–6, 58]

Renaturated

(reversible thermal denaturation [14], crystallographic studies of denaturation and renaturation [30, 31], reversible thermal denaturation conditions: acid pH, 0.2 M NaCl, 0.10 mM dithiothreitol, 0.01 mM T4 lysozyme [38]) [14, 30, 31, 38]

---

5 STABILITY

pH

6–8 (stable in presence of 0.01% bovine serum albumin, Pseudomonas aeruginosa) [26]; 6.5 (stable at high enzyme concentrations, for at least 2 months at 4°C) [26]; 3.5 (stability at acidic pH, stable for weeks at 4°C, Ceratitis capitata) [33]; 8.7 (lability at alkaline pH, Ceratitis capitata) [33]; More (very stable at acid pH's even at 100°C [57], lability at alkaline pH [57]) [33, 57]

Temperature (°C)

21 (trout: half life 1–2 days, in presence of another protein (5mg/ml) stable for at least 1 week) [1]; 77 (transition temperature, high thermal stability) [14]; 20–70 (unaffected, 10 minutes, pH 5.0) [17]; 45 (inactivation above, 10 minutes, phage T4 ghost) [20]; 70 (80% loss of activity after 10 minutes) [21]; 37 (stable below, Pseudomonas aeruginosa) [26]; 40 (unstable above) [26]; 80 (83% of activity of that at 25°C, tortoise, more stable than hen lysozyme) [29]; 53 (50% inactivation at, phage T4) [41]; 40 (no loss of activity up to, 5 minutes) [45]; 70 (90% loss of activity after 5 minutes, phage P22) [45]; 37 (35% residual activity after 4 days, pH 4) [56]; More (phage T4, mutant and wilde-type, study of structural basis of thermal stability [10], tortoise lysozyme more stable than hen lysozyme [29], human lysozyme more resistant to heat than chicken [43]) [10, 29, 43, 14, 22, 28, 33, 49, 57]

Oxidation

Inactivated by photooxidation [57]

Organic solvent

Ethanol (denaturant) [14]; Methanol (denaturant) [14]; Isopropanol (denaturant) [14]; n-Propanol (denaturant) [14]; Glycerol (nondenaturant) [14]; Ethylene glycol (nondenaturant) [14]; More [14]

General stability information

At high concentrations or in presence of albumin, lyophilized or desalted without loss of activity [9]; Ruminant and monkey lysozymes C unusually resistant to inactivation by pepsin [13]; Guanidine hydrochloride denaturates [14]; Effects of freezing, thawing and freeze-drying are neglegible [17]; Na^+ and Mg^{2+} exert a stabilizing effect (phage T4) [41]; Denaturation by ultraviolet irradiation [57]; Inactivated by photooxidation and action of X-rays [57]; More [41]

Storage

Several weeks at –20°C [8]; Several months, at –20°C, 0.1% bovine serum albumin [23]; For weeks at 4°C, pH 3.5, 0.1 M ionic strength; More [41, 33]

6 CROSSREFERENCES TO STRUCTURE DATABANKS

PIR/MIPS code

LZHU (human); LZBA (baboon); LZBO (c 2, bovine); LZRT (rat); LZCH (c, precursor, chicken); LZQJEC (c, California quail); LZQJEB (c, common bobwhite, tentative sequence); LZQJE (c, precursor, Japanese quail, tentative sequence); LZFER (c, precursor, ring-necked pheasant); LZTK (c, precursor, turkey tentative sequence); LZUH (c, helmeted guineafowl, tentative sequence); LZDK (c, precursor, duck); LZDK3 (c III, duck); LZOVE (c, plain chachalaca); LZPY (c, pigeon); LZWK (c, precursor, cecropia moth,

Enzyme Handbook © Springer-Verlag Berlin Heidelberg 1991
Duplication, reproduction and storage in data banks are only
allowed with the prior permission of the publishers

fragment); LZWSG (g, black swan); LZGSG (g, goose); LZOSG (g, ostrich); LZBPT4 (bacteriophage T4); WMBPP9 (bacteriophage phi-29); A34277 (1, precursor, bovine); A35558 (1, bovine); B34277 (2a, precursor, bovine, fragment); B35558 (2, bovine); C34277 (2b, precursor, bovine, fragment); C35558 (3, bovine); D34277 (2c, precursor, bovine, fragment); D35558 (1, sheep); E34277 (2d, precursor, bovine, fragment); E35558 (2, sheep); F34277 (3, precursor, bovine); F35558 (3, sheep); G35558 (1, axis deer); H35558 (2, axis deer); PL0163 (common ring-tailed possum, fragment); S05657 (chicken); S01692 (tail, bacteriophage T4); A11762 (starfish, fragment); S00489 (rainbow trout, fragment); B34047 (g, black swan, fragments); JT0526 (c, Indian peafowl); A29736 (Hanuman langur); S04938 (precursor, human); A31240 (precursor, human); S01661 (c, donkey); JX0104 (rabbit); A11541 (mouse, fragment); A31239 (M, precursor, mouse)

Brookhaven code

0FDL (mouse - Mus musculus, hen - Gallus gallus, egg white) ; 0GLM (embden goose - Anser anser) ; 0LZ6 (Streptomyces erythreaus) ; 0LZE (hen - Gallus gallus, egg white) ; 0LZT (hen - Gallus gallus, egg white) ; 0TEL (tortoise - Trionyx gangeticus, egg white) ; 1L01 (bacteriophage T4, double mutant Thr 155 replaced by Ala and Thr 157 replaced by Ile, mutant gene is derived from the M13 plasmid by cloning of the T4 lysozyme gene) ; 1L02 (bacteriophage T4, mutant with Thr 157 replaced by Ala, mutant gene is derived from the M13 plasmid by cloning of the T4 lysozyme gene) ; 1L03 (bacteriophage T4, mutant with Thr 157 replaced by Cys, mutant gene is derived from the M13 plasmid by cloning of the T4 lysozyme gene) ; 1L04 (bacteriophage T4, mutant with Thr 157 replaced by Asp, mutant gene is derived from the M13 plasmid by cloning of the T4 lysozyme gene) ; 1L05 (bacteriophage T4, mutant with Thr 157 replaced by Asp, mutant gene is derived from the M13 plasmid by cloning of the T4 lysozyme gene) ; 1L06 (bacteriophage T4, mutant with Thr 157 replaced by Glu, mutant gene is derived from the M13 plasmid by cloning of the T4 lysozyme gene) ; 1L07 (bacteriophage T4, mutant with Thr 157 replaced by Phe, mutant gene is derived from the M13 plasmid by cloning of the T4 lysozyme gene) ; 1L08 (bacteriophage T4, mutant with Thr 157 replaced by Gly, mutant gene is derived from the M13 plasmid by cloning of the T4 lysozyme gene) ; 1L09 (bacteriophage T4, mutant with Thr 157 replaced by His, mutant gene is derived from the M13 plasmid by cloning of the T4 lysozyme gene) ; 1L10 (bacteriophage T4, mutant with Thr 157 replaced by Ile, mutant gene is derived from the M13 plasmid by cloning of the T4 lysozyme gene) ; 1L11 (bacteriophage T4, mutant with Thr 157 replaced by Leu, mutant gene is derived from the M13 plasmid by cloning of the T4 lysozyme gene) ; 1L12 (bacteriophage T4, mutant with Thr 157 replaced by Asn, mutant gene is derived from the M13 plasmid by cloning of the T4 lysozyme gene) ; 1L13 (bacteriophage T4, mutant with Thr 157 replaced by Arg, mutant gene is derived from the M13 plasmid by cloning of the T4 lysozyme gene) ; 1L1 4 (bacteriophage T4, mutant with Thr 157 replaced by Ser, mutant gene is derived from the M13 plasmid by cloning of the T4 lysozyme gene) ; 1L15 (bacteriophage T4, mutant with Thr 157 replaced by Val, mutant gene is derived from the M13 plasmid by cloning of the T4 lysozyme gene) ; 1L16 (bacteriophage T4, mutant with Gly 156 replaced by Asp, mutant gene is derived from the M13 plasmid by cloning of the T4 lysozyme gene) ; 1LYM (hen - Gallus gallus, egg white) ; 1LYZ (hen - Gallus gallus, egg white) ; 1LZ1 (human - Homo sapiens) ; 1LZH (hen - Gallus gallus, egg white) ; 1LZT (hen - Gallus gallus, egg white, triclinic crystal form) ; 2HFL (balb - Slash, mouse - Mus musculus, chicken - Gallus gallus) ; 2HFM (mouse - Mus musculus, hen - Gallus gallus, egg white, antibody-antigen complex, IgA FV fragment and lysozyme complex / model) ; 2LYM (hen - Gallus gallus, egg white, 1 atmosphere, 1.4 M NaCl) ; 2LYZ (hen - Gallus gallus, egg white) ; 2LZ2 (turkey - Meleagris gallopavo) ; 2LZH (hen - Gallus gallus, egg white, orthorhombic) ; 2LZM (E. coli infected with bacteriophage T4) ; 4LYZ (hen - Gallus gallus, egg white) ; 3LYM (hen - Gallus gallus, egg white, 1000 atmospheres, 1.4 M NaCl) ; 3LYZ (hen - Gallus gallus, egg white) ; 6LYZ (hen - Gallus gallus, egg white) ; 7LYZ (hen - Gallus gallus, egg white, triclinic crystall) ; 5LYZ (hen - Gallus gallus, egg white) ; 8LYZ (hen - Gallus gallus, egg white, iodine-inactivated) ; 9LYZ (hen - Gallus gallus, egg white, NAM-NAG-NAM substrate only)

7 LITERATURE REFERENCES

[1] Van De Guchte, M., Van Der Vossen, J.M.B.M., Kok, J., Venema, G.: Appl. Environ. Microbiol., 55 (1) , 224–228 (1989)
[2] Castanon, M.J., Spevak, W., Adolf, G.R., Chlebowicz-Sledziewska, E., Sledtiewski, A.: Gene, 66, 223–234 (1988)
[3] Miki, T., Yasukochi, T., Nagatani, H., Furuno, M., Orita, T., Yamada, H., Imoto, T., Horiuchi, T.: Protein Eng., 1 (4) , 327–332 (1987)
[4] Hayakawa, T., Toibana, A., Marumoto, R., Nakahama, K., Kikuchi, M., Fujimoto, K., Ikehara, M.: Gene, 56, 53–59 (1987)
[5] Oberto, J., Davison, J.: Gene, 40, 57–65 (1985)
[6] Matthias, P.D., Renkawitz, R., Grez, M., Schütz, G.: EMBO J., 1 (10) , 1207–1212 (1982)
[7] Ries-Kautt, M.M., Ducruix, A.F.: J. Biol. Chem., 264 (2) , 745–748 (1989)
[8] Audy, P., Trudel, J., Asselin, A.: Plant Sci., 58, 43–50 (1988)
[9] Grinde, B., Jolles, J., Jolles, P.: Eur. J. Biochem., 173, 269–273 (1988)
[10] Alber, T., Matthews, W.: Methods Enzymol., 154 (Recomb. DNA, Pt. E) , 511–533 (1987)
[11] Doucet, J., Benoit, J.P.: Nature, 325, 643–646 (1987)
[12] Cho, R.K., Okitani, A., Kato, H.: Agric. Biol. Chem., 48 (12) , 3081–3089 (1984)
[13] Dobson, D.E., Prager, E.M., Wilson, A.C.: J. Biol. Chem., 259 (18) , 11607–11616 (1984)
[14] Imoto, T., Johnson, L.N., North, A.C.T., Phillips, D. C., Rupley, J.A. in "The Enzymes", 3rd Ed. (Boyer, P.D., Ed.) 7, 665–868 (1972)

Enzyme Handbook © Springer-Verlag Berlin Heidelberg 1991
Duplication, reproduction and storage in data banks are only
allowed with the prior permission of the publishers

[15] Wang, Ch.-S., Kloer, H.-U.: Anal. Biochem., 139, 224–227 (1984) (Review)
[16] Szewczyk, B., Skorko, R.: Eur. J. Biochem., 133, 717–722 (1983)
[17] Pahud, J.-J., Widmer, F.: Biochem. J., 201, 661–664 (1982)
[18] Kondo, K., Fujio, H., Amano, T.: J. Biochem., 91, 571–587 (1982)
[19] Nerurkar, L.S. in "Methods Stud. Mononucl. Phagocytes" (Adams, D.O., Edelson, P.J., Koren, H.S., Eds.), 667–683 (1981) (Review)
[20] Szewczyk, B., Skorko, R.: Biochim. Biophys. Acta, 662, 131–137 (1981)
[21] Jensen, H.B., Kleppe, K.: Eur. J. Biochem., 28, 116–122 (1972)
[22] Sidhan, V., Gurnani, S.: Agric. Biol. Chem., 45 (8), 1817–1823 (1981)
[23] Carroll, St. F., Martinez, R.J.: Infect. Immun., 24 (2), 460–467 (1979)
[24] Perkins, St. J., Johnson, L.N., Machin, P.A., Phillips, D.C.: Biochem. J., 173, 607–616 (1978)
[25] Fernandez-Sousa, J.M., Perez-Castells R., Rodriguez, R.: Biochim. Biophys. Acta, 523, 430–434 (1978)
[26] Ochi, N., Azegami, M., Ishi, S.: J. Biochem., 83, 727–736 (1978)
[27] Morita, T., Hara, S., Matsushima, Y.: J. Biochem., 83, 893–903 (1978)
[28] Elwell, M.L., Schellman, J.A.: Biochim. Biophys. Acta, 494, 367–383 (1977)
[29] Gayen, S.K., Som, S., Sinha, N.K., Sen, A.: Arch. Biochem. Biophys., 183, 432–442 (1977)
[30] Yonath, A., Sielecki, A., Moult, J., Podjarny, A., Traub, W.: Biochemistry, 16 (7), 1413–1417 (1977)
[31] Yonath, A., Podjarny, A., Honig, B., Sielecki, A., Traub, W.: Biochemistry, 16 (7), 1418–1430 (1977)
[32] White, Th. J., Mross, G.A., Osserman, E.F., Wilson, A.C.: Biochemistry, 16 (7), 1430–1436 (1977)
[33] Fernandez-Sousa, J.M., Gavilanse, J.G., Municio, A.M., Perez-Arando, A., Rodriguez, R.: Eur. J. Biochem., 72, 25–33 (1977)
[34] Yoshimoto, T., Tobiishi, M., Tsuru, D.: J. Biochem., 80, 703–709 (1976)
[35] Raghunathan, R., Shantoo, G.: Int. J. Pept. Protein Res., 8, 349–356 (1976)
[36] Junowicz, E., Charm, St.E.: FEBS Lett., 57 (2), 219–221 (1975)
[37] Cornelius, D.A., Brown, W.H., Shrake, A.F., Rupley, J.A.: Methods Enzymol., 34 (Affinity Tech. Enzyme Purif., Part B), 639–645 (1974)
[38] Elwell, M., Schellman, J.: Biochim. Biophys. Acta, 386, 309–323 (1975)
[39] Guenther, H.L., Sorgente, N., Guenther, H.E., Eisenstein, R., Kuettner, K.E.: Biochim. Biophys. Acta, 372, 321–334 (1974)
[40] Perin, J.-P., Jolles, P.: Mol. Cell. Biochem., 2 (2), 189–195 (1973)
[41] Tsugita, A. in "The Enzymes", 3rd Ed. (Boyer, P.D., Ed.) 5, 343–411 (1971) (Review)
[42] Jauregui-Adell, J., Cladel, G., Ferraz-Pina, C., Rech, J.: Arch. Biochem. Biophys., 151, 353–355 (1972)
[43] Barel, A.O., Prieels, J.P., Maes, E., Looze, Y., Leonis, J.: Biochim. Biophys. Acta, 257, 288–296 (1972)
[44] Hermann, J., Jolles, J., Jolles, P.: Eur. J. Biochem., 24 (1), 12–17 (1971)
[45] Koteswara Rao, G.R., Burma, D.P.: J. Biol. Chem., 246 (21), 6474–6479 (1971)
[46] Buss, D.H.: Biochim. Biophys. Acta, 236, 587–592 (1971)
[47] Bernier, I., Van Leemputten, E., Horisberger, M., Bush, D.A., Jolles, P.: FEBS Lett., 14 (2), 100–104 (1971)
[48] Howard, J.B., Glazer, A.N.: J. Biol. Chem., 242 (24), 5715–5723 (1967)
[49] Hawiger, J.: J. Bacteriol., 95 (2), 376–384 (1968)
[50] LaRue, J.N., Speck, J.C.: J. Biol. Chem., 245 (8), 1985–1991 (1970)

[51] Hermann, J., Jolles, J.: Biochim. Biophys. Acta, 200, 178–179 (1970)
[52] Canfield, R.E., Kammerman, S., Sobel, J.H., Morgan, F.J.: Nature (New Biol.) , 232, 16–17 (1971)
[53] Balekjian, A.Y., Hoerman, K.C., Berzinskas, V.J.: Biochem. Biophys. Res. Commun., 35 (6) , 887–894 (1969)
[54] Glazer, A.N., Barel, A.O., Howard, J.B., Brown, D.M.: J. Biol. Chem., 244 (13) , 3583–3589 (1969)
[55] Black, L.W., Hogness, D.S.: J. Biol. Chem., 244 (8) , 1968–1975 (1969)
[56] Parry, R.M., Chandan, R.C., Shahani, K.M.: Arch. Biochem. Biophys., 103, 59–65 (1969)
[57] Jolles, P. in "The Enzymes", 2nd Ed. (Boyer, P.D., Cardy, H., Myrbach, K., Eds.) 4, 431–445 (1960) (Review)
[58] Vasavada, H.A., Murthy, I., Padayatty, J.D.: Gene, 34, 9–15 (1985)

Enzyme Handbook © Springer-Verlag Berlin Heidelberg 1991
Duplication, reproduction and storage in data banks are only
allowed with the prior permission of the publishers

1 NOMENCLATURE

EC number
3.2.1.18

Systematic name
Acyneuraminyl hydrolase

Recommended name
Sialidase

Synonymes
Neuraminidase
Alpha-neuraminidase
Acetylneuraminidase

CAS Reg. No.
9001-67-6

2 REACTION AND SPECIFICITY

Catalysed reaction
N-Acetylneuraminosyloligosaccharide + H_2O →
→ N-acetylneuraminic acid + oligosacharide (hydrolysis of sialic acid residues alpha-ketosidically linked to oligosaccharides, glycoproteins, glycopeptides and glycolipids)

Reaction type
O-Glycosyl bond hydrolysis

Natural substrates
Glycoproteins + H_2O [1, 2]
Glycolipids + H_2O
More (in spleen: removal of sialyl residues from red blood cells [31], viral enzyme: binding to cell surface, penetration into host cell, release of progeny virus [5], relationship to pathogenicity [37], role in influenza infection [13], cleavage of alpha-N-acetylneuraminic acid from glomerular polyanion in renal cortex [9])

Enzyme Handbook © Springer-Verlag Berlin Heidelberg 1991
Duplication, reproduction and storage in data banks are only
allowed with the prior permission of the publishers

Substrate spectrum

1 N-Acetylneuraminyllactose + H_2O (preference for substrates containing alpha-(2–3)-linked sialic acid [2, 8, 10, 11, 15, 17, 26, 27, 28, 30, 34, 47, 50], hydrolysis of substrates containing alpha-(2–6)-linked sialic acid [1, 2, 8, 10, 11, 17, 26, 28, 29, 30, 32, 34, 36, 47, 50], slow hydrolysis of substrates containing alpha-(2–8)-linked sialic acid [10, 28, 30, 34], no hydrolysis of substrates containing alpha-(2–8)-linked sialic acid [26, 29], prerequisite: carboxy-group vicinal to anomeric center [18])
2 4-Methylumbelliferyl-alpha-N-acetylneuraminic acid + H_2O [1, 2, 17]
3 Di-(and tri-) gangliosides + H_2O (in presence of detergents [6], hydrolysis by cytosolic form only [15]) [1, 2, 6, 12, 15, 21, 41, 45]
4 Sialyllactosylceramide + H_2O [45]
5 Colominic acid + H_2O [1, 2, 11, 48]
6 Glycopeptides + H_2O (or glycoproteins, e.g. fetuin, mammalian submaxillay mucin, orosomucoid) (specific for erythrocyte sialoglycoprotein [31]) [1, 2, 10, 11, 12, 15, 17, 21, 24, 27, 28, 29, 48, 50]
7 More (transglycosylation with methanol as nucleophilic reagent [7], esterase activity [10])

Product spectrum

1 N-Acetylneuramic acid + lactose [47]
2 ?
3 N-Acetylneuramic acid + monosialoganglioside [45]
4 Sialic acid + lactosylceramide [45]
5 ?
6 ?
7 ?

Inhibitor(s)

Hg^{2+} [1, 2, 9, 10, 17, 25, 40, 47]; 2-Deoxy-2, 3-didehydro-N-acetylneuraminic acid [1, 2, 8, 10]; N-(4-Nitrophenyl)oxamic acid [1, 10, 17]; N-Acetylneuraminic acid (product inhibition) [1, 9, 17, 47]; Cu^{2+} (plasma membrane-bound form [3]) [3, 8, 10, 11, 36, 38, 40, 41, 47]; Dithiothreitol (0.25 mM) [5, 10]; Glutathione [8, 47]; Merthiolate [8]; Triton X-100 (0.5%) [9]; Mercaptoethanol (enzyme III [49]) [10, 49]; Sodium cholate (substrates sialyllactose or orosomucoid) [11]; Mn^{2+} [11, 24, 36]; Bivalent cations (depending on substrate) [11]; EDTA [17, 19, 24, 26, 27, 47]; Zn^{2+} [19, 24, 40]; Mg^{2+} [24, 36, 40]; Acetate (0.2 M/l, 89% activity) [33]; Ca^{2+} [36, 40]; Co^{2+} [36, 40]; NH_4^+ [40, 49]; Detergents [41]; Fe^{3+} [41, 47]; 1-(4-Methoxyphenoxymethyl)-3, 4-dehydroisoquinoline [41]; p-Chloromercuribenzoate [41]; Fe^{2+} [47, 49]; Thioglycolic acid [47]; Sr^{2+} [49]

Cofactor(s)/prosthetic group(s)

Sodium cholate (activation with substrate gangliosides [11], activation of lysosomal form [3]) [3, 6, 11]; Non-ionic detergents (activation of plasmamembrane bound form) [3]

Metal compounds/salts

Mg^{2+} (activation) [5, 19, 39, 47, 49]; Ca^{2+} (activation) [5, 10, 17, 19, 24, 26, 32, 39, 47, 49]; Fe^{2+} (activation) [8]; Mn^{2+} (activation) [19, 39, 47]; Ba^{2+} (activation) [24, 49]; Co^{2+} (activation) [24, 39]

Turnover number (min⁻¹)

360000 [38]

Specific activity (U/mg)

520 (extracellular) [1]; 680 (cell bound) [1]; 950 [4]; More (assay method [16], comparison of assay methods [28]) [2, 3, 5–7, 11, 12, 14–17, 19–29, 31, 33, , 34, 35, 38–43, 45, 47, 48, 50]

K_m-value (mM)

0.2 (4-methylumbelliferyl-N-acetylneuraminic acid, similar values [1, 5, 6, 9, 11]) [2]; 1.67–7.9 (N-acetylneuraminyl-alpha-(2–3)lactose) [1, 2, 6–8, 10–12, 17, 20, 26, 30]; 1.91–7.5 (N-acetylneuraminyl-alpha-(2–6) lactose) [1, 2, 6–8, 10–12, 17, 20, 26]; 0.096–6.7 (fetuin) [2, 6, 8, 11, 12, 17, 25, 33]; 1.4 (colominic acid, alpha-(2–8)) [10]; 0.9 (disialyllactose, NeuAc-alpha 2 → 8 NeuAc-alpha 2 → 3) [20]; 4.1 (3'-methoxyphenyl-N-acetylneuraminic acid, similar values [8, 33]) [28]; 0.3 (ganglioside, molar ratio of di- to monoganglioside is 95:5) [31]; 0.36–0.4 (bovine submaxillary mucin) [32, 39], 4.0 (porcine submaxillary mucin) [39]; 0.0017 (Cowper's gland mucin) [36]; 0.8 (glycopeptides from edible birds' nests) [38]; 0.4 (porcine thyroglobulin) [39]; 0.043 (gangliosides) [41]; More [2, 6, 10, 11, 12, 23, 30, 33, 34, 36, 39, 42, 46–48]

pH-optimum

3.1–4.0 (depending on substrate) [40]; 3.5–3.8 (substrate: ganglioside GM3 (II³Neu-AcLacCer)) [12]; 3.8 [31]; 4.0 [23]; 4–4.5 [21]; 4.0–5.6 (depending on buffer [48]) [6, 48]; 4.1 (substrate gangliosides [10]) [10]; 4.2 [2]; 4.2–4.6 (lysosome-bound [15]) [15, 33]; 4.3–5.2 (substrates 4-methylumbelliferyl-N-acetylneuraminic acid and fetuin [12], depending on substrate [36]) [12, 36, 38]; 4.4 (substrate sialyllactose [10]) [10, 45]; 4.5 [22]; 4.5–7.0 (depending on type of virus and substrate) [47]; 4.6 (acetate buffer [5]) [5, 17]; 4.7 (intralysosomal [15]) [15, 46]; 5.0 (enzyme I [49]) [1, 25, 41, 49]; 5.4 [7]; 5.5 (phosphate buffer [5]) [5, 8, 24, 39]; 5.5–5.7 [32]; 5.8 (cytosolic) [15]; 6.0 (substrates orosomucoid and gangliosides [11]) [11, 20]; 6.4 [19]; 6.5 (substrate sialyllactose) [11]; 6.6 (enzyme III) [49]; 6.6–7.0 [27]; 6.7 [42]; 8.3 (enzyme II) [49]

pH-range

3–6 [1]; 3–8 [48]; 5–6 [39]; 5.4 (less than 20% activity) [33]; 5.3–7.8 (50% activity, extracellular form) [27]; 5.9 (50% activity, intracellular form) [27]

Temperature optimum (°C)

37 (enzyme III [49]) [1, 2, 24, 26, 49]; 40 (enzymes I and II) [49]; 70 [40]

Enzyme Handbook © Springer-Verlag Berlin Heidelberg 1991
Duplication, reproduction and storage in data banks are only
allowed with the prior permission of the publishers

Temperature range (°C)
10–60 [1]; 47 (50% activity) [26]

3 ENZYME STRUCTURE

Molecular weight
240000–270000 (Influenza virus, gel filtration) [13, 28]
150000–180000 (Actinomyces viscosus, FPLC-gel filtration, SDS-PAGE [1],
Bacteroides fragilis, gel filtration [20], bull, gel filtration [22]) [1, 20, 22]
88000–125000 (Bacteroides fragilis, gel filtration [19], Streptococcus group
B type II, gel filtration, SDS-PAGE [29], Arthrobacter sp., gel filtration, SDS-
PAGE [30, 34], Streptococcus group A, gel filtration [32], Streptococcus
group K, gel filtration, SDS-PAGE [39]) [19, 29, 30, 32, 34, 39]
60000–70600 (Sendai virus, unreduced enzyme, SDS-PAGE [5], Clostridium
perfringens, gel filtration, SDS-PAGE [7, 38], ultracentrifugation, electrophoresis in acid urea [38] , rat, cytosolic + lysosomal form, gel filtration
[15], Pneumococcus, SDS-PAGE [42], Diplococcus pneumoniae, 4 isoenzymes, multiple forms, gel filtration [44]) [5, 7, 15, 38, 42, 44]
38000–49000 (Sendai virus, reduced form, SDS-PAGE [5], rat, sucrose
density centrifugation [11], man, gel filtration, SDS-PAGE [17], Bifidobacterium lactentis, gel filtration [26]) [5, 11, 17, 26]
23000 (Asterias rubens, gel filtration, FPLC gel filtration) [2]

Subunits
Tetramer (4 × 50000–60000, Influenza virus, value depending on type of
virus and analytical method, comparison of primary and secondary
structures of different subtypes) [13]
? (? × 76000 (neuraminidase subunit) + ? × 32000 (beta-galactosidase
protective protein) + ? × 66000 (beta-galactosidase), man, SDS-PAGE
[14], ? × 70000, man, SDS-PAGE [21], Corynebacterium ulcerans, SDS-
PAGE [24]) [14, 21, 24]

Glycoprotein/Lipoprotein
Glycoprotein (carbohydrate composition [13], no glycoprotein [7]) [13, 33]

4 ISOLATION/PREPARATION

Source organism
Actinomyces viscosus [1]; Asterias rubens (starfish) [2]; Man [3, 6, 10, 12,
14, 17, 21, 23, 33]; Trypanosoma cruzi [4]; Sendai virus [5]; Clostridium perfringens [7, 38, 48]; Influenza virus [8, 13, 28, 47, 50–53]; Rat [9, 11, 15, 41];
Vibrio cholerae [18, 50]; Bacteroides fragilis [19, 20]; Bull [22, 31];
Corynebacterium ulcerans [24]; Pasteurella haemolytica [25]; Bifidobacterium lactentis [26]; Pseudomonas aeruginosa [27]; Streptococcus

type III, group B [29]; Streptococcus group A [32]; Streptococcus group K [39]; Streptococcus group A, B, C, G, L [56]; Arthrobacter sialophilus [30]; Arthrobacter sp. (distribution in [35]) [34, 35]; Rabbit [36]; Calf [40, 45]; Pneumococcus [42]; Diplococcus pneumoniae [43, 44]; Pig [46]; Bufo arenarum (toad) [49]; Clostridium welchii [50]; Pseudomonas fluorescens [50]; Viruses (of myxo group [50], overview [37]) [37, 50, 55]; Bacteria (overview) [37, 55]; Animals (overview) [37]

Source tissue
Culture medium [1, 7, 19, 24–27, 29, 30, 32, 34, 35, 38, 39, 42–44, 48]; Cell [1, 20]; Whole animals (without stomach and intestines) [2]; Fibroblasts [3, 33]; Promastigotes [4]; Glycoproteins (viral envelope) [5]; Placenta [6, 12, 14]; Surface components (virus) [8]; Kidney (glomeruli of renal cortex [9]) [9, 46]; Liver [10, 11, 15, 21, 23]; Leukocytes [17]; Testis [22]; Spleen [31]; Spermatozoal acrosomes [36]; Brain [40, 45]; Heart [41]; Oviduct [49]; Tissues, organs, body fluids (overview) [37]

Localisation in source
Extracellular (bacteria) [1, 7, 16, 19, 24–27, 29, 30, 32, 34, 35, 38, 39, 42–44, 48]; Envelope (virus) [5, 16]; Cell-bound [1]; Soluble part of cell [2, 23]; Plasma membrane (external surface [3], inner membrane [36]) [3, 31, 36, 40, 45]; Cell membrane [4, 20]; Lysosomes (membranes of [10]) [3, 10, 14, 15, 17, 21, 22, 23, 54]; Microsomes [10, 11, 31]; Nucleus [10]; Particulate parts of cell [46]; Golgi apparatus [16, 54]; Cytoplasm [11, 15]

Purification
Actinomyces viscosus [1]; Asterias rubens [2]; Trypanosoma cruzi [4]; Sendai virus [5]; Clostridium perfringens [7, 38, 48]; Influenza virus [8, 47]; Rat [11, 41]; Man (copurification with beta-galactosidase [14]) [12, 14 , 21]; Bacteroides fragilis [19, 20]; Bull (copurification with beta-galactosidase) [22]; Bifidobacterium lactentis (partial) [26]; Streptococcus group B type III [29]; Streptococcus group A (partial) [32]; Arthrobacter ureafaciens (2 isoenzymes [35]) [34, 35]; Rabbit [36]; Streptococcus group K [39]; Pneumococcus [42]; Diplococcus pneumoniae (multiple forms) [43]; Calf (partial) [45]; Pig [46]; Bufo arenarum (3 isoenzymes) [49]

Crystallization
(crystal structure [13]) [13, 43, 50]

Cloned
−

Renaturated
−

Enzyme Handbook © Springer-Verlag Berlin Heidelberg 1991
Duplication, reproduction and storage in data banks are only
allowed with the prior permission of the publishers

5 STABILITY

pH
4–11 [38]; 6–9 [42]; 4.6 (best value for storage) [15]; 5.3 (best value for storage) [46]

Temperature (°C)
–20 (inactivation enzyme II and II) [49]; 24 (15 minutes, 78% activity) [33]; 37 (15 minutes, 15% activity [33], 24 hours [27] , 20 hours in different buffer solutions [38], 10 hours, 65% activity [40], inactivation above [46]) [27, 33, 38, 40, 46]; 40 (30 minutes, 50% activity) [10]; 44 (5 hours, 25% activity) [40]; 50 (below [24, 39], 30 minutes, 30% activity [8], 5 minutes, 70% activity [28]) [8, 24, 28, 39]; 52 (inactivation) [33]; 55 (15 minutes, 20 % activity) [40]; 56 (30 minutes, 20% activity) [27]; 60 (inactivation enzyme I and II) [49]; 65 (1 minute, 50% activity) [40]; 70 (1 minute, 15 % activity) [40]

Oxidation

Organic solvent

General stability information
Freezing/thawing (inactivation, not sensible [40]) [1, 2, 10 , 11, 21, 33, 39, 41]; Leupeptin (stabilization) [15]; Bovine serum albumin (stabilization) [15, 48]; Divalent cations (stabilization) [15]; Proteinase inhibitor (stabilization of crude enzyme) [17]; Sonication (inactivation) [33]; Dilute solutions (inactivation) [48]; Lysosomal enzyme extremely unstable [15]

Storage
–80°C, whole human liver, up to 1 year [23]; –70°C [14, 22]; –70°C, cytosolic enzyme, at least 1 week [15]; –70°C, 0. 1 M acetate buffer, pH 5.5, 1 mM Ca^{2+}, at least 4 weeks [17]; –20°C, 10 mM acetate buffer, pH 6.0, 0.2% n-octylglucoside [5]; –20°C, lysosomal suspension, 1 week, 80% activity [10]; –20°C, culture filtrate, several months [27]; –20°C, 10 mM sodium citrate buffer, pH 6.1 [29]; –20 °C, 10 mM citrate phosphate buffer, pH 6.0, 0.1 mM phenylmethanesulfonylfluoride, more than 1 year [33]; –20 °C, 20 mM KCN [38]; 0°C, 0.2% BSA, 2 days, 80% activity [11]; Liquid N_2, soluble form, up to 15 days [23]; Lyophilized, several months [36, 47]; 4°C, 0.3 mg/ml albumin, purified enzyme, more than 2 years [48]; More (similar conditions) [1, 2, 18, 21, 23, 36, 39, 42, 45, 47]

6 CROSSREFERENCES TO STRUCTURE DATABANKS

PIR/MIPS code

NMIV (Influenza A virus, strain A/PR/8/34); NMIVU7 (Influenza A virus, strain USSR/90/77); NMIV3 (Influenza A virus, strain A/WSN/33 [H1N1]); NMIV42 (Influenza A virus, strain A/Memphis/1/71H-A/Bellamy/42N [H3N1], fragment); NMIVEK (Equine influenza A virus, strain A/Ken/1/81[N8]); NMIVAA (Avian influenza A virus, strain A/shearwater/Australia/1/72 [H6N5]); NMIV9 (Influenza A virus, strain A/tern/Australia/G70C/75); NMIVW8 (Influenza A virus, strain A/whale/Maine/1/84 [H13N9]); NMIVEA (Equine influenza A virus, strain A/Cor/16/74[N7]); NMIV2 (Influenza A virus, strain A/NT/60/68); NMIV72 (Influenza A virus, strain A/Memphis/102/72 [H3N2], fragment); NMIV71 (Influenza A virus, strain A/turkey/Oregon/71 [H7N3], fragment); NMIV68 (Influenza A virus, strain A/turkey/Ontario/6118/68 [H8N4], fragment); NMIVN5 (Influenza A virus, strain A/shearwater/Australia/72 [H6N5], fragment); NMIVN6 (Influenza A virus, strain A/duck/Alberta/28/76 [H4N6], fragment); NMIVN7 (Influenza A virus, strain A/duck/Germany/49 [H10N7], fragment); NMIV78 (Influenza A virus, strain A/black duck/Australia/702/78, fragment); NMIVN2 (Influenza A virus, 2 strains); NMIV27 (Influenza A virus, strain A/RI/5-/57 [H2N2]); NMIV4 (Influenza B virus, strain B/Lee/40); A35264 (Bacteroides fragilis, fragment); S04801 (Influenza A virus, strain Chile/1/83 [H1N1]); A27734 (precursor, Vibrio cholerae, fragment); S01339 (Clostridium perfringens)

Brookhaven code

7 LITERATURE REFERENCES

[1] Teufel, M., Roggentin, P., Schauer, R.: Biol. Chem. Hoppe-Seyler, 370, 435–443 (1989)

[2] Schauer, R., Wember, M.: Biol. Chem. Hoppe-Seyler, 370, 183–190 (1989)

[3] Lieser, M., Harms, E., Kern, H., Bach, G., Cantz, M.: Biochem. J., 260, 69–74 (1989)

[4] Harth, G., Haidaris, C.G., So, M.: Proc. Natl. Acad. Sci. USA, 84, 8320–8324 (1987)

[5] Barnes, J.A., Allen, T.M.: Enzyme Microb. Technol., 9, 553–558 (1987)

[6] Hiraiwa, M., Uda, , Y., Nishizawa, M., Miyatake, T.: J. Biochem., 101, 1273–1279 (1987)

[7] Bouwstra, J.B., Deyl, C.M., Vliegenthart, J.F.G.: Biol. Chem. Hoppe-Seyler, 368, 269–275 (1987)

[8] Arora, D.J.S., Gabriel, L.F.: Biochim. Biophys. Acta, 884, 73–83 (1986)

[9] Baricos, W.H., Cortez-Schwartz, S., Shah, S.V.: Biochem. J., 239, 705–710 (1986)

[10] Michalski, J.-C., Corfield, A.P., Schauer, R.: Biol. Chem. Hoppe-Seyler, 367, 715–722 (1986)

[11] Miyagi, T., Tsuiki, S.: J. Biol. Chem., 260, 6710–6716 (1985)

[12] Hiraiwa, M., Nishizawa, M., Uda, Y., Nakajima, T., Miyatake, T.: J. Biochem., 103, 86–90 (1988)

[13] Colman, P.M., Ward, C.W.: Curr. Top. Microbiol. Immunol., 114, 177–255 (1988) (Review)

Enzyme Handbook © Springer-Verlag Berlin Heidelberg 1991
Duplication, reproduction and storage in data banks are only allowed with the prior permission of the publishers

[14] Verheijen, F.W., Palmeri, S., Galjaard, H.: Eur. J. Biochem., 162, 63–67 (1987)

[15] Miyagi, T., Tsuiki, S.: Eur. J. Biochem., 141, 75–81 (1984)

[16] Schauer, R., Nöhle, U. in "Methods Enzym. Anal.", 3rd. Ed. (Bergmeyer, H.V., Ed.) 4, 195–208 (1984)

[17] Schauer, R., Wember, M.: Hoppe-Seyler's Z. Physiol. Chem., 365, 419–426 (1984)

[18] Eschenfelder, V., Brossmer, R., Wachter, M.: Hoppe-Seyler's Z. Physiol. Chem., 364, 1411–1417 (1983)

[19] Berg, J.O., Lindqvist, L., Andersson, G., Nord, C.E.: Appl. Environ. Microbiol., 46, 75–80 (1983)

[20] Nicolai, H., Von, Hammann, R., Werner, H., Zilliken, F.: FEMS Microbiol. Lett., 17, 217–220 (1983)

[21] Michalsky, J.-C., Corfield, A.P., Schauer, R.: Hoppe-Seyler's Z. Physiol. Chem., 363, 1097–1102 (1982)

[22] Verheijen, F., Brossmer, R., Galjaard, H.: Biochem. Biophys. Res. Commun., 108, 868–875 (1982)

[23] Meyer, D.M., Lemonnier, M., Bourrillon, R.: Biochem. Biophys. Res. Commun., 103, 1302–1309 (1981)

[24] Vertiev, Y.V., Ezepchuk, Y.V.: Hoppe-Seyler's Z. Physiol. Chem., 362, 1339–1344 (1981)

[25] Tabatabai, L.B., Frank, G.H.: Curr. Microbiol., 5, 203–206 (1981)

[26] Nicolai, H. Von, Esser, P., Lauer, E.: Hoppe-Seyler' S Z. Physiol. Chem., 362, 153–162 (1981)

[27] Leprat, R., Michel-Briand, Y.: Ann. Microbiol., 131, 209–222 (1980)

[28] Cabezas, J.A., Calvo, P., Eid, P., Martin, J., Perez, N., Reglero, A., Hannoun, C.: Biochim. Biophys. Acta, 616, 228–238 (1980)

[29] Milligan, T.W., Mattingly, S.J., Straus, D.C.: J. Bacteriol., 144, 164–172 (1980)

[30] Wang, P., Schafer, D., Hohm, C.E., Tanenbaum, S.W., Flashner, M. in "Glycoconjugate Res." (Gregory, J.D., Jeanloz, R.W., Eds.) 2, 955–957 (1977)

[31] Schengrund, C.-L., Repman, M.A., Nelson, J.T.: Biochim. Biophys. Acta, 568, 377–385 (1979)

[32] Davis, L., Braig, M.M., Ayoub, E.M.: Infect. Immun., 24, 780–786 (1979)

[33] Thomas, G.H., Reynolds, L.W., Miller, C.S.: Biochim. Biophys. Acta, 568, 39–48 (1979)

[34] Wang, P., Tanenbaum, S.W., Flashner, M.: Biochim. Biophys. Acta, 523, 170–180 (1978)

[35] Uchida, Y., Tsukada, Y., Sugimori, T.: J. Biochem., 82, 1425–1433 (1977)

[36] Srivastava, P.N., Abou-Issa, H.: Biochem. J., 161, 193–200 (1977)

[37] Ray, P.K.: Adv. Appl. Microbiol., 21, 227–267 (1977) (Review)

[38] Nees, S., Veh, R.W., Schauer, R.: Hoppe-Seyler's Z. Physiol. Chem., 356, 1027–1042 (1975)

[39] Kiyohara, T., Terao, T., Shioiri-Nakano, K., Osawa, T.: Arch. Biochem. Biophys., 164, 575–582 (1974)

[40] Preti, A., Lombardo, A., Cestaro, B., Zambotti, S., Tettamanti, G.: Biochim. Biophys. Acta, 350, 406–414 (1974)

[41] Tallman, J.F., Brady, R.O.: Biochim. Biophys. Acta, 293, 434–443 (1973)

[42] Stahl, W.L., O'Toole, R.D.: Biochim. Biophys. Acta, 268, 480–487 (1972)

[43] Tanenbaum, S.W., Gulbinsky, J., Katz, M., Sun, S.-C.: Biochim. Biophys. Acta, 198, 242–254 (1970)

[44] Tanenbaum, S.W., Sun, S.-C.: Biochim. Biophys. Acta, 229, 824–828 (1971)

[45] Gatt, S., Leibovitz, Z.: Methods Enzymol., 14, 149–152 (1969)

[46] Tuppy, H., Palese, P.: Hoppe-Seyler's Z. Physiol. Chem., 349, 1169–1178 (1968)

[47] Rafelson, M.E., Gold, S., Priede, I.: Methods Enzymol., 8, 677–680 (1966)

[48] Cassidy, J.T., Jourdian, G.W., Roseman, S.: Methods Enzymol., 8, 680–685 (1966)

[49] De Martinez, N.R., Olavarria, J.M.: Biochim. Biophys. Acta, 320, 301–310 (1973)

[50] Gottschalk, A. in "The Enzymes", 2nd. Ed. (Boyer, P.D., Ed.) 4, 461–473 (1960) (Review)

[51] Bucher, D.J., Palese, P. in "Influenza Virus And Influenza" (Kilbourne, E.D., Ed.) 83–123, Academic Press, N.Y. (1975) (Review)

[52] Colman, P.M. in "The Structure And Function Of Neuraminidases", Peptide Protein Rev., 4, 215–255 (1984) (Review)

[53] Rosenberg, A., Schengrund, C.-L. in "The Biological Roles Of Sialic Acid" (Rosenberg, A., Schengrund C.-L., Eds.) 295–395, Plenum, N.Y. (1976) (Review)

[54] Kishore, G.S., Tulsiani, D.R.P., Bhavanandan, V.P., Carubelli, R.: J. Biol. Chem., 250, 2655–2659 (1975)

[55] Drzeniek, R.: Curr. Top. Microbiol. Immunol., 59, 35–72 (1972) (Review)

[56] Hayano, S., Tanaka, A.: J. Bacteriol., 97, 1328–1333 (1969)

Enzyme Handbook © Springer-Verlag Berlin Heidelberg 1991
Duplication, reproduction and storage in data banks are only
allowed with the prior permission of the publishers

9

1 NOMENCLATURE

EC number
3.2.1.20

Systematic name
Alpha-D-glucoside glucohydrolase

Recommended name
Alpha-glucosidase

Synonymes
Maltase
Glucoinvertase
Glucosidosucrase
Maltase-glucoamylase
Alpha-glucopyranosidase
Glucosidoinvertase
Alpha-D-glucosidase
Alpha-glucoside hydrolase
Alpha-1, 4-glucosidase

CAS Reg. No.
9001-42-7

2 REACTION AND SPECIFICITY

Catalysed reaction
Maltose + H_2O →
→ 2 alpha-D-glucose (hydrolysis of terminal, non-reducing 1, 4-linked alpha-D-glucose residues with release of alpha-D-glucose)

Reaction type
O-Glycosyl bond hydrolysis (exohydrolysis)

Natural substrates
Alpha-D-glucooligosaccharides + H_2O (produced by alpha-and beta-amylases [12])
Starch + H_2O [12]
Glycogen + H_2O [62]
More (constituent of endogenous alpha-glucan metabolism [54], synthesis of glucosides containing pyridoxin, esculin, or rutin [43])

Enzyme Handbook © Springer-Verlag Berlin Heidelberg 1991
Duplication, reproduction and storage in data banks are only
allowed with the prior permission of the publishers

Substrate spectrum

1 Maltose + H_2O (absolute requirement for alpha-D-glucosidic linkage [31])
2 Maltotriose + H_2O
3 Maltotetraose + H_2O
4 Maltopentaose + H_2O
5 Starch (soluble) + H_2O
6 Amylose + H_2O
7 p-Nitrophenyl-alpha-D-glucoside + H_2O
8 Isomaltose + H_2O (hydrolysis of alpha-(1–6)-linkage only [11], alpha-(1–4)-linkage preferred to alpha-(1–6) [16])
9 Kojibiose (O-alpha-D-glucopyranosyl-(1–2)-D-glucopyranose) + H_2O [15, 18, 23]
10 Sucrose (beta-D-fructofuranosyl-alpha-D-glucopyranoside) + H_2O [33]
11 Phenyl-alpha-maltoside + H_2O
12 Nigerose (O-alpha-D-glucopyranosyl-(1–3)-D-glucopyranose) + H_2O
13 Turanose (O-alpha-D-glucopyranosyl-(1–3)-D-fructofuranose) + H_2O (enzymes II and III [35])
14 Amylopectin + H_2O
15 Glycogen + H_2O
16 Melizitose (O-alpha-D-glucopyranosyl-(1–3)-O-beta-D-fructofuranosyl-(2–1)-alpha-D-glucopyranoside) + H_2O [11]
17 4-Methylumbelliferyl-alpha-D-glucopyranoside + H_2O [62]
18 More (hydrolysis of alpha-(1–6)linkages in glucosides [11, 23], hydrolysis of alpha-(1–4)-linkages [23, 24, 26, 43, 47], hydrolysis of alpha-(1–3)-linkages [23, 43, 47], weak hydrolysis of alpha-(1–2)-linkages [23, 43], no hydrolysis of alpha-(1–6)-linkages [43], trans-glucosidase activity [64])

Product spectrum

1 Alpha-D-glucose (transglucosylation: panose [4, 12, 18, 29, 34 , 39], maltotriose [13, 18, 25, 26, 32, 34, 38, 39, 43, 47], isomaltose [18, 34, 39, 45])
2 Alpha-D-glucose
3 Alpha-D-glucose
4 Alpha-D-glucose
5 Alpha-D-glucose
6 Alpha-D-glucose
7 p-Nitrophenol + glucose
8 Alpha-D-glucose + ?
9 Alpha-D-glucose
10 D-Glucose + D-fructose
11 Phenol + alpha-D-glucose (or phenyl-alpha-glucoside + glucose [17, 18, 26, 32, 34, 38, 39, 43, 47]) [18, 23, 35]
12 Alpha-D-glucose

13 Alpha-D-glucose + D-fructose
14 ?
15 Alpha-D-glucose
16 ?
17 4-Methylumbelliferone + alpha-D-glucose
18 ?

Inhibitor(s)

Starch (hydrolysis of maltose, non-competetive) [1]; Sucrose (hydrolysis of starch) [1]; N-Ethylmaleimide (membrane-bound form [3]) [1, 3]; p-Chloromercuribenzoate [1]; Hg^{2+}; Inhibitor (from 4 types of green algae) [6]; Cu^{2+} [8, 20, 26, 32, 34, 38, 39, 43, 44, 45, 47]; Iodoacetamide [8, 20]; UO^{2+} [8]; EDTA [8]; Glucono-delta-lactone [9, 17]; Tris; Castanospermine [12], Ag^+ [13, 18]; Turanose [13, 16, 25, 26, 30, 38, 39, 43, 45, 47, 59]; Erythritol [17, 40, 51]; Xylose [17]; Inositol [17]; Sn^{2+} [18, 25]; Zn^{2+} [20, 38, 43, 44, 45, 47, 59]; Ni^{2+} [20, 39, 47]; Rose Bengal [20]; Pb^{2+} [25, 32, 34, 39, 43, 44 , 45]; Glycerol [30]; Methyl-alpha-D-glucopyranoside [30]; Fructose [30]; Ribose [30]; Al^{3+} [33]; 6-Bromo-3, 4, 5-trihydroxycyclohex-1-ene (bromoconduritol) [35, 36]; Histidine [44, 45]; Carbohydrates (depending on concentration) [38, 44, 50]; 5-Amino-5-deoxy-D-glucopyranose (nojirimycin) [57]; Mn^{2+} [59]; Maltose (substrate inhibition) [64]

Cofactor(s)/prostethic group(s)

EDTA (activation) [5]; o-Phenanthroline (activation) [5]; Alpha, alpha'-dipyridyl (activation) [5]; Albumin (activation) [62]

Metal compounds/salts

Ba^{2+} [5]; Sr^{2+} [5]; Mn^{2+} [5, 20]; Ca^{2+} [8]; Mg^{2+} [8, 20]; K^+ [8, 40]

Turnover number (min^{-1})

Specific activity (U/mg)

69 [1]; 71.4 [5]; 114–140 [18]; More (assay method [19]) [3–5, 7–10, 12, 13, 16–19, 22–24, 26–28, 30–35, 37, 39, 40, 42, 44–52, 55–58, 60, 61, 63]

K_m-value (mM)

0.166–6.94 (maltose) [1, 2, 4, 7, 8, 12, 13, 16, 22, 23, 25–27, 30 , 31, 38, 45–47, 50–52, 55, 56, 61]; 0.21–13.2 (maltotriose) [1, 2, 13, 25, 26, 30, 38, 60]; 0.1–2.0 (maltotetraose) [1, 2, 60]; 0.14–1.0 (maltopentaose) [1, 2, 60]; 0.93–3.7 (starch) [1, 2]; 0.8–3.6 (phenyl-alpha-glucoside) [35]; 2.813 (phenyl-alpha-maltoside) [38]; 0.7–11.8 (amylose) [13, 26]; 0.134–11.5 (p-nitrophenyl-alpha-D-glucopyranoside) [16, 20, 27, 30, 31]; 9.25–33 (isomaltose) [16, 30, 38, 60]; 8.9–13.1 (sucrose) [28]; 8.3 (kojibiose) [60]; 3.3 (nigerose) [60]; More [2, 23, 38, 39, 40, 60]

Enzyme Handbook © Springer-Verlag Berlin Heidelberg 1991
Duplication, reproduction and storage in data banks are only
allowed with the prior permission of the publishers

pH-optimum
3.0–5.0 [4, 45]; 3.5 (depending on buffer [50]) [39, 50]; 3. 5–4.0 [7]; 3.6 [23, 46]; 3.7–4.0 (enzyme II) [51]; 4.0 [12, 16, 25]; 4.0–4.3 (enzyme I) [51]; 4.0–6.0 [37]; 4.3 (alpha-glucosides) [11]; 4.5 (glycogen, addition af albumin [62]) [10, 48, 55, 62]; 4.6 [42]; 4.6–5.0 [49]; 5.0 (glycogen [62]) [14, 20, 26, 27, 32, 34, 52, 62, 64]; 5.0–6.0 [44]; 5.5–6.0 [5]; 5–6.5 (depending on substrate and isoenzyme) [28]; 5.6 [57]; 6.0–6.5 [13]; 6.0 [9, 17]; 6–7 [61, 63]; 6.3–7.1 [35]; 6.25 [60]; 6.5 (maltose [11]) [8, 11]; 6.5–7.0 [29]; 6.7–6.8 [31]; 6.8 [33]; 6.8–7.3 [47]; 7.0 [21, 24]; 7.0–7.5 [53]

pH-range
2.5–5 [48]; 3.0–6.0 [16]; 4.0–6.5 [37]; 4.3–7.6 [44]; 4.5–7.5 [8]; 4.5–9.5 [17]; 5–7 [28]; 5.5–6.5 [9]; 6–8 [47]; 9. 5 (15% activity) [21]

Temperature optimum (°C)
30 (maltase [11]) [8, 11]; 36 (enzyme III) [35]; 37 [20]; 37–55 (enzyme I) [25]; 37–50 (enzyme II) [25]; 38 [33]; 40 (alpha-glucosidase [11]) [11, 21, 24, 47, 49]; 42 (enzymes I and II) [35]; 50 (starch [51]) [17, 31, 37, 39, 45, 51]; 50–60 [5, 50]; 50–55 (enzymes II and III [12]) [12]; 52 [52]; 55 (enzyme I [12], en-zymes Ia and IIa [13], maltose [51]) [4, 12, 13, 34, 42, 51]; 60 (enzyme I [26]) [7, 10, 26, 63]; 65 (enzymes Ib and IIb [13], enzymes II-1 and II-2 [26]) [13, 14, 26]; 75 [44]

Temperature range (°C)
25–40 [8]; 50–75 [44]

---

3 ENZYME STRUCTURE

Molecular weight
590000 (rabbit, gel filtration, SDS-PAGE) [1]
500000 (rat, polyacrylamide gel electrophoresis, density gradient centrifugation, complex of maltase/glucoamylase, dissociation by heat or low pH yields fragments of MW 134000–480000) [61]
280000 (horse, gel filtration) [60]
270000 (pig, gel filtration, SDS-PAGE) [63]
120000–160000 (Bacillus licheniformis, maltase, gel filtration [9], pig, en-zymes Ia and IIa, SDS-PAGE [13], rat, gel filtration [22], Phaseolus vidis-simus, enzyme I, SDS-PAGE [26], potato, gel filtration [29], Aspergillus awamori, enzymes I, II, III, SDS-PAGE [34], Mucor javanicus, sedimentation and diffusion constants [41], Penicillium purpurogenum, SDS-PAGE [45]) [9, 13, 22, 26, 29, 34, 41, 45]

95000–115000 (Oryza sativa, SDS-PAGE [2], Oryza sativa, enzyme I, gel filtration [12], Glycine max, enzyme II, gel filtration [15], Tetrahymena pyriformis, gel filtration, SDS-PAGE [16], Phaseolus vidissimus, enzyme II-1, SDS-PAGE [26], Aspergillus niger, intracellular form, gel filtration, SDS-PAGE [27], Vitis vinifera, gel filtration [30], Mucor racemosus, calculation from sedimentation and diffusion constants, sedimentation equilibrium centrifugation [37], pig, SDS-PAGE [55], man, gel filtration [59]) [2, 12, 15, 16, 26, 27, 30, 37, 55, 59]

70000–84000 (human, soluble form, immunoblotting analysis [3], Oryza sativa, enzyme II, gel filtration [12], enzyme I, sedimentation equilibrium centrifugation [25], pig, enzymes Ib and IIb, SDS-PAGE [13], Beta vulgaris, SDS-PAGE [14], Glycine max, enzyme I, gel filtration [15], Drosophila melanogaster, gel filtration, SDS-PAGE [28], Fagopyrum esculentum, gel filtration, SDS-PAGE [40]) [3, 12–15, 25, 28, 40]

40000–68000 (Oryza sativa, gel filtration [4], enzyme III [12], enzyme II, sedimentation equilibrium centrifugation [25], Bacillus licheniformis, alpha-glucosidase, gel filtration [9], Bacillus amyloliquefaciens, maltase, gel filtration [11], Allium tistulosum, enzymes I -III, SDS-PAGE [18], Phaseolus vidissimus, enzyme II-2, SDS-PAGE [26], Saccharomyces carlsbergensis, gel filtration, SDS-PAGE [31, 35], Lentinus edodes, SDS-PAGE [32], Bacillus thermoglucosidius, gel filtration [44], flint corn, gel filtration [46], Saccharomyces cerevisiae, gel filtration [53]) [4, 9, 11, 12, 25, 26, 31, 32, 35, 44, 46, 53]

27000–30000 (Bacillus caldotenax, gel filtration, SDS-PAGE [5], Bacillus amyloliquefaciens, alpha-glucosidase, gel filtration [11], SDS-PAGE [33], Saccharomyces losos, gel filtration [49], man [57]) [5, 11, 33, 49, 57]

Subunits
Octamer (8 × 79000, rabbit, SDS-PAGE with reducing agents) [1]
Tetramer (4 × 33500, rat, SDS-PAGE with mercaptoethanol [22])
Dimer (2 × 93000, Drosophhila melanogaster, enzyme I, SDS-PAGE) [28]
Monomer (1 × 75000–77000, Drosophila melanogaster, enzymes II and III, SDS-PAGE [28], 1 × 80000–88000, Fagopyrum esculentum, SDS-PAGE [40], 1 × 30000–33000, man [57]) [28, 40, 57]
? (x × 60000 + x × 40000, Bacillus licheniformis, maltase, SDS-PAGE) [9]

Glycoprotein/Lipoprotein
Glycoprotein (carbohydrate content [16, 27]) [15, 16, 22, 25, 27, 28, 38, 42, 48–50, 60, 61, 63]

Enzyme Handbook © Springer-Verlag Berlin Heidelberg 1991
Duplication, reproduction and storage in data banks are only
allowed with the prior permission of the publishers

4 ISOLATION/PREPARATION

Source organism

Rabbit [1]; Oryza sativa (rice) [2, 4, 12, 25, 51]; Human [3, 57, 59, 62, 65–67]; Bacillus caldovelox [5]; Paecilomyces varioti [7]; E. coli (enteropathogenic) [8]; Bacillus licheniformis [9, 17]; Lipomyces starkeyi [10]; Bacillus amyloliquefaciens [11, 33]; Pig [13, 55, 56, 63]; Beta vulgaris (edible beet) [14]; Glycine max (soybean) [15]; Tetrahymena pyriformis [16]; Allium tistulosum (Welsh onion) [18]; Lactobacillus acidophilus [20]; Bacillus sp. (acidophilic) [21]; Rat [22, 58]; Zea mays [23]; Bacillus amylolyticus [24]; Phaseolus vidissimus (green gram) [26]; Aspergillus niger [27]; Drosophila melanogaster [28]; Potato [29]; Vitis vinifera [30]; Saccharomyces carlsbergensis [31, 35]; Lentinus edodes [32]; Aspergillus awamori [34]; Mucor racemosus [37, 38]; Penicillium oxalicum [39]; Fagopyrum esculentum (buckwheat) [40, 52]; Mucor javanicus [41–43]; Bacillus thermoglucosidius [44]; Penicillium purpurogenum [45]; Flint corn [46]; Bacillus cereus [47]; Aspergillus fumigatus [48]; Saccharomyces losos [49]; Mucor rouxii [50]; Horse [60]; Cattle [64]

Source tissue

Kidney (cortex) [1, 22, 57, 60, 62, 65]; Seed [2, 14, 23, 25, 40, 46, 51, 52]; Seedlings [26]; Placenta [3]; Cell culture [4]; Cell [5, 11, 17, 20, 31, 33, 47, 49, 50, 53]; Culture medium [7, 9, 10, 16, 17, 21, 44, 48]; Duodenal mucosa [13]; Callus [15]; Leaves [18]; Mycelia [27, 34, 37, 38, 41–43, 45]; Whole body [28]; Tuber [29]; Berries [30]; Fruit body [32]; Liver [55, 58, 58, 64, 66, 67]; Small intestine [56, 61]; Serum [63]

Localisation in source

Brush border membrane [1, 22, 60]; Membranes (microvillous membranes of enterocytes [56]) [3, 33, , 56, 61]; Cell wall [4, 15]; Intracellular [5, 10, 17, 24, 27]; Extracellular [7, 9, 10, 12, 16, 17, 21, 24, 27, 44, 48]; Cytoplasm [11, 20]; Cell bound [10, 24]; Lysosomes [16, 58]

Purification

Rabbit [1]; Human (partial, 2 isoenzymes [59]) [3, 57, 59, 66, 67]; Oryza sativa (3 isoenzymes [12], 2 isoenzymes [25, 51]) [4, 12, 25, 51]; Bacillus caldovelox [5]; Paecilomyces varioti [7]; E. coli (partial) [8]; Bacillus licheniformis [9]; Lipomyces starkeyi [10]; Bacillus amyloliquefaciens [11, 33]; Pig (4 isoenzymes [13]) [13, 55, 56, 63]; Beta vulgaris [14]; Glycine max (2 isoenzymes) [15]; Tetrahymena pyriformis [16]; Bacillus licheniformis [17]; Allium tistulosum (3 isoenzymes) [18]; Lactobacillus acidophilus (partial) [20]; Rat [22, 58]; Zea mays [23]; Phaseolus vidissimus [26]; Aspergillus niger [27]; Drosophila melanogaster (3 isoenzymes) [28]; Potato (partial) [29]; Vitis vinifera (partial) [30]; Saccharomyces carlsbergensis [31, 35];

Aspergillus awamori [34]; Mucor racemosus [37]; Penicillium oxalicum [39];
Fagopyrum esculentum [40]; Mucor javanicus [42], Bacillus
thermoglucosidius [44]; Penicillium purpurogenum [45]; Flint corn [46];
Bacillus cereus [47]; Aspergillus fumigatus [48]; Saccharomyces losos [49];
Mucor rouxii [50]; Saccharomyces cerevisiae [53]; Horse [60]

Crystallization
[41–43]

Cloned
–

Renaturated
–

---

5 STABILITY

pH
3.4–7.3 [46]; 3.5–6.5 [26]; 3.5–7.0 (enzyme I [18]) [18, 39]; 3.5–8.0 (enzymes II
and III) [18]; 3.6–6.6 [49]; 4–7.3 [52]; 4.0–5.5 [25]; 4.0–6.0 [32]; 4.0–7.0 [3, 12,
37, 42]; 4.0–8.0 [55]; 4.0–9.0 [5]; 4.5–6.0 [34]; 5.0–7.0 [9]; 5.0–9.5 [57];
5.0–11.0 (at 31°C) [44]; 5.5–7.0 [45]; 6.0–8.5 (at 55°C) [44]; 6.0–9.0 (alpha-glucosidase [11]) [11, 29, 60, 63]; 6.0–9.5 (maltase) [11]; 6.2–7.0 [33];
6.8–7.3 [47]

Temperature (°C)
4–30 [9]; 30 (up to) [25, 45]; 35 (up to) [47]; 37 (1 hour, 100% activity [30],
up to [39]) [30, 39]; 40 (up to) [32]; 45 (below, enzyme I and II [12]) [12, 18,
33, 34, 37, 42, 52]; 50 (below [4, 55, 57], inactivation [17, 49]) [4, 17, 49, 55,
57]; 55 (below, enzyme II [12], enzymes Ia and IIa [13]) [12, 13, 61]; 60 (1
hour, 100% activity [5], 2 hours, 86% activity [7], inactivation [8, 44], 10
minutes, 20% activity [46]) [5, 7, 8, 44, 46]; 65 (up to [26], intracellular form
more sensitive than extracellular form [27]) [26, 27]; 70 (1 hour, 50% activity
[5], below, enzymes Ib and IIb [13]) [5, 13]; More (Ca^{2+}, K^+, protection from
heat inactivation) [8]

Oxidation

Organic solvent
Ethanol (45% v/v, stable) [44]

General stability information
Mn^{2+} (stabilization) [5]; Cysteine (stabilization) [5], Histidine (stabilization)
[5]; Urea (7.2 M, stable) [44]; SDS (0.06%, stable) [44]; Freezing/thawing
(inactivation) [56]

Enzyme Handbook © Springer-Verlag Berlin Heidelberg 1991
Duplication, reproduction and storage in data banks are only
allowed with the prior permission of the publishers

Storage

−20°C, 0.1 M potassium phosphate buffer, pH 6.8, 6 months [29]; −20°C, at least 2 months [30]; −20°C, 0.15 M potassium phosphate buffer, pH 6.8, 1 mM EDTA, at least 1 year [44]; −20°C [17]; −20°C, 1 year, more than 90% activity [58]; 4°C, 0.1 M phosphate buffer, pH 6.0 [5]; 4°C, 6 days, 88% activity [33]; 4°C, 1 month [56]; Lyophilized [63]

6 CROSSREFERENCES TO STRUCTURE DATABANKS

PIR/MIPS code

Brookhaven code

7 LITERATURE REFERENCES

[1] Pereira, B., Sivakami, S.: Biochem. J., 261, 43–47 (1989)
[2] Matsui, H., Ito, H., Chiba, S.: Agric. Biol. Chem., 52, 1859–1860 (1988)
[3] Tsuji, A., Suzuki, Y.: Biochem. Biophys. Res. Commun., 151, 1358–1363 (1988)
[4] Yamasaki, Y., Konno, H.: Agric. Biol. Chem., 51, 3239–3244 (1987)
[5] Giblin, M., Kelly, C.T., Fogarty, W.M.: Can. J. Microbiol., 33, 614–618 (1987)
[6] Cannel, R.J.P., Walker, J.M.: Biochem. Soc. Trans., 15, 521 (1987)
[7] O'Mahony, M.R., Kelly, C.T., Fogarty, W.M.: Biotechnol. Lett., 9, 317–322 (1987)
[8] Olusanya, O., Olutiola, P.O.: FEMS Microbiol. Lett., 36, 239–244 (1986)
[9] Kelly, C.T., Giblin, M., Fogarty, W.M.: Can. J. Microbiol., 32, 342–347 (1986)
[10] Kelly, C.T., Moriarty, M.E., Fogarty, W.M.: Appl. Microbiol. Biotechnol., 22, 352–358 (1985)
[11] Fogarty, W.M., Kelly, C.T., Kadam, S.K.: Can. J. Microbiol., 31, 670–674 (1985)
[12] Yamasaki, Y., Konno, H.: Agric. Biol. Chem., 49, 3383–3390 (1985)
[13] Suzuki, Y., Uchida, K.: Agric. Biol. Chem., 49, 1573–1581 (1985)
[14] Suzuki, Y., Uchida, K.: Agric. Biol. Chem., 49, 863–865 (1985)
[15] Yamasaki, Y., Konno, H.: Agric. Biol. Chem., 49, 849–850 (1985)
[16] Banno, Y., Nozawa, Y.: J. Biochem., 97, 409–418 (1985)
[17] Thirunavukkarasu, M., Priest, F.G.: J. Gen. Microbiol., 130, 3135–3141 (1984)
[18] Suzuki, Y., Uchida, K.: Agric. Biol. Chem., 48, 1343–1345 (1984)
[19] Dahlqvist, A. in "Methods Enzym. Anal.", 3rd. Ed. (Bergmeyer, H.U., Ed.) 4, 208–217 (1984)
[20] Li, K.-B., Chan, K.-Y.: Appl. Environ. Microbiol., 46, 1380–1387 (1983)
[21] Kelly, C.T., O'Reilly, F., Fogarty, W.M.: FEMS Microbiol. Lett., 20, 55–59 (1983)
[22] Reiss, U., Sacktor, B.: Arch. Biochem. Biophys., 209, 342–348 (1981)
[23] Matsui, H., Yazawa, I., Chiba, S.: Agric. Biol. Chem., 45, 887–894 (1981)
[24] Kelly, C.T., Heffernan, M.E., Fogarty, W.M.: Biotechnol. Lett., 2, 351–356 (1980)
[25] Yamasaki, Y., Suzuki, Y.: Agric. Biol. Chem., 44, 707–715 (1980)
[26] Yamasaki, Y., Suzuki, Y.: Agric. Biol. Chem., 43, 481–489 (1979)
[27] Rudick, M.J., Fitzgerald, Z.E., Rudick, V.L.: Arch. Biochem. Biophys., 193, 509–520 (1979)
[28] Tanimura, T., Kitamura, K., Fukuda, T., Kikuchi, T.: J. Biochem., 85, 123–130 (1979)
[29] Killilea, S.D., Clancy, M.J.: Phytochemistry, 17, 1429–1431 (1978)

[30] Peruffo, A.D.B., Renosto, F., Pallavicini, C.: Planta, 142, 195–201 (1978)
[31] Needleman, R.B., Federoff, H.J., Eccleshall, T.R., Buchferer, , B., Marmur, J.:
     Biochemistry, 17, 4657–4661 (1978)
[32] Yamasaki, Y., Suzuki, Y.: Agric. Biol. Chem., 42, 971–980 (1978)
[33] Urlaub, H., Wöber, G.: Biochim. Biophys. Acta, 522, 161–173 (1978)
[34] Yamasaki, Y., Suzuki, Y., Ozawa, J.: Agric. Biol. Chem., 41, 2149–2161 (1977)
[35] Matsusaka, K., Chiba, S., Shimomura, T.: Agric. Biol. Chem., 41, 1917–1923 (1977)
[36] Legler, G.: Methods Enzymol., 46, 368–381 (1977)
[37] Yamasaki, Y., Suzuki, Y., Ozawa, J.: Agric. Biol. Chem., 41, 1553–1558 (1977)
[38] Yamasaki, Y., Suzuki, Y., Ozawa, J.: Agric. Biol. Chem., 41, 1559–1565 (1977)
[39] Yamasaki, Y., Suzuki, Y., Ozawa, J.: Agric. Biol. Chem., 41, 1451–1458 (1977)
[40] Kanaya, K.-I., Chiba, S., Shimomura, T., Nishi, K.: Agric. Biol. Chem., 40, 1929–1936
     (1976)
[41] Yamasaki, Y., Suzuki, Y., Ozawa, J.: Agric. Biol. Chem., 40, 1909–1915 (1976)
[42] Yamasaki, Y., Miyake, T., Suzuki, Y.: Agric. Biol. Chem., 37, 131–137 (1973)
[43] Yamasaki, Y., Miyake, T., Suzuki, Y.: Agric. Biol. Chem., 37, 251–259 (1973)
[44] Suzuki, Y., Yuki, T., Kishigami, T., Abe, S.: Biochim. Biophys. Acta, 445, 386–397
     (1976)
[45] Yamasaki, Y., Suzuki, Y., Ozawa, J.: Agric. Biol. Chem., 40, 669–676 (1976)
[46] Chiba, S., Shimomura, T.: Agric. Biol. Chem., 39, 1033–1040 (1975)
[47] Yamasaki, Y., Suzuki, Y.: Agric. Biol. Chem., 38, 443–454 (1974)
[48] Rudick M.J., Elbein, A.D.: Arch. Biochem. Biophys., 161, 281–290 (1974)
[49] Chiba, S., Saeki, T., Shimomura, T.: Agric. Biol. Chem., 37, 1823–1829 (1973)
[50] Flores-Carreon, A., Ruiz-Herrera, J.: Biochim. Biophys. Acta, 258, 496–505 (1972)
[51] Takahashi, N., Shimomura, T., Chiba, S.: Agric. Biol. Chem., 35, 2015–2024 (1971)
[52] Takahashi, M., Shimomura, T.: Agric. Biol. Chem., 32, 929–939 (1968)
[53] Khan, N.A., Eaton, N.R.: Biochim. Biophys. Acta, 146, 173–180 (1967)
[54] Urlaub, H., Wöber, G.: Biochem. Soc. Trans., 3, 1076–1078 (1975)
[55] Matsui, H., Chiba, S.: Agric. Biol. Chem., 47, 707–713 (1983)
[56] Sorenson, S.H., Noren, O., Sjöström, H., Danielen, E.M.: Eur. J. Biochem., 126,
     559–568 (1982)
[57] Minamiura, N., Matoba, K., Nishinaka, H., Yamamoto, T.: J. Biochem., 91, 809–816
     (1982)
[58] Dissous, C., Ansart, J.F., Cheron, A., Krembel, J.: Anal. Biochem., 116, 35–38 (1981)
[59] Martiniuk, F., Hirschhorn, R.: Biochim. Biophys. Acta, 658, 248–261 (1981)
[60] Giudicelli, J., Emiliozzi, R., Vannier, C., De Burlet, G., Sudaka, P.: Biochim. Biophys.
     Acta, 612, 85–96 (1980)
[61] Flanagan, P.R., Forster, G.G.: Biochem. J., 173, 553–563 (1978)
[62] Koster, J.F., Slee, R.G.: Biochim. Biophys. Acta, 482, 89–97 (1977)
[63] Hibi, N., Chiba, S., Shimomur, T.: Agric. Biol. Chem., 40, 1805–1812 (1976)
[64] Bruni, C.B., Sica, V., Auricchio, F., Covelli, I.: Biochim. Biophys. Acta, 212, 470–477
     (1970)
[65] De Burlet, G., Sudaka, P.: Biochimie, 58, 621–623 (1976)
[66] Koster, J.F., Slee, , R.G., Van Der Klei-Moorsel, J.M., Rietra, P.J.G.M., Lucas, L.J.:
     Clin. Chim. Acta, 68, 49–58 (1976)
[67] Auricchhio, F., Bruni, C.B., Sicer, V.: Eur. J. Biochem., 31, 156–165 (1968)

Enzyme Handbook © Springer-Verlag Berlin Heidelberg 1991
Duplication, reproduction and storage in data banks are only
allowed with the prior permission of the publishers

1 NOMENCLATURE

EC number
   3.2.1.21

Systematic name
   Beta-D-glucoside glucohydrolase

Recommended name
   Beta-glucosidase

Synonymes
   Emulsin
   Gentiobiase
   Cellobiase
   Elaterase
   Aryl-beta-glucosidase
   Beta-D-glucosidase
   Beta-glucoside glucohydrolase
   Arbutinase
   Amygdalinase
   p-Nitrophenyl beta-glucosidase
   Primeverosidase
   Amygdalase
   Limarase [6]
   Salicilinase [15]
   Beta-1,6-glucosidase [58]
   Beta-D-glycosidase [75]
   Novozyme 188
   Maxazyme Cl [39]

CAS Reg. No.
   9001-22-3

---

2 REACTION AND SPECIFICITY

Catalysed reaction
   Beta-D-oligosaccharides + H_2O →
   → n D-glucose (hydrolysis of terminal non-reducing beta-D-glucose
   residues with release of beta-D-glucose)

Reaction type
   O-Glycosyl bond hydrolysis

Enzyme Handbook © Springer-Verlag Berlin Heidelberg 1991
Duplication, reproduction and storage in data banks are only
allowed with the prior permission of the publishers

Natural substrates

Beta-D-oligosaccharides (e.g. cellobiose in cellulose-degrading organisms)
More (biological function of mammalian enzyme unclear [77])

Substrate spectrum

1 Beta-D-oligosaccharides + H_2O (strict requirement for beta-D-configuration [16], hydrolysis of alpha(1–4) and beta(1–4) configuration [1], requirement of trans equatorial oxygen at position 1, 2, and 3 of monosaccharide [57], oligosaccharides with beta(1–6) linkages [11, 16, 58], beta(1–3) linkages, e.g. laminarose [16], beta(1–2) linkages e.g. sophorose [11, 16], beta-glucosyltransferase activity [27, 52], influence of conformation at C-4 and C-5 of saccharide on substrate recognition [76])

2 p-Nitrophenyl-beta-D-glucopyranoside + H_2O [1, 4, 6, 14, 19, 20, 22, 28, 32, 37, 39, 45, 48, 49, 57, 78]

3 p-Nitrophenyl-beta-D-galactopyranoside + H_2O [1, 4, 6, 27, 48, 49]

4 o-Nitrophenyl-D-glucopyranoside (or galactopyranoside, or xylopyranoside) + H_2O [1, 11, 27, 46]

5 p-Nitrophenyl-beta-D-xylopyranoside + H_2O [1, 6, 27, 34, 39, 48, 49]

6 Cellobiose + H_2O [1, 8, 14, 20, 22, 27, 28, 32, 34, 37, 45, 49, 52, 53]

7 2-(Beta-D-glucopyranosyloxy)-2-methyl-propanenitrile (linamarin) + H_2O [6]

8 Cellotriose + H_2O [11, 49, 56]

9 Cellotetraose + H_2O [49, 56]

10 Cellopentaose + H_2O [49, 56]

11 4-Methylumbelliferyl derivatives of beta-D-glucose + H_2O [57, 76–78]

12 2-(Hydroxymethyl)phenyl-beta-D-glucopyranoside (salicin) + H_2O [6, 20, 27]

13 Cellulose + H_2O [39]

14 4-(Beta-D-galactosido)-D-gluconic acid (lactobionic acid) + H_2O [78]

15 Estrone 3-beta-glucoside + H_2O [78]

16 More [1, 11, 27, 49, 78]

Product spectrum

1 D-Glucose
2 p-Nitrophenol + D-glucose
3 p-Nitrophenol + D-galactose
4 o-Nitrophenol + D-glucose
5 p-Nitrophenol + D-xylose
6 D-Glucose
7 ?
8 D-Glucose (or cellobiose + D-glucose)
9 Cellobiose + D-glucose [56]
10 Cellobiose + D-glucose [56]
11 4-Methylumbelliferone + D-glucose
12 2-Hydroxybenzenemethanol (saligenol) + D-glucose

13 ?
14 ?
15 ?
16 ?

Inhibitor(s)

Glucose (competetive, non competetive [46], not inhibitory [31, 33], glucose derivatives [38]); Glucono-1, 5-lactone (not inhibitory [20]) [6, 7, 15, 16, 20, 29, 34, 37, 45, 49, 55, 57, 58, 75, 77, 78]; Cu^{2+} [1, 29, 32, 40, 48, 53, 58]; Ni^{2+} [1, 26, 27]; Hg^{2+} [1, 7, 22, 26, 27, 32, 40, 48, 49, 55, 58]; Fe^{3+} [1, 20, 32, 58]; Fe^{2+} [29, 48]; Ca^{2+} [1, 20]; Co^{2+} [1, 20, 40]; Ag^{2+} [6, 29]; Pb^{2+} [26]; Zn^{2+} [40]; Norjirimycin [16]; 1-Deoxynorijimycin [38]; D-Glucono-1, 5-lactam [38]; Detergents (reversible [22], ionic [57]) [22, 27, 32, 40, 57]; p-Chloromercuribenzoate (reversed by dithiothreitol [32]) [27, 32, 40, 48, 49, 57, 58]; Cello-oligosaccharides [28]; EDTA [29, 45]; Al^{3+} [27]; Cd^{2+} [27]; Cellobiose (high concentrations [64]) [29, 64]; Gentiobiose [29]; Laminaribose [29]; Gamma-galactono-lactone [34, 45]; p-Nitrophenyl-beta-D-glucopyranoside (more than 0.4 mM) [44]; Melibiose [45]; Maltose [45]; p-Nitrophenyl-alpha-D-glucopyranoside [45]; Iodoacetamide [49]; N-Ethylmaleimide [57, 58]; Iodoacetic acid [57]; Glutathione [57]; p-Nitrothiophenyl-beta-D-glucopyranoside (and other substrate analogs) [75]; Galactosylsphingosine [76]; p-Nitrophenyl-beta-D-xylopyranoside [77]; p-Nitrophenyl-beta-D-mannopyranoside [77]; Cetyltriammoniumbromide [77]; Sodium taurocholate [77]; N-Bromoacetyl-beta-D-galactosylamine [79]; Phloridzin [80]; 2, 5-Dihydroxymethyl-3, 4-dihydroxypyrrolidine [81]; Phosphatidylserine [78]; Phosphatidylinositol [78]

Cofactor(s)/prostethic group(s)

Ethanol (activation) [1, 37]; Reducing agents (activation) [22, 41, 57]

Metal compounds/salts

Mn^{2+} [1, 24]; Co^{2+} [24]; Divalent cations [22]

Turnover number (min^{-1})

Specific activity (U/mg)

11.0–2300 (depending on growth medium) [3]; 825 [24]; 33.3–127 (depending on substrate) [49]; More [1, 2, 4, 7, 8, 10, 11, 14–16, 18, 20, 23, 25–28, 30, 35, 36, 46–49, 52, 56–59, 75–78]

Enzyme Handbook © Springer-Verlag Berlin Heidelberg 1991
Duplication, reproduction and storage in data banks are only
allowed with the prior permission of the publishers

K_m-value (mM)

0.21 (p-nitrophenyl-beta-D-glucoside, similar values [6, 13, 16, 22, 26, 29, 30, 35, 37, 40, 44–48, 59]) [4]; 1.17 (p-nitrophenyl-beta-D-glucoside, similar values [7, 16, 32, 33, 42, 49, 55]) [24]; 0.27 (p-nitrophenol-beta-D-fucoside) [4]; 6.7 (p-nitrophenol-beta-D-galactoside, similar values [7, 45]) [4]; 0.2 (p-nitrophenol-beta-D-mannoside) [4]; 2.8 (p-nitrophenyl-beta-D-xyloside) [4]; 0.2 (cellobiose, , similar value [45]) [37]; 2.0 (cellobiose, similar values [7, 8, 11, 16, 24, 31, 53, 56]) [4]; 33 (cellobiose, similar values [26, 32, 33, 40, 49]) [22]; 5.0 (salicin, similar values [6, 24, 26]) [7]; 0.86 (salicin, similar values [4, 55]) [53]; 0.7–1.98 (cellotriose) [16, 55]; 24.8 (cellotriose, similar value [40]) [32]; 50.0 (laminarin) [7]; 0.06–0.15 (methylumbelliferyl derivatives of glucopyranoside, galactopyranoside, xylopyranoside, alpha-L-arabinopyranoside) [76]; More (comparison of values from different fungi and bacteria [50]) [4, 6, 16, 24, 26, 40, 42, 44, 45, 50, 55, 75]

pH-optimum

3.5–5.0 (p-nitrophenyl-beta-d-glucopyranoside) [42]; 4.0 [11, 20, 37]; 4.0–4.5 [32]; 4.0–5.0 [33]; 4.1 (enzyme I) [18]; 4.2 (p-nitrophenyl-beta-D-glucopyranoside) [16]; 4.3 [35]; 4.5 (cellobiose [16]) [16, 31, 56]; 4.5–5.0 [28]; 4.5–5.5 (cellobiose) [42]; 4.5–6.5 [4]; 4.6 [29]; 4.8 [47]; 4.8–5.5 [59]; 5.0 [2, 7, 8, 14, 24, 30, 44, 46, 53]; 5.5 [55]; 5.7 (enzyme IV [18]) [18, 77]; 5–6 [78]; 5.5–6.0 [76]; 6.0–6.5 (depending on buffer [1], cloned enzyme [21]) [1, 23, 49, 54]; 6.0 [21, 42]; 6–7.3 [6]; 6.2 [27, 45]; 6.4 [48]; 6.5 [34, 40]; 8.0 [52]

pH-range

3.0–5.2 [20]; 3.0–6.5 [35]; 3.9–5.5 [30]; 4–5 [56]; 4–7 [2]; 4.8–8.0 [48]; 5–7 [1, 22]

Temperature optimum (°C)

35 (enzyme IV [18]) [18, 21]; 37 (cloned enzyme) [21]; 40 [58]; 45 [35]; 49 (p-nitrophenyl-beta-D-glucopyranoside) [39]; 50 [1, 7, 8, 14, 40, 53]; 50–55 [47]; 50–60 [20]; 52 (cellulose [39]) [39, 48]; 55 (linamarin [6], intracellular form [44]) [6, 26, 44 , 46, 55]; 60 (soluble form [2], p-nitrophenyl-beta-D-glucopyranoside [6]) [2, 6, 29, 52]; 60–65 [42]; 60–70 [31]; 65 (cellobiose [16]) [16, 22]; 65 [32, 54]; 68 (p-nitrophenyl-beta-D-glucopyranoside) [16]; 70 (enzyme I [18]) [13, 18, 37]; 70–75 [11]; 75 (extracellular form [44]) [44, 56]; 80–85 [4]; More (comparison of values from different fungi and bacteria) [50]

Temperature range (°C)

20–70 [35]; 40–70 [42]; 40–80 [2]; 50–75 [22, 56]

3 ENZYME STRUCTURE

Molecular weight

300000–380000 (Kluyveromyces fragilis, gel filtration [26] , Aspergillus fumigatus, gel filtration [32], Coriolus versicolor, SDS-PAGE [35], Candida pelliculosa, gel filtration [40], Lezentis trabea, gel filtration [56]) [26 , 32, 35, 40, 56]

250000–295000 (Humicola insolens, SDS-PAGE [14, 53], Hanseniaspora vineae, gel filtration [1], Aspergillus terres, gel filtration [47]) [1, 14, 47, 53]

180000–240000 (Aspergillus fumigatus, gel filtration [33], Sclerotinia sclerotium, gel filtration [8], Pyricularia oryzae, SDS-PAGE, sucrose density centrifugation [55], Penicillium funiculosum, enzyme I, gel filtration [37], Botrydiplodia theobromae, gel filtration [65]) [8, 33, 37, 55, 65]

90000–160000 (Erinnyis ello, density gradient centrifugation, gel electrophoresis [34], Alternaria alternata, gel filtration [7], Sporotrichum pulverulentum, SDS-PAGE [15], Aspergillus niger [11], Sclerotium rolfsii, gel filtration, SDS-PAGE [16, 51], Schizophyllum commune [17], Clostridium stercorarium, FPLC gel filtration [22], Rhynchosciara americana, glycerol gradient centrifugation [45], Talaromyces emersonii, SDS-PAGE [18]) [7, 11, 15, 16, 17, 22, 34, 45, 51]

50000–85000 (Thermoascus aurantiacus, gel filtration, SDS-PAGE [13], Talaromyces emersonii, intracellular form [18], E. coli, containing gene from Caldocellum saccharolyticum, SDS-PAGE, calculation from DNA sequence [23], Ruminococcus albus, gel filtration [21, 41], Carica papaya, gel filtration [30], Trichoderma reesei [28], Trichoderma viride, SDS-PAGE [39], Rhizobium trifolii, gel filtration [48], Clostridium thermocellum, gel filtration [54], rat, gel filtration, sedimentation equilibrum centrifugation [57], Flavobacterium, gel filtration, SDS-PAGE [58], pig, gel filtration [75], man, gel filtration [76, 78], man, denaturing and non-denaturing gel electrophoresis [78]) [18, 21, 23, 30, 39, 41, 48, 57, 58, 75, 76, 78]

18000–48000 (Thermus, sp. Z-1, gel filtration, SDS-PAGE [4], Evernia prunastri, gel filtration [20], thermophilic anaerobic bacterium, HPLC gel filtration [27], Penicillium funicolosum, enzyme II, gel filtration [37], Clostridium thermocellum, gel filtration, SDS-PAGE [49]) [4, 20, 27, 37, 49]

Enzyme Handbook © Springer-Verlag Berlin Heidelberg 1991
Duplication, reproduction and storage in data banks are only
allowed with the prior permission of the publishers

Subunits
Monomer [4, 13, 16, 27, 49, 51, 75, 78]
Dimer (2 × 27000, Carica papaya, SDS-PAGE [30], 2 × 94000, Candida wickerhamii, SDS-PAGE [33], 2 × 120000, Penicillium funiculosum, SDS-PAGE [37]) [30, 33, 37]
Tetramer (4 × 79000, Hanseniaspora vineae, SDS-PAGE [1], 4 × 90000–95000, Kluyveromyces fragilis, DNA sequence, SDS-PAGE [26], Aspergillus fumigatus, SDS-PAGE [32], Candida pelliculosa, SDS-PAGE [40]) [1, 26, 32, 40]
? (60000–70000 + 80000, Alternaria alternata [7], Evernia prunastri [20], x × 69000–74000, Scytalidium lignicola, enzyme I and II, [46], SDS-PAGE) [7, 20, 46]
Octamer (8 × 45000–47000, Botrydiplodia theobromae, each subunits consits of 4 inactive subunits of MW 10000–12000) [65]

Glycoprotein/Lipoprotein
Glycoprotein (11% galactose, 4% mannose, 1% glucose [7], carbohydrate content 2.5% [14], 160 units monosaccharide per subunit [33], no glycoprotein [75]) [7, 14, 28, 39, 47]

4 ISOLATION/PREPARATION

Source organism
Hanseniaspora vineae [1]; Myceliopthora thermophila [2]; Streptomyces lividans [3]; Thermus sp. Z-1 [4, 19]; Cellulomonas thermocellum [5]; Manihot esculenta (cassava) [6]; Alternaria alternata [7]; Sclerotinia sclerotiorum [8]; Aspergillus niger [11, 31]; Humicola insolens [14, 53]; Humicola grisea [14]; Sporotrichum pulverulentum [15]; Sclerotium rolfsii [16, 51]; Schizophyllum commune [17]; Talaromyces emersonii [18]; Evernia prunastri [20]; Ruminococcus albus [21, 41]; Clostridium stercorarium [22]; E. coli (containing gene from thermophilic anaerobic bacterium Tp8) [23]; Aspergillus japonicus [24]; Rhynchosciara americana [45]; Scytalidium lignicola [46]; Aspergillus terreus [47]; Rhizobium trifolii [48]; Clostridium thermocellum [49, 54]; Mucor miehei [52]; Pyricularia oryzae [55]; Lenzites trabea [56]; Rat [57]; Flavobacterium [58]; Chaetomium thermophile [59]; Aspergillus aculeatus [12]; Thermoascus aurantiacus [13]; Trichoderma viride [9]; Trichoderma koningii [10]; Hevea [60]; Trifolium [61]; Phaseolus [62]; Phytophora infestans [63]; Aspergillus orizae [50]; Saccharomyces cerevisiae [50]; Botrydiplodia theobromae [65]; Erwinia [66]; Thermobacter ethanolicus [67]; Kluyveromyces fragilis [27]; Trichoderma reesei [28]; Aspergillus ornatus [29]; Carica papaya (papaya) [30]; Aspergillus fumigatus [32]; Candida wickerhamii [33]; Erinnyis ello (cassava hornworm) [34]; Coriolus versicolor [35]; Penicillium funiculosum [36, 37]; Trichoderma viride [39]; Candida pelliculosa var. acethericus [40]; Aspergillus wentii [42]; Termitomyces [43]; Humicola spp. [44]; Aspergillus foetidus [64]; Bac-

teroides succinogenes [68]; Cellulomonas fimi [69]; Stachybotrys atra [70]; Kluyveromyces dobzhanskii [71]; Kluyveromyces lactis [72]; Rhodotorula minuta [73]; Dekkera intermedia [74]; Man [76–79]; Pig [75]

Source tissue
Cell [1, 4, 20, 22, 23, 26, 27, 40, 44, 48, 49, 58, 59]; Culture medium [2, 7, 8, 12–18, 22, 24, 28, 29, 32, 33, 35, 36, 44, 46, 51, 53, 59]; Leaf [6]; Peel [6]; Tuber cortex [6]; Mycelia [18]; Fruit [30]; Midgut [34, 45]; Conidiophores [43]; Fruit body [43]; Kidney (cortex) [57, 75]; Liver [76, 78, 79]; Spleen [77]

Localisation in source
Cytoplasm [1, 57, 75–78]; Extracellular (with cellulose as carbon source [15, 37]) [2, 7, 8, 12–18, 22, 24, 28, 29, 32, 33, 35–37, 44, 46, 55, 56, 59]; Cell bound (with cellobiose as carbon source [15, 37]) [15, 37, 41, 59]; Intracellular [18, 20, 40, 44, 58]; Periplasm [22, 48, 54]; Plasma membrane [45]

Purification
Hanseniaspora vinea [1]; Myceliopthora thermophila [2]; Thermus sp Z-1 [4]; Manihot erculenta (partial) [6]; Alternaria alternata [7]; Sclerotinia sclerotium [8]; Aspergillus niger (from Novozym 188, commercially available preparation) [11]; Humicola insolens [14, 53]; Sporotrichum pulverulentum (5 isoenzymes) [15]; Sclerotium rolfsii (4 isoenzymes) [16, 51]; Aspergillus aculeatus (3 isoenzymes) [12]; Thermoascus aurantiacus [13]; Schizophyllum commune [17]; Talaromyces emersonii (extracellular and intracellular forms) [18]; Evernia prunastri [20]; Ruminococcus albus [21, 41]; Clostridium stercorarium [22]; E.coli [23]; Aspergillus japonicus [24]; Kluyveromyces fragilis [26]; Saccharomyces cerevisiae [26]; Trichoderma reesei [28]; Aspergillus fumigatus [32]; Candida wickerhamii [33]; Coriolus versicolor [35]; Penicillium funiculosum (2 isoenzymes) [36], Trichoderma viride (from Maxazyme Cl, commercilally available cellulase preparation) [39]; Termitomyces (partial) [43]; Candida pelliculosa [40]; Rhynchosciara americana [45]; Scytalidium lignicola (2 isoenzymes) [46]; Aspergillus terreus [47]; Rhizobium trifolii [48]; Clostridium thermocellum [54]; Mucor miehei [52]; Pyricularia oryzae [55]; Lenzites trabea [56]; Rat [57]; Flavobacterium [58]; Chaetomium thermophile [59]; Dekkera intermedia [74]; Aspergillus foitidus [64]; Pig [75]; Man [76–78]

Crystallization
–

Cloned
[3, 21, 23, 25, 26]

Renaturated
–

Enzyme Handbook © Springer-Verlag Berlin Heidelberg 1991
Duplication, reproduction and storage in data banks are only
allowed with the prior permission of the publishers

5 STABILITY

pH

2–8 [32]; 2.5–9 [11]; 2.5–10.0 [52]; 3–8 [7]; 3.5–7.5 [2]; 4.0–4.5 [31]; 4.0–8.0 [14]; 4.5–6.5 [19]; 5.0 (optimal stability [28], 24 hours [46]) [28, 46]; 5.0–7.0 [58]; 6–8 [13]

Temperature (°C)

0–60 [7]; 35 (up to) [33]; 30 (up to) [46]; 40 (up to) [45 , 77]; 45 (unstable below and above [8], 20 hours, 50% activity [35]) [8, 35]; 50 (denaturation above [1], 5 minutes, 70% activity [58], 10 minutes, 50% activity [78]) [1, 40, 48, 58, 78]; 53 (up to) [34]; 55 (stable [42], 20 minutes [47], 30h, 74% activity [49, 54]) [42, 47, 49, 54]; 58–85 (comparison of stability values from different organisms) [27]; 60 (1 hour [2], stable below [20], crude enzyme, 3 hours, 50% activity, purified enzyme 5 hours, 50% activity [22]) [2, 20, 22, 31, 32, 53]; 60–70 (1 hour) [52]; 65 (6. 3 hours, 50% activity [43]) [24, 43]; 70 (1 hour, immobilized enzyme [2], enzyme I: 410 minutes, 50% activity, enzyme IV: 2 minutes, 50% activity [18], several days [23]) [2, 18, 23]; 75 (1 hour, 70% activity [13], 5 minutes [52]) [13, 52]; 80 (30 minutes , immobilized enzyme, 30% activity [2]) [2]; 90 (11 minutes) [23]; 95 (5 minutes, 56% activity) [52]

Oxidation

Organic solvent

Ethanol (80% activity in 5% w/v solution) [27]; Acetone (crude enzyme stable, purified enzyme inactivation) [23]

General stability information

Bovine serum albumin (stabilization) [37]

Storage

4°C, 0.02% w/v NaN_3, 50 mM sodium acetate buffer, pH 4.8 [2]; –10°C, more than 1 year [6]; 4°C, or frozen, 20 mM sodium acetate buffer, pH 4, propylene container [11]; 0–4 °C, 50 mM acetate buffer, pH 5.0, at least 2 weeks [14]; –15°C, 50 mM citrate buffer, pH 4.5, several months [16]; –70°C, at least 6 months [23]; Liquid N_2 [27]; –20°C, more than 6 months [33, 36]; –20°C, or 4°C, at least 1 year [49]

---

6 CROSSREFERENCES TO STRUCTURE DATABANKS

PIR/MIPS code

GLHQ (precursor, yeast, Hansenula anomala); GLVK (precursor, yeast Kluyveromyces fragilis); GLAG (Agrobacterium sp.); A28571 (Schizophyllum commune, fragment); S03813 (Caldocellum saccharolyticum); S08243 (Ruminococcus albus); S04381 (B, Clostridium thermocellum); A29171 (A-3, Aspergillus wentii, fragment)

Brookhaven code

7 LITERATURE REFERENCES

[1] Vasserot, Y., Christiaens, H., Chemardin, P., Arnaud, A., Galzy, P.: J. Appl. Bacteriol., 66, 271–279 (1989)

[2] Roy, S.K., Raha, S.K., Dey, S.K., Chakrabarty, S.L.: Enzyme Microb. Technol., 11, 431–435 (1989)

[3] Jaurin, B., Granström, M.: Appl. Microbiol. Biotechnol., 30, 502–508 (1989)

[4] Takase, M., Horikoshi, K.: Agric. Biol. Chem., 53, 559–560 (1989)

[5] Kim, H.K., Pack, M.Y.: Enzyme Microb. Technol., 11, 313–316 (1989)

[6] Yeoh, H.-H.: Phytochemistry, 28, 721–724 (1989)

[7] Martinez, M.J., Vazquez, C., Guillen, F., Reyes, F.: FEMS Microbiol. Lett., 55, 263–268 (1988)

[8] Waksman, G.: Biochim. Biophys. Acta, 967, 82–86 (1988)

[9] Voragen, A.G.J., Beldman, G., Rombouts, F.M.: Methods Enzymol., 160, 243–251 (1988)

[10] Wood, T.M.: Methods Enzymol., 160, 221–233 (1988)

[11] McCleary, B.V., Harrington, J.: Methods Enzymol., 160, 575–583 (1988)

[12] Murao, S., Sakamoto, R., Arai, M.: Methods Enzymol., 160, 274–299 (1988)

[13] Shephard, M.G., Cole, A.L., Tong, C.C.: Methods Enzymol., 160, 300–307 (1988)

[14] Hayashida, S., Ohta, K., Mo, K.: Methods Enzymol., 160, 323–352 (1988)

[15] Deshpande, V., Eriksson, K.-E.: Methods Enzymol., 160, 415–424 (1988)

[16] Sadana, J.C., Patil, R.V., Shewale, J.G.: Methods Enzymol., 160, 424–431 (1988)

[17] Lo, A.C., Willick, G., Bernier, R., Desrochers, M.: Methods Enzymol., 160, 432–437 (1988)

[18] Coughlan, M.P., McHale, A.: Methods Enzymol., 160, 437–443 (1988)

[19] Takase, M., Horikoshi, K.: Appl. Microbiol. Biotechnol., 29, 55–60 (1988)

[20] Yagüe, E., Estevez, M.P.: Eur. J. Biochem., 175, 627–632 (1988)

[21] Honda, H., Saito, T., Iijima, S., Kobayashi, T.: Enzyme Microb. Technol., 10, 559–562 (1988)

[22] Bronnemeier, K., Staudenbauer, W.L.: Appl. Microbiol. Biotechnol., 28, 380–386 (1988)

[23] Plant, A.R., Oliver, J.E., Patchett, M.L., Daniel, R. M., Morgan, H.W.: Arch. Biochem. Biophys., 262, 181–188 (1988)

[24] Sanyal, A., Kundu, R.K., Dube, S., Dube, D.K.: Enzyme Microb. Technol., 10, 91–99 (1988)

[25] Kadam, S., Demain, A.L., Millet, J., Beguin, P., Aubert, J.-P.: Enzyme Microb. Technol., 10, 9–13 (1988)

[26] Leclerc, M., Chemardin, P., Arnaud, A., Ratomahenina, R., Galzy, P., Gerbaud, C., Raynal, A., Guerineau, M.: Biotechnol. Appl. Biochem., 9, 410–422 (1987)

[27] Patchett, M.L., Daniel, R.M., Morgan, H.W.: Biochem. J., 243, 779–782 (1987)

[28] Chirco, W.J., Brown, R.D.: Eur. J. Biochem., 165, 333–341 (1987)

[29] Yeoh, H.H., Tan, T.K., Koh, , S.K.: Appl. Microbiol. Biotechnol., 25, 25–28 (1986)

[30] Hartmann-Schreier, J., Schreier, P.: Phytochemistry, 25, 2271–2274 (1986)

[31] Dekker, R.F.H.: Biotechnol. Bioeng., 28, 1438–1442 (1986)

[32] Kitpreechavanich, V., Hayashi, M., Nagai, S.: Agric. Biol. Chem., 50, 1703–1711 (1986)

Enzyme Handbook © Springer-Verlag Berlin Heidelberg 1991
Duplication, reproduction and storage in data banks are only
allowed with the prior permission of the publishers

[33] Freer, S.N.: Arch. Biochem. Biophys., 243, 515–522 (1985)

[34] Santos, C.D., Terra, W.R.: Biochim. Biophys. Acta, 831, 179–185 (1985)

[35] Evans, C.S.: Appl. Microbiol. Biotechnol., 22, 128–131 (1985)

[36] Kantham, L., Vartak, H.G., Jagannathan, V.: Biotechnol. Bioeng., 27, 781–785 (1985)

[37] Kanthan, L., Jagannathan, V.: Biotechnol. Bioeng., 27, 786–791 (1985)

[38] Dale, M.P., Ensley, H.E., Kern, K., Sastry, K.A.R., Byers, L.D.: Biochemistry, 24, 3530–3539 (1985)

[39] Beldman, G., Searle-Van Leeuwen, M.F., Rombouts, F.M., Voragen, F.G.J.: Eur. J. Biochem., 146, 301–308 (1985)

[40] Kohchi, C., Hayashi, M., Nagai, S.: Agric. Biol. Chem., 49, 779–784 (1985)

[41] Ohmiya, K., Shira, M., Kurachi, Y., Shimizu, S.: J. Bacteriol., 161, 432–434 (1985)

[42] Srivastava, S.K., Gopalkrishnan, K.S., Ramachandran, K.B.: Enzyme Microb. Technol., 6, 508–512 (1984)

[43] Osore, H., Okech, M.A.: J. Appl. Biochem., 5, 172–179 (1983)

[44] Araujo, E.F., Barros, E.G., Caldas, R.A., Silva, D.O.: Biotechnol. Lett., 5, 781–784 (1983)

[45] Ferreira, C., Terra, W.R.: Biochem. J., 213, 43–51 (1983)

[46] Desai, J.D., Ray, R.M., Patel, N.P.: Biotechnol. Bioeng., 25, 307–313 (1983)

[47] Workman, W.E., Day, D.F.: Appl. Environ. Microbiol., 44, 1289–1295 (1982)

[48] Abe, M., Higashi, S.: J. Gen. Appl. Microbiol., 28, 551–562 (1982)

[49] Ait, , N., Creuzet, N., Cattaneo, J.: J. Gen. Microbiol., 128, 569–577 (1982)

[50] Woodward, J., Wiseman, A.: Enzyme Microb. Technol., 4, 73–79 (1982) (Review)

[51] Shewale, J.G., Sadana, J.: Arch. Biochem. Biophys., 207, 185–196 (1981)

[52] Yoshioka, H., Hayashida, S.: Agric. Biol. Chem., 44, 2817–2824 (1980)

[53] Yoshioka, H., Hayashida, S.: Agric. Biol. Chem., 44, 1729–1735 (1980)

[54] Ait, N., Creuzet, N., Cattaneo, J.: Biochem. Biophys. Res. Commun., 90, 537–546 (1979)

[55] Hirayama, T., Horie, S., Nagayama, H., Matsuda, K.: J. Biochem., 84, 27–37 (1978)

[56] Herr, D., Baumer, F., Dellweg, H.: Eur. J. Appl. Microbiol. Biotechnol., 5, 29–36 (1978)

[57] Glew, R.H., Peters, S.P., Christopher, A.R.: Biochim. Biophys. Acta, 422, 179–199 (1976)

[58] Sano, K., Amemura, A., Harada, T.: Biochim. Biophys. Acta, 377, 410–420 (1975)

[59] Lusis, A.J., Becker, R.R.: Biochim. Biophys. Acta, 329, 5–16 (1973)

[60] Selmar, D., Lieberei, R., Biehl, B., Voigt, J.: Plant Physiol., 83, 557 (1987)

[61] Hughes, M.A.: J. Exp. Bot., 19, 427 (1968)

[62] Itoh-Nashida, T., Hiraiwa, M., Uda, Y.: J. Biochem., 101, 847 (1987)

[63] Bodenmann, J., Heininger, U., Hohl, H.R.: Can. J. Microbiol., 31, 75–82 (1985)

[64] Gusakov, A.V., Sinitsyn, A.P., Goldsteins, G.H., Klyosov, A.A.: Enzyme Microb. Technol., 6, 275–282 (1984)

[65] Umezurike, G.M.: Biochem. J., 145, 361–368 (1975)

[66] Barras, F., Chambost, J.P., Chippeaux, M.: Mol. Gen. Genet., 197, 490–496 (1984)

[67] Mitchell, R.W., Hägerdal, B., Ferchak, J.D., Pye, E. K.: Biotechnol. Bioeng. Symp., 12, 461–467 (1982)

[68] Forsberg, C.W., Groleau, D.: Can. J. Microbiol., 28, 144–148 (1982)

[69] Wakarchuk, W.W., Kilburn, D.G., Miller, R.C., Warren, R.A.: J. Gen. Microbiol., 130, 1385–1389 (1984)

[70] DeGussen, R.L., Aerts, G.M., Glaeyssens, M., DeBruyne, C.K.: Biochim. Biophys. Acta, 525, 142 (1978)

[71] Fleming, L.W., Duerksen, J.D.: J. Bacteriol., 93, 142–150 (1967)

[72] Marchin, G.L., Duerksen, J.D.: J. Bacteriol., 96, 1187–1190 (1968)

[73] Duerksen, J.D., Halvorson, H.: J. Biol. Chem., 233, 1113–1120 (1958)

[74] Blondin, B., Ratomahenina, R., Arnaud, A., Galzy, P.: Eur. J. Appl. Microbiol. Biotechnol., 17, 1–6 (1983)

[75] Pocsi, I., Kiss, L.: Biochem. J., 256, 139–146 (1988)

[76] LaMarco, K.L., Glew, R.H.: Arch. Biochem. Biophys., 236, 669–676 (1985)

[77] Maret, A., Salvayre, R., Negre, A., Douste-Blazy, L.: Eur. J. Biochem., 133, 283–287 (1983)

[78] Daniels, L.B., Coyle, P.J., Chiao, Y.-B., Glew, R.H.: J. Biol. Chem., 256, 13004–13013 (1981)

[79] Meisler, M.H.: Biochim. Biophys. Acta, 410, 347–352 (1975)

[80] Abrahams, H.E., Robinson, D.: Biochem. J., 111, 749–755 (1969)

[81] Chinchetru, M.A., Calvo, P., Cenci Di Bello, I., Winchester, B.: Comp. Biochem. Physiol. B Comp. Biochem., 84, 623–628 (1986)

Enzyme Handbook © Springer-Verlag Berlin Heidelberg 1991
Duplication, reproduction and storage in data banks are only
allowed with the prior permission of the publishers

1 NOMENCLATURE

EC number
3.2.1.22

Systematic name
Alpha-D-galactoside galactohydrolase

Recommended name
Alpha-galactosidase

Synonymes
Melibiase
Galactosidase, .alpha.
.alpha.-D-Galactosidase
.alpha.-Galactosidase
.alpha.-Galactosidase A
.alpha.-D-Galactoside galactohydrolase
. alpha.-Galactoside galactohydrolase

CAS Reg. No.
9025-35-8

---

2 REACTION AND SPECIFICITY

Catalysed reaction
Alpha-D-galactoside (galactose-ROH) + H_2O →
→ galactose + R (mechanism [1], hydrolysis of terminal, non-reducing
alpha-D-galactose residues in alpha-D-galactosides, including galactose
oligosaccharides, galactomannans and galactolipids)

Reaction type
O-Glycosyl bond hydrolysis

Natural substrates
Melibiose + H_2O [1]
Raffinose + H_2O [1]

Substrate spectrum
1 Alpha-D-galactoside (e.g. o-nitrophenyl-alpha-D-galactoside, ethyl
alpha-D-galactoside) + H_2O [1–62]
2 Alpha-D-fucosides (e.g. p-nitrophenyl-alpha-D-fucoside) + H_2O [1, 57]
3 Melibiose + H_2O [1, 15, 33, 53]
4 Raffinose + H_2O [1, 2, 4, 15, 20, 29, 33, 44, 53]
5 Stachyose + H_2O [2, 15, 20, 29, 33]

Enzyme Handbook © Springer-Verlag Berlin Heidelberg 1991
Duplication, reproduction and storage in data banks are only
allowed with the prior permission of the publishers

  6 Glycolipids + H_2O [40]
  7 Galabiosylceramide + H_2O [40]
  8 Globotriglycosylceramide + H_2O [40]
  9 Disaccharides (and derivatives) + H_2O (overview) [5]
10 Trisaccharides (reducing and nonreducing) + H_2O (overview) [5]
11 Oligosaccharides + H_2O (overview) [5]
12 Polysaccharides + H_2O (overview) [5]
13 More (galactosyl transferase activity [1, 5, 24, 37, 49, 55], alpha-
     fucosidase activity [33], hydrolysis of alpha-D-galactopyranosidic
     linkages in glycosides of methanol, phenol and p-nitrophenol [5], lectin
     activity [33], not: substances with beta-galactosidic linkages [2], not:
     arabinosides [57]) [1, 2, 5, 24, 33, 37, 49, 55]

Product spectrum

  1 Galactose + ? (e.g. p-nitrophenol + D-galactose)
  2 Fucose + ? (e.g. p-nitrophenol + D-fucose)
  3 D-Galactose + D-glucose
  4 Galactose + sucrose [2]
  5 Galactose [2]
  6 ?
  7 ?
  8 ?
  9 Monosaccharides
10 ?
11 ?
12 ?
13 ?

Inhibitor(s)

Sulfhydryl reagents (e.g. p-chloromercuribenzoate, N-ethylmaleimide, iodoacetamide, no inhibition of all alpha-galactosidases [1]) [1, 53]; Carbohydrates (inhibition of lectin activity) [23]; Ag^+ [1, 7, 15, 19, 24, 25, 29, 30, 33,
44, 53, 61]; Cu^{2+} [1, 15, 29]; Hg^{2+} [1, 7, 15, 19, 24, 25, 27, 29, 30, 33, 35, 44,
53, 61]; D-Galactose (no effect [13]) [1, 5, 7, 18, 20, 29, 30, 33, 36, 37, 38, 39,
44, 52]; L-Arabinose [1, 18, 52]; D-Fucose [1, 52]; D-Galactal [1, 19, 24];
Aldono-(1 ---> 5)-lactones [1]; N-(N-Benzoylcarbonyl-Epsilon-amino-
caproyl)alpha-D-galactopyranoside [3]; N-Epsilon-aminocaproyl-alpha-D-
galactopyranoside [3]; Melibiose [7, 18, 24, 33, 36, 39]; Myo-inositol [18, 36,
47, 54]; D-Glucose-6-phosphate [18]; Iodoacetic acid [18]; Galactonolactone [19]; Sucrose [20]; Fructose [20]; Ca^{2+} [29]; Mg^{2+} [29]; Mn^{2+} [29];
Ni^{2+} [29]; Zn^{2+} [29]; Co^{2+} [29]; Triethanolamine-hydrochloride-NaOH (10
mM) [29]; Hepes-NaOH (10 mM) [29]; L-Ascorbic acid [33]; Stachyose [33,
26]; Glycerol-3-phosphate [35]; Raffinose [36]; N-6-Aminohexanoyl-alpha-
D-galactosyl amine; Sugars [37, 57]; p-Nitrophenyl-alpha-D-galactoside
[38]; Tris [39]; Cellobiose [39]; Lactose [39]; Sodium taurocholate [40]

Cofactor(s)/prostethic group(s)
Mn^{2+} (E. coli: activity depends on) [1]; NAD^+ (E. coli: activity depends on) [1]

Metal compounds/salts
K^+ (Vicia faba: activates) [1]

Turnover number (min^{-1})

Specific activity (U/mg)
0.29 [5]; 24.5 [16]; 41 [21]; 89.15 [28]; More [6, 9, 10, 13, 14, 15, 18, 19, 20, 21, 22, 25, 27, 29 , 31, 33, 34, 35, 36, 37, 38, 40, 41, 44, 45, 47, 48, 53, 54, 56]

K_m-value (mM)
0.38 (p-nitrophenyl-alpha-D-galactoside) [1]; 4.76 (p-nitrophenyl-alpha-D-fucoside) [1]; 14.3 (p-nitrophenyl-beta-L-arabinoside) [1]; 7.13 (methyl alpha-D-galactoside) [1]; 8.93 (ethyl alpha-D-galactoside) [1]; 6.13 (n-propyl alpha-D-galactoside) [1]; 1.11 (phenyl alpha-D-galactoside) [1]; 0.38 (p-nitrophenyl alpha-D-galactoside) [1]; 0.96 (melibiose) [21]; 4.0 (raffinose) [1]; 2.0 (alpha-galactosidase I, alpha-phenylgalactoside) [5]; 1.0 (alpha-galactosidase II, alpha-phenylgalactoside) [5]; More (of immobilized enzyme [7, 8]) [1, 7, 8, 9, 13, 14, 15, 17, 18, 20, 21, 24 , 25, 27, 29, 30, 31, 33, 37, 38, 39, 40, 41, 42, 43, 45, 46, 47, 48, 50, 54, 55, 56, 57, 58, 59, 60, 61]

pH-optimum
5.3 (alpha-galactosidase I [5], Vigna unguiculata, II2 [9] , Coffea sp., phenyl alpha-D-galactoside [1], Plantago ovata, II, phenyl alpha-D-galactoside [1], Spinacia oleracea, p-nitrophenyl alpha-D-galactoside [1], Polyplastron multivesculatus, melibiose [1]) [1, 5, 9]; 6.3 (alpha-galactosidase II) [5]; 4.5–5.0 (guar [6]) [6, 7, 15]; 4.5–5.5 (Medicago sativa) [6]; 5.0 (Vigna unguiculata, I) [9]; 5.9 (Vigna unguiculata, II1) [9]; 6.1 (enzyme I) [17]; 4.7 (enzyme II) [17]; 7.5 [18]; 3.0–5.0 (Calvatia cyanthiformis, o-nitrophenyl alpha-D-galactoside, Mortiella vinacea, raffinose) [1]; 6.0 (Malolontha melolontha, phenyl alpha-D-galactoside [1], enzyme II [34] , alpha-galactosidase I [39]) [1, 34, 39]; 3.8–5.5 (Aspergillus niger, p-nitrophenyl alpha-D-galactoside) [1]; 4.2–4.8 (Aspergillus niger, melibiitol) [1]; 4.6 (Aspergillus paxillus) [1]; 4.0–5.0 (Canavalia ensiformis, p-nitrophenyl alpha-D-galactoside) [1]; 4.2 (Citrullus vulgaris, melibiose, raffinose) [1]; 4.0 (Candida javanica) [13]; 4.5 (natural substrates [19], enzyme I [34], raffinose [44]) [19, 33, 34, 44]; 5.9 (artificial substrates) [19]; 5.9–6.4 (galactosidase I) [21]; 6.3–6.5 (galactosidase II) [21]; 5.0 [24]; 4.8 [25]; 5.2 [37]; 7 (alpha-galactosidase II) [39]; 4.0–5.5 (p-nitrophenyl alpha-D-galactoside) [44]; 6.5 (Bacillus) [53]; 7.5 (Micrococcus) [53]; 5.5–5.7 [62]; 2.5–4.5 [61]; 6.8 [55]; 5.5–6.0 [58]; More [1, 14, 28, 29, 30, 40, 41, 42, 45, 46, 49, 59, 60]

Enzyme Handbook © Springer-Verlag Berlin Heidelberg 1991
Duplication, reproduction and storage in data banks are only
allowed with the prior permission of the publishers

pH-range

2–6 (2: about 70% of activity maximum, 6: about 25% of activity maximum) [13]; 3.5–6.5 (3.5, 6.5: about 10% of activity maximum) [33]; 2.5–7.0 (2.5: about 10% of activity maximum, 7.0: about 20% of activity maximum) [15]; 3–7 (3/7: about 20% of activity maximum) [37]; 1.1–2.0 (fairly active at pH 1.1–2.0) [61]; More [9, 20, 41, 53]

Temperature optimum (°C)

70 (above, 2-nitrophenyl-alpha-D-galactoside) [56]; 55 (stachyose, raffinose [56]) [15, 56]; 37 [55]; 60 [33]; 50 [44]; 35–40 [53]; 45 (guar) [6]; 75 (immobilized enzyme, Pycnoporas cinnabarinus) [7, 24]; 70 (Candida javanica) [13]

Temperature range (°C)

30–80 (30°C: about 10% of activity maximum, 80°C: about 15 % of activity maximum) [13]; 30–70 (30°C: about 30% of activity maximum, 70°C: about 10% of activity maximum) [15]; 25–50 (25°C: about 60% of activity maximum, 50°C: about 15% of activity maximum) [53]; 20–55 (pH 5.8, 20°C, 55°C: less than 10% of activity maximum) [20]; 40–90 (40 °C: about 10% of activity maximum, 90°C: about 20% of activity maximum) [24]; 35–75 (35°C: about 15% of activity maximum, 75°C: about 20% of activity maximum) [33]

---

3 ENZYME STRUCTURE

Molecular weight

210000 (Medicago sativa, enzyme II, SDS-PAGE) [6]
33000 (Medicago sativa, enzyme I, SDS-PAGE) [6]
111000 (Vigna unguiculata, I, gel filtration) [9]
29000 (Vigna unguiculata, II1, gel filtration) [9]
30000 (Vigna unguiculata, II2, gel filtration) [9]
150000 (Monascus pilosus, gel filtration) [15]
240000 (Cephalosporium acremonium, gel filtration) [10]
50000 (E. coli, SDS-PAGE) [11]
160000 (Lens culinaris, enzyme I, gel filtration) [17]
40000 (Lens culinaris, enzyme II, gel filtration) [17]
23000 (coconut, enzyme A, gel filtration) [18]
26000 (coconut, enzyme B, gel filtration) [18]
45000 (Citrullus battich, gel filtration, SDS-PAGE) [19]
250000 (Bacteroides ovatus, alpha-galactosidase I /II, gel filtration) [21]
210000 (Pycnoporus cinnabarinus, gel filtration) [24]
56000 (Aspergillus tamarii, SDS-PAGE) [25]
254000 (Aspergillus tamarii, disc gel electrophoresis) [26]
47000 (Saccharum officinarum, gel filtration) [33]
215000 (Castanea sativa, gel filtration, enzyme I) [34]
53000 (Castanea sativa, gel filtration, enzyme II) [34]

360000 (Poterioochromonas malhamensis, sucrose density-gradient centrifugation) [35]
101000 (human, gel filtration) [36]
42000 (Stachys affinis, gel filtration) [37]
280000 (Bacillus stearothermophilus, alpha-galactosidase I, disc gel electrophoresis) [39]
325000 (Bacillus stearothermophilus, alpha-galactosidase II, disc gel electrophoresis) [39]
57000 (Medicago sativa, gel filtration) [44]
103000 (human, alpha-galactosidase A, gel filtration) [48]
117000 (human, alpha-galactosidase B, gel filtration) [48]
More [1, 27, 30, 31, 38, 45, 47, 50, 54]

Subunits
Tetramer (Bacillus stearothermophilus, alpha-galactosidase I: 4 × 81000, alpha-galactosidase II: 4 × 84000 [39], Pycoporus cinnabarinus, 4 × 52000, SDS-PAGE [24] , Vigna unguiculata, enzyme I is a tetramer of enzyme II2, 4 × 31500, SDS-PAGE [9]) [9, 24, 39]
Trimer (Bacteroides ovatus, alpha-galactosidase I: 3 × 85000, alpha-galactosidase II: 3 × 80500, SDS-PAGE) [21]
Dimer (2 × 49800, human, SDS-PAGE [36], 2 × 45700, human, alpha-galactosidase A, SDS-PAGE, 2 × 47700, human, alpha-galactosidase B, SDS-PAGE [48]) [36, 48]
Monomer (Trifolium repens, SDS-PAGE, 1 × 41000) [49]
More (monomer-dimer-tetramer transition) [54]

Glycoprotein/Lipoprotein
Glycoprotein (Lens culinaris [17], Mortiella vinacea [1], Cephalosporium acremonium [10], coconut: 12% carbohydrate [16], Aspergillus tamarii [25, 26], Saccharomyces carlsbergensis [46], human [47], Aspergillus niger [50]) [1, 10, 16, 17, 25, 26, 46, 47, 50]

---

4 ISOLATION/PREPARATION

Source organism
Guar (Cyamopsis tetragonolobus) [6]; Medicago sativa [6, 44]; Pycnoporus cinnabarinus (immobilized enzyme [7]) [7, 22, 24]; Vigna unguiculata (3 forms: I, II1, II2) [9]; Fungi imperfecti (various strains) [10]; Cephalosporium acremonium [10]; E. coli [11, 55]; Candida javanica [13]; Lens culinaris (enzyme I, enzyme II) [14, 17]; Monascus pilosus [15]; Coconut [16, 18]; Citrullus battich [19]; Lactobacillus fermenti [20]; Bacteroides ovatus (alpha-galactosidase I and II) [21]; Vicia faba (galactosidase II2) [23]; Aspergillus tamarii (2 forms [26]) [25, 26]; Lycopersicon esculentum (alpha-galactosidase I and II) [27]; Vigna radiata [28]; Cucurbita pepo (LIV) [29]; Lupinus angustifolius [30]; Glycine max [31]; Human (normal and patients

Enzyme Handbook © Springer-Verlag Berlin Heidelberg 1991
Duplication, reproduction and storage in data banks are only
allowed with the prior permission of the publishers

with Fabry disease [43]) [32, 36, 40, 41, 42, 43, 47, 48, 60]; Saccharum officinarum [33]; Castanea sativa [34]; Poteriochroomonas malhamensis [35]; Stachys affinis [37]; Cucumis sativus [38]; Bacillus stearothermophilus (alpha-galactosidase I and II) [39]; Trifolium repens [45, 49]; Saccharomyces carlsbergensis [46]; Aspergillus niger [50]; Dictyostelium discoideum [51]; Streptomyces [52, 57]; Micrococcus sp. No. 31–2 (alkalophilic) [53]; Bacillus sp. No. 7–5 (alkalophilic) [53]; Mouse [54]; Penicillium duponti [56]; Lens esculanta [58]; Cajanus indicus (2 enzymes: I and II) [59]; Corticium rolfsii [61]; Sweet almond [62]; Coffea sp. (2 alpha-galactosidases: I and II [5]) [3, 5, 8]; Microorganisms [1, 4]; Plants [1]; Animals [1]; More (occurs widly in microorganisms, plants and animals [1], multimolecular forms [1]) [1, 4]

Source tissue

Parathyroid gland [1]; Intestine [1]; Kidney [1]; Thyroid gland [1]; Seeds [1, 6, 14, 17, 23, 30, 31, 34, 44, 45, 49]; Mycelium [4, 26]; Beans [5, 8]; Cotyledons [14]; Endosperm [16, 18]; Fruit [19, 27]; Leaves [29, 38]; Stalks [33]; Spleen [36]; Placenta [36, 48, 60]; Plasma [36, 47]; Tuber [37]; Liver [40, 42, 54]; Leukocytes [43]; Culture medium [4, 13, 25, 50]

Localisation in source

Soluble [1]; Extracellular [4, 25, 51]; Intracellular [15]; Lysosomes [54]; Microsomes [54]

Purification

Coffea sp. (affinity chromatography [3], 2 alpha-galactosidases: I and II [5], partial [5], immobilized enzyme [8]) [3, 5, 8]; Guar (Cyamopsis tetragonolobus) [6]; Medicago sativa [6, 44]; Vigna unguiculata [9]; Cephalosporium acremonium [10]; E. coli [11, 55]; Candida javanica [13]; Lens culinaris (partial [14]) [14, 17]; Monascus pilosus [15]; Coconut [16, 18]; Citrullus battich [19]; Lactobacillus fermenti [20]; Bacteroides ovatus (alpha-galactosidase I and II) [21]; Pycnoporus cinnabarinus [22, 24]; Vicia faba [23]; Aspergillus tamarii [25, 26]; Lycopersicon esculentum (alpha-galactosidase I and II) [27]; Vigna radiata [28]; Cucurbita pepo (LIV) [29]; Lupinus angustifolius [30]; Glycine max [31]; Saccharum officinarum [33]; Poterioochromonas malhamensis [35]; Human [36, 40, 41, 42, 47, 48, 60]; Stachys affinis [37]; Cucumis sativus (partial) [38]; Bacillus stearothermophilus (alpha-galactosidase I and II) [39]; Trifolium repens [45]; Aspergillus niger [50]; Streptomyces [52]; Micrococcus sp. No. 31–2 (partial) [53]; Bacillus sp. No. 7–5 (partial) [53]; Mouse [54]; Penicillium duponti [56]; Lens esculanta [58]; Cajanus indicus [59]; Corticium rolfsii [61]; Sweet almond [62]

Crystallization

(Mortiella vinacea [1], Pycnoporus cinnabarinus [22]) [1, 22]

Cloned
  (E. coli) [12]

Renaturated
  –

---

5 STABILITY

pH
  3–9 (stable [7, 19], Pycoporus cinnabarinus, immobilized enzyme [7]) [7, 19]; 3.0–7.2 (stable) [19]; 6.0–8.5 (alkalophilic Bacillus) [53]; 7.5–8.5 (alkalophilic Micrococcus) [53]; More [9, 24, 46]

Temperature (°C)
  75 (Vicia sativa, 40 minutes, 10% loss of activity) [1]; 60 (bovine: 5 minutes, complete loss of activity, Prunus amygdalus, 20 minutes, 15% loss of activity, Vicia faba: 30 minutes, 42% loss of activity) [1]; 50 (guar, unstable above) [6]; 45 (Medicago, unstable above) [6]; 80 (Pycnoporus cinnabarinus, immobilized enzyme, stable below, 15 minutes, pH 5.0) [7]; 70 (15 minutes, 30% loss of activity) [13]; 80 (15 minutes, complete loss of activity) [13]; 37 (45–50 hours, 50% loss of activity) [21]; 65 (half-life: 2 hours, alpha-galactosidase I, 3 minutes, alpha-galactosidase II) [39]; 35 (unstable above) [53]; 55 (Aspergillus niger: 60 minutes, 35% loss of activity [1], bovine: 5 minutes, 20% loss of activity [1], Streptomyces olivaceus, 15 minutes, 90% loss of activity [1], Monascus pilosus, stable below [15]) [1, 15]; More (human: thermolabile alpha-galactosidase A and thermostable galactosidase B) [9, 25, 37, 43, 45, 59, 60]

Oxidation

Organic solvent

General stability information
  Mn^{2+} (stabilizes during enzyme assay) [1]; Reducing agents (stabilize during enzyme assay) [1]; Thawing (loss of activity) [5]; Immobilization [7, 8]; Lyophilized (–20 °C, enzyme I stable enzyme, II not stable) [21]; Freeze-drying (60 % loss of activity [25], stable [35]) [25, 35]; Freezing and thawing (stable) [35]; Dilute solutions (very unstable even at 0–5°C) [55]

Storage
  4°C, 30 days (36% loss of activity) [20]; –20°C, 20% glycerol (indefinitely stable) [20]; –20°C, pH 6.5, 6 months [22]; 4°C, several months [25]; –20°C, 30% glycerol, 10% sucrose, 3 months [35]; –20°C, 0.02 M sodium acetate buffer, pH 5.2 [54]

Enzyme Handbook © Springer-Verlag Berlin Heidelberg 1991
Duplication, reproduction and storage in data banks are only
allowed with the prior permission of the publishers

6 CROSSREFERENCES TO STRUCTURE DATABANKS

PIR/MIPS code

GBHUA (precursor, human fragment); GBBYAG (precursor, yeast, Saccharomyces cerevisiae); GBECAG (Escherichia coli); B35160 (Escherichia coli, fragment); S04081 (A, precursor, human); A30214 (A, precursor, human, fragment); A29608 (A, precursor, human, fragment)

Brookhaven code

7 LITERATURE REFERENCES

[1] Dey, P.M., Pridham, J.B.: Adv. Enzymol. Relat. Areas Mol. Biol., 36, 91–130 (1972) (Review)

[2] Beutler, H.-O. in "Methods Enzym. Anal.", 3rd. Ed. (Bergmeyer, H.U., Ed.) 6, 90–96 (1984) (Review)

[3] Harpaz, N., Flowers, H.M.: Methods Enzymol., 34, 347–350 (1974) (Review)

[4] Crueger, A., Crueger, W. in "Biotechnology" (Kieslich, K., Ed.) 6a, 421–457, Verlag Chemie, Weinheim (1984) (Review)

[5] Courtois, J.E., Petek, F.: Methods Enzymol., 8, 565–571 (1966) (Review)

[6] McCleary, B.V.: Methods Enzymol., 160, 627–632 (1988) (Review)

[7] Mitsutomi, M., Uchida, Y., Ohtakara, A.: J. Ferment. Technol., 63, 325–329 (1985)

[8] Kuo, J.-Y., Goldstein, J.: Enzyme Microb. Technol., 5, 286–290 (1983)

[9] Alani, S.R., Markakis, P.: Phytochemistry, 28, 2047–2051 (1989)

[10] Zaprometova, O.M., Ulezlo, I.V.: Biotechnol. Appl. Biochem., 10, 232–241 (1988)

[11] Nagao, Y., Nakada, T., Imoto, M., Shimamoto, T., Sakai, S., Tsuda, M., Tsuchiya, T.: Biochem. Biophys. Res. Commun., 151, 236–241 (1988)

[12] Hanatani, M., Yazu, H., Shiota-Niiya, S., Moriyama, Y., Kanazawa, H., Futai, M., Tsuchiya, T.: J. Biol. Chem., 259, 1807–1812 (1984)

[13] Cavazzoni, V., Adami, A., Craveri, R.: Appl. Microbiol. Biotechnol., 26, 555–559 (1987)

[14] Corchete, M.P., Guerra, H.: Phytochemistry, 26, 927–932 (1987)

[15] Wong, H.-C., Hu, C.-A., Yeh, H.-L., Su, W., Lu, H.-C., Lin, C.-F.: Appl. Environ. Microbiol., 52, 1147–1152 (1986)

[16] Balasubramaniam, K., Mathew, C.D.: Phytochemistry, 25, 1819–1821 (1986)

[17] Dey, P.M., Del Campillo, E.M., Lezica, R.P.: J. Biol. Chem., 258, 923–929 (1983)

[18] Mujer, C.V., Ramirez, D.A., Mendoza, E.M.T.: Phytochemistry, 23, 1251–1254 (1984)

[19] Itoh, T., Uda, Y., Nakagawa, H.: J. Biochem., 99, 243–250 (1986)

[20] Schuler, R., Mudgett, R.E., Mahoney, R.R.: Enzyme Microb. Technol., 7, 207–211 (1985)

[21] Gherardini, F., Babcock, M., Salyers, A.A.: J. Bacteriol., 161, 500–506 (1985)

[22] Mitsutomi, M., Ohtakara, A.: Agric. Biol. Chem., 48, 3153–3155 (1984)

[23] Dey, P.M., Naik, S., Pridham, J.B.: Planta, 167, 114–118 (1986)

[24] Ohtakara, A., Mitsutomi, M., Uchida, Y.: Agric. Biol. Chem., 48, 1319–1327 (1984)

[25] Civas, A., Eberhard, R., Le Dizet, P., Petek, F.: Biochem. J., 219, 857–863 (1984)

[26] Civas, A., Eberhard, R., Le Dizet, P., Petek, F.: Biochem. J., 219, 849–855 (1984)

[27] Pressey, R.: Phytochemistry, 23, 55–58 (1984)

[28] Dey, P.M.: Eur. J. Biochem., 140, 385–390 (1984)

[29] Gaudreault, P.-R., Webb, J.A.: Plant Physiol., 71, 662–668 (1983)
[30] Plant, A.R., Moore, K.G.: Phytochemistry, 21, 985–989 (1982)
[31] Del Campillo, E., Shannon, L.M.: Plant Physiol., 69, 628–631 (1982)
[32] Schram, A.W., Tager, J.M.: Trends Biochem. Sci., December 1981, 6, 328–329 (1981) (Review)
[33] Chinen, I., Nakamura, T., Fukuda, N.: J. Biochem., 90, 1453–1461 (1981)
[34] Dey, P.M.: Phytochemistry, 20, 1493–1496 (1981)
[35] Dey, P.M., Kauss, H.: Phytochemistry, 20, 45–48 (1981)
[36] Bishop, D.F., Desnick, R.J.: J. Biol. Chem., 256, 1307–1316 (1981)
[37] Ueno, Y., Ikami, T., Yamauchi, R., Kato, K.: Agric. Biol. Chem., 44, 2623–2629 (1980)
[38] Smart, E.L., Pharr, D.M.: Plant Physiol., 66, 731–734 (1980)
[39] Pederson, D.M., Goodman, R.E.: Can. J. Microbiol., 26, 978–984 (1980)
[40] Dean, K.J., Sweeley, C.C.: J. Biol. Chem., 254, 9994–10000 (1979)
[41] Dean, K.J., Sweeley, C.C.: J. Biol. Chem., 254, 10001–10005 (1979)
[42] Dean, K.J., Sweeley, C.C.: J. Biol. Chem., 254, 10006–10010 (1979)
[43] Salvayre, R., Maret, A., Negre, A., Douste-Blazy, L.: Eur. J. Biochem., 100, 377–383 (1979)
[44] Itoh, T., Shimura, S., Adachi, S.: Agric. Biol. Chem., 43, 1499–1504 (1979)
[45] Williams, J., Villarroya, H., Petek, F.: Biochem. J., 175, 1069–1077 (1978)
[46] Lazo, P.S., Ochoa, A.G., Gascon, S.: Arch. Biochem. Biophys., 191, 316–324 (1978)
[47] Bishop, D.F., Sweeley, C.C.: Biochim. Biophys. Acta, 525, 399–409 (1978)
[48] Kusiak, J.W., Quirk, J.M., Brady, R.O.: J. Biol. Chem., 253, 184–190 (1978)
[49] Williams, J., Villarroya, H., Petek, F.: Biochem. J., 161, 509–515 (1977)
[50] Adya, S., Elbein, A.D.: J. Bacteriol., 129, 850–856 (1977)
[51] Kilpatrick, D.C., Stirling, J.L.: Biochem. J., 158, 409–417 (1976)
[52] Oishi, K., Aida, K.: Agric. Biol. Chem., 39, 2129–2135 (1975)
[53] Akiba, T., Horikoshi, K.: Agric. Biol. Chem., 40, 1851–1855 (1976)
[54] Lusis, A.J., Paigen, K.: Biochim. Biophys. Acta, 437, 487–497 (1976)
[55] Kawamura, S., Kasai, T., Tanusi, S.: Agric. Biol. Chem., 40, 641–648 (1976)
[56] Arnaud, N., Bush, D.A., Horisberger, M.: Biotechnol. Bioeng., 18, 581–585 (1976)
[57] Oishi, K., Aida, K.: Agric. Biol. Chem., 40, 57–65 (1976)
[58] Dey, P.M., Wallenfels, K.: Eur. J. Biochem., 50, 107–112 (1974)
[59] Dey, P.M., Dixon, M.: Biochim. Biophys. Acta, 370, 269–275 (1974)
[60] Beutler, E., Kuhl, W.: J. Biol. Chem., 247, 7195–7200 (1972)
[61] Kaji, A., Yoshihara, O.: Agric. Biol. Chem., 36, 1335–1342 (1972)
[62] Malhotra, O.P., Dey, P.M.: Biochem. J., 103, 508–513 (1967)

Enzyme Handbook © Springer-Verlag Berlin Heidelberg 1991
Duplication, reproduction and storage in data banks are only
allowed with the prior permission of the publishers

1 NOMENCLATURE

EC number
3.2.1.23

Systematic name
Beta-D-galactoside galactohydrolase

Recommended name
Beta-galactosidase

Synonymes
Lactase
Galactosidase, .beta.
.beta.-Galactosidase
. beta-Lactosidase
Maxilact
Hydrolact
.beta.-D-Lactosidase
S 2107
Lactozym
.beta.-D-Galactoside galactohydrolase
Trilactase
.beta.-D-Galactanase
Oryzatym
Sumiklat
p-Nitrophenyl.beta.-galactosidase
Galactosyl
Lactosylceramidase II
Beta-galase [18]

CAS Reg. No.
9031-11-2

2 REACTION AND SPECIFICITY

Catalysed reaction
A beta-D-galactoside + H_2O →
→ hydrolyzed beta-D-galactoside + beta-D-galactose (mechanism [1, 3],
hydrolysis of terminal non-reducing beta-D-galactose residues in beta-D-
galactosides) [1, 3]

Enzyme Handbook © Springer-Verlag Berlin Heidelberg 1991
Duplication, reproduction and storage in data banks are only
allowed with the prior permission of the publishers

Reaction type
O-Glycosyl bond hydrolysis
Cleavage of : C-O, C-S, C-F and C-N bonds [1]

Natural substrates
Lactose + H_2O (intestine: absorption of dietary lactose) [1]
Delta-D-galactosides + H_2O (key enzyme in degradation of glycolipids, mucopolysaccharides and glycoproteins, plant enzyme related to catabolism of galactolipids) [1]

Substrate spectrum
1 Lactose + H_2O [36]
2 Beta-1, 3-linked galactobiose + H_2O [18]
3 Galactooligosaccharides + H_2O (beta-1,3-linked, beta-1,6-linked, overview) [18]
4 o-Nitrophenyl-beta-galactoside + H_2O (low activity [14, 15]) [3, 19, 24, 27, 33]
5 Cellobiose + H_2O [24, 25, 27]
6 Gentiobiose + H_2O [25]
7 p-Nitrophenyl-beta-D-galactoside + H_2O [27]
8 p-Nitrophenyl-beta-D-glucoside + H_2O [27]
9 o-Nitrophenyl-beta-D-fucoside + H_2O [27]
10 Monogalactosyldiacylglycerol + H_2O [30]
11 O-Beta-D-galactosyl-1, 3-D-arabinoside + H_2O [36]
12 Lactosyl ceramide + H_2O [49]
13 4-Methylumbelliferyl-beta-D-galactoside + H_2O [56]
14 p-Nitrophenyl-beta-D-fucoside + H_2O [56]
15 More (specificity to sugar moiety and to anomeric character of the linkage but not to a particular aglycon [1], D-pyranose ring essential [1], substitution at C-2, C-3, C-4 and C-6 by groups that are bulkier than the hydroxyl group prevents binding to active site [1], cleavage of : C-O, C-S, C-F and C-N bonds [1], aglycon: another sugar residue, alkyl or aryl group [1], transfer-reaction of free galactose and alcohols [1], not: alkyl beta-D-thiogalactosides [1], not: phenyl beta-D-thiogalactosides [1], not: p-beta-galactosidase activity [34], acts on galactan from citrus pectic polysaccharides in exo-fashion [28], polysaccharides: slight [36], synthesis of galactosyl-N-acetylgalactosamine with immobilized enzyme [1, 2], synthesis of (2R)-glycerol-o-beta-D-galactopyranoside [16], synthesis of galactosyl-N-acetylgalactosamine with immobilized enzyme [12], some enzymes in this group: alpha-L-arabinosides, some animal enzymes: beta-D-fucosides and beta-D-glucosides, cf. E.C. 3.2.1.108) [1, 3, 4, 18, 25, 26, 27, 30, 33, 36, 49, 51, 52, 56, 58]

Product spectrum
1 Glucose + galactose
2 Galactose
3 More (enzyme splits off beta-1, 3-and beta-1, 6-linked D-galactosyl residues from the nonreducing ends, rate increases with increasing chain lengths) [18]
4 Galactose + o-nitrophenol
5 Glucose
6 Glucose
7 Galactose + p-nitrophenol
8 Glucose + p-nitrophenol
9 Fucose + p-nitrophenol
10 Galactose + diacylglycerol
11 ?
12 ?
13 Galactose + 4-methylumbelliferone
14 Fucose + p-nitrophenol
15 ?

Inhibitor(s)
Sulfhydryl inhibitors [31, 39]; Urea (irreversible inactivation [3, 42], uneffected by short incubation with 2 M urea [43]) [3, 42]; Alkali ions (at very high concentrations [1]) [1, 19]; Metal chelators (which bind Mg^{2+} and Ca^{2+}) [1]; Alcohols (various, e.g. 2-mercaptoethanol, n-propanol) [1]; Amines (various, e.g. ethanolamine) [1]; Citrate [1]; EDTA (not [24, 55], activates [31]) [1, 48, 52]; 2-Mercaptoethanol [1]; n-Propanol [1]; Ethylenediamine [1]; Mercaptoethylamine [1]; N-Bromoacetylgalactosylamine [5]; D-Galactal [14, 18, 48, 51]; Hg^{2+} [18, 23, 25, 33, 35, 36, 39, 44, 52, 54, 55, 60, 63]; p-Chloromercuribenzoate (no effect [55]) [18, 19, 23, 24, 35, 36, 57, 63]; Transition metal ions [19, 31]; Ca^{2+} [19, 34, 39]; Mg^{2+} [19, 39]; Zn^{2+} [19, 35, 36, 39, 52]; Fe^{2+} [19, 25, 36, 52, 54]; Co^{2+} [19]; Ni^{2+} [19, 25]; Cu^{2+} [19, 23, 25, 33, 35, 36, 41, 52, 54, 55, 60]; Fe^{3+} [19]; Phenylmethylsulfonyl fluoride [19]; Ag^+ [23]; Monoiodoacetic acid [23]; Trisaminomethane [23]; Glucono-delta-lactone [24, 25]; Glucosamine [24, 25]; Monosaccharides (e.g. glucose [32, 40]) [31, 32, 40]; p-Hydroxymercuribenzoate [24, 25, 29, 51]; Ba^{2+} [25]; p-Nitrophenyl-beta-D-thiogalactopyranoside [29]; D-Galactono-1, 4-lactone (chloroplast enzyme insensitive vacuolar enzyme inhibited [30]) [18, 30, 57]; Cd^{2+} [35]; Pb^{2+} [35, 52, 54]; Galactose-6-phosphate [32]; Lactate [32, 38]; p-Aminophenyl beta-D-thiogalactopyranoside (not [55]) [33]; Thiol group reagents [48]; Sphingosine [49]; Fatty acids [49]; Ceramide [49]; Gamma-galactonolactone [49]; Methyl beta-D-galactopyranoside [51]; Ag^{2+} [54]; Phenyl-beta-D-thiogalactoside [57]; N-Bromosuccinimide [60]; Sodium laurylsulfate [60]; Be^{2+} [3]; Cysteine [3, 34]; Glutathione [3, 34]; Thiogalactoside (competitive) [3]; Disaccharides (e.g. galactose (poor

Enzyme Handbook © Springer-Verlag Berlin Heidelberg 1991
Duplication, reproduction and storage in data banks are only
allowed with the prior permission of the publishers

inhibition [29], chloroplast enzyme insensitve, vascular enzyme inhibited
[30]) [29, 30, 32, 38, 40, 41, 48, 49, 51, 54, 59]), melibiose [54]) [29, 30, 32, 38,
40, 41, 48, 49, 51, 54, 59];
More (strong product inhibition [31], no substrate inhibition [55], no
product inhibition [55]) [26, 31, 55]

Cofactor(s)/prostethic group(s)
More (Kluyveromyces fragilis: Mn^{2+}, K^+, Kluyveromyces lactis: Mn^{2+}, Na^+,
E. coli: Na^+, K^+) [6]

Metal compounds/salts
Mn^{2+} (activates) [1, 2, 39, 44, 52, 63]; Rb^+ (activates) [1]; Cs^+ (activates)
[1]; Na^+ (synergistic activation with either Mg^{2+} and K^+ or Mg^{2+} and Na^+
[26], activates [19]) [19, 26]; K^+ (synergistic activation with either Mg^{2+} and
K^+ or Mg^{2+} and Na^+ [26], activates [63]) [19, 63]; Li^+ (activates) [19]; Ca^{2+}
(activates) [52]; Mg^{2+} (synergistic activation with either Mg^{2+} and K^+ or
Mg^{2+} and Na^+ [26], activates [34, 52, 63], Mg^{2+} and reducing agent required for activity [29]) [26, 29, 34, 52, 63]; Alkali ions (activate) [1, 19]; KCl
(activates lactose hydrolysis) [1]; Co^{2+} (activates) [63]; More (no cation required [40], divalent metals not necessary for enzymatic activity [1]) [1, 40]

Turnover number (min⁻¹)

Turnover number (min^{-1})
358000 (o-nitrophenyl beta-D-galactopyranoside) [1]; 25860 (o-nitrophenyl-
beta-D-galactopyranoside) [19]; 720000 [5]; 53280 (o-nitrophenyl-beta-D-
galactopyranoside) [20]

Specific activity (U/mg)
5.62 [18]; 99 [19]; 279.2 [23]; 466 [26]; 1126 [35]; 2382 [52]; More [19, 25, 28,
29, 31, 32, 34, 36, 41, 43, 46, 48, 49, 50, 51, 55, 57, 59, 63]

K_m-value (mM)
0.95 (o-nitrophenyl-beta-D-galactoside, without NaCl) [3]; 0.161
(o-nitrophenyl-beta-D-galactoside, with NaCl) [3]; 0.445
(p-nitrophenyl-beta-D-galactoside, without NaCl) [3]; 0.05
(p-nitrophenyl-beta-D-galactoside, with NaCl) [3]; 2.5 (methylsalicylate, with
NaCl) [3]; 3.23 (phenyl-beta-D-galactoside, without NaCl) [3]; 7.52 (al-
pha-lactose, without NaCl) [3]; 8.32 (allo-lactose, without NaCl) [3]; 0.120
(thio-(o-nitrophenyl)-beta-D-galactoside) [3]; 85 (lactose, Aspergillus niger)
[6]; 50 (lactose, Aspergillus oryzae) [6]; 7.79 (beta-1, 3-galactobiose) [18];
4.87 (beta-1, 3-galactotriose) [18]; 4.3
(o-nitrophenyl-beta-D-galactopyranoside) [23]; 2.0
(p-nitrophenyl-beta-galactoside) [23]; 30 (lactose) [23]; More (K_m of
immobilized and free enzyme is similar [9], influence of pH and temperature
[1]) [1, 3, 6, 8, 9, 18, 20, 21, 24, 25, 28, 29, 31, 32, 33, 34, 35, 36, 38, 39, 40, 41,
42, 43, 44, 47, 48, 49, 50, 51, 52, 53, 54, 55, 57, 60, 62]

pH-optimum

7.5 (Neurospora crassa, 2 enzymes: one with optimum 4.2, another with 7.5)
[50]; 6. 0–6.5 [53]; 6.5 [54]; 4.5 (o-nitrophenyl beta-D-galactopyranoside
[60]) [55, 58, 60]; 3.5 [57]; 6.0–6.4 [59]; 4.8 (lactose) [60]; 7.3 (40°C, veronal
buffer) [2]; 7.7 (20°C, 0.05 M, Tris-acetic acid buffer) [3]; 6.6 (20°C, 0. 05 M,
Tris-acetic acid buffer, NaCl) [3]; 3.0–4.0 (Aspergillus niger) [6]; 5.0
(Aspergillus oryzae) [6]; 6.6 (Kluyveromyces fragilis) [6]; 6.5–7.0
(Klyuveromyces lactis) [6]; 7.2 (E. coli) [6]; 3.5–4 (Aspergillus niger) [10];
4.5–5 (Aspergillus oryzae, soluble and immobilized) [10]; 3–3.5 (Aspergillus
niger, immobilized) [10]; 6.5 [32]; 4.0 (Raphanus sativus, p-nitrophenyl
beta-D-galactoside, beta-1, 3-linked galactobiose) [18] [18, 33, 62]; 6.0
(Thermoaerobacter, o-nitrophenyl-beta-D-galactopyranoside [19], Thermus
[27], human [31]) [19, 27, 31]; 7.0 (Saccharomyces lactis [20], Corynebacterium murisepticum [21]) [20, 21, 34]; 6.5 (Bacillus macerans) [23]; 3.5
(Paecilomyces varioti) [24, 25]; 3.4 (Machantia polymorpha, McIlvaine buffer) [28]; 2.6 (Machantia, glycine-HCl buffer) [28]; 6. 8 [29]; 4.8 (chloroplast,
Triticum) [30]; 3.5–4.3 (vacuole, Triticum) [30]; 6.5 (Treponema phagedenis)
[35]; 6.4 [36]; 7 (intracellular enzyme, Medicago sativa) [38]; 4 (cell-wall enzyme, Medicago sativa) [38]; 7.0–7.5 [39]; 2. 4 [41]; 4.3 [42, 51]; 5.0 [43]; 4.4
[44]; 3.6–5.5 [46]; 3.0 [47]; 5.5–5.8 [48]; 4.2 (Neurospora crassa, 2 enzymes:
one with optimum 4.2, another with 7.5) [50]

pH-range

5.0–8.0 (5.0, 8.0: about 30% of activity maximum) [19]; 4. 0–8.9 (4.0, 8.0:
about 25% of activity maximum) [27]; 5.4–7.8 (5.4: 7.4% of activity maximum, 7.8: 16.2% of activity maximum) [29]; More [17, 20, 36, 41, 59]

Temperature optimum (°C)

78 (enzyme from thermophilic anaerobe strain, cloned in E. coli) [13]; 70
(Bacillus acidocaldarius 11–10, soluble enzyme) [17]; 75 (Bacillus
acidocaldarius 11–10, immobilized enzyme) [17]; 65 (Bacilllus
acidocaldarius HM-1 [17]) [17, 59]; 45 [29, 34, 54]; 42 (Medicago sativa,
cell-wall [38] [32, 38]; 60 [33]; 40 [36]; 46 (Medicago sativa, intracellular
[38]) [38, 60]; 58 (Medicago sativa, extracellular) [38]; 30 [39]; 55 [40, 42]; 80
[43]; 65 [46]; 37 [48]; 55 (Aspergillus niger, Aspergillus oryzae) [6]; 60
(Bacillus circulans, lactose) [8]; 30–35 (Kluyveromyces fragilis,
Kluyveromyces lactis) [6]; 55–60 (Aspergillus niger) [10]; 35 (E. coli) [6];
50–55 (Aspergillus oryzae) [10]; 50 (Paecilomyces varioti) [24, 25]

Temperature range (°C)

15–50 (15°C, 50°C: about 20% of activity maximum) [29]; 35–80 (35°C:
about 40% of activity maximum, 80°C: about 20% of activity maximum)
[46]; More [17, 36]

Enzyme Handbook © Springer-Verlag Berlin Heidelberg 1991
Duplication, reproduction and storage in data banks are only
allowed with the prior permission of the publishers

3 ENZYME STRUCTURE

Molecular weight

630000 (Saccharomyces lactis, 4 enzyme forms, gel filtration, 630000, 55000, 41000 and 19000) [20]

380000 (human, gel filtration) [27]

580000 (Treponema phagedenis, gel filtration) [35]

450000 (Actinomyces viscosus, molecular exclusion chromatography) [53]

430000 (Bacillus coagulans, gel filtration) [40]

250000 (Lactobacillus helveticus, gel filtration) [32]

260000 (Medicago sativa, intracellular, gel filtration) [38]

320000 (Bacillus macerans, gel filtration) [23]

215000 (Bacillus stearothermophilus, disc gel electrophoresis) [59]

203000 (Saccharomyces fragilis, sedimentation equilibrium) [63]

200000–233000 (yeast, gel filtration) [39]

185000 (Bacillus C-125, method of Andrews) [54]

126000 (Penicillium multicolor, sedimentation equilibrium) [33]

154000 (Thermoaerobacter, gel permeation) [19]

170000 (Lactobacillus murinus, gel filtration) [34]

125000 (Medicago sativa, cell-wall, gel filtration) [38]

124000–173000 (Aspergillus niger, gel filtration, 3 multiple forms, differ in carbohydrate content) [47]

122000 (Rhizobium trifolii, gel filtration) [36]

110000 (Pycnoporus cinnabarinus, gel filtration) [41]

105000 (Aspergillus oryzae, gel filtration) [60]

104000 (Daucus carota, gel filtration) [22]

100000 (Penicillium citrinum, gel filtration) [55]

95000 (Scopulariopsis sp., gel filtration) [46]

94000 (Paecilomyces varioti, gel filtration, polyacrylamide gel electrophoresis) [24, 25]

68000 (bovine, gel filtration) [51]

62000 (Marchantia polymorpha, gel filtration) [28]

55000 (Saccharomyces lactis, 4 enzyme forms, gel filtration, 630000, 55000, 41000 and 19000) [20]

50000 (Medicago sativa, extracelllular, gel filtration) [38]

41000 (Saccharomyces lactis, 4 enzyme forms, gel filtration, 630000, 55000, 41000 and 19000) [20]

19000 (Saccharomyces lactis, 4 enzyme forms, gel filtration, 630000, 55000, 41000 and 19000) [20]

More [1, 4, 45, 50, 55]

Subunits

Hexamer (6 × 90000, Treponema phagedenis, SDS-PAGE) [35]
Tetramer (4 × 135000, E. coli, 6 M guanidine hydrochloride, low speed sedimentation (without reaching equilibrium) [1], 4 × 78000, Bacillus macerans, SDS-PAGE [23], 4 × 65000, Lactobacillus helveticus [32], E. coli [61]) [1, 23, 32, 61]
Dimer (2 × 100000, Corynebacterium murisepticum, SDS-PAGE [21], 2 × 50000, Daucus carota, SDS-PAGE [22], 2 × 60000, Penicillium citrinum, SDS-PAGE [55]) [21, 22, 55]
Monomer (1 × 45000, Raphanus sativus, SDS-PAGE) [18]
More [1]

Glycoprotein/Lipoprotein

Glycoprotein (Aspergillus niger: form 1 /13.5%, form II /19.2%, form III /28.5% [47], Neurospora crassa [50], bovine [51], Penicillium citrinum: 6% carbohydrate [55], animal enzymes: large amounts of carbohydrate [1], Saccharomyces lactis [20], Paecilomyces varioti [24, 25], Aspergillus niger [47], Penicillium multicolor: 15% mannose [33]) [1, 20, 24, 25, 47, 33, 47, 50, 51, 55]

---

4 ISOLATION/PREPARATION

Source organism

More (universal occurence, overview) [1]; Aspergillus niger (immobilized [10]) [6, 10]; Aspergillus oryzae [6, 10, 58, 60]; E. coli [4]; Bacillus megaterium [4]; Bacillus circulans (immobilized) [8]; Kluyveromyces fragilis [15]; Bacillus acidocaldarius (strains: HM-1 and 11–10) [17]; Raphanus sativus [18]; Thermoaerobacter sp. [19]; Saccharomyces lactis [20]; Corynebacterium murisepticum (inducible) [21]; Daucus carota [22]; Bacillus macerans [23]; Paecilomyces varioti [24, 25]; Streptococcus thermophilus [26]; Human [27]; Marchantia polymorpha [28]; Bacteroides polypragmatus [29]; Triticum aestivum [30]; Thermus [31]; Lactobacillus helveticus [32]; Penicillium multicolor [33]; Lactobacillus murinus [34]; Treponema phagedenis [35]; Rhizobium trifolii [36]; Bovine [37, 51]; Medicago sativa [38]; Torulopsis versatilis [39]; Torulopsis sphaerica [39]; Candida pseudotropicalis [39]; Bacillus coagulans [40]; Pycnoporus cinnabarinus [41]; Petunia hybrida (multiple forms) [42]; Caldariella acidophila [43]; Sugar cane [44]; Bacillus (various species) [43]; Scopulariopsis sp. [46]; Aspergillus niger (3 multiple forms) [47]; Xanthomonas campestris [48]; Rat [49, 57]; Neurospora crassa [50, 62]; Streptococcus 6646K [52]; Actinomyces viscosus [53]; Bacillus C-125 (alkalophilic) [54]; Penicillium citrinum [55]; Cat [56]; Bacillus stearothermophilus [59]; E. coli (multiple forms) [61]; Saccharomyces fragilis [63]

Enzyme Handbook © Springer-Verlag Berlin Heidelberg 1991
Duplication, reproduction and storage in data banks are only
allowed with the prior permission of the publishers

Source tissue

More (overview, various mammalian organs) [1]; Intestine [1, 27]; Seeds [1],
Leaves [1]; Brain [1, 49]; Cell suspension [22, 28]; Culture medium [33, 52,
62]; Testis [37, 51]; Liver [56]; Mammary gland [57]; Small intestine [61]

Localisation in source

Intracellular [18, 38, 44]; Extracellular [24, 25, 38, 62]; Membrane (brush-
border) [27]; Chloroplast (80% stroma) [30]; Vacuole [30]; Periplasmic
space [36]; Cell wall [38, 44]; More [53]

Purification

E. coli [1, 4]; Bacillus megaterium [4]; Raphanus sativus [18];
Thermoaerobacter sp. [19]; Saccharomyces lactis (4 forms) [20];
Corynebacterium murisepticum [21]; Daucus carota [22]; Bacillus
macerans [23]; Paecilomyces varioti [24, 25]; Streptococcus thermophilus
[26]; Human [27]; Marchantia polymorpha [28]; Bacteroides polypragmatus
[29]; Triticum aestivum [30]; Thermus (partial) [31]; Lactobacillus murinus
[34]; Treponema phagedenis [35]; Rhizobium trifolii [36]; Bovine
(neuramidase and beta-galactosidase form an enzyme complex) [37];
Medicago sativa (3 enzymes: intracellular, extracellular, cell-wall) [38];
Bacillus coagulans [40]; Pycnoporus acidophila [43]; Sugar cane [44];
Scopulariopsis sp. [46]; Aspergillus niger [47]; Xanthomonas campestris
(partial) [48]; Rat [49, 57]; Neurospora crassa [50]; Streptococcus 6646K
[52]; Bacillus C-125 [54]; Penicillium citrinum [55]; Cat [56]; Aspergillus
oryzae (partial [58]) [58, 60]; E. coli [61]; Saccharomyces fragilis [63]

Crystallization

(Bacillus macerans [23], Saccharomyces fragilis [63]) [23, 63]

Cloned

(Rhizobium meliloti gene in E. coli [7], enzyme from an thermophilic
anaerobe strain NA10 in E. coli [13]) [7, 13]

Renaturated

[1, 61]

---

5 STABILITY

pH

6–8 (stability decreases sharply below pH 6 and slowly above pH 8) [3]; 3.5
(completely dissociates into inactive monomers) [1]; 11.5 (completely dissociates into inactive monomers) [1]; 2.5–8.0 (Aspergillus niger) [6]; 6.5–7.5
(Kluyveromyces) [6]; 2.5–7.0 (Aspergillus oryzae) [6]; 6.0–8.0 (E. coli) [6]; 3–8
(Aspergillus niger) [10]; 3.5–8 (Aspergillus oryzae) [10]; 3.5–8.8 (24 hours,
4°C, radish) [18]; 6–9 [23]; 6.0–7.0 [28]; 4.0 (low stability below) [28]; 3.5–7.5
[3]; 3–6 (37°C, 2 hours) [41]; 5.5–9. 0 [54]; 4.5–7 (40°C, 3 hours) [55]; 4.0–9.0
[60]; More (high stability of immobilized enzyme [9]) [9, 28]

Temperature (°C)

80 (37% loss of activity after 3 hours, presence of 2-mercaptoethanol, enzyme from thermophilic anaerobe cloned in E. coli) [13]; 85 (half-life: 55 hours) [43]; 70 (half-life: 550 hours) [43]; 75 (half-life: 155 minutes) [19]; 60 (10 minutes, 60% loss of activity [59], stable up to [47], 10 minutes, complete loss of activity [23], 3 hours, 5% loss of activity [25]) [23, 25, 47, 59]; 55 (pH 4.0, 10 minutes, completely inactivated, radish [18], 30 minutes, 2 mg/ml bovine serum albumin, 42% loss of activity [40], complete loss of activity [54]) [18, 40, 54]; 50 (half-life: 32–42 days, immobilized enzyme [17], 20 days, 50% loss of activity [59], rapidly inactivated above [52]) [17, 52, 59]; 45 (stable below [33], 15 minutes, pH 2.4 [41]) [33, 41]; 40 (pH 4.0, 10 minutes, radish) [18]; 37 (relatively stable below [23], several days [58]) [23, 58]; 38 (10 minutes) [54]; 40 (pH 5.6, 15 minutes, stable below) [55]; More (of free and immobilized enzyme [9], of enzyme immobilized in a capillary bed reactor [11], chloroplast enzyme much more stable than vacuolar enzyme [30]) [3, 8, 9, 11, 30]

Oxidation

Frozen or lyophilized (solutions which are either frozen or lyophilized show markedly increased loss of activity upon thawing or dissolution) [4]; Immobilization [8, 9, 10, 11, 17, 31]; 2-Mercaptoethanol (stabilizes) [13]; Glycerol (stabilizes) [29, 61]; $MgCl_2$ (protects from thermal denaturation) [32, 34]; Bovine serum albumin (protects from thermal denaturation) [40, 59]; Aqueous solution [58]

Organic solvent

General stability information

0°C, ammonium sulfate precipitate, weeks [4]; 0°C, buffered solution, 2–3 days, loss of activity [4]; –20°C, 2 months [28]; –20°C, 10% glycerol (optimum storage conditions) [29]; –20°C, 6 months [49]; 4°C, 30 days [50]; 4°C, several months [51, 58]; 4°C, 2 months [52]

Storage

---

6 CROSSREFERENCES TO STRUCTURE DATABANKS

PIR/MIPS code

GBEC (Escherichia coli); GBECE (ebgA, Escherichia coli); A32688 (human); B32241 (Streptococcus thermophilus, fragment); S06878 (Escherichia coli, fragment); A24925 (Klebsiella pneumoniae); A27233 (Staphylococcus aureus); A29836 (I Bacillus stearothermophilus); A30093 (Lactobacillus delbrueckii subsp. bulgaricus); S06762 (Sulfolobus solfataricus); A05063 (Kluyveromyces marxianus, fragment); A29357 (yeast, Kluyveromyces marxianus, fragment); A31673 (precursor, human); A32611 (precursor, human)

Enzyme Handbook © Springer-Verlag Berlin Heidelberg 1991
Duplication, reproduction and storage in data banks are only
allowed with the prior permission of the publishers

Brookhaven code

7 LITERATURE REFERENCES

[1] Wallenfels, K., Weil, R. in "The Enzymes", 3rd. Ed. (Boyer, P.D., Ed.) 7, 617–663 (1972) (Review)

[2] Wallenfels, K., Zarnitz, M.L., Laule, G., Bender, H., Keser, M..: Biochem. Z., 331, 459 (1959)

[3] Wallenfels, K., Malhotra, O.M. in "The Enzymes", 2nd Ed. (Boyer, P.D., Ed.) 4, 409–430 (1960) (Review)

[4] Steers, E., Cuatrecasas, P.: Methods Enzymol., 34, 350–358 (1974) (Review)

[5] Yariv, J.: Methods Enzymol., 46, 398–403 (1977) (Review)

[6] Crueger, A., Crueger, W. in "Biotechnology" (Kieslich, K., Ed.) 60, 421–457, Verlag Chemie, Weinheim (1984) (Review)

[7] Fanning, S., O'Gara, F.: Gene, 71, 57–64 (1988)

[8] Nakanishi, K., Matsuno, R., Torii, K., Yamamoto, K., Kamikubo, T.: Enzyme Microb. Technol., 5, 115–120 (1983)

[9] Sarto, V., Marzetti, A., Focher, B.: Enzyme Microb. Technol., 7, 515–520 (1985)

[10] Baret, J.L.: Methods Enzymol., 136, 411–423 (1987)

[11] Peterson, R.S., Hill, C.G., Amundson, C.H.: Biotechnol. Bioeng., 34, 429–437 (1989)

[12] Larsson, P.-O., Hedbys, L., Svensson, S., Mosbach, K. : Methods Enzymol., 136, 230–233 (1987)

[13] Saito, T., Honda, H., Iijima, S., Kobayashi, T.: Enzyme Microb. Technol., 11, 302–305 (1989)

[14] Viratelle, O.M., Yon, J.M.: Biochemistry, 19, 4143–4149 (1980)

[15] Fenton, D.M.: Enzyme Microb. Technol., 4, 229–232 (1982)

[16] Boos, W.: Methods Enzymol., 89, 59–64 (1982)

[17] Kobayashi, T., Hirose, Y., Ohmiya, K., Shimizu, S., Uchino, F.: J. Ferment. Technol., 56, 309–314 (1978)

[18] Sekimata, M., Ogura, K., Tsumuraya, Y., Hashimoto, Y., Yamamoto, S.: Plant Physiol., 90, 567–574 (1989)

[19] Lind, D.L., Daniel, R.M., Cowan, D.A., Morgan, H.W.: Enzyme Microb. Technol., 11, 180–186 (1989)

[20] Mbuyi-Kalala, A., Schek, A.G., Leonis, J.: Eur. J. Biochem., 178, 437–443 (1988)

[21] Priyolkar, M.R., Nair, C.K.K., Pradhan, D.S.: Arch. Microbiol., 151, 49–53 (1989)

[22] Konno, H., Katoh, K., Kubota, I.: Phytochemistry, 27, 1301–1302 (1988)

[23] Miyazaki, Y.: Agric. Biol. Chem., 52, 625–631 (1988)

[24] O'Mahony, M., Kelly, C.T., Fogarty, W.M.: Biochem. Soc. Trans., 624th Meeting, 16, 183–184 (1988)

[25] Kelly, C.T., O'Mahony, M.R., Fogarty, W.M.: Appl. Microbiol. Biotechnol., 27, 383–388 (1988)

[26] Smart, J., Richardson, B.: Appl. Microbiol. Biotechnol., 26, 177–185 (1987)

[27] Lau, H.K.F.: Biochem. J., 241, 567–572 (1987)

[28] Konno, H., Yamasaki, Y., Katoh, K.: Plant Sci., 44, 97–104 (1986)

[29] Patel, G.B., Mackenzie, C.R., Agnew, B.J.: Appl. Microbiol. Biotechnol., 22, 114–120 (1985)

[30] Bhalla, P.L., Dalling, M.J.: Plant Physiol., 76, 92–95 (1984)

[31] Cowan, D.A., Daniel, R.M., Martin, A.M., Morgan, H.W. : Biotechnol. Bioeng., 26, 1141–1145 (1984)

[32] Nadder De Macias, M.E., Manca De Nadra, M.C., Strasser De Saad, A.M., Pesce De Ruiz Holgado, A.A., Oliver, G.: J. Appl. Biochem., 5, 275–281 (1983)

[33] Takenishi, S., Watanabe, Y., Miwa, T., Kobayashi, R.: Agric. Biol. Chem., 47, 2533–2540 (1983)

[34] Nader De Macias, M.E., Manca De Nadra, M.C., Strasser De Saad, A.M., Pesce De Riuz Holgado, A.A., Oliver, G.: Curr. Microbiol., 9, 99–104 (1983)

[35] Takahshi, T., Sugahara, T., Yamaya, S.: Curr. Microbiol., 8, 341–345 (1983)

[36] Abe, M., Higashi, S.: J. Gen. Appl. Microbiol., 28, 551–562 (1982)

[37] Verheijen, F., Brossmer, R., Galjaard, H.: Biochem. Biophys. Res. Commun., 108, 868–875 (1982)

[38] Chaubet, N., Pareilleux, A.: Z. Pflanzenphysiol., 106 , 401–407 (1982)

[39] Itoh, T., Suzuki, M., Adachi, S.: Agric. Biol. Chem., 46, 899–904 (1982)

[40] Levin, R.E., Mahoney, R.R.: Antonie Leeuwenhoek, 47, 53–64 (1981)

[41] Ohtakara, A., Hayashi, N., Mitsutomi, M.: J. Ferment. Technol., 59, 325–328 (1981)

[42] Komp, M., Hess, D.: Phytochemistry, 20, 973–976 (1981)

[43] Buonocore, V., Sgambati, O., De Rosa, M., Esposito, E., Gambacorta, A.: J. Appl. Biochem., 2, 390–397 (1980)

[44] Etcheberrigaray, J.L., Vattuone, M.A., Sampietro, A. R.: Phytochemistry, 20, 49–51 (1986)

[45] Schulein, M.: Enzyme Eng., 5, 79–83 (1980)

[46] Pastore, G.M., Park, Y.K.: J. Ferment. Technol., 58, 79–81 (1980)

[47] Widmer, F., Leuba, J.-L.: Eur. J. Biochem., 100, 559–567 (1979)

[48] Frank, J.F., Somkuti, G.A.: Appl. Environ. Microbiol. , 38, 554–556 (1979)

[49] Gatt, S.: Methods Enzymol., 14, 156–161 (1969)

[50] Stephens, R., DeBusk, G.: Methods Enzymol., 42, 497–503 (1975)

[51] Distler, J.J., Jourdian, G.W.: Methods Enzymol., 50, 514–520 (1978)

[52] Kiyohara, T., Terao, T., Shioiri-Nakano, K., Osawa, T.: J. Biochem., 80, 9–17 (1976)

[53] Kiel, R.A., Tanzer, J.M., Woodiel, F.N.: Infect. Immun., 16, 81–87 (1977)

[54] Ikura, Y., Horikoshi, K.: Agric. Biol. Chem., 43, 1359–1360 (1979)

[55] Watanabe, Y., Kibesaki, Y., Takenishi, S., Sakai, K., Tsujisaka, Y.: Agric. Biol. Chem., 43, 943–950 (1979)

[56] Holmes, E.W., O'Brien, J.S.: Biochemistry, 18, 952–958 (1979)

[57] Kuo, C.-H., Wells, W.W.: J. Biol. Chem., 253, 3550–3556 (1978)

[58] Akasaki, M., Suzuki, M., Funakoshi, I., Yamashina, I. : J. Biochem., 80, 1195–1200 (1976)

[59] Goodman, R.E., Pederson, D.M.: Can. J. Microbiol., 22 , 817–825 (1976)

[60] Tanaka, Y., Kagamiishi, A., Kiuchi, A., Horiuchi, T.: J. Biochem., 77, 241–247 (1975)

[61] Marchesi, S.L., Steers, E., Shifrin, S.: Biochim. Biophys. Acta, 181, 20–34 (1969)

[62] Comp, P.C., Lester, G.: J. Bacteriol., 107, 162–167 (1971)

[63] Uwajima, T., Yagi, H., Terada, O.: Agric. Biol. Chem. , 36, 570–577 (1972)

1 NOMENCLATURE

EC number
3.2.1.24

Systematic name
Alpha-D-mannoside mannohydrolase

Recommended name
Alpha-mannosidase

Synonymes
Mannosidase, .alpha.-
.alpha.-Mannosidase
.alpha.-D-Mannosidase
p-Nitrophenyl-.alpha.-mannosidase
.alpha.-D-Mannopyranosidase
1, 2-Alpha-mannosidase [4, 6, 9]
1, 2-Alpha-D-mannosidase [7]
Exo-alpha-mannosidase [12]

CAS Reg. No.
9025-42-7

---

2 REACTION AND SPECIFICITY

Catalysed reaction
Hydrolysis of terminal, non-reducing alpha-D-mannose residues in alpha-D-mannosides

Reaction type
O-Glycosyl bond hydrolysis

Natural substrates
Oligosaccharides + H_2O (mannose containing, enzyme may be involved in processing of oligosaccharide chains of mammalian glycoproteins [26]) [23, 26, 29, 30]
GlcNAc(Man)$_5$ GlcNAc + H_2O (biological substrate for Golgi mannosidase II) [29]
Man9 precursor (rat: mannosidase IA, IB convert Man$_9$ precursor to Man$_5$ intermediate) [31]
GlcNAcMan$_5$ + H_2O (rat: mannosidase II is the GlcNAcMan$_5$-cleaving enzyme in glycoprotein biosynthesis) [31]
More [35]

Enzyme Handbook © Springer-Verlag Berlin Heidelberg 1991
Duplication, reproduction and storage in data banks are only
allowed with the prior permission of the publishers

Substrate spectrum

1 Alpha-D-mannosides + H_2O
2 Alpha-D-lyxosides (with the same configuration at C-2, C-3 and C-4 as mannose) + H_2O
3 Heptopyranosides (with the same configuration at C-2, C-3 and C-4 as mannose) + H_2O
4 Disaccharides and oligosaccharides + H_2O (r) (reversed alpha-mannosidase reaction at high concentration of mannose) [10]
5 Ovalbumin + H_2O (not [4]) [1]
6 Ovalbumin-glycopeptide + H_2O [1, 3, 13, 20, 51]
7 Ovomucoid + H_2O [1]
8 Orosomucoid + H_2O [1]
9 p-Nitrophenyl-alpha-D-mannoside + H_2O (slowly [5], not [6, 8, 14]) [1, 2, 3, 18, 21, 28, 30, 32, 35, 45, 46, 52]
10 Glycopeptides (mannose containing) + H_2O [1, 3, 6, 20, 21, 32, 51]
11 2-O-Alpha-D-mannobiose + H_2O [4]
12 2-O-Alpha-mannotriose + H_2O [4]
13 4-O-Alpha-mannobiose + H_2O [4]
14 6-O-Alpha-mannobiose + H_2O [4]
15 2-Acetamido-4-O-alpha-D-mannopyranosyl-2-deoxy-D-glucose + H_2O
16 2-Acetamido-6-O-alpha-D-mannopyranosyl-2-deoxy-D-glucose + H_2O
17 Yeast mannan + H_2O (slowly [46]) [6, 19, 20, 27, 32, 46, 51]
18 Glycoproteins + H_2O (mannose containing glycoproteins) [6, 32]
19 Taka-amylase A + H_2O [7]
20 Arylmannosides + H_2O [46]
21 Alkylmannosides + H_2O [46]
22 $Man_9GlcNAc$ + H_2O (not [36], the only substrate cleaved [14]) [14]
23 $Man_9GlcNAc$ + H_2O [28]
24 4-Methylumbelliferyl-alpha-D-mannoside + H_2O [16, 18]
25 GDP-mannose + H_2O [20, 32]
26 1, 2-Alpha-mannobiose + H_2O [6]
27 $(Man)_5GlcNAc$ + H_2O [29]
28 $(Man)_4GlcNAc$ + H_2O [29]
29 $GlcNAc(Man)_5GlcNAc$ + H_2O
30 Mannose oligosaccharides + H_2O [30, 39]
31 More (rat brain: cleaves alpha1, 2-, alpha1, 3-and alpha1, 6-linked mannosyl residues in high mannose oligosaccharides [29], cleaves (Man)alpha1-->3(Man) linkages more than 10 times faster than (Man)alpha1-->6(Man) and (Man)alpha1-->2(Man) linkages [8], specific for hydrolysis of alpha-1, 2-mannosyl-mannose [6, 26, 27, 47], EC 3.2.1.24 can be divided into 2 classes: 1. enzymes hydrolyzing nonreducing terminal alpha-D-mannosidic linkages regardless of the aglycon moiety, 2. enzymes hydrolyzing a specifically linked alpha-D-mannosidic bond [5], not: yeast mannan [2], not: 3-O-alpha-, 4-O-alpha-and 6-O-alpha-mannobioses,

2-acetamido-2-deoxy-3-O-alpha-, 4-O-alpha-and
6-O-alpha-D-mannopyranosyl-D-glucoses [4]) [2, 4, 5, 6, 8, 18, 20, 26,
29, 29, 30, 31, 34, 35, 36, 39, 46, 47, 51]

Product spectrum
1 Mannose + ?
2 ?
3 ?
4 Mannose (r) (reversed alpha-mannosidase reaction at high concentrations of mannose) [10]
5 Mannose + ovalbumin
6 Mannose + ovalbumin-glycopeptide
7 Mannose + ovomucoid [1]
8 Mannose + orosomucoid
9 Mannose + p-nitrophenol
10 Mannose + glycopeptide
11 Mannose
12 Mannose
13 Mannose
14 Mannose
15 ?
16 ?
17 Mannose + yeast mannan
18 Mannose + glycoprotein
19 Mannose + Taka-amylase
20 Mannose + aromatic alcohol
21 Mannose + alkyl alcohol
22 $Man_8GlcNAc$ + mannose [14]
23 $Man_5GlcNAc$ + mannose [28]
24 4-Methylumbelliferone + mannose
25 GDP + mannose
26 Mannose
27 $Man_3GlcNAc$ + mannose [29]
28 $Man_3GlcNAc$ + $Man_2GlcNAc$ + mannose [29]
29 $GlcNAc(Man)_3GlcNAc$ + mannose [29]
30 Mannose + oligosaccharides [30, 39]
31 ?

Inhibitor(s)
Methyl-alpha-D-mannoside [1, 7]; D-Mannono-(1 --> 5)-lactone (weak [4],
strong [5]) [4, 5, 6, 18]; Hg^{2+} [1, 4, 5, 11, 19, 23, 51]; Ag^+ [1, 5, 11, 19, 23];
Zn^{2+} [1, 5, 6, 14, 19, 44, 52]; Mn^{2+} [5, 11, 14, 19]; Mg^{2+} [5, 14, 44]; Fe^{3+} [5];
Ba^{2+} [5]; Sr^{2+} [5]; Ca^{2+} (low [11]) [5, 11]; Cu^{2+} [5, 8, 11, 14, 16, 19, 21, 23,
26, 31, 34, 44, 51]; Cd^{2+} [5]; Heavy metal ions [6]; EDTA (slight [7], not [8,
12, 34, 45]) [6, 7, 11, 14, 19, 20, 23, 24, 26, 28, 32, 51];

Enzyme Handbook © Springer-Verlag Berlin Heidelberg 1991
Duplication, reproduction and storage in data banks are only
allowed with the prior permission of the publishers

$MnCl_2$ [7]; $CuSO_4$ [7]; $FeSO_4$ [7]; $CoCl_2$ [7]; Monoiodoacetic acid (slight) [7];
1-Cyclohexyl-3-(2-morpholine-4-ethyl)carbodiimide [7]; Fe^{2+} (slight [44])
[11, 19, 21, 34, 44, 51]; D-Mannose (partial [12]) [12, 17, 21 , 23, 34, 45];
Co^{2+} (moderate activator [21]) [14, 24]; $AgNO_3$ [18]; Swaisonine (relatively
insensitive to [30]) [18]; Pb^{2+} [19]; Ni^{2+} [19]; p-Chloromercuribenzoate [19];
p-Chloromercuriphenylsulfonic acid [21, 31, 32, 34, 45]; Ag^{2+} [24];
Phosphatidylinositol (at phosphatidyl concentrations which provide optimal
activity) [26]; Phosphatidylglycerol (at phosphatidylcholine concentrations
which provide optimal activity) [26]; Glycerol [26]; Mannoseoligo-
saccharides (alpha1,2- and alpha 1, 3-linked) [28]; Mannosamine [28]; Tris
(mannosidase IA /IB, not II) [31]; Bovine serum albumin [34]; Mannosylamine [35]; Li^{2+} [44]; NaN_3 [46]; Monoiodoacetate [46]; Methyl alpha-D-mannoside [46]; Chelating agents [49]

Cofactor(s)/prosthethic group(s)
More (requirement for nonionic detergents or for specific phospholipids)
[26]

Metal compounds/salts
Ca^{2+} (EDTA inhibition reversed completely by Ca^{2+} [14, 28] and partially by
Mg^{2+} [14], not by other divalent cations [14], activates [19], alpha-mannosidase activated by Ca^{2+}, 1, 2-alpha-mannosidase not [8]) [8,
14, 19, 28]; Zn^{2+} (contains 2 mol zinc per mol protein [38], jack bean:
0.470–0.565 mg zinc per g of protein, metalloenzyme [49], activity restored
after inactivation by chelating agents [49], required to retain activity even at
room temperature [3], not Zn^{2+}-dependent [4, 5], absolute requirement
[11], increases activity [12], required [13, 37], stimulates [14, 24, 34, 51],
restores activity after EDTA inhibition [23, 32, 52]) [3, 4, 5, 11, 12, 13, 14, 23,
24, 32, 34, 37, 49, 51]; Mg^{2+} (EDTA inhibition reversed completely by Ca^{2+}
and partially by Mg^{2+}, not by other divalent cations [14], activates [19]) [14,
19]; Co^{2+} (activates) [34]; More (divalent cations not required [14], no metal ion requirement [18]) [14, 18]

Turnover number (min^{-1})
61.62 (mannobiose) [7]; 7800 [17]

Specific activity (U/mg)
35.5 [1]; 12 [2]; 19 [3]; 1.3 [4]; More [5, 6, 8, 11, 14, 16, 18, 19, 20, 22, 23, 25,
26, 28, 31, 35, 36 , 39, 44, 47, 52]

K_m-value (mM)
1.2 (p-nitrophenyl-alpha-D-mannopyranoside) [5]; 8.2 (methyl
4-O-alpha-mannopyranosyl-alpha-D-mannopyranoside) [5]; 2.3 (alpha-D-mannoside) [2]; 0.91 (p-nitrophenyl-alpha-D-mannoside) [3]; 2
(2-O-alpha-D-mannobiose) [4]; 0.57 (1, 2-alpha-mannobiose) [6]; 0.67
(mannobiose) [7]; 1.25 (Man₃GlcNAcGlcNAc (NaB₃H₄-reduced)) [8]; 3.7
(p-nitrophenyl-alpha-D-mannobiosides) [11]; 0.27
(p-nitrophenyl-alpha-D-mannopyranoside) [12]; 0.140
(4-methylumbelliferyl-alpha-D-mannopyranoside) [16]; 2.2
(4-methylumbelliferyl-alpha-D-mannopyranoside) [18]; More [13, 16, 18, 19,
20, 21, 23, 30, 32, 34, 35, 39, 44, 45, 46, 52]

pH-optimum
4.0 (alpha-mannosidase A, broad) [23]; 4.6 (alpha-mannosidase B, sharp)
[23]; 5.6 (golgi) [32]; 5.5–5.9 [35]; 6.0–6.4 (tissue, human) [41]; 5.2–5.8
(serum, human) [41]; 5.5–6.0 [43]; 4.2–4.5 [44]; 6.3 [45]; 4.5–5.0 (free and insolubilized enzyme) [48]; 3.4 [51]; 4.1 [52]; 4.0–5. 0 [1]; 4.0 [2, 14]; 5.5–6.0
[28]; 4.6 [3, 20, 32]; 4.8 [4]; 5.5 [34]; 4.5 [37]; 4.2 [5]; 4.9–5.3 [6]; 5.0 [8, 27];
4.4 [11]; 3.9–4.3 [18]; 5.8 [19]; 5.5 [21]; 4.6–4.8 [24]

pH-range
4.5–8.0 (4.5: about 20% of maximal activity, 8.0: about 75 % of maximal activity) [13]; 4.9–8.2 (50% of maximal activity at 4.9 and 8.2) [19]; 3.8–5.5
(50% of maximal activity at 3.8 and 5.5) [20]; 4.0–8.0 (very little activity at pH
4.0 and 8.0) [29]; 3–6 (3: about 20% of maximal activity, 6: about 10% of
maximal activity) [2]; 3.5–6.0 (6.0: 50% of maximal activity, 3.5: 70% of maximal activity) [8]; 3.3–5.4 (over 50% of maximal activity between) [18]; 3.0–8.0
(3.0: about 60% of maximal activity, 8.0: about 40% of maximal activity) [5]

Temperature optimum (°C)
45 [5, 19]; 48 [24]; 50 [11]; 40 [20]; 37–40 [44]

Temperature range (°C)
15–65 (very low activity at 15°C and 65°C) [5]; 35–55 (half maximal activity)
[24]; More (increase in activity when incubated between 20°C and 62°C
before assay) [52]

Enzyme Handbook © Springer-Verlag Berlin Heidelberg 1991
Duplication, reproduction and storage in data banks are only
allowed with the prior permission of the publishers

3 ENZYME STRUCTURE

Molecular weight

60000 (SDS-PAGE under nonreducing conditions, Saccharomyces cerevisiae) [14]

217000 (sucrose density gradient centrifugaation, Canavalia ensiformis) [15]

560000 (gel filtration, Saccharomyces cerevisiae) [16]

200000 (gel filtration, Canavalia ensiformis) [17, 48]

150000 (gel filtration, Prunus serotina) [18]

30000 (SDS-polyacrylamide gel electrophoresis, Rhodococcus erythopolis) [19]

32000 (gel filtration, Rhodococcus erythropolis) [19]

335000 (molecular sieve chromatography, rat, lysosomal [32]) [20, 32]

229000 (molecular sieve chromatography, rat) [21]

200000 (sucrose density centrifugation, rat, lysosomal [32]) [20, 32]

110000 (sucrose density centrifugation, , rat) [21]

300000 (Lupinus angustifolius, gel filtration) [23]

197000 (gel filtration, wheat) [24]

188000 (sucrose gradient centrifugation, wheat) [24]

280000 (gel filtration, human, mannosidase A) [25]

260000 (gel filtration, human, mannosidase B) [25]

51000 (Aspergillus satoi, gel filtration) [27]

372000–490000 (different methods, rat liver, cytoplasm) [35]

220000 (gel filtration, rat epididymis [36], sedimentation equilibrium analysis, Phaseolus vulgaris [38]) [36, 38]

42000 (SDS-PAGE, hog) [39]

45000 (SDS-PAGE, monkey) [40]

500000–600000 (alpha-D-mannosidase I_1 /I_3, gel filtration, human) [42]

130000–140000 (alpha-mannosidase I_2 /I_4, gel filtration, human) [42]

450000 (gel filtration, baker's yeast) [46]

275000 (gel chromatography, Styela plicata) [51]

230000 (disc-electrophoresis, Medicago sativa) [52]

49000 (gel exclusion chromatography, Aspergillus oryzae) [6]

64000 (SDS-gel electrophoresis, Aspergillus satoi) [7]

260000 (gel filtration, Helianthus annuus) [11]

270000 (gel filtration, watermelon) [13]

75000 (gel filtration, Saccharomyces cerevisiae) [14]

More (enzyme concentration above 0.1 mg/ml: enzyme associates into aggregates with MW 1000000 [19]) [19, 20, 32]

Subunits

Dimer (SDS-PAGE in presence of reducing agent, 2 subunits, 1 × 44500,
1 × 225000, Saccharomyces cerevisiae) [14]
Tetramer (2 × 44000, 2 × 66000, Canavalia ensiformis, heating prior to
electrophoresis) [17]
Polymer (Prunus serotina, SDS-PAGE) [18]
Polymer (human, SDS-PAGE after reductive carboxymethylation, man-
nosidase A: x × 62000, x × 26000, equimolar proportion of subunits, man-
nosidase B: a small subunit of MW 26000 and variable mixture of larger
subunits 58000 and 62000) [25]
Tetramer (rat [34, 35], SDS-PAGE, 4 × 110000 [35]) [34, 35]
Dimer (2 non-covalently bound subunits, SDS-PAGE, Phaseolus vulgaris)
[38]
Tetramer (heating in SDS and mercaptoethanol, SDS-PAGE, 2 × 66000,
2 × 44000, jack bean) [46]
Tetramer (SDS-PAGE, Styela plicata, 4 × 70000) [51]
Tetramer (Medicago sativa, SDS-PAGE, 75000, 60000, 50000, 45000) [52]
Dimer (non-denaturing and SDS-electrophoresis, Helianthus annuus,
mannosidase I: 2 × 55000, mannosidase II: 2 × 60000) [11]

Glycoprotein/Lipoprotein

Glycoprotein (Helianthus annuus L. [11], Rhodococcus erythropolis: 15%
carbohydrate, rhamnose, glucose, galactose [19], Canavalia ensiformis,
larger subunit: 6.05 %, smaller subunit: 0.4% [17], wheat: D-glucose,
D-mannose, D-galactose, N-acetyl-D-glucosamine [24], rat [33 , 34, 35, 36],
5.6% hexose, 3.1% mannose, 2.0% glucosamine and small amounts of
glucose, fucose, galactose [36], Phaseolus vulgaris: alpha-mannosidase I:
8.3% carbohydrate [38], alpha-mannosidase II: 16.5% carbohydrate [38],
hog: 8% mannose, 3.3% glucosamine [39]) [11, 17, 19, 20, 24, 32, 34, 35, 36,
37, 38, 39]

---

4 ISOLATION/PREPARATION

Source organism

Jack bean [1, 10, 48, 49]; Turbo cornutus [2]; Hog [3, 39]; Aspergillus
oryzae (1, 2-alpha-mannosidase [6, 9]) [6, 9]; Aspergillus satoi (1, 2-al-
pha-mannosidase [7], alpha-mannosidase [8]) [7, 8, 27]; Aspergillus niger
(1, 2-alpha-mannosidase [4], alpha-mannosidase [5]) [4, 5]; Helianthus annuus (2 forms: I /II) [11]; Water melon [13]; Yarrowia lipolytica [12];
Saccharomyces cerevisiae [14, 16]; Canavalia ensiformis [15, 17]; Prunus
serotina [18]; Rhodococcus erythropolis [19]; Rat (alpha-mannosidase: IA
/IB [31], mannosidase II [31]) [20, 21, 29, 30, 31, 32, 33, 34, 35, 36]; Monkey
(neutral and acid alpha-D-mannosidase [22]) [22, 40]; Lupinus
angustifolius (mannosidase A and B) [23]; Wheat [24]; Human (man-

Enzyme Handbook © Springer-Verlag Berlin Heidelberg 1991
Duplication, reproduction and storage in data banks are only
allowed with the prior permission of the publishers

nosidase A /B [25], serum [41], plasma: 4 forms /I_1 /I_2 /I_3 /I_4 [42]) [25, 41–43]; Rabbit [26]; Mung bean [28]; Phaseolus vulgaris (3 isoenzymes: I /II /II [37], alpha-mannosidase I /II [38]) [37, 38]; Aspergillus flavus [44]; Saccharomyces cerevisiae [45, 46, 50]; Acinetobacter sp. [42]; Styela plicata [51]; Medicago sativa [52]

Source tissue

Hypocotyl [11]; Fruits [13]; Leaf [15]; Seeds [18, 52]; Culture fluid [19]; Cotyledons [23, 37]; Aleurone layer [24]; Seedlings [28]; Brain [29]; Muscle [33]; Epididymis [36]; Brain [40]; Serum [41]; Plasma [42]; Red blood cells [43]; Tissue [41]; Growth medium [44, 47]; Internal organs [51]; Meal [1]; Liver [2, 20, 21, 25, 26, 30, 31, 32, 34, 35]; Cell [5, 6, 12, 44]; Kidney [3, 39]; Culture filtrate [7, 27]; Mycelium [8]

Localisation in source

Cell wall [37]; Protein bodies [37]; Extracellular [44, 47]; Membrane (peripheral membrane protein or ecto-type integral membrane protein) [16]; Lysosomes [20, 32]; Golgi membranes [21, 31, 32, 34]; Cytoplasm [32, 35]; Microsomes [26, 28, 29]; Extracellular [27]; Endoplasmic reticulum [30, 37]; More (particulate fraction [12, 45], purified soluble alpha-mannosidase represents the catalytically active domain of the enzyme which has been proteolytically released from its membrane-bound form [14]) [12, 14, 45]

Purification

Helianthus annuus [11]; Watermelon [13]; Saccharomyces cerevisiae [14, 16, 46]; Prunus serotina (partial) [18]; Rhodococcus erythropolis [19]; Rat [20, 21, 29, 31, 32, 34, 35, 36]; Monkey [22, 40]; Lupinus angustifolius [23]; Wheat [24]; Human [25]; Rabbit [26]; Mung bean [28]; Aspergillus flavus [44]; Acinetobacter sp. [47]; Styela plicata [51]; Medicago sativa [52]; Jack bean [1, 48]; Turbo cornutus [2]; Hog [3, 39]; Aspergillus niger [4]; Aspergillus oryzae [6, 9]; Aspergillus saitoi [8, 27]

Crystallization

–

Cloned

–

Renaturated

–

---

5 STABILITY

pH

5.0–8.0 (stable) [4]; 5.0–7.0 (maximum stability) [5]; 4.5–6.0 (4°C, stable for at least 6 months) [6]; 7.0–9.0 (stable) [13, 34]; 4.8–8.0 (20°C, 5 hours, stable) [19]; 7 (stable) [52]; 6 (unstable below) [46]; 4–6 (stable) [51]

Temperature (°C)

60 (15 minutes [3], inactivation in 5 minutes [19]) [3, 19]; 45 (rapid loss of activity above [5], pH 5.0, stable for 10 minutes [27]) [5, 27]; 50 (1 hour, 10% loss of activity [6], complete loss of activity after 15 minutes [44]) [6, 44]; 75 (24 hours, 83% mannose) [10]; 30 (retains activity for 3 days) [11]; 70 (2 hours, alpha-mannosidase I: 68% loss of activity, mannosidase II: 80% loss of activity [11], 90% loss of activity after 10 minutes [27], in-activated after 5 minutes [46], stable up to, free and insolubilized [48]) [11, 27, 46, 48]; 55 (no loss of activity after 20 minutes) [33]; 44 (stable up to 5 minutes) [46]; 40 (rapid inactivation after 15 minutes) [20]; More (temperature at which the enzyme was fully stable for 3 hours and 24 hours appears to be approximately 20°C higher in 83% mannose than in 0% mannose [10], $CaCl_2$ increases thermostability) [10]

Oxidation

Organic solvent

General stability information

Protein concentration (stability depends on [3, 20], unstable at low con-centration [35]) [3, 20, 35]; Bovine serum albumin (stabilizes) [20, 34]; Sucrose (slight stabilization) [20]; Glycerol (slight stabilization) [20]; Co^{2+} (cytoplasmic enzyme, partial stabilization) [32, 35]; Zn^{2+} (stabilizes) [34]; Dithiothreitol (stabilizes) [35]; Mn^{2+} (stabilizes) [35]; Fe^{2+} (stabilizes) [35]; Freezing and thawing (very unstable upon [1], stable [19, 21]) [1, 19, 21]; Freezing (stable) [3]; Lyophilization (loss of activity) [3]; More [17]

Storage

4°C, 10 mM sodium actetate, pH 6.0, enzyme concentration 0.1 mg/ml (stable up to 1 month) [10]; 0–4°C, Tris/acetate buffer, pH 7.5, 4 weeks (stable) [22]; –70°C, 24 hours (complete loss of activity) [29]; 0–4°C, 2–3 days (no loss of activity) [29]; –20°C (no loss of activity after 2–3 months [1], indefinitely stable [2]) [1, 2]; 4°C or –20°C (less than 5% loss of activity after 4 months) [7]; 4°C (activity retained for 9 months) [11]; –20°C (stable for at least 1 year) [19]; More [19, 20, 27, 31, 32, 35, 44]

---

6 CROSSREFERENCES TO STRUCTURE DATABANKS

PIR/MIPS code

A33511 (yeast, Saccharomyces cerevisiae)

Brookhaven code

7 LITERATURE REFERENCES

[1] Yu-Teh Li: J. Biol. Chem., 214, 1010–1012 (1966)

[2] Maramatsu, T., Egami, F.: J. Biochem., 62, 700–709 (1967)

[3] Okumura, T., Yamashina, I.: J. Biochem., 68, 561–571 (1970)

[4] Swaminathan, N., Matta, K.L., Donoso, L.A., Bahl, O.P. : J. Biol. Chem., 247, 1775–1779 (1972)

[5] Matta, K.L., Bahl, O.P.: J. Biol. Chem., 247, 1780–1787 (1972)

[6] Yamamoto, K., Hitomi, J., Kobatake, K., Yamaguchi, H.: J. Biochem., 91, 1971–1979 (1982)

[7] Shigematsu, Y., Tsukahara, K., Tanaka, T., Takeuchi, M., Ichishima, E.: Curr. Microbiol., 13, 43–46 (1986)

[8] Amano, J., Kobata, A.: J. Biochem., 99, 1645–1654 (1986)

[9] Tanimoto, K., Katsuragi, T., Yamaguchi, H.: Agric. Biol. Chem., 53, 1083–1088 (1989)

[10] Johansson, E., Hedbys, L., Mosbach, K., Larsson, P.-O.: Enzyme Microb. Technol., 11, 347–352 (1989)

[11] Lopez-Valbuena, R., Jorrin, J., Polanco, A., Tena, M. : Plant Sci., 62, 11–19 (1989)

[12] Vega, R., Dominguez, A.: J. Basic Microbiol., 28, 371–379 (1988)

[13] Nakagawa, H., Enomoto, N., Asakawa, M., Uda, Y.: Agric. Biol. Chem., 52, 2223–2230 (1988)

[14] Jelinek-Kelly, S., Herscovics, A.: J. Biol. Chem., 263, 14757–14763 (1988)

[15] Niyogi, K., Singh, M.: Phytochemistry, 27, 2737–2741 (1988)

[16] Yoshihisa, T., Ohsumi, Y., Anraku, Y.: J. Biol. Chem., 263, 5158–5163 (1988)

[17] Einhoff, W., Rüdiger, H.: Biol. Chem. Hoppe-Seyler, 369, 165–169 (1988)

[18] Waln, K.T., Poulton, J.E.: Plant Sci., 53, 1–10 (1987)

[19] Zacharova, I.Y., Tamm, V.Y., Pavlova, I.N.: Methods Enzymol., 160, 620–626 (1988)

[20] Opheim, D.J., Touster, O.: Methods Enzymol., 50, 494–500 (1978)

[21] Ram, D., Tulsiani, P., Touster, O.: Methods Enzymol., 50, 500–505 (1970)

[22] Mathur, R., Balasubramanian, A.: Biochem. J., 222, 261–264 (1984)

[23] Plant, A.R., Moore, K.G.: Phytochemistry, 21, 985–989 (1982)

[24] Conti, S., Carratu, G., Giannattasio, M.: Phytochemistry, 26, 2909–2912 (1988)

[25] Cheng, S.H., Malcolm, S., Pemble, S., Winchester, B.: Biochem. J., 233, 65–72 (1986)

[26] Forsee, W.T., Schutzbach, J.S.: J. Biol. Chem., 256, 6577–6582 (1981)

[27] Ichishima, E., Arai, M., Shigematsu, Y., Kumagai, H., Sumida-Tanaka, R.: Biochim. Biophys. Acta, 658, 45–53 (1981)

[28] Szumilo, T., Kaushal, G.P., Hori, H., Elbein, A.D.: Plant Physiol., 81, 383–389 (1986)

[29] Tulsiani, D.R.P., Touster, O.: J. Biol. Chem., 260, 13081–13087 (1985)

[30] Bischoff, J., Kornfeld, R.: J. Biol. Chem., 258, 7907–7910 (1983)

[31] Tulsiani, D.R.P., Hubbard, S.C., Robbins, P.W., Touster, O.: J. Biol. Chem., 257, 3660–3668 (1982)

[32] Opheim, D.J., Touster, O.: J. Biol. Chem., 253, 1017–1023 (1978)

[33] Wallace, R.R., Lewis, M.H.R.: Biochem. Soc. Trans., 566th Meeting, 5, 231–233 (1977)

[34] Tulsiani, D.R.P., Opheim, D.J., Touster, O.: J. Biol. Chem., 252, 3227–3233 (1977)

[35] Shoup, V.A., Touster, O.: J. Biol. Chem., 251, 3845–3852 (1976)

[36] Tulsiani, D.R.P., Coleman, V.D., Touster, O.: Arch. Biochem. Biophys., 267, 60–68 (1988)

[37] Van Der Wilden, W., Chrispeels, M.J.: Plant Physiol., 71, 82–87 (1983)

[38] Paus, E.: Eur. J. Biochem., 73, 155–161 (1977)

[39] Okumura, T., Yamashina, I.: J. Biochem., 73, 131–138 (1973)

[40] Mathur, R., Panneerselvam, K., Balasubramanian, A.S.: Biochem. J., 253, 677–685 (1988)
[41] Hultberg, B., Masson, P.K., Sjöblad, S.: Biochim. Biophys. Acta, 445, 398–405 (1976)
[42] Hirani, S., Winchester, B.: Biochem. J., 179, 583–592 (1979)
[43] Poenaru, L., Dreyfus, J.C.: Biochim. Biophys. Acta, 566, 67–71 (1979)
[44] Augustin, J., Sikil D., Zemek, J.: Collect. Czech. Chem. Commun., 43, 2812–2816 (1978)
[45] Opheim, D.J.: Biochim. Biophys. Acta, 524, 121–130 (1978)
[46] Kaya, T., Aikawa, M., Matsumoto, T.: J. Biochem., 82, 1443–1449 (1977)
[47] Kathoda, D., Sawaya, Y., Asatsuma, K., Suzuki, F., Hayashibe, M.: Agric. Biol. Chem., 41, 331–337 (1977)
[48] Shepherd, V., Montgomery, R.: Biochim. Biophys. Acta, 429, 884–894 (1976)
[49] Snaith, S.M.: Biochem. J., 147, 83–90 (1975)
[50] Kaya, T., Shibano, M., Kutsumi, T.: J. Biochem., Short Comm., 73, 181–182 (1973)
[51] Shigeta, S., Kubota, H., Tamura, H., Oka, S.: J. Biochem., 94, 1827–1832 (1983)
[52] Curdel, A., Petek, F.: Biochem. J., 185, 455–462 (1980)

Enzyme Handbook © Springer-Verlag Berlin Heidelberg 1991
Duplication, reproduction and storage in data banks are only
allowed with the prior permission of the publishers

1 NOMENCLATURE

EC number
3.2.1.25

Systematic name
Beta-D-mannoside mannohydrolase

Recommended name
Beta-mannosidase

Synonymes
Mannanase
Mannosidase, .beta.-
Mannase
.beta.-Mannosidase
.beta.-D-Mannosidase
Beta-D-Mannosidase
.beta.-Mannoside mannohydrolase
Exo-beta-D-mannanase [4]

CAS Reg. No.
9025-43-8

---

2 REACTION AND SPECIFICITY

Catalysed reaction
Hydrolysis of terminal, non-reducing beta-D-mannose residues in
beta-D-mannosides

Reaction type
O-Glycosyl bond hydrolysis (exohydrolysis)

Natural substrates
Beta-D-mannosides + H_2O (degradation of glycoprotein [1], key role in
hydrolysis of galactomannan in germinating guar seed [4]) [1, 4]

Substrate spectrum
1 4-Methylumbelliferyl beta-D-mannopyranoside + H_2O [1, 4, 6, 12, 13, 15, 18, 19, 20, 21]
2 p-Nitrophenyl-beta-D-mannopyranoside + H_2O [3, 4, 6, 9]
3 Naphtyl-beta-D-mannopyranoside + H_2O [3]
4 Beta-D-mannotriitol + H_2O [3–5]
5 Beta-D-mannotetraitol + H_2O [3, 4]
6 Beta-D-mannopentaitol + H_2O [3, 4]

Enzyme Handbook © Springer-Verlag Berlin Heidelberg 1991
Duplication, reproduction and storage in data banks are only
allowed with the prior permission of the publishers

7 Beta-D-mannohexaitol + H_2O [3, 4]
8 Mannobiose + H_2O [5, 8, 12, 15]
9 Mannotriose + H_2O (slowly [12]) [5, 8, 12, 15]
10 Mannopentaose + H_2O [5]
11 Mannotetraose + H_2O (slowly [12]) [5, 12]
12 Man(beta) 1-->4GlcNAc(beta) 1–4Man(beta) 1–4GlcNAc(beta) 1–4GlcNAc + H_2O
13 Galactomannans + H_2O [11]
14 Manno-oligosaccharides + H_2O [11]
15 4-O-Beta-D-mannosyl-D-mannose + H_2O [12]
16 Core-glycopeptide + H_2O [15]
17 Man(GlcNAc)$_2$-Asn + H_2O [15, 21]
18 Taka-amylase A + H_2O [15]
19 Alpha$_1$-acid glycoprotein + H_2O [15]
20 4-O-Beta-D-mannopyranosyl-L-rhamnose + H_2O [15]
21 6-O-Beta-D-mannopyranosyl-D-galactose + H_2O [15]
22 4-O-Beta-D-mannopyranosyl-N-acetylglucosamine + H_2O [15]
23 More (acts on Man/beta)1-->2Man bond very slowly [2], relatively specific for the mannosyl beta-1, 4-linkages [13]) [2, 11, 13, 14, 15, 16, 19, 21]

Product spectrum
1 Mannose + methylumbelliferone
2 Mannose + nitrophenol
3 Mannose + alpha-naphthol
4 Mannose
5 Mannose
6 Mannose
7 Mannose
8 Mannose
9 Mannose
10 Mannose
11 Mannose
12 Mannose + GlcNAc(beta)1–4Man(beta)1–4GlcNAc(beta)1–4 GlcNAc
13 ?
14 ?
15 Mannose
16 Mannose + core-glycopeptide
17 Mannose + (GlcNAc)$_2$-Asn
18 ?
19 ?
20 Mannose + L-rhamnose
21 Mannose + D-galactose
22 Mannose + N-acetylglucosamine
23 ?

Inhibitor(s)

Hg^{2+} [3, 4, 5, 12, 15, 17, 19, 21]; Ag^+ [3, 4, 5, 19, 21]; Ca^{2+} [3]; Cu^{2+} (partial [4]) [4, 5, 15]; Zn^{2+} (partial [4]) [4, 5, 15]; D-Mannose (not [5]) [4, 13]; Cd^{2+} [5, 12]; SDS [5]; N-Bromosuccinimide [5]; p-Chloromercuribenzoate [5, 12]; N-Dodecylbenzene sulfonate [5]; Mannosylamine [8]; Beta-mannono-1, 4-lactone (not [12]) [9, 20]; Sodium taurocholate [10]; $AgNO_3$ [15, 19]; $Fe_2(SO_4)_3$ [15]; p-Chloromercuriphenylsulfonic acid [15]; 2-Amino-2-deoxy-D-mannose [18]; p-Nitrophenyl-alpha-D-mannopyranoside (activates at low concentrations, inhibits at high concentrations) [18]; Heavy metal salts [19]; p-Nitrophenyl-beta-galactoside [20]; Mannono(1 -- > 5) lactone [21]

Cofactor(s)/prostethic group(s)

Metal compounds/salts

More (not metal dependent [9, 15], not stimulated by Co^{2+} and Zn^{2+} [10]) [9, 10, 15]

Turnover number (min^{-1})

Specific activity (U/mg)

0.996 [1]; 5.7 [4]; 38.8 [5]; 1.1 [6]; 32.86 [1]; More [9, 12, 13, 14, 16, 18, 21]

K_m-value (mM)

1.43 (p-nitrophenyl-beta-D-mannopyranoside) [3]; 2.33 (o-nitrophenyl-beta-D-mannopyranoside) [3]; 0.91 (methyl-umbelliferyl-beta-D-mannopyranoside) [3]; 3.22 (naphthyl-beta-D-mannopyranoside) [3]; 12.5 (beta-D-mannotriitol, beta-D-mannotetraitol, beta-D-mannopentaitol, beta-D-mannohexaitol) [3]; 80.0 (beta-D-mannotriitol) [4]; 12.0 (beta-D-mannotetraitol) [4]; 2.8 (beta-D-mannopentaitol, beta-D-mannohexaitol) [4]; 0.5 (beta-nitrophenyl-beta-D-mannopyranoside, methylumbel-liferyl-beta-D-mannopyranoside) [4]; 3.0 (o-nitrophenyl-beta-D-mannopyranoside) [4]; 0.56 (4-methylumbelliferyl beta-D-mannopyranoside) [1]; 50 (Man(beta)1–4Man) [2]; 100 (Man(beta)1–2Man) [2]; More [6, 9, 10, 11, 12, 13, 14, 15, 16, 17, 18, 20]

pH-optimum

2.4–3.4 [15]; 4.5 [1, 21]; 4 [3]; 5–6 [4]; 5.1–5.6 [11]; 5.0 [12]; 3.5 [13]; 3.5–4.0 [20]; 5.2 [16]; 6.0 [5]; 7.0 [7]; 4.2 [9]; 4.6 [19]; 5. 0–5.5 (acidic form) [10]; 6.0–8.0 (neutral form) [10]; 3.5–4.5 [14]

pH-range

3–7 (3: about 20% of maximal activity, 7: about 10% of maximal activity) [1]; 5.5–9.5 (no activity at pH 5 and 10) [5]; 3.7–4.6 (more than 95% of maxi-mal activity at pH 3. 7–4.6) [9]; 3.5–6.6 (3.5: 58% of maximal activity at, 6.5: 33% of maximal activity at) [16]; 3–6 [18]; 3.5–5.5 (more than 50% of maxi-mal activity at) [19]

Enzyme Handbook © Springer-Verlag Berlin Heidelberg 1991
Duplication, reproduction and storage in data banks are only
allowed with the prior permission of the publishers

Temperature optimum (°C)
52 [4]; 50 [5, 16]; 55 [7, 13, 18]

Temperature range (°C)
55–70 (55°C: maximum, 70°C: significant drop in activity) [18]

---

3 ENZYME STRUCTURE

Molecular weight
43000 (beta-mannanase I, Trifolium repens, SDS-PAGE, disc gel electrophoresis) [11]
38000 (beta-mannanase II, Trifolium repens, SDS-PAGE, disc gel electrophoresis) [11]
140000 (Sephadex gel exclusion chromatography, Tremella fuciformis) [12]
130000 (Aspergillus niger, gel filtration) [13]
120000 (Aspergillus niger, gel filtration) [14]
64000 (gel filtration, Polyporus sulfureus) [15]
40000 (disc gel electrophoresis, SDS-PAGE, Alfalafa) [16]
20000 (gel filtration, various legumes) [17]
88000 (barley, acrylamide gel electrophoresis) [18]
100000 (gel exclusion, hen) [21]
110000 (gel filtration, human) [1]
94000 (SDS-PAGE, Helix pomatia [3], Bacillus sp. [5]) [3, 5]
59000 (SDS-gel electrophoresis, gel filtration, Cyamopsis tetragonolobus) [4]
79000 (gel filtration, human) [9]
More [17]

Subunits
More (single band on disc electrophoresis) [15]

Glycoprotein/Lipoprotein
Glycoprotein (human: high mannose type oligosaccharide chains [1], Helix pomatia: 3% carbohydrate [3], Cyamopsis tetragonolobus: 7% [4], Aspergillus niger: 17.8% neutral sugar, 2.6% N-acetylhexosamine [13]) [13, 14]

4 ISOLATION/PREPARATION

Source organism
Human [1, 9, 20]; Turbo cornutus [2, 23]; Helix pomatia [3]; Cyamopsis
tetragonolobus [4]; Bacillus sp. [5, 7]; Latuca sativa [8]; Goat (liver, 2 forms:
lysosomal /acidic form and nonlysosomal /neutral form [10]) [6, 10]; Trifolium repens (beta-mannanase I/II) [11]; Tremella fuciformis [12];
Aspergillus niger [13, 14]; Polyporus sulfureus [15]; Lucerne (Alfalfa) [16,
17]; Carob [17]; Honey locust [17]; Guar [17]; Soybean [17]; Barley [18]; Rat
[19]; Achatina fulica [2, 21]; Hen [22]

Source tissue
Placenta [1, 9]; Seed [4, 8, 11, 16, 17]; Oviduct [22]; Cell [5]; Kidney [6];
Endosperm [17]; Synovial fluid [20]; Liver [10, 19, 23]; Culture filtrate [13,
14]; Viscera [21]

Localisation in source
Lysosomes [1, 10, 19]; More (in association with a cellulosic cell-wall fraction) [8]

Purification
Human (partial [9]) [1, 9]; Helix pomatia [3]; Cyamopsis tetragonolobus [4];
Bacillus sp. [5]; Goat [6]; Trifolium repens [11]; Turbo cornutus [23];
Tremella fuciformis [12]; Aspergillus niger [13, 14]; Polyporus sulfureus [15];
Alfalfa [16]; Barley [18]; Achatina fulica [21]; Hen [22]

Crystallization
–

Cloned
–

Renaturated
–

5 STABILITY

pH
7.5–9.0 [1]; 7.0 (activity decreases below [1], unstable above [3]) [1, 3]; 4.5
(45°C) [3]; 5–8 (40°C, 30 minutes) [4]; 6.5–8.0 (40°C, 30 minutes) [5]; 5.0–5.5
[12]; 3.5 (activity lost rapidly below) [12]; 2.5–7.5 (2 hours, 37°C) [13]; 10
(5–20% loss of activity after 60 minutes) [19]

Enzyme Handbook © Springer-Verlag Berlin Heidelberg 1991
Duplication, reproduction and storage in data banks are only
allowed with the prior permission of the publishers

Temperature (°C)

50 (rapid loss of activity above) [3]; 45 (stable up to 45 °C, 15 minutes at pH 5.5) [4]; 40 (stable up to, pH 6.5–8. 0, 30 minutes [4], 15 minutes stable up to 40°C) [9]; 37 (activity remains constant, 2 hours [9], stable up to 60 minutes [11]) [9, 11]; 50 (readily heat inactivated at) [10]; 65 (1 hour, 30–40% loss of activity) [19]

Oxidation

Inactivated in presence of oxygen, reactivated by dithiothreitol [12]

Organic solvent

General stability information

Freezing and thawing (stable) [11]; Freeze-drying (stable) [11]; Albumin (stabilizes) [1]; EDTA (stabilizes) [9]; 2-Mercaptoethanol (stabilizes) [9]

Storage

–20°C, [4]; –20°C, freeze-dried enzyme (no loss of activity) [11]

---

6 CROSSREFERENCES TO STRUCTURE DATABANKS

PIR/MIPS code

Brookhaven code

---

7 LITERATURE REFERENCES

[1] Iwasaki, Y., Tsuji, A., Omura, K., Suzuki, Y.: J. Biochem., 106, 331–335 (1989)
[2] Takegawa, K., Miki, S., Osaka, F., Jikibara, T., Iwahara, S.: Agric. Biol. Chem., 53, 1179–1180 (1989)
[3] McCleary, B.V.: Methods Enzymol., 160, 614–619 (1988)
[4] McCleary, B.V.: Methods Enzymol., 160, 589–595 (1988)
[5] Akino, T., Nakamura, N., Horikoshi, K.: Agric. Biol. Chem., 52, 1459–1464 (1988)
[6] Frei, J.I., Cavanagh, K.T., Fisher, R.A., Hausinger, J. R.P., Dupuis, M., Rathke, E.J.S., Jones, M.Z.: Biochemistry, 249, 871–875 (1988)
[7] Akino, T., Nakamura, N., Horikoshi, K.: Appl. Microbiol. Biotechnol., 26, 323–327 (1987)
[8] Ouellette, B.F.F., Bewley, J.D.: Planta, 169, 333–338 (1986)
[9] Noeske, C., Mersmann, G.: Hoppe-Seyler's Z. Physiol. Chem., 364, 1645–1651 (1983)
[10] Dawson, G.: J. Biol. Chem., 257, 3369–3371 (1982)
[11] Villarroya, H., Williams, J., Dey, P., Villarroya, S. , Petek, F.: Biochem. J., 175, 1079–1087 (1978)
[12] Sone, Y., Misaki, A.: J. Biochem., 83, 1135–1144 (1978)
[13] Bouquelet, S., Spik, G., Montreuil, J.: Biochim. Biophys. Acta, 522, 521–530 (1978)
[14] Elbein, A.D., Adya, S., Lee, Y.C.: J. Biol. Chem., 252, 2026–2031 (1977)
[15] Wan, C.C., Muldrey, J.E., Li, S.-C., Li, Y.-T.: J. Biol. Chem., 251, 4384–4388 (1976)
[16] Villarroya, H., Petek, F.: Biochim. Biophys. Acta, 438 , 200–211 (1976)
[17] McCleary, B.V., Matheson, N.K.: Phytochemistry, 14, 1187–1194 (1975)

[18] Houston, C., Latimer, S.B., Mitchell, E.D.: Biochim. Biophys. Acta, 370, 276–282 (1974)
[19] LaBadie, J.H., Aronson, N.N.: Biochim. Biophys. Acta, 321, 603–614 (1973)
[20] Bartholomew, B.A., Perry, A.L.: Biochim. Biophys. Acta, 315, 123–127 (1973)
[21] Sugahara, K., Okumura, T., Yamashina, I.: Biochim. Biophys. Acta, 268, 488–496 (1972)
[22] Sukeno, T., Tarentino, A.L., Plummer, T.H., Maley, F.: Biochemistry, 11, 1493–1501 (1972)
[23] Muramatsu, T., Egami, F.: J. Biochem., 62, 700–709 (1957)

Enzyme Handbook © Springer-Verlag Berlin Heidelberg 1991
Duplication, reproduction and storage in data banks are only
allowed with the prior permission of the publishers

1 NOMENCLATURE

EC number
3.2.1.26

Systematic name
Beta-D-fructofuranoside fructohydrolase

Recommended name
Beta-fructofuranosidase

Synonymes
Fructofuranosidase, .beta.-
Glucosucrase
.beta.-Fructofuranosidase
.beta-h-Fructosidase
.beta.-Fructosidase
Invertase
Saccharase
Invertin
Sucrase
Maxinvert L 1000
Fructosylinvertase
Alkaline invertase [18]
Acid invertase [21, 38, 39]
More (Sigma) [28, 29]

CAS Reg. No.
9001-57-4

2 REACTION AND SPECIFICITY

Catalysed reaction
Hydrolysis of terminal non-reducing beta-D-fructofuranoside residues in
beta-D-fructofuranosides (mechanism) [1]

Reaction type
O-Glycosyl bond hydrolysis

Natural substrates
Sucrose + H_2O

Enzyme Handbook © Springer-Verlag Berlin Heidelberg 1991
Duplication, reproduction and storage in data banks are only
allowed with the prior permission of the publishers

Substrate spectrum

1 Sucrose + H_2O [1–51]
2 Raffinose + H_2O (not [18], low activity [40]) [1, 5, 6, 12, 26, 27, 30, 32, 34, 37, 38, 39, 40, 43, 44, 48, 49]
3 Beta-fructofuranosides + H_2O (beta-fructofuranosides of simple alcohols [1]) [1]
4 Gentianose + H_2O [1]
5 Sucrose phosphate + H_2O [1]
6 Inulin + H_2O (low rate [1, 19]) [1, 19, 20, 34]
7 Disaccharides (of sucrose type) + H_2O [1]
8 Methyl-beta-D-fructofuranoside + H_2O [3, 5, 6]
9 Stachyose + H_2O [6, 12, 27, 39, 49]
10 1-Kestose + H_2O [7]
11 Nystose + H_2O [7]
12 Sucrose + H_2O [7]
13 More (also catalyses fructotransferase reaction [1, 5, 7, 8], not: melizitose [1, 3], not: trehalose [3], no appreciable transferase activity [11], external transferase: transferring activity with methanol, ethanol, n-propanol, butyl alcohol, aniline, nitroaniline, 2-mercaptoethanol, allyl alcohol, 2-chloroethanol, internal invertase: transferring activity with methanol, ethanol propanol, isopropanol, benzyl alcohol [8]) [1, 3, 5, 7, 8, 11]

Product spectrum

1 Fructose (beta-configuration) + glucose + fructooligosaccharides
2 Galactose [6] + sucrose + melibiose [6, 37] + fructose [6, 37] + glucose [6]
3 Fructose + alcohol
4 ?
5 ?
6 Fructose [34]
7 Monosaccharides
8 Fructose + ?
9 ?
10 Nystose [7]
11 1^F-Fructosyl-nystose [7]
12 1-Kestose [7]
13 ?

Inhibitor(s)

I_2 [1, 5, 8, 26, 49]; Hg^+ [1]; Hg^{2+} [1, 2, 7, 12, 18, 26, 27, 32, 40, 43, 49]; Organomercury compounds [1]; KI (slightly: isoform Ib, not: Ia, IIa, IIb) [12]; $AgNO_3$ (invertase II [49]) [13, 26, 49]; K[Fe(CN)6] [13]; Phenyl mercuri acetate [1]; o-Hydroxymethyl-phenyl mercury chloride [1]; Ag^+ (polyacrylamide embedded enzyme less inhibited than CM-cellulose adsorbed enzyme [35]) [1, 32, 35, 43]; Cu^{2+} [1, 6, 18, 40, 43]; Cd^{2+} [1, 43];

Zn^{2+} (slight [18]) [1, 3, 6]; Pb^{2+} [1]; UO_2^{2+} [1]; Suramin [1]; SDS [1, 32];
Heparin [1]; Aniline [1, 3, 5]; Aromatic amines [1]; 3, 4-Xylidine [1]; 3, 5-
Xylidine [1]; m-Toluidine [1]; Iodoacetate (very weak [1], not [2]) [1]; FeCN
(very weak) [1]; 1-Fluoro-2, 4-dinitrobenzene [1]; Antibiotics [1]; Tyrosinase
[1]; Peroxidase [1]; Pyridoxal [3, 5, 12, 26, 49]; Pyridoxal phosphate [3]; Urea
[3]; 2-Amino-2-hydroxymethyl propane-1 , 3-diol [5]; Fructose (slight [18],
not [27]) [6, 18, 39, 41]; Melezitose (at high concentration) [6]; Beta-methyl-
fructoside (at high concentration [6]) [6, 39]; Inulin (at high concentration)
[6]; Co^{2+} [6]; Mn^{2+} [6]; Pyridoxine [6 , 12, 26, 49]; Pyridoxamine [6, 26, 49];
Pyridine [6]; Heptamolybdate [6]; NH_4^+ [8]; CNBr [8]; K_2PtCl_4 [8]; p-Chloromercuribenzoate (not: 0.010 mM or lower [48], polyacrylamide embedded
enzyme less inhibited than CM-cellulose adsorbed enzyme [35]) [12, 13, 26,
27, 32, 35, 40, 49]; N-Ethylmaleimide [12]; p-Chloromercuriphenyl sulfonic
acid [26]; Fructose-6-phosphate (slight) [27]; Lauryl sulfate [33]; Metasili-
cate [33]; o-Phenanthroline [36]; Alpha, alpha-biphyridyl [36]; Sufhydryl
reagents [40]; Turanose [39]; Maltose [39]; Glucose (not [6]) [39, 41]; Fructose 1, 6-diphosphate [41]; Maleic acid [41]; Trans-aconitic acid [41]; Malic
acid [41]; Ascorbic acid [41]; 2, 5-Anhydro-D-mannitol [43]; Aniline (invertase I) [49]; Sugars (overview) [39]; p-Substituted mercuribenzoate (not:
Neurospora crassa enzyme [3]) [1, 2]; m-Substituted mercuribenzoate [1]

Cofactor(s)/prostethic group(s)
More (phosphate and arsenate ions required for activity) [40]

Metal compounds/salts
Mg^{2+} (activates) [6, 36]; Sr^{2+} (activates) [6]; Ba^{2+} (activates) [6]; KCN
(stimulates) [40]; NaN_3 (stimulates) [40]; $Cu(NO_3)_2$ (0.001 mM–1 mM, stimulates) [48]; Co^{2+} (reactivation after inhibition by o-phenanthroline) [36]; Arsenate (required for activity) [40]; Phosphate (required for activity) [40]

Turnover number (min⁻¹)
1555 [41]

Specific activity (U/mg)
2700 [2]; 28.9 [6]; 2810 [8]; 2900 [10]; 455 [32]; 2560 [34]; More [2, 3, 8, 13,
18, 19, 25, 26, 27, 37, 38, 39, 40, 42, 43, 47, 48]

K_m-value (mM)
44 (sucrose, yeast, soluble) [3]; 0.33 (beta-methyl-fructoside, Neurospora)
[3]; 6.1 (sucrose, Neurospora) [3]; 25–26 (sucrose, yeast, external enzyme)
[5]; 6.5 (raffinose, Neurospora) [3]; 150 (raffinose, external enzyme) [5]; 4.2
(sucrose, self transferase reaction) [6]; 290 (sucrose, self-transferase reaction) [7]; 800 (1-kestose) [7]; 140 (nystose, self-transferase reaction) [7]; 200
(raffinose, self transferase reaction) [7]; More (K_m of micelle entrapped invertase [16], of immobilized enzyme [29, 45], enzyme II: biphasic curve in
Lineweaver-Burk plot [18]) [8, 11, 13, 16, 18, 20, 21, 26, 27, 28, 29, 30, 33, 35,
39, 40, 42, 43, 44, 45, 48, 49, 51]

Enzyme Handbook © Springer-Verlag Berlin Heidelberg 1991
Duplication, reproduction and storage in data banks are only
allowed with the prior permission of the publishers

pH-optimum
3.5 (isoform IIb [12], 2 optima: 3.5 and 5.5 [41]) [12, 41]; 5.0 (isoform Ia [12],
P-II [32], invertase II [49]) [12, 32, 49]; 5.0–6.0 [7]; 4.3 (invertase I) [49]; 8.0
[18]; 5.5 (soluble [26], 2 optima: 3.5 and 5.5 [41]) [26, 41]; 4.0 (invertase 1
[27]) [27, 34, 38]; 5.6 (invertase 2) [27]; 3–6 (immobilized enzyme) [29];
4.6–5.0 [30]; 11.8 [36]; 6.8 [40]; 6.5 [42]; 6.2 [48]; 3.5–5.0 (yeast, external enzyme) [5]; 5.1 (raffinose) [13]; 4.5 (isoform: Ib, IIa [12], cell-wall bound enzyme [26], sucrose [13], P-I [32], polyacrylamide embedded and
carboxymethyl-cellulose adsorbed enzyme [35]) [6, 11, 12, 13, 26, 32, 35, 37,
43]; 4–4.5 [50]; More [1, 3]

pH-range
4.0–10.0 (4.0: 25% of maximal activity, 10.0: 10% of maximal activity) [27];
2.5–7.5 (2.5: about 25% of maximal activity, 7.5: about 5% of maximal activity) [11]; 3.0–7.0 (3.0: about 50–60% of maximal activity, 7.0: less than
10% of maximal activity) [30]; 3.0–6.0 (3.0: about 50% of maximal activity,
6.2: less than 5% of maximal activity) [13]; More [1, 26, 27, 30]

Temperature optimum (°C)
50–60 [7]; 50 (immobilized enzyme [29]) [26, 30]; 40 [40]; 47 (P-I) [32]; 52
(P-2) [32]; 75 [34]; 37 [36]; 60 [38]; 70 [50]

Temperature range (°C)
30–85 (at 30°C and 85°C about 10% of maximal activity) [34]; 30–70 (30°C:
about 25% of maximal activity, 70°C: about 40% of maximal activity) [7];
10–70 (10°C: about 10% of maximal activity, native and immobilized enzyme, 70°C: about 25% of maximal activity, immobilized enzyme) [29]; More
[30]

---

3 ENZYME STRUCTURE

Molecular weight
270000 (sedimentation equilibrium, Saccharomyces) [3]
52000 (Carica papaya) [6]
340000 (gel filtration, Aspergillus niger) [7]
100000 (SDS-PAGE, Aspergillus niger) [7]
210000 (sedimentation equilibrium centrifugation, Neurospora crassa) [9]
135000 (yeast, gel filtration) [10]
58000 (SDS-PAGE, Helianthus tuberosus) [17]
30000 (SDS-PAGE, Helianthus tuberosus) [17]
280000 (gel filtration, sugar beet, I /II) [18]
84000 (gel filtration, Aspergillus ficuum) [19]
60000 (gel filtration, Citrus sinensis, acid phosphatase [21], gel filtration,
Saccharomyces cerevisiae [25]) [21, 25]
158000 (Triticum aestivum, DEAE-cellulose chromatography) [26]
28000 (gel filtration, sugar beet) [30]

202000 (sedimentation equilibrium, Fusarium oxysporum, P-1) [32]
140000 (sedimentation equilibrium, Fusarium oxysporum, P-2) [32]
66000 (gel filtration, sugar cane, neutral invertase) [33]
380000 (gel filtration, sugar cane, acid invertase) [33]
22000 (gel filtration, Corticium rolfsii) [34]
300000 (gel filtration, Candida utilis) [37]
73000 (SDS-PAGE, Nicotiana tabacum) [38]
76000 (gel filtration, Tropaeolum majus) [39]
92000 (gel filtration, Brevibacterium divaricatum) [40]
11000 (gel filtration in presence of SDS, Ricinus communis) [41]
78000 (gel filtration, Ricinus communis) [41]
35400 (gel filtration, Pseudomonas fluorescens) [42]
48500 (polyacrylamide gradient gel electrophoresis, Raphanus sativus) [44]
59000 (gel filtration, invertase I) [49]
108000 (gel filtration, invertase II) [49]
95000 (SDS-disc electrophoresis, Candida utilis) [50]
More (Saccharomyces cerevisiae: ultracentifuge analysis, molecular weight
depends on pH, pH 9.4: 75800, pH 8.3: 112500, pH 7.5: 180500, pH 4.9:
409000 [24], Fusarium oxysporum: molecular weight and subunit structure
are temperature dependent [32]) [24, 32]

Subunits

Tetramer (4 × 51000, may be composed of more than one type of subunit,
sedimentation equilibrium centifugation /guanidine hydrochloride, 2-
mercaptoethanol, Neurospora crassa) [9]
Monomer (pH 9.4, Saccharomyces cerevisiae) [24]
Dimer (pH 8.3, Saccharomyces cerevisiae) [24]
Octamer (pH 4.9, Saccharomyces cerevisiae) [24]
Monomer (column chromatography, Saccharomyces cerevisiae, internal
enzyme exists in native state as monomer) [25]
Oligomer (column chromatography, Saccharomyces cerevisiae, external
enzyme: oligomer, inactive when converted to monomer) [25]
Heptamer (7 × 110000, identical, molecular weight determination in
presence and absence of activator, point of optimal activation, gel filtration
in presence of SDS, Ricinus communis) [41]
More (subunit structure: pH-dependent [24], temperature-dependent, P-1:
trimer/dimer/monomer/, P-2: tetramer/dimer/monomer [32]) [24, 25, 32, 42]

Glycoprotein/Lipoprotein

Glycoprotein (yeast [1], Saccharomyces: 50 % carbohydrate, mannan with
3% glucosamine [2], yeast: less than 3%–80% carbohydrate [3], external
invertase: outside cell membrane contains 50% carbohydrate depending
on the species of yeast and purification method, internal invertase: within
cell, little or no carbohydrate [5], Aspergillus niger: 20% [7], Neurospora
crassa: glycoprotein containing 11% mannose, 3% glucosamine [9],
Saccharomyces: internal enzyme has little or no carbohydrate, external

Enzyme Handbook © Springer-Verlag Berlin Heidelberg 1991
Duplication, reproduction and storage in data banks are only
allowed with the prior permission of the publishers

enzyme is glycoprotein containing 50% mannan and 3% glucosamine [11], tomato, E: 5.3%, L: 9.1% [13], Fusarium oxysporum, mycelial: 36% conidial: 23%, predominantly mannose with smaller percentage of glucose, galactose, N-acetylglucosamine, N-acetylgalactosamine [32], sugar cane, acid invertase: 23.5%, neutral invertase: 22% [33], Raphanus sativus: sugar content 7.7% [44], carbohydrate structure of yeast invertase [46], Phytophtora megasperma: protein to carbohydrate ratio: 3.2, major components: glucose + mannose [48]) [1, 2, 3, 5, 7, 9, 11, 13, 31, 32, 33, 38, 43, 44, 46, 47, 48, 50]

---

4 ISOLATION/PREPARATION

Source organism
Carica papaya [6]; Aspergillus niger [7]; Neurospora crassa [9]; Vitis vinifera (isoforms: Ia, Ib, IIa, IIb) [12]; Tomato (2 forms: E, L [13], embedded within polyacrylamide gel /adsorbed on carboxymethyl-cellulose [35]) [13, 35]; Helianthus tuberosus [17]; Sugar beet (2 forms: I, II [18], alkaline and acid invertase [51]) [18, 30, 51]; Aspergillus ficuum [19]; Clostridium acetobutylicum [20]; Citrus sinensis [21]; Triticum aestivum (2 invertases: soluble, cell wall bound) [26]; Lilium auratum (invertase I: cytoplasmic, invertase II: bound to pollen wall) [27]; Fusarium oxysporum (P-1: mycelial, P-2 : conidial [32]) [32, 47]; Sugar cane (acid invertase and neutral invertase) [33]; Corticium rolfsii [34]; Neurospora sitophila [36]; Nicotiana tabacum [38]; Tropaeolum majus [39]; Brevibacterium divaricatum [40]; Ricinus communis [41]; Pseudomonas fluorescens [42]; Kluyveromyces fragilis [43]; Raphanus sativus [44]; Phytophthora megasperma (var. sojae) [48]; Avena sativa (invertase I, II) [49]; Yeast (internal and external invertase [5], internal [8, 10], membrane associated isozyme precursor of external glycoprotein invertase [31]) [1, 3, 5, 8, 10, 14, 15, 31]; Saccharomyces cerevisiae [1, 3, 4, 23, 24, 25, 46]; Bacteria [1]; Saccharomyces (strain FH4C) [2]; Neurospora [3]; Candida utilis [3, 4, 37, 50]; More [1]

Source tissue
Fruits [6, 13]; Pollen [27]; Shoots [17]; Cell [16, 38, 42, 47]; Berries [11]; Roots [18, 30]; Coleoptiles [26, 48 , 49]; Mycelium [32, 48]; Conidia [32]; Juice (sugar cane) [33]; Culture filtrate [34]; Leaves [21, 39, 41]; Seedlings [44, 48, 49]; Cultured cells (single cells from leaf explant) [51]

Localisation in source
Intracellular [3, 5, 8, 10, 11, 20]; Soluble [3, 11, 17, 26, 27]; Cell wall (bound [26, 30], acid invertase [50]) [13, 26, 27, 30, 50]; Membrane [31]; Cytoplasm (alkaline and acid invertase) [51]; Extracellular (acid invertase) [51]; More (outside permeability barrier [1, 2, 5, 11]) [1, 2, 5, 11, 31]

Purification
Saccharomyces cerevisiae [8, 24, 46]; Tomato (2 forms: E, L) [13];
Helianthus tuberosus [17]; Sugar beet (partial, 2 forms: I, II) [18, 30];
Aspergillus ficuum [19]; Clostridium acetobutylicum [20]; Citrus sinensis
[21]; Triticum aestivum [26]; Lilium auratum [27]; Fusarium oxysporum [32,
47]; Corticium rolfsii [34]; Candida utilis [37]; Nicotiana tabacum [38];
Tropaeolum majus (partial) [39]; Brevibacterium divaricatum [40]; Pinus
commumis [41]; Pseudomonas fluorescens (partial) [42]; Kluyveromyces
fragilis [43]; Raphanus sativus [44]; Phytophthora megasperma [48];
Saccharomyces (strain FH4C) [2]; Yeast (internal and external invertase [5,
28], internal [10], membrane associated isozyme-precursor of external
glycoprotein invertase) [31]; Carica papaya [6]; Aspergillus niger [7]

Crystallization
(Fusarium oxysporum) [47]

Cloned
More (amplification) [24]

Renaturated
(yeast, no effect of carbohydrate moiety on renaturation [14], external invertase: complete renaturation, beta-N-acetylglucosaminidase H-treated
external invertase: 40% renaturation, internal invertase: 2% renaturation
[25]) [14, 25]

---

5 STABILITY

pH
5 (most stable at) [2]; 3.5 (low stability at and below, invertase II) [49]; 3–7
(lability not pH-dependent between) [6]; 6 (very unstable above [5], unstable at acid pH [40]) [5, 40]; 4.4–10.0 (stable, purified enzyme) [7]; 6.5–7.0
(stablility optimum) [43]; 10 (30°C, stability optimum) [36]; 3–7.5 (30°C,
stable between, external invertase) [11]; 6–9 (30°C, stable between, internal
invertase) [11]; 1.5–8.0 (2°C, stable) [34]; 3–6 (stable) [37]; More [11, 49, 50]

Temperature (°C)
56 (pH 4.9, stable for 15 minutes, yeast, external enzyme) [5]; 55 (5 minutes,
complete inactivation [13], micelle entrapped enzyme quite stable at [14],
10 minutes: P-I, complete loss of activity, P-II, 20% loss of activity [32], 5
minutes, complete loss of activity [44]) [13, 14, 32, 44]; 60 (substitution of
D_2O for H_2O: half-life is 4 times greater) [22]; 65 (10 minutes, stable below)
[34]; 70 (10 minutes, 10 % loss of activity) [34]; 30 (14% loss of activity after
90 minutes [36], less than 5% loss of activity [40]) [36, 40]; 30 (pH 3–7,
stable for at least 2 hours, yeast, external enzyme) [5]; 50 (30 minutes, less
than 20% loss of activity [7], 4 minutes: soluble, 40% loss of activity, cell-
wall bound, 80% loss of activity [26], 20 minutes, invertase 1: 60% loss of

Enzyme Handbook © Springer-Verlag Berlin Heidelberg 1991
Duplication, reproduction and storage in data banks are only
allowed with the prior permission of the publishers

activity, invertase 2: 8% loss of activity [27], 8.2% loss of activity after 90 minutes [36], stable below [37], 15 minutes, more than 85% loss of activity [40], stable at [43]) [7, 26, 27, 36, 37, 40, 43]; More (no significant effect of carbohydrate moiety on heat stability [15], thermostability of immobilized enzyme [23], themostability of polyacrylamide embedded and carboxymethyl-cellulose adsorbed enzyme [35]) [23, 26, 35, 49, 50]

Oxidation
Oxidation of sugar moiety with periodate (stabilization decreased) [50]

Organic solvent
Benzene (stabilization of invertase entrapped in reversed micelles of sodium lauryl sulfate and sodium tauroglycolate in organic solvents) [16]; n-Decanol (stabilization of invertase entrapped in reversed micelles of sodium lauryl sulfate and sodium tauroglycolate in organic solvents) [16]; Carbon tetrachloride (stabilization of invertase entrapped in reversed micelles of sodium lauryl sulfate and sodium tauroglycolate in organic solvents) [16]; Hexane (stabilization of invertase entrapped in reversed micelles of sodium lauryl sulfate and sodium tauroglycolate) [16]

General stability information
Immobilized enzyme (on bead DEAHP-cellulose, half life: 215 days) [28]; Stability as salt-free lyophilized powder [31]; Repeated freezing and thawing (no loss of activity) [44]; Labile [6]; Invertase entrapped in reversed micelles of sodium lauryl sulfate and sodium tauroglycolate in organic solvents (stabilization) [16]

Storage
For months in frozen or lyphilized state [2]; 21°C or −20 °C, 6 days (little loss of activity) [31]; More [5]

---

6 CROSSREFERENCES TO STRUCTURE DATABANKS

PIR/MIPS code
IFBY (precursor, yeast); A27748 (yeast, Saccharomyces cerevisiae); A05306 (7, precursor, yeast, Saccharomyces cerevisiae, fragments)

Brookhaven code

---

7 LITERATURE REFERENCES

[1] Myrbäck, K. in "The Enzymes", 2nd Ed. (Boyer, P, D., Ed.) 4, 379–396 (1960) (Review)
[2] Neumann, N.P., Lampen, J.O.: Biochemistry, 6, 468–475 (1967)
[3] Lampen, J.O. in "The Enzymes", 3rd Ed. (Boyer, P.D., Ed.) 5, 291–305 (1971) (Review)

[4] Shiomi, N., Onodera, S.: Agric. Biol. Chem., 52, 2347–2348 (1988)
[5] Goldstein, A., Lampen, J.O.: Methods Enzymol., 42C, 504–511 (1975)
[6] Lopez, M.E., Vattuone, M.A., Sampietro, A.R.: Phytochemistry, 27, 3077–3081 (1988)
[7] Hirayama, M., Sumi, N., Hidaka, H.: Agric. Biol. Chem., 53, 667–673 (1989)
[8] Baseer, A., Shall, S.: Biochim. Biophys. Acta, 250, 192–202 (1971)
[9] Meachum, Z.D., Colvin, H.R., Braymer, H.J.: Biochemistry, 10, 326–332 (1971)
[10] Gascon, S., Lampen, J.O.: J. Biol. Chem., 243, 1567–1572 (1968)
[11] Gascon, S., Neuman, N.P., Lampen, J.O.: J. Biol. Chem., 243, 1573–1577 (1968)
[12] Ishikawa, N., Nakagawa, H., Ogura, N.: Agric. Biol. Chem., 53, 837–838 (1989)
[13] Nakagawa, H., Kawasaki, Y., Ogura, N., Takehana, H.: Agric. Biol. Chem., 36, 18–26 (1971)
[14] Schülke, N., Schmidt, F.X.: J. Biol. Chem., 263, 8832–8837 (1988)
[15] Schülke, N., Schmidt, F.X.: J. Biol. Chem., 263, 8827–8831 (1988)
[16] Madamwar, D.B., Bhatt, J.P., Ray, R.M.: Enzyme Microb. Technol., 10, 302–305 (1988)
[17] Goupil, P., Croisille, Y., Croisille, F., Ledoigt, G.: Plant Sci., 54, 45–54 (1988)
[18] Masuda, H., Takahshi, T., Sugawara, S.: Agric. Biol. Chem., 51, 2309–2314 (1987)
[19] Ettalibi, M., Baratti, J.C.: Appl. Microbiol. Biotechnol., 26, 13–20 (1987)
[20] Looten, Ph., Blanchet, D., Vandecasteele, J.P.: Appl. Microbiol. Biotechnol., 25, 419–425 (1987)
[21] Schaffer, A.A.: Phytochemistry, 25, 2275–2277 (1986)
[22] Combes, D., Monsan, P. in "Eur. Congr. Biotechnol.", 3rd Ed., 1, 233–237 (1984)
[23] Mansfeld, J., Schellenberger, A.: Acta Biotechnol., 6, 89–99 (1986)
[24] Williams, R.S., Trumbly, R.J., MacColl, R., Trimble, R.B., Maley, F.: J. Biol. Chem., 260, 13334–13341 (1985)
[25] Chu, F.K., Takase, K., Guarino, D., Maley, F.: Biochemistry, 24, 6125–6132 (1985)
[26] Krishnan, H.B., Blanchette, J.T., Okita, T.W.: Plant Physiol., 78, 241–245 (1985)
[27] Singh, M.B., Knox, R.B.: Plant Physiol., 74, 510–515 (1984)
[28] Hradil, J., Svec, F.: Enzyme Microb. Technol., 3, 336–340 (1981)
[29] Hradil, J., Svec, F.: Enzyme Microb. Technol., 3, 331–335 (1981)
[30] Masuda, H., Sugawara, S.: Plant Physiol., 66, 93–96 (1980)
[31] Babczinski, P.: Biochim. Biophys. Acta, 614, 121–133 (1980)
[32] Nishizawa, M., Maruyama, Y., Nakamura, M.: Agric. Biol. Chem., 44, 489–498 (1980)
[33] Del Rosario, E.J., Santisopasri, V.: Phytochemistry, 16, 443–445 (1977)
[34] Sato, M., Kaji, A.: Agric. Biol. Chem., 40, 2107–2108 (1976)
[35] Nakagawa, H., Arao, T., Matsuzawa, T., Ito, S., Ogura, N., Takehana, H.: Agric. Biol. Chem., 39, 1–5 (1975)
[36] Dixon, M.M., Fogarty, W.M.: Biochem. Soc. Trans., 552nd Meeting, Vol. 2, 1339–1341 (1974)
[37] Iizuka, M., Tsuji, Y., Yamamoto, T.: Agric. Biol. Chem., 38, 213–215 (1974)
[38] Nakamura, M., Hagimori, M., Matsumoto, T.: Agric. Biol. Chem., 52, 3157–3158 (1988)
[39] Isla, M.I., Vattuone, M.A., Gutierrez, M.I., Sampietro, A.R.: Phytochemistry, 27, 1993–1998 (1988)
[40] Yamamoto, K., Kitamoto, Y., Ohata, N., Isshiki, S., Ichikawa, Y.: J. Ferment. Technol., 64, 285–291 (1986)
[41] Prado, F.E., Vattuone, M.A., Fleischmacher, O.L., Sampietro, A.R.: J. Biol. Chem., 260, 4952–4957 (1985)
[42] Bugbee, W.M.: Can. J. Microbiol., 30, 1326–1329 (1984)

Enzyme Handbook © Springer-Verlag Berlin Heidelberg 1991
Duplication, reproduction and storage in data banks are only
allowed with the prior permission of the publishers

[43] Workman, W.E., Day, D.F.: FEBS Lett., 160, 16–20 (1983)
[44] Faye, L., Berjonneau, C., Rollin, P.: Plant Sci. Lett., 22, 77–87 (1981)
[45] Shiomi, T., Tohyama, M., Satoh, M., Miya, M., Imai, K.: Biotechnol. Bioeng., 32, 664–668 (1988)
[46] Lehle, L., Cohen, R.E., Ballou, C.E.: J. Biol. Chem., 254, 12209–12218 (1979)
[47] Onodera, K., Mariyama, Y.: J. Gen. Appl. Microbiol., 25, 335–337 (1979)
[48] West, C., Wade, M., McMillan III, C., Albersheim, P.: Arch. Biochem. Biophys., 201, 25–35 (1980)
[49] Pressey, R., Avants, J.K.: Plant Physiol., 65, 135–140 (1980)
[50] Iizuka, M., Yamamoto, T.: Agric. Biol. Chem., 43, 217–222 (1979)
[51] Masuda, H., Takahashi, T., Sugawara, S.: Plant Physiol., 86, 312–317 (1988)

1 NOMENCLATURE

EC number
3.2.1.28

Systematic name
Alpha, alpha-trehalose glucohydrolase

Recommended name
Alpha, alpha-trehalase

Synonymes
Alpha, alpha'-trehalase
Trehalase

CAS Reg. No.
9025-52-9

---

2 REACTION AND SPECIFICITY

Catalysed reaction
Alpha, alpha-trehalose + $H_2O \rightarrow$
$\rightarrow$ 2 D-glucose

Reaction type
O-Glycosyl bond hydrolysis

Natural substrates
Alpha, alpha-trehalose + H_2O (function of trehalose in Saccharomyces cervisiae [31])

Substrate spectrum
1 Alpha, alpha-trehalose (alpha-D-glucopyranosyl-alpha-D-glucopyranoside) + H_2O (absolute specificity for trehalose)

Product spectrum
1 D-Glucose

Enzyme Handbook © Springer-Verlag Berlin Heidelberg 1991
Duplication, reproduction and storage in data banks are only
allowed with the prior permission of the publishers

Inhibitor(s)

$HgCl_2$ (protection from inactivation by NaCl [2]) [1, 2, 7, 11, 17, 18, 20]; Tris [2, 7, 8, 11, 16, 17, 19, 20]; Phlorizin [2, 7, 8, 17]; Sucrose [6, 16, 19, 20]; Mannose [6]; Glucose [6]; Maltose [6]; Fructose [6]; Mannitol [6]; Sorbitol [6]; Methyl-beta-glucoside [7]; EDTA (c-trehalase) [10]; Acetic acid/acetate buffer (v-trehalase) [10]; $ZnCl_2$ (c-trehalase) [10]; $MgCl_2$ [11, 20]; $CuSO_4$ [11]; $CaCl_2$ [11]; Urea [11]; ATP [12, 21]; p-Nitrophenyl-beta-D-glucoside [16, 19]; SDS [17]; N-Ethylmaleimide [17]; p-Chloromercuribenzene sulfonic acid [17]; Phenyl-beta-glucoside [19]; p-Aminophenyl-beta-D-glucoside [19]; p-Chloromercuribenzoate [19]; Iodoacetamide [19]

Cofactor(s)/prostethic group(s)

Metal compounds/salts

Ca^{2+} (activation of c-trehalase) [10]; Mn^{2+} (activation of c-trehalase) [10]; KCl (activation of v-trehalase) [10]; $NaNO_3$ (activation of v-trehalase) [10]

Turnover number (min^{-1})

Specific activity (U/mg)

1056 [5]; 343 [8]; 99 [1]; More (assay method [9]) [2, 4, 7, 8, 10, 11, 13, 14, 15, 17–20, 22]

K_m-value (mM)

0.13–5.7 (trehalose) [1–8, 10, 11, 15, 16, 18–20, 23]; 5.5–11 (trehalose, depending on buffer) [17]; 55 (trehalose) [21]

pH-optimum

4–5 (v-trehalase) [10]; 4.4 [14]; 4.5 [1]; 4.5–5.3 [6]; 4.7 (acetate buffer) [17]; 5.5 [15, 18, 23]; 5.5–5.7 [20]; 5.5–6.0 [8]; 5.7 (phosphate or histidine buffer) [17]; 5.8–6.0 [2]; 5.9 [7]; 6.0 [16]; 6.5 [19]; 6.7 (c-trehalase) [10]

pH-range

3–7 [6]; 4.5–6.8 (more than 30% activity) [2]

Temperature optimum (°C)

50 [11, 14, 15]

Temperature range (°C)

3 ENZYME STRUCTURE

Molecular weight
   210000–240000 (yeast suc 2 deletion mutant, gel filtration, SDS-PAGE [1],
   Saccharomyces cerevisiae, v-trehalase, gel filtration [10], rat, gel filtration
   [17], Phycomyces blakesleeanus, gel filtration [21]) [1, 10, 17, 21]
   160000–175000 (rabbit, membrane bound form, gel filtration [3, 7],
   Saccharomyces cerevisiae, gel filtration, glycerol gradient centrifugation
   [4], c-trehalase, gel filtration [10], Thermomyces lanuginosus, gel filtration
   [15]) [3, 4, 7, 10, 15]
   48000–105000 (pig, SDS-PAGE [2], rabbit amnionic fluid, gel filtration, calculation from sedimentation coefficient and Stoke's radius, radiation inactivation [3], rabbit intestine, gel filtration of detergent solubilized and
   protease treated enzyme [5], rabbit kidney, gel filtration, SDS-PAGE [8], Dictyostelium discoideum, gel filtration, pore gradient electrophoresis, non-
   denaturing polyacrylamide gel electrophorresis, SDS-PAGE, comparison
   with values from other organisms [11], Trichoderma reesei, SDS-PAGE [14],
   Apis mellifera, gel filtration at pH 6.5 [19], rat, gel filtration [20]) [2, 3, 5, 8,
   11, 14, 19, 20]
   33000 (Apis mellifera, gel filtration at pH 3.5, SDS-PAGE) [19]

Subunits
   Monomer [11]
   Dimer (2 × 86000, rabbit, SDS-PAGE [4], 2 × 75000, rabbit, SDS-PAGE [5])
   [4, 5]
   Oligomer (x × 30000, rat, SDS-PAGE) [17]

Glycoprotein/Lipoprotein
   Glycoprotein (86% w/w neutral sugar [1]) [1, 3, 7, 11]

4 ISOLATION/PREPARATION

Source organism
   Rabbit [3, 5, 7, 8, 24]; Yeast (suc 2 deletion mutant 2.64.1 Calpha) [1]; Pig
   [2, 25]; Saccharomyces cerevisiae [4, 10, 18, 22, 31]; Frankia (strain Ar I3)
   [6]; Dictyostelium discoideum [11]; Pichia pastoris [12]; Drosophila
   melanogaster [13]; Trichoderma reesei [14]; Thermomyces lanuginosus
   (formerly Humicola lunuginosa) [15]; Rhynchosciara americana [16]; Rat
   [17, 20]; Apis mellifera (honey bee) [19]; Phycomyces blakesleeanus [21];
   Galleria mellonella [23]; Blaberus discoidales (cockroach) [26]; Pseudomonas fluorescens [27]; Cecropia [33]; Streptomyces hygroscopicus [28];
   Aspergillus oryzae [29]; Ant [30]; Yeast [32]; Fungi [32]; Insects [32]; Plants
   [32]; Bacteria [32]

Source tissue
Kidney cortex [2, 7, 8, 24]; Amnionic fluid [3]; Small intestine [5, 8, 17, 20, 25]; Intestines [26]; Muscle [26, 33]; Conidia [29]; Cell [11, 18, 22]; Ascospores [12]; Culture filtrate [14]; Mycelia (acetone/butanol dried) [15]; Larvae [16, 23]; Thorax [19]; Spores [21]

Localisation in source
Vacuoles [1]; Membranes (brush border [2] microvillar [6, 17, 20]) [2, 6–8, 17, 19, 20]; Soluble parts of cell [10, 13, 19]

Purification
Yeast (suc 2deletion mutant) [1]; Saccharomyces cerevisae (partial, cryptic and active trehalase [4], cryptic trehalase and activating factor protein [22]) [4, 18, 22]; Rabbit (amphiphilic form [5], 4 forms of enzyme [7]) [5, 7, 8]; Drosophila melanogaster (partial) [13]; Trichoderma reesei (partial) [14]; Thermomyces lanuginosus (partial) [15]; Rhynchosciara americana (partial) [16]; Rat (partial) [17, 20]; Apis mellifera (solube and membrane bound form) [19]; Galleria mellonella [23]; Streptomyces hygroscopicus (partial) [28]

Crystallization
–

Cloned
–

Renaturated
–

---

5 STABILITY

pH
2.0 (above) [18]; 4.5–7.0 [8]; 5.1 (best stability) [14]

Temperature (°C)
45–50 (up to, comparison with enzyme from other sources) [11]; 50 (up to [8], 34 hours, 50% activity [14]) [8, 14]; 55 (up to [3], slight inactivation [8]) [3, 8]; 60 (1.8 minutes, 50% activity [14]); 70 (0.35 minutes, 50% activity [14]); More (influence of protection factor [15], comparison of stability of enzyme isolated from activated and dormand spores) [21]

Oxidation

Organic solvent

General stability information
Detergents (e.g. Tween 80, required for stability) [2]; Glyerol (stabilization) [11]; Freezing/thawing (no inactivation) [11]; Urea (8 M, 85% activity) [19]

Storage

−80°C, stability differs with form of enzyme [7]; −60°C, at least 6 months [3]; −20°C, [1, 22]; Liquid N_2, 5 mM sodium phosphate buffer, pH 6.8, 0.2% Tween 80 [2]; −15°C [6]; −15°C, 3 months, 82% activity [18]; 4°C, membrane enzyme 3 days, soluble enzyme 3 months [3]; 4°C, 20% glycerol, 5 months, 80% activity [4]; −12°C, at least 2 months, 4–6°C, 2 days [11]; 4°C, 3 months [14]; 0°C, at least 20 days [15]; −10°C, water pH 7, more than 2 months [16]; −10°C, 1 week, 70–80% activity [20]; 0°C, 0.05 M maleate buffer pH 6.5, 1 month [23]

6 CROSSREFERENCES TO STRUCTURE DATABANKS

PIR/MIPS code
S04782 (precursor, Escherichia coli)

Brookhaven code

7 LITERATURE REFERENCES

[1] Mittenbühler, K., Holzer, H.: J. Biol. Chem., 263, 8537–8543 (1988)
[2] Yoneyama, Y.: Arch. Biochem. Biophys., 255, 168–175 (1987)
[3] Morin, P.-R., Potier, M.: Biochim. Biophys. Acta, 923, 371–380 (1987)
[4] Dellamora-Ortiz, G.M., Ortiz, C.H.D., Maia, J.C.C., Panek, A.D.: Arch. Biochem. Biophys., 251, 205–214 (1986)
[5] Yokota, K., Nishi, Y., Takesue, Y.: Biochim. Biophys. Acta, 881, 405–414 (1986)
[6] Lopez, M.F., Torrey, J.G.: Arch. Microbiol., 143, 209–215 (1985)
[7] Nakano, M., Sacktor, B.: J. Biochem., 97, 1329–1335 (1985)
[8] Galand, G.: Biochim. Biophys. Acta, 789, 10–19 (1984)
[9] Dahlqvist, A. in "Methods Enzym Anal.", 3rd. Ed. (Bergmeyer, H.U., Ed.) 4, 208–217 (1984)
[10] Londesborough, J., Varimo, K.: Biochem. J., 219, 511–518 (1984)
[11] Killick, K.A.: Arch. Biochem. Biophys., 222, 561–573 (1983)
[12] Thevelein, J.M., Den Hollander, J.A., Shulman, R.C.: Proc. Natl. Acad. Sci. USA, 79, 3503–3507 (1982)
[13] Bargiello, T.A., Grossfield, J.: Anal. Biochem., 101, 131–137 (1980)
[14] Vijayakumar, P., Ross, W., Reese, E.T.: Can. J. Microbiol., 24, 1280–1283 (1978)
[15] Prasad, A.R.S., Maheshwari, R.: Biochim. Biophys. Acta, 525, 162–170 (1978)
[16] Terra, W.R., Ferreira, C., De Bianchi, A.G.: Biochim. Biophys. Acta, 524, 131–141 (1978)
[17] Nakano, M., Sumi, Y., Miyakawa, M.: J. Biochem., 81, 1041–1049 (1977)
[18] Kelly, P.J., Catley, B.J.: Anal. Biochem., 72, 353–358 (1976)
[19] Talbot, B.G., Muir, J.G., Huber, R.E.: Can. J. Biochem., 53, 1106–1117 (1975)
[20] Sasajima, K., Kawachi, T., Sato, S., Sugimura, T.: Biochim. Biophys. Acta, 403, 139–146 (1975)
[21] Van Assche, J.A., Carlier, A.R.: Biochim. Biophys. Acta, 391, 154–161 (1975)
[22] Van Solingen, P., Van Der Plaat, J.B.: Biochem. Biophys. Res. Commun., 62, 553–560 (1975)

Enzyme Handbook © Springer-Verlag Berlin Heidelberg 1991
Duplication, reproduction and storage in data banks are only
allowed with the prior permission of the publishers

[23] Kalf, G.F., Rieder, S.V.: J. Biol. Chem., 230, 691–698 (1958)
[24] Nakano, M.: Biochim. Biophys. Acta, 707, 115–120 (1982)
[25] Dahlqvist, A.: Acta Chem. Scand., 14, 9–16 (1960)
[26] Gilby, A.R., Wyatt, S.S., Wyatt, G.R.: Acta Biochim. Pol., 14, 83–100 (1967)
[27] Gouilloux, E., Acrila, M.A., Courtois, J.E., Mourmrikoff, V.: Biochimie, 53, 853–857 (1971)
[28] Hey, A., Elbein, A.D.: J. Bacteriol., 96, 105–110 (1968)
[29] Horikoshi, K., Ikeda, Y.: J. Bacteriol., 91, 1883–1887 (1966)
[30] Paulsen, R.: Arch. Biochem. Biophys., 142, 170–172 (1971)
[31] Panek, A.: Arch. Biochem. Biophys., 100, 422–425 (1963)
[32] Elbein, A.: Adv. Carbohydr. Chem. Biochem., 30, 227–256 (1974)
[33] Gussin, A.E.S., Wyatt, G.R.: Arch. Biochem. Biophys., 112, 626–634 (1965)

1 NOMENCLATURE

EC number
3.2.1.30

Systematic name
N-Acetyl-beta-D-glucosaminide N-acetylglucosaminohydrolase

Recommended name
N-Acetyl-beta-glucosaminidase

Synonymes
Beta-N-acetylglucosaminidase
Beta-acetylaminodeoxyglucosidase
Beta-acetamidodeoxyglucosidase
Beta-acetylglucosaminidase
N-Acetyl-beta-glucosaminidase
N-Acetyl-beta-D-glucosaminidase
Chitobiase
EC 3.2.1.29 (now included with EC 3.2.1.30)
N-Acetyl-beta-glucosaminidase
Acetyl-beta-glucosaminidase
Beta-D-glucosaminidase
Beta-N-acetyl-D-glucosaminidase
Beta-N-acetylaminodeoxyglucosidase
Exo-N-acetyl-beta-D-glucosaminidase
p-Nitrophenyl-beta-N-acetylglucosaminidase
Exochitinase
Beta-D-N-acetylglucosaminidase
More (discussion of IUB nomenclature [1])

CAS Reg. No.
9012-33-3

---

2 REACTION AND SPECIFICITY

Catalysed reaction
Hydrolysis of terminal, non-reducing N-acetyl-beta-D-glucosamine residues
in chitobiose and higher analogs and in glycoproteins (exoglycosidase,
reaction mechanism [12])

Reaction type
O-Glycosyl bond hydrolysis

Enzyme Handbook © Springer-Verlag Berlin Heidelberg 1991
Duplication, reproduction and storage in data banks are only
allowed with the prior permission of the publishers

Natural substrates

N, N'-Diacetylchitobiose (and higher oligomers) + H_2O [13]
Ovalbumin + H_2O [7, 14]
Asialo-agalactofetuin + H_2O [7]
Acid glycoprotein + H_2O [14]
Ovomucoid glycopeptide + H_2O [14]
Kappa-casein glycopeptide + H_2O [14]
Chitin + H_2O [18]
Murein-lipoprotein complex + H_2O [17]
Hylauronic acid + H_2O [18]
Chondroitin sulfate + H_2O [18]

Substrate spectrum

1 p-Nitrophenyl-2-acetamido-2-deoxy-beta-D-glucopyranoside
  (pNp-beta-GlcNAc, with respect to substrate specificity some of the cited
  enzymes probably belong to EC 3.2.1.52, so far analysed only those enzymes cited in [17, 21, 22, 27] clearly belong to EC 3.2.1.30) + H_2O
2 p-Nitrophenyl-2-acetamido-2-deoxy-beta-D-glactopyranoside
  (pNp-beta-GalNAc) + H_2O (no hydrolysis [17, 21]) [4, 7, 14]
3 N, N'-Diacetylchitobiose (and higher oligomers [4]) + H_2O [4, 5, 16, 23]
4 4-Methylumbelliferyl-beta-D-glucosaminide (or galactoaminide) + H_2O
  [8, 13, 20, 26]

Product spectrum

1 N-Acetylglucosamine + p-nitrophenol
2 N-Acetylgalactosamine + p-nitrophenol
3 N-Acetylglucosamine
4 4-Methylumbelliferone + N-acetylglucosamine (or
  N-acetylgalactosamine)

Inhibitor(s)

N-Acetylglucosamine [2, 8, 13, 17, 21, 25, 27]; N-Acetylgalactosamine [2];
p-Nitrophenyl-2-deoxy-2-thioacetamido-beta-D-gluco(or
galacto)pyranoside [7]; Hg^{2+} [3, 4, 7, 12, 14, 18, 20, 21, 24]; $CaCl_2$ [4];
$MgSO_4$ [4]; Zn^{2+} [7, 18, 21]; Fe^{3+} [7, 20, 24]; Ca^{2+} [12, 21]; Cu^{2+} (only
together with L-ascorbic acid [20]) [12, 20]; Ag^+ [12, 18, 20, 21, 24, 27];
2-Deoxy-2-acetamido-D-glucono-1,5-lactone [13, 17, 21, 23]; Iodoacetamide
[14]; p-Chloromercuribenzoate [14, 24, 25, 27];
p-Aminophenyl-1-thio-beta-L-fucopyranoside [15]; N-Acetylmuramic acid
[17]; Cd^{2+} [21]; Acetate [25]

Cofactor(s)/prostethic group(s)

EDTA (11% activation) [4]

Metal compounds/salts

No metal ions required [7]; No Zn^{2+} or Mg^{2+} required [12]

Turnover number (min^{-1})

Specific activity (U/mg)
233 [22]; 2.7 [17]; More [3–7, 9–18, 22–24, 26]

K$_m$-value (mM)
0.15 (p-nitrophenyl-2-acetamido-2-deoxy-beta-D-glucose, similar values [2, 3, 5, 12, 13, 14, 17, 20, 26]) [21]; 0.11 (4-methyl-umbelliferyl-2-acetamido-2-deoxy-beta-D-glucose, similar values [8, 13, 25]) [20, 21]; 0.018 (O-2-acetamido-2-deoxy-beta-D-glucopyranosyl-(1–4)-2-acetamido-3-O-(D -1-carboxyethyl)-2-deoxy-D-glucose) [21]; 0.31 (N, N'-diacetylchitobiose, similar values [5, 13, 23]) [21]; 0.39 (N, N', N'-triacetylchitotriose) [21]; 0.38 (N, N', N', N'-tetraacetylchitotetraose) [21]; 0.35 (N, N', N', N', N'-pentaacetylchitopentaose) [21]; 2.3 (p-nitrophenyl-2-acetamido-2-deoxy-beta-D-galactopyranoside, similar values [13]) [5]; 0.86 (p-nitrophenyl-2-acetamido-2-deoxy-beta-D-galactopyranoside, similar values [20]) [12]; 1.3 (p-nitrophenyl-N-thioacetyl-beta-D-glucosaminide) [12]; 0.27 (2, 4-dinitrophenyl-N-acetyl-beta-D-glucosaminide) [12]; 1.23 (4-methylumbelliferyl-2-acetamido-2-deoxy-beta-D-galactose) [13]; 0.58 (4-methylumbelliferyl-2-acetamido-2-deoxy-beta-D-galactose) [20]; 2.58 (glycopeptide monomer) [17]

pH-optimum
3.4 (galactosaminides) [14]; 4.5 (glucosaminides [14]) [14, 25]; 4.5–5.0 [20]; 4.6 [26]; 4.8 (p-nitrophenyl-2-acetamido-2-deoxy-beta-D-glucose [7]) [7, 15]; 5.0 [5]; 5.0–6.0 [3]; 5.2 [27]; 5.4 (p-nitrophenyl-2-acetamido-2-deoxy-beta-D-galactose [7], synthetic substrate [13]) [7, 13]; 5.5–6.0 [4]; 5.9 [21]; 6.0 [2]; 6.5 (natural substrates) [13]; 7.7 [17]

pH-range
4.0–5.0 [25]; 4–6 [26]

Temperature optimum (°C)
45 [4]; 55 [3]

Temperature range (°C)

Enzyme Handbook © Springer-Verlag Berlin Heidelberg 1991
Duplication, reproduction and storage in data banks are only
allowed with the prior permission of the publishers

3 ENZYME STRUCTURE

Molecular weight
234000 (cow, enzyme B, gel filtration) [6]
190000 (bull, gel filtration) [14]
136000–149000 (Aspergillus niger, sedimentation data, amino acid composition [12], Sclerotinia fructigena, gel filtration [23], bull, sedimentation equilibrium centrifugation [24], Patella vulgata, sedimentation equilibrium centrifugation [19]) [12, 19, 23, 24]
105000–125000 (Bombyx mori, gel filtration [3, 4 , 16], cow, enzyme A, gel filtration [6], Turbatix aceti, gel filtration, gel electrophoresis [7], Octopus vulgaris, gel filtration [9], human plasma, gel filtration, sedimentation equilibrium centrifugation [20]) [3, 4, 7, 9, 16, 20]
92000 (Hordeum vulgare, sedimentation equilibrium centrifugation, SDS-PAGE) [26]
33500–38000 (E. coli, gel filtration, SDS-PAGE) [17]

Subunits
Monomer [17]
Dimer (1 × 68900 + 1 × 65900, Turbatix aceti, SDS-PAGE [7], 2 × 61000, Bombyx mori, SDS-PAGE [16] , 2 × 67000, Bombyx mori, enzyme Ex1, 1 × 67500 + 1 × 57500, Bombyx mori, enzyme Ex2, SDS-PAGE [3, 4]) [3, 4, 7, 16]
Tetramer (alpha$_2$-beta$_2$, alpha: 27000, beta: 34000, Octopus vulgaris, SDS-PAGE) [9]
? (x × 70000, Conidiobolus lamprauges, SDS-PAGE [10], x × 55000 + x × 25000, cow, SDS-PAGE [6], bull, sedimentation equilibrium centrifugation with dithiothreitol or guanidium chloride gives two bands [24]) [6, 10, 24]

Glycoprotein/Lipoprotein
Glycoprotein (15.9% carbohydrate [7]) [6, 7, 14, 16, 19, 20]

---

4 ISOLATION/PREPARATION

Source organism
E. coli (K-12) [17]; Bacillus subtilis [21, 22]; Calf [27]; Streptomyces sp. [2]; Bombyx mori (silkworm) [3, 4, 16]; Phycomyces blakesleeanus [5]; Cow [6]; Turbatix aceti [8]; Pig [9]; Octopus vulgaris [9]; Conidiobolus lamprauges [10]; Mycobacterium leprae [11]; Aspergillus niger [12]; Cupiennius salei [13]; Bull [14, 24]; Limulus polyphemus (horseshoe crab) [15]; Dictyostelium discoideum [18]; Patella vulgata [19]; Human [20, 25]; Sclerotinia fructigena [23]; Hordeum vulgare (barley) [26]

Source tissue
   Mycelia [2]; Integuments [3]; Alimentary canal [4]; Sporangiophores [5];
   Mammary gland [6]; Culture medium [7, 10, 23]; Gastric mucosa [8];
   Hepatopancreas [9]; Armadillo liver (infected wtih Mycobacterium leprae)
   [11]; Digestive fluid [13]; Sperm [14]; Serum (maternal [25]) [15, 25];
   Haemolymph [16]; Cell [17, 21, 22]; Digestive gland [20]; Plasma [20];
   Spleen [24]; Seed (malted) [26]; Brain [27]; Acetone extract [12]

Localisation in source
   Surface (of mycelia) [2]; Extracellular [7, 10, 23]; Lysosomes [8]; Cytoplasm
   (soluble) [17, 27]; Soluble and particulate parts of cell (subcellular distribution) [22]

Purification
   Bombyx mori [4, 16]; Phycomyces blakesleeanus (partial) [5]; Cow (2 forms)
   [6]; Turbatix aceti [7]; Pig (2 forms) [8]; Octopus vulgaris [9]; Conidiobolus
   lamprauges (partial) [19]; Aspergillus niger [12]; Cupiennius salai [13]; Bull
   (together with EC 3.2.1.52) [14]; Limulus polyphemus [15]; E. coli [17]; Dictyostelium discoideum [18]; Patella vulgata [20]; Human plasma [20];
   Bacillus subtilis [22]; Sclerotinia fructigena [23]; Bull (2 isoenzymes) [24];
   Maternal serum (2 forms) [25]; Hordeum vulgare [26]; Calf [27]

Crystallization
   –

Cloned
   –

Renaturated
   –

---

5 STABILITY

pH
   2.6 (60 minutes, 20% activity) [7]; 4.0 (inactivation below) [27]; 4–11 [23];
   4–8 [2]; 5.0–7.0 [13]; 5.0–8.5 [4]; 5.0–9.0 (Ex2) [3]; 6.0–8.0 (Ex1) [3]; 8.5 (best
   value for stability) [21]

Temperature (°C)
   30 (below) [4, 13]; 48 (below) [3]; 50 (10 minutes, 50% inactivation [23],
   form A: inactivation, form P: stable [25]) [23, 25]

Oxidation

Organic solvent
   Acetone (inactivation) [27]

Enzyme Handbook © Springer-Verlag Berlin Heidelberg 1991
Duplication, reproduction and storage in data banks are only
allowed with the prior permission of the publishers

General stability information

Dithiothreitol (stabilization) [27]; Cysteine (stabilization) [27]; Glycerol (stabilization) [17]; Extremly unstable [17]; Freezing/thawing (inactivation) [7]

Storage

−70°C, 20% glycerol, at least 3 months [17]; −20°C, 50 mM sodium phosphate buffer, pH 6.0, 0.1 M NaCl, 27 days, 87.1% activity, lyophilized 34.5% activity, unstable at 4°C [4]; 4°C, 30% $(NH_4)_2SO_4$, 50 mM phosphate buffer, pH 7.0, several months [7]; 50 mM citrate buffer, pH 4.6 [12]; Lyophilized [20]; 2°C, $(NH_4)_2SO_4$ [24]; Frozen, 10 mM sodium acetate buffer, pH 6 [26]

---

6 CROSSREFERENCES TO STRUCTURE DATABANKS

PIR/MIPS code

Brookhaven code

---

7 LITERATURE REFERENCES

[1] Cabezas, J.A.: Biochem. J., 261, 1059–1061 (1989)
[2] Iwamoto, T., Okiura, T., Sasaki, T., Inaoka, M.: J. Ferment. Technol., 5, 593–596 (1987)
[3] Koga, D., Shimazaki, C., Yamamoto, K., Inoue, K., Kimura, S., Ide, A.: Agric. Biol. Chem., 51, 1679–1681 (1987)
[4] Koga, D., Nakashima, M., Matsukara, T., Kimura, S., Ide, A.: Agric. Biol. Chem., 50, 2357–2368 (1986)
[5] Cohen, R.J.: Plant Sci., 43, 93–101 (1986)
[6] Kitchen, B.J., Masters, C.J.: Biochim. Biophys. Acta, 831, 125–132 (1985)
[7] Bedi, G.S., Shah, R.H., Bahl, O.P.: Arch. Biochem. Biophys., 233, 237–250 (1984)
[8] Devery, R., Collins, P., Johnson, A., Watson, G.: Biochem. Soc. Trans., 12, 460 (1984)
[9] Ceccarini, C., D'Anniello, A., Cacace, M.G., Atkinson, P.H.: Eur. J. Biochem., 132, 469–476 (1983)
[10] Ishikawa, F., Oishi, K., Aida, K.: Agric. Biol. Chem., 47, 149–151 (1983)
[11] Wheeler, P.R., Bharadwaj, V.P., Gregory, D.: J. Gen. Microbiol., 128, 1063–1071 (1982)
[12] Jones, C.S., Kosman, D.J.: J. Biol. Chem., 255, 11861–11869 (1980)
[13] Mommsen, T.P.: Biochim. Biophys. Acta, 612, 361–372 (1980)
[14] Khar, A., Anand, S.R.: Biochim. Biophys. Acta, 483, 141–151 (1977)
[15] Jain, R.S., Walz, C., Buck, C.A., Warren, L.: J. Chromatogr., 136, 141–146 (1977)
[16] Kimura, S.: Biochim. Biophys. Acta, 446, 399–406 (1976)
[17] Yem, D.W., Wu, H.C.: J. Bacteriol., 125, 324–331 (1976)
[18] Dimond, R.L., Loomis, W.F.: J. Biol. Chem., 249, 5628–5632 (1974)
[19] Phizackerley, P.J.R., Bannister, J.V.: Biochim. Biophys. Acta, 362, 129–135 (1974)
[20] Verpoorte, J.A.: Biochemistry, 13, 793–799 (1974)

[21] Berkeley, R.C.W., Brewer, S.J., Ortiz, J.M., Gillespie, J.B.: Biochim. Biophys. Acta, 309, 157–168 (1973)
[22] Ortiz, J.M., Gillespie, J.B., Berkeley, R.C.W.: Biochim. Biophys. Acta, 289, 174–186 (1972)
[23] Reyes, F., Byrde, R.J.: Biochem. J., 131, 381–388 (1973)
[24] Verpoorte, J.A.: J. Biol. Chem., 247, 4787–4793 (1972)
[25] Stirling, J.L.: Biochim. Biophys. Acta, 271, 154–162 (1972)
[26] Mitchell, E.D., Houston, C.W., Latimer, S.B.: Phytochemistry, 15, 1869–1871 (1976)
[27] Frohwein, Y.-Z., Gatt, S.: Biochemistry, 6, 2775–2782 (1967)

1 NOMENCLATURE

EC number
3.2.1.31

Systematic name
Beta-D-glucuronoside glucuronosohydrolase

Recommended name
Beta-glucuronidase

Synonymes
Glucuronidase, beta-glucuronide glucuronohydrolase
Exo-beta-D-glucuronidase
Ketodase

CAS Reg. No.
9001-45-0

---

2 REACTION AND SPECIFICITY

Catalysed reaction
A beta-D-glucuronoside + ·H_2O →
→ an alcohol + D-glucuronate

Reaction type
O-Glycosyl bond hydrolysis (exoglycosidic)

Natural substrates
Luteolin triglucoronide + H_2O [1]
Dermatan sulfate + H_2O [3]
Heparan sulfate + H_2O [3]
Oestrone 3-glucuronide + H_2O [11]
More (role in physiology, in tissues, in body fluids [32]) [2, 12]

Enzyme Handbook © Springer-Verlag Berlin Heidelberg 1991
Duplication, reproduction and storage in data banks are only
allowed with the prior permission of the publishers

Substrate spectrum

1 p-Nitrophenyl-beta-D-glucuronide + H_2O [2, 4, 9, 12, 14, 23, 24, 26]
2 4-Methylumbelliferyl-beta-D-glucuronide + H_2O [5, 14]
3 Phenolphthalein-beta-D-glucuronide + H_2O [4, 12, 22]
4 Luteolin 7-O-diglucuronide 4'-O-glucuronide + H_2O [1]
5 Luteolin 7-O-diglucuronide + H_2O [1]
6 Luteolin 7-O-glucuronide + H_2O [1]
7 Apigenin 7, 4'-diglucuronide + H_2O [1]
8 Androstendione-enol-beta-D-glucuronide + H_2O [3]
9 Glucuronic acid 1-phosphate + H_2O [3]
10 Naphthol AS-BI-D-glucuronide + H_2O [3]
11 Ammonium 1-deoxy-1-(6-thiopurinyl)-beta-D-glucopyranosidurate + H_2O [7]
12 1-Deoxy-1-(6-thiopurinyl)-beta-D-glucopyranosiduronamide + H_2O [7]
13 1-Deoxy-1-(6-thiopurinyl)-beta-D-glucopyranoside + H_2O [7]
14 More (transfer of beta-glucuronosyl residues from aryl and alicyclic glucuronides to aliphatic alcohols and glycerols) [33]

Product spectrum

1 p-Nitrophenol + D-glucuronic acid
2 4-Methylumbelliferone + D-glucuronic acid
3 Phenolphthalein + D-glucuronic acid
4 ?
5 ?
6 ?
7 Apigenin 7-O-glucuronide + glucuronic acid (from 4-position)
8 ?
9 Glucuronic acid + phosphate
10 ?
11 6-Mercaptopurine + ammonium 1-deoxy-beta-D-glucopyranosiduronate
12 6-Mercaptopurine + 1-deoxy-beta-D-glucopyranosiduronamide
13 6-Mercaptopurine + 1-deoxy-beta-D-glucopyranoside
14 ?

Inhibitor(s)

D-Saccharic acid 1, 4-lactone [1, 2, 4, 31, 32]; Cu^{2+} [1, 2, 4, 12, 30, 31]; Ag^+ [1, 4, 30, 31]; Methanol [1]; Delta-D-glucuronolactone [2]; Heparin [2, 12]; Hg^{2+} (reversed by EDTA [4, 11]) [4, 11, 12, 26, 30, 31]; Ni^{2+} [4, 12]; Ascorbic acid [4, 12]; Sodium desoxycholate [11]; Phenolic and alcoholic glucuronides [11]; Tris [12]; D-Glucaro-1, 4-lactone [12]; Glucuronic acid [12, 31]; Potassium saccharate [26]; p-Chloromercuribenzoate [30]; Citric acid [32]; Inhibitor (heat stable, competetive, in plasma [28], from endogenous tissue [32]) [28, 32]

Cofactor(s)/prostethic group(s)
Ethyleneglycol (activation) [1]; Bovine serum albumin (activation) [11, 28]; Plasma serum (acivation) [28]

Metal compounds/salts
NaCl [11, 26]; Mg^{2+} [11]; Ca^{2+} [11]

Turnover number (min⁻¹)

Turnover number (min^{-1})
12000 [20]

Specific activity (U/mg)
100–533 [2]; 95 [17]; More (assay method [3, 25], use of different substrates in assay) [1, 3–5, 10–18, 20–26, 28]

K_m-value (mM)
0.13–2.9 (p-nitrophenyl-beta-D-glucoronide) [2, 4, 9, 12, 14, 23, 24]; 0.007 (luteolin 7-O-diglucuronide 4'-O-glucuronide) [1]; 0.018–3.08 (phenolphthalein glucuronide, dependency on pH [4, 28], dependency on form of enzyme [12]) [4, 12, 22]; 0.2 (estriol-3-glucuronide, estriol-16-alpha-glucuronide) [9]; 0.02 (oestrone-3-glucuronide) [11]; 0.041–1.3 (4-methylumbelliferyl-beta-D-glucuronide) [12, 28]; 0.132 (ammonium 1-deoxy-1-(6-thiopurinyl)-beta-D-glucopyranosidurate) [7]; 9.25 (1-deoxy-1-(6-thiopurinyl)-beta-D-glucopyranosiduronamide) [7]; 1.49 (1-deoxy-1-(6-thiopurinyl)-beta-D-glucopyranoside) [7]; More [18, 31, 32]

pH-optimum
2.6 (form III) [2]; 3 (substrate 4-methylumbelliferyl-beta-D-glucuronide) [28]; 3.2 (form II) [2]; 3.4 (form I [2], form II [12]) [2, 12]; 3.8 [14]; 4.0–4.4 (acetate buffer) [11]; 4.1 (form II) [12]; 4.3 [1, 4]; 4.4 (form I [12]) [12, 17, 24]; 4.5 [5, 23]; 4.6 [18]; 4.7 [4, 22]; 5.0 [18, 26, 28]; 5.4 (citrate buffer) [11]; 6.8 (soluble and immobilized) [9]; More (comparison of values for different substrates and organisms) [29, 31, 32]

pH-range
2.1–5.8 (half maximal activity) [1]; 3.5–5.3 [5]; 4.0 (70% activity) [13]; 6.0 (90% activity) [13]; 4.0–5.5 [18]

Temperature optimum (°C)
55 [1]; 56 [11]; 70 [4, 13]

Temperature range (°C)

Enzyme Handbook © Springer-Verlag Berlin Heidelberg 1991
Duplication, reproduction and storage in data banks are only
allowed with the prior permission of the publishers

3 ENZYME STRUCTURE

Molecular weight
250000–310000 (man, gel filtration [4, 14], sedimention equilibrium
centrifugation [14], rat, gel filtration [10], sedimentation equilibrium
centrifugation [17, 21, 23], gel electrophoresis [18, 23], pig, gel filtration
[11], Littorina littorea, gel filtration [12], mouse, gel filtration [15, 20],
gradient polyacrylamide electrophoresis [22], rabbit, sucrose gradient
centrifugation [26]) [4, 10–12, 14, 15, 17, 18, 20, 21, 23, 26]
113500 (Littorina littorea, sedimentation equilibrium centrifugation of form I)
[12]
70000 (Kobayasia nipponica, gel filtration, SDS-PAGE) [2]

Subunits
Tetramer (4 × 70000–80000, man, 2 identical 2 non-identical chains [4],
SDS-PAGE [4, 14], rat, SDS-PAGE [10, 17, 18, 21, 23], mouse, SDS-PAGE
[15, 20, 22], rabbit, SDS-PAGE [26]) [4, 10, 14, 15, 17, 18, 20–23]

Glycoprotein/Lipoprotein
Glycoprotein (oligosaccharide structure [8], carbohydrate composition [14,
23]) [8, 14, 19, 20, 23, 24]

---

4 ISOLATION/PREPARATION

Source organism
Secale cereale (rye) [1]; Kobayasia nipponica [2]; Man [4, 14, 16];
Drosophila melanogaster [5]; Mycobacterium leprae [6]; Ox [7, 24, 29]; Rat
[8, 10, 13, 17–19, 21, 23]; E. coli [9, 29, 30]; Pig [11]; Littorina littorea (marine
mollusc) [12]; Mouse [15, 20, 22]; Rabbit [26, 27]; Patella vulgata [29]; Helix
pomatia (snail) [29]; Mammals (distribution in [32]) [31, 32, 33]; Fish [32];
Bacteria [32]

Source tissue
Leaves [1]; Whole animals [5]; Cell (from armadillo liver infected with
Mycobacterium leprae) [6]; Liver [7, 12, 17–20, 24, 26, 27, 29, 31]; Preputial
gland [8, 13, 21, 23, 31]; Basophil leukemia tumor [10]; Kidney [11, 22, 31];
Placenta [14, 16, 28]; Urine [1525]; Serum [25]; Spinal fluid [25]; Duodenal
juice [25]; Bile [25]; Digestive juice [29]; More (nearly all mammalian tissues
and body fluids [3, 25, 31, 32], distribution in tissues and organs) [32]

Localisation in source
Membrane bound (form I and II) [5]; Soluble part of cell [5]; Lysosomes [11,
17, 19, 21, 22, 27, 28, 31, 32]; Microsomes [17, 18, 22, 27, 28]; More (subcellular distribution) [27]

Purification
Secale cereale (partial) [1]; Kobayasia nipponica (3 forms) [2]; Man [4, 14, 16, 28]; Drosophila melanogaster (2 forms, partial) [5]; Rat (2 forms [10], 3 forms [13]) [8, 10, 13, 17, 18, 21, 23]; Littorina littorea (2 forms) [12]; Mouse [15, 20 , 22]; Ox [24]; Rabbit [26]; E. coli [30]; Calf [32]

Crystallization
[23]

Cloned
–

Renaturated
[19]

5 STABILITY

pH
3.0–10.5 [26]; 4–11 [22]; 4.2–7.0 (form III) [2]; 5.0 (inactivation) [18]; 5.0–7.0 (form I) [2]; 5.4–7.0 (form II) [2]; 6.5 (inactivation) [18]; 7.0 (or 7.8, stable at, depending on buffer) [23]; 8.0 (stable at) [18]

Temperature (°C)
–20 (inactivation) [18]; 37 (presence of oestrone-3-glucuronide, 45 minutes) [11]; 40 (10 minutes) [2]; 50 (inactivation above [26], more stable in neutral buffers [23], depending on buffer [24]) [23, 24, 26]; 55 (denaturation above [1], depending on pH, substrate protects against inactivation [4], 30 minutes, form I: inactivation, form II: stable [5], 10 minutes, inactivation at pH 5.0, not at pH 7.0 [22]) [1, 4, 5, 22]; 60 (form I: 60 minutes, form II: 5 minutes) [12]; 70 (stable at [14], absence of substrate, 10 minutes, inactivation [13], lysosomal form: 60 minutes, 80% activity, plasma enzyme: 15 minutes, 0% activity [28]) [13, 14, 28]; 71 (30 minutes, pH 5.0, 80% activity, 30 minutes, pH 8.0, 20% activity) [20]

Oxidation

Organic solvent
More (effect of solvents on soluble and immobilized enzyme) [9]

General stability information
Freezing (inactivation due to dissociation) [12]; Sodium acetate buffer (inactivation) [13]; Freezing/thawing (no inactivation) [20]; Bovine serum albumin (stabilization) [23]

Enzyme Handbook © Springer-Verlag Berlin Heidelberg 1991
Duplication, reproduction and storage in data banks are only
allowed with the prior permission of the publishers

Storage

−20°C, 50% glycerol, at least 2 months [1]; −20°C, form I: unstable, form II: 6 months, both forms stable in whole flies [5]; −20°C, neutral pH, 50% glycerol [26]; 0°C, 0.2 M potassium phosphate buffer, pH 8.0, 3 days, unstable at −20°C or 4°C or in sodium acetate buffer pH 4.0 [11]; 4°C, immobilized enzyme, 1 year, 40% activity [9]; 4°C, at least 1 month [12]; 0–4°C, Tris-HCl buffer, pH 8.0, more than 1 month [13]; 4°C, 2 weeks, 90% activity, or −20°C, 50% glycerol [18]; −20°C or 4°C, 0.02 M sodium acetate buffer, pH 5.2, 0.15 M NaCl, 1 mg/ml protein, several months, sucrose prevents aggregation [20]

6 CROSSREFERENCES TO STRUCTURE DATABANKS

PIR/MIPS code

GBECGC (Escherichia coli); A32576 (B, mouse); B32576 (H, mouse); A24983 (human, fragment); A26581 (precursor, placental, human); A29977 (precursor, mouse); A28954 (precursor, mouse); A25047 (precursor, rat); S00345 (precursor, rat, fragment)

Brookhaven code

7 LITERATURE REFERENCES

[1] Schulz, M., Weissenböck, G.: Phytochemistry, 26, 933–937 (1987)
[2] Tsuchihashi, H., Yadome, T., Miyazaki, T.: J. Biochem., 96, 1789–1797 (1984)
[3] Stahl, P.D., Fishman, W.H. in "Methods Enzym. Anal.", 3rd. Ed. (Bergmeyer, H.U., Ed.) 4, 246–256 (1984)
[4] Gupta, G.S., Singh, G.P.: Biochim. Biophys. Acta, 748, 398–404 (1983)
[5] Langley, S.D., Wilson, S.D., Gross, A.S., Warner, C.K., Finnerty, V.: J. Biol. Chem., 258, 7416–7424 (1983)
[6] Wheeler, P.R., Bharadwaj, V.P., Gregory, D.: J. Gen. Microbiol., 128, 1063–1071 (1982)
[7] Parker, A., Maw, B., Fedor, L.: Biochem. Biophys. Res. Commun., 103, 1390–1294 (1981)
[8] Byrd, J.C., Touster, O.: Biochim. Biophys. Acta, 677, 69–78 (1981)
[9] Bowlers, L.D., Johnson, P.R.: Biochim. Biophys. Acta, 661, 100–105 (1981)
[10] Schwartz, L.B., Austen, K.F.: Biochem. J., 193, 663–670 (1981)
[11] Gowers, H.M., Breuer, H.: J. Steroid Biochem., 13, 1021–1027 (1980)
[12] Diez, T., Cabezas, J.A.: Eur. J. Biochem., 93, 301–311 (1979)
[13] Tulsiani, D.R.P., Touster, O.: Methods Enzymol., 50, 510–514 (1978)
[14] Brot, F.E., Bell, C.E., Sly, W.S.: Biochemistry, 17, 385–391 (1978)
[15] Mills, N.C., Gupta, C., Bardin, C.W.: Arch. Biochem. Biophys., 185, 100–107 (1978)
[16] Contractor, S.F., Oakey, M.: Anal. Biochem., 78, 279–282 (1977)
[17] Himeno, M., Nishimura, Y., Tsuji, H., Kato, K.: Eur. J. Biochem., 70, 349–359 (1976)
[18] Owens, J.W., Stahl, P.: Biochim. Biophys. Acta, 438, 474–486 (1976)
[19] Potier, M., Gianetto, R.: Can. J. Biochem., 54, 321–326 (1976)

[20] Tomino, S., Paigen, K.: J. Biol. Chem., 250, 8503–8509 (1975)
[21] Keller, R.K., Touster, O.: J. Biol. Chem., 250, 4765–4769 (1975)
[22] Lin, C.-W., Orcutt, M.L., Fishman, W.H.: J. Biol. Chem., 250, 4737–4743 (1975)
[23] Himeno, M., Ohhara, H., Arakawa, Y., Kato, K.: J. Biochem., 77, 427–438 (1975)
[24] Himeno, M., Hashiguchi, Y., Kato, K.: J. Biochem., 766, 1243–1252 (1974)
[25] Fishman, W.H. in "Method. Enzym. Anal" (Bergmeyer, H. U., Ed.) 1, 964–979 (1974)
[26] Dean, R.T.: Biochem. J., 138, 395–405 (1974)
[27] Dean, R.T.: Biochem. J., 138, 407–413 (1974)
[28] Contractor, S.F., Shane, B.: Biochem. J., 128, 11–18 (1972)
[29] Wakabayashi, M., Fishman, W.H.: J. Biol. Chem., 236, 996–1001 (1961)
[30] Doyle, M.L., Katzman, P.A., Doisy, E.A.: J. Biol. Chem., 217, 921–930 (1955)
[31] Levy, G.A., Marsh, C.A. in "The Enzymes", 2nd Ed. (Boyer, P.D., Lardy, H., Myrbäck, K., Eds.) 4, 397–407 (1960) (Review)
[32] Fishman, W.H.: Adv. Enzymol. Relat. Subj. Biochem., 16, 361–409 (1955) (Review)
[33] Fishman, W.H., Green, S.: J. Biol. Chem., 225, 435 (1957)

Enzyme Handbook © Springer-Verlag Berlin Heidelberg 1991
Duplication, reproduction and storage in data banks are only
allowed with the prior permission of the publishers

1 NOMENCLATURE

EC number
3.2.1.32

Systematic name
1, 3-Beta-D-xylan xylanohydrolase

Recommended name
Xylan endo-1, 3-beta-xylosidase

Synonymes
Endo-1, 3-.beta.-xylanase
Xylanase, endo-1, 3-
Endo-1, 3-xylanase
1, 3-.beta.-Xylanase
Xylanase
Endo-1, 3-beta-xylanase
1, 3-Xylanase
Endo-1, 3-.beta.-xylanase
.beta.-1 , 3-Xylanase
Endo-beta-1, 3-xylanase [1]

CAS Reg. No.
9025-55-2

---

2 REACTION AND SPECIFICITY

Catalysed reaction
Random hydrolysis of 1, 3-beta-D-xylosidic linkages in 1, 3-beta-D-xylans

Reaction type
O-Glycosyl bond hydrolysis

Natural substrates
Beta-1, 3-xylan + H_2O [1]

Substrate spectrum
1 Beta-1, 3-xylan + H_2O [1, 3, 4, 5]
2 Xylotriose + H_2O [1]
3 Xylotetraose + H_2O [1]
4 Xylopentaose + H_2O [1]
5 Rhodymenan + H_2O [2]
6 More (not: xylobiose [1], p-nitrophenyl-beta-D-xyloside [1], beta-1, 4-xylan [1, 3, 5], cellulose [3, 5], laminaran [3], beta-1, 3-xylotriose [3], beta-1, 3-xylobiose [3]) [1, 3, 5]

Enzyme Handbook © Springer-Verlag Berlin Heidelberg 1991
Duplication, reproduction and storage in data banks are only
allowed with the prior permission of the publishers

Product spectrum
1 Xylose + xylooligosaccharides [1, 3, 4, 5]
2 Xylose + xylobiose [1]
3 Xylose + xylotriose + xylobiose (small amount) [1]
4 Xylose + xylotetraose + xylobiose (small amount) + xylotriose (small amount) [1]
5 Beta-1, 4-xylotriose + xylooligosaccharides (beta-1, 4-linked, trace amount) [2]
6 ?

Inhibitor(s)
Fe^{3+} [1]; Hg^{2+} [1, 3, 4]; Mn^{2+} [1, 3]; Cu^{2+} [1]; Zn^{2+} [1]; Pb^{2+} [1]; Ag^+ [1];
p-Chloromercuribenzoate [1]; N-Bromosuccinimide [3]; Ag^{2+} [4]; Cd^{2+} [4]

Cofactor(s)/prostethic group(s)

Metal compounds/salts

Turnover number (min^{-1})

Specific activity (U/mg)
13.6 [1]; 13.3 [3]; More [4]

K_m-value (mM)

pH-optimum
6.0–6.5 [1]; 4.0 (Aspergillus terreus, EF-4) [3]; 4.5 (Aspergillus terreus, EF-5, EF-6) [3]; 5.0 (Aspergillus terreus, EF-1, EF-3) [3]; 5.5 (Aspergillus terreus, EF-2) [3]; 6.0–7.5 [5]; 6–8 [4]

pH-range
3–10 (3: less than 10 % of maximal activity, 10: 50% of maximal activity) [4]; 3–9 (5% of maximal activity at pH 3 and 9) [5]

Temperature optimum (°C)
40 (Aspergillus terreus, EF-4, EF-5) [3]; 45 (Aspergillus terreus, EF-2) [3]; 50 (Aspergillus terreus, EF-3, EF-6) [3]; 55 (Aspergillus terreus, EF-1) [3]; 60 [4]

Temperature range (°C)

3 ENZYME STRUCTURE

Molecular weight
53000 (gel filtration, Vibrio sp.) [1]
11000 (SDS-PAGE, Aspergillus terreus, EF-1, EF-5) [3]
13500 (SDS-PAGE, Aspergillus terreus, EF-6) [3]
14500 (SDS-PAGE, Aspergillus terreus, EF-4, EF-3) [3]
20000 (SDS-PAGE, Aspergillus terreus, EF-2) [3]

Subunits

Glycoprotein/Lipoprotein

–

---

4 ISOLATION/PREPARATION

Source organism
Vibrio sp. AX-4 [1]; Aspergillus terreus [2]; Aspergillus terreus A-07 (6 different enzymes) [3]; Bacillus No. C-59–2 [4]; Vibrio [5]

Source tissue
Culture medium [1, 3. 4]

Localisation in source
Extracellular [2, 5]

Purification
Vibrio sp. AX-4 [1]; Aspergillus terreus (6 different enzymes: EF-1 -EF-6) [3]; Bacillus No. C-59–2 [4]

Crystallization
–

Cloned
–

Renaturated
–

---

5 STABILITY

pH
7.5 (highest stability) [4]; 4.5–10 (below 40°C) [1]; 3–6 (30°C) [3]; 8 (inactivation) [3]

Temperature (°C)
60 (pH 4.5, Aspergillus terreus, EF-1, EF-2, EF-3, EF-4) [3]; 70 (rapid decrease of activity, EF-1, EF-2, EF-3, EF-4) [3]; 50 (pH 4.5, stable up to, Aspergillus terreus, EF-5 , EF-6) [3]; 60 (rapid decrease of activity, Aspergillus terreus, EF-5, EF-6) [3]; 30–60 (10 minutes, presence of Ca^{2+}, no loss of activity) [4]; 40 (15 minutes, stable up to 40°C) [1]; 37 (20 hours, 5% loss of activity) [1]; 55 (15 minutes, complete loss of activity) [1]

Oxidation

Organic solvent

Enzyme Handbook © Springer-Verlag Berlin Heidelberg 1991
Duplication, reproduction and storage in data banks are only
allowed with the prior permission of the publishers

General stability information

Storage

---

6 CROSSREFERENCES TO STRUCTURE DATABANKS

PIR/MIPS code
A31842 (precursor, Clostridium thermocellum)

Brookhaven code

---

7 LITERATURE REFERENCES

[1] Aoki, T., Araki, T., Kitamikado, M.: Nippon Suisan Gakkaishi, 54, 277–281 (1988)
[2] Chen, W.P., Matsuo, M., Yasui, T.: Agric. Biol. Chem., 50, 1195–1200 (1986)
[3] Wen Pin Chen, Matsuo, M., Yasui, T.: Agric. Biol. Chem., 50, 1183–1194 (1986)
[4] Horikoshi, K., Atsukawa, Y.: Agric. Biol. Chem., 37, 2097–2103 (1973)
[5] Araki, T., Aoki, T., Kitamikado, M.: Nippon Suisan Gakkaishi, 53, 2077–2081 (1987)

1 NOMENCLATURE

EC number
3.2.1.33

Systematic name
Dextrin 6-alpha-D-glucanohydrolase

Recommended name
Amylo-1, 6-glucosidase

Synonymes
Dextrin 6-alpha-D-glucosidase
Glucosidase, amylo-1, 6-
Amylo-1, 6-glucosidase
Amylopectin 1, 6-glucosidase
Dextrin-1, 6-glucosidase
E.C. 3.2.1.9 (formerly)
More (in mammals and yeast the enzyme is linked to a glycosyltransferase similar to EC 2.4.1.25, together these 2 activities constitute the glycogen debranching system with the following names: Amylo-1, 6-glucosidase/Oligo-1, 4 -- > 1, 4-glucantransferase [3], Glycogen debranching enzyme [12], Oligo-alpha-1, 4-glucan: alpha-1, 4-glucan-4-glycosyltransferase-amylo-1, 6-glucosidase [12], Amylo-1, 6-glucosidase/4-alpha-glucanotransferase [5, 7, 8], Amylo-1, 6-glucosidase/1, 4-alpha-glucan: 1, 4-alpha-glycosyltransferase [11]) [3, 5, 7, 8, 11, 12]

CAS Reg. No.
9012-47-9

2 REACTION AND SPECIFICITY

Catalysed reaction
Endohydrolysis of 1, 6-alpha-D-glucoside linkages at points of branching in chains of 1, 4-linked alpha-D-glucose residues

Reaction type
O-Glycosyl bond hydrolysis (endoglycosidic)

Natural substrates
Glycogen + H_2O (regulation of glycogen metabolism in liver) [3]

Enzyme Handbook © Springer-Verlag Berlin Heidelberg 1991
Duplication, reproduction and storage in data banks are only
allowed with the prior permission of the publishers

Substrate spectrum

1 Glycogen + H_2O (reversion reaction: synthesis of (1 --> 6)-bound side chains, incorporation of glucose into polysaccharides [11], glucose incorporation into glycogen [15, 16]) [2, 3, 7, 11, 14–18, 19]
2 Limit dextrin + H_2O (reversion reaction: glucose incorporation into limit dextrin [15]) [10, 13, 14, 15, 18]
3 Amylopectin + H_2O [11, 14, 18]
4 Alpha-dextrins + H_2O [13]
5 More [21]

Product spectrum

1 Glycogen (partially debranched) + glucose (glucose incorporation into glycogen [15]) [11, 15]
2 Limit dextrin (partially debranched) + glucose (glucose incorporation into limit dextrin [15]) [10]
3 Amylopectin (partially debranched) + glucose [11]
4 Alpha-dextrin (partially debranched) + glucose
5 ?

Inhibitor(s)

p-Hydroxymercuribenzoate [21]; 1-S-Dimethylarsino-1-thio-beta-D-gluconopyranoside [5]; 2, 2-Bis(hydroxymethyl)-2, 2', 2'-nitrilotriethanol [5]; Urea [10, 12]; Guanidine [10, 12]; Amines (protonated, hydroxyalkylsubstituted) [23]; Tris (buffer) [12, 20, 21]; Imidazole (buffer) [12]; Cationic buffers [23]; NH_4^+ (slight) [23]

Cofactor(s)/prostethic group(s)

Metal compounds/salts

Turnover number (min⁻¹)

Specific activity (U/mg)

8.4 [18]; More [1, 4, 6, 10, 12, 13, 21, 23]

K_m-value (mM)

1.6 (shellfish glycogen, phi-dextrin, yeast) [11]; 3.8 (shellfish glycogen, beta-dextrin, yeast) [11]; 4.3 (amylopectin beta-dextrin, yeast) [11]; 7.2 (shellfish glycogen, beta-dextrin, rabbit) [11]; 11 (amylopectin beta-dextrin, rabbit) [11]; More (function of pH [23]) [21, 23]

pH-optimum

6 (limit dextrin, 37°C) [12]; 6.0–6.6 [18]; 6.6 (anionic buffer, limit dextrin) [20, 23]; 7.2 (cationic buffer, limit dextrin) [20, 23]; 6.4 (phosphate buffer, glycosyl incorporation into glycogen) [20]; 6–6.5 (phosphate buffer, limit dextrin, glycosyl incorporation into glycogen) [20]; 6.1–6.4 (sodium citrate/mercaptoethanol buffer) [21]; More (pH-optimum depends on type of buffer [20], 2 enzymes: one with a acid and one with a neutral pH-optimum [22]) [20, 22]

pH-range

5.0–8.0 (5.0: about 25% of maximum activity, 8.0: about 40 % of maximum activity) [18]

Temperature optimum (°C)

37 [21]

Temperature range (°C)

---

3 ENZYME STRUCTURE

Molecular weight

166000 (polyacrylamide gel electrophoresis in absence and presence of SDS, chicken) [6]
160000–170000 (analytical gel chromatography under denaturing and non-denaturing conditions, rabbit) [7]
170000 (SDS-PAGE, rabbit) [8]
179250 (sucrose density centrifugation, rabbit) [18, 21]
280000 (sedimentation equilibrium, Saccharomyces cerevisiae) [18]
210000 (Sephadex chromatography, polyacrylamide gel electrophoresis, Saccharomyces cerevisiae) [18]
267000–279000 (equilibrium method, rabbit) [21]
More [9]

Subunits

Oligomer (3 basic subunits, 120000, 85000, 70000, carboxymethylation prior to SDS-PAGE, yeast) [9]
Oligomer (2 or 3 different subunits, SDS-PAGE, Saccharomyces cerevisiae, 74000, 94000, 175000) [18]
Monomer (rabbit: analytical gel chromatography under denaturing conditions [7], SDS-PAGE, alkylated enzyme [8]) [7, 8]
More (evidence for a subunit of 120000, rabbit) [9]

Glycoprotein/Lipoprotein

–

Enzyme Handbook © Springer-Verlag Berlin Heidelberg 1991
Duplication, reproduction and storage in data banks are only
allowed with the prior permission of the publishers

4 ISOLATION/PREPARATION

Source organism
　Rabbit (2 enzymes: acid and neutral [22]) [1, 2, 4, 7, 8, 9, 10, 11, 12, 13, 15, 16, 19, 20, 21, 22, 23]; Dog [3]; Chicken [6]; Yeast [9, 11]; Human (type III storage disease) [17]; Saccharomyces cerevisiae [18]

Source tissue
　Muscle [1, 2, 4, 6, 7, 8, 9, 10, 11, 13, 15, 16, 20, 21, 22, 23]; Cell [18]; Liver [3, 10, 12]; Fibroblasts [17]

Localisation in source

Purification
　Rabbit [1, 4, 10, 12, 13, 21, 23]; Chicken [6]; Saccharomyces cerevisiae [18]

Crystallization
　(rabbit) [2]

Cloned
　–

Renaturated
　–

---

5 STABILITY

pH

Temperature (°C)

Oxidation

Organic solvent

General stability information

Storage

---

6 CROSSREFERENCES TO STRUCTURE DATABANKS

PIR/MIPS code

Brookhaven code

7 LITERATURE REFERENCES

[1] Scopes, R.K., Stoter, A.: Methods Enzymol., 90, 479–490 (1982)
[2] Osterlund, B.R., Hayakawa, K., Madsen, N.B., James, M. N.G.: J. Mol. Biol., 174, 557–559 (1984)
[3] Palmer, T.N., Ryman, B.E.: FEBS Lett., 18, 277–279 (1971)
[4] White, R.C., Ruff, C.J., Nelson, T.E.: Anal. Biochem., 115, 388–390 (1981)
[5] Gillard, B.K., White, R.C., Zingaro, R.A., Nelson, T. E.: J. Biol. Chem., 255, 8451–8457 (1980)
[6] Heizmann, C.W., Eppenberger, H.M.: FEBS Lett., 105, 35–39 (1979)
[7] White, R.C., Nelson, T.E.: Biochim. Biophys. Acta, 400, 154–161 (1975)
[8] White, R.C., Nelson, T.E.: Biochim. Biophys. Acta, 365, 274–280 (1974)
[9] Lee, E.Y.C., Carter, J.H.: FEBS Lett., 32, 78–80 (1973)
[10] Brown, D.H., Gordon, R.B., Illingworth Brown, B.: Ann. N. Y. Acad. Sci., 210, 238–253 (1973)
[11] Lee, E.Y.C., Carter, J.H.: Arch. Biochem. Biophys., 154, 636–641 (1973)
[12] Gordon, R.B., Brown, D.H., Illingworth Brown, B.: Biochim. Biophys. Acta, 289, 97–107 (1972)
[13] Watts, T.E., Nelson, T.E.: Anal. Biochem., 49, 479–491 (1972)
[14] Manners, D.J.: Nature (New Biol.), 234, 150–151 (1971)
[15] Stark, J.R., Thambyrajah, V.: Biochem. J., 120, 17–18 (1970)
[16] Huijing, F., Lee, E.Y.C., Carter, J.H., Whelan, W.J.: FEBS Lett., 7, 251–254 (1970)
[17] Justice, P., Ryan, C., Yi-Yung Hsia, D.: Biochem. Biophys. Res. Commun., 39, 301–306 (1970)
[18] Lee, E.Y.C., Carter, J.H., Nielsen, L.D., Fischer, E. H.: Biochemistry, 9, 2347–2355 (1970)
[19] Nelson, T.E., Larner, J.: Biochim. Biophys. Acta, 198, 538–545 (1970)
[20] Nelson, T.E., Kolb, E., Larner, J.: Biochim. Biophys. Acta, 151, 212–215 (1968)
[21] Brown, D.H., Illingworth Brown, B.: Methods Enzymol., 8, 515–524 (1966)
[22] Taylor, P.M., Whelan, W.J. in "Contol Of Glycogen Metabolism" (FEBS Proc.4th Meeting, Oslo, 1967) (Whelan, W.J., Ed.), 101–114 (1968)
[23] Nelson, T.E., Kolb, E., Larner, J.: Biochemistry, 8, 1419–1428 (1969)

Enzyme Handbook © Springer-Verlag Berlin Heidelberg 1991
Duplication, reproduction and storage in data banks are only
allowed with the prior permission of the publishers

1 NOMENCLATURE

EC number
3.2.1.35

Systematic name
Hyaluronate 4-glucanohydrolase

Recommended name
Hyaluronoglucosaminidase

Synonymes
Hyaluronoglucosidase
Hyaluronidase
Chondroitinase
Chondroitinase I
E.C. 3.2.1.34 (now included with EC 3.2.1.35)

CAS Reg. No.
37326-33-3; 9012-81-1

---

2 REACTION AND SPECIFICITY

Catalysed reaction
Random hydrolysis of 1, 4-linkages between N-acetyl-beta-D-glucosamine and D-glucuronate residues in hyaluronate; More (mechanism [3], also hydrolyzes 1, 4-beta-D-glycosidic linkages between N-acetylgalactosamine or N-acetylgalactosamine sulfate and glucuronic acid in chondroitin, chondroitin 4-and 6-sulfates and dermatan)

Reaction type
O-Glycosyl bond hydrolysis

Natural substrates
Hyaluronic acid $+$ H_2O (microorganisms: nutrient role, lowering of connective tissue barrier) [3]

Enzyme Handbook © Springer-Verlag Berlin Heidelberg 1991
Duplication, reproduction and storage in data banks are only
allowed with the prior permission of the publishers

Substrate spectrum

1 Chondroitin + H_2O (testicular and snake venom enzyme) [3]
2 Chondroitin + H_2O (bacterial enzyme) [3]
3 Hyaluronic acid + H_2O (testicular and snake venom enzyme) [3, 4]
4 Hyaluronic acid + H_2O (bacterial enzyme) [3, 4]
5 Chondroitin sulfates + H_2O (A and C, testicular enzyme) [3, 4]
6 Chondroitin sulfates + H_2O (A and C, bacterial enzyme) [3, 4]
7 Chondroitin 4-sulfate + H_2O [2, 3]
8 Chondroitin 6-sulfate + H_2O [3]
9 More (testicular enzyme: transglycosylation [3, 5], bacterial enzyme: no transglycosalation [3]) [3, 5]

Product spectrum

1 Tetrasaccharides (testicular and snake venom enzyme: (glucuronic acid)-(N-acetylglucosamine)-(glucuronic acid)-(N-acetylglucosamine)) [3]
2 Glucuronido-N-acetylhexosamine (Delta4,5-unsaturated, bacterial enzyme) [3]
3 Tetrasaccharides (testicular and snake venom enzyme: (glucuronic acid)-(N-acetylglucosamine)-(glucuronic acid)-(N-acetylglucosamine)) [3]
4 Glucuronido-N-acetylhexosamine (Delta4,5-unsaturated, bacterial enzyme) [3]
5 Tetrasaccharides (testicular and snake venom enzyme: (glucuronic acid)-(N-acetylglucosamine)-(glucuronic acid)-(N-acetylglucosamine)) [3]
6 Glucoronido-N-acetylhexosamine (Delta4,5-unsaturated, bacterial enzyme) [3]
7 ?
8 ?
9 ?

Inhibitor(s)

Alcohols (sulfated, aliphatic) [3]; Hexylresorcinol [3]; Aurin tricarboxylic acid [3]; Anionic dyes [3]; Bile salts [3]; Steroids (sulfated) [3]; Polyphenols [3]; Heavy metals (di- and trivalent) [3]; Fe^{3+} [3, 4]; Cu^{2+} [3, 4]; Fe^{2+} [3, 4]; Zn^{2+} [3, 4]; Serum (of most mammals [3]) [3, 4]; Anticoagulants [4]; Chondroitin sulfates [4]; o-Quinones [4]; p-Quinones [4]; Quinols [4]; Heparin [3, 4]; Hyaluronic acid (sulfated, nitrated or acylated [3]) [3, 4]; Chitin esters (sulfated) [3]; Cellulose esters (sulfated) [3]; Carboxycellulose (sulfated) [3]; Xylan (sulfated) [3]

Cofactor(s)/prostethic group(s)

Metal compounds/salts

Turnover number (min^{-1})

Specific activity (U/mg)
  7.47 [1]

K$_m$-value (mM)

pH-optimum
  5.4 (testis: broad optimum below) [4]

pH-range

Temperature optimum (°C)
  50 (testis) [3]

Temperature range (°C)
  8–38 (testis, Q$_{10}$: 2 (between 18°C and 38°C)) [4]

---

3 ENZYME STRUCTURE

Molecular weight
  70000 (commercial preparation, ovine testicular hyaluronidase) [1]

Subunits

Glycoprotein/Lipoprotein
  –

---

4 ISOLATION/PREPARATION

Source organism
  Dog [2]; Bacteria [3, 4]; Mammalia [3, 4]; Buffalo [7]; Goat [7]; Flavobacterium [3]; Clostridium perfringens [3]; Proteus vulgaris [6]; Pneumococci
  [3]; Streptococci [3]; Staphylococci [3]; Clostridia [3]; Proteus [3]

Source tissue
  Commercial preparation [1]; Testis [3–5]; Blood [3]; Liver [2]; Spermatozoa
  [7]

Localisation in source
  Soluble [7]; More (associated with denuded sperm) [7]

Purification
  Clostridium perfringens [4]; More (commercial preparation, ovine testicular
  hyaluronidase [1]) [1, 4]

Crystallization
  –

Enzyme Handbook © Springer-Verlag Berlin Heidelberg 1991
Duplication, reproduction and storage in data banks are only
allowed with the prior permission of the publishers

Cloned
–

Renaturated
–

5 STABILITY

pH

Temperature (°C)
100 (testis: pH 5.6, 5 minutes, 80% loss of activity, bacteria: heat labile) [3, 4]

Oxidation

Organic solvent

General stability information

Storage

6 CROSSREFERENCES TO STRUCTURE DATABANKS

PIR/MIPS code
A30566 (Streptococcus bacteriophage H4489A)

Brookhaven code

7 LITERATURE REFERENCES

[1] Morton, D.B.: Biochem. Soc. Trans., 534th Meeting, Vol.1, 385 (1973)
[2] Hayashi, S.: J. Biochem., 83, 149–157 (1978)
[3] Meyer, K., Hoffman, P., Linker, A. in "The Enzymes", 2nd. Ed. (Boyer, P.D., Ed.) 4, 447–460 (1960) (Review)
[4] Meyer, K., Rapport, M.M.: Adv. Enzymol. Relat. Subj. Biochem., 13, 199–236 (1952) (Review)
[5] Weissmann, B.: J. Biol. Chem., 216, 783–794 (1955)
[6] Thurston, C.F.: J. Gen. Microbiol., 80, 515–522 (1974)
[7] Anand, S.R., Kaur, S.P., Chaudhry, P.S.: Hoppe-Seyler' S Z. Physiol. Chem., 358, 685–688 (1977)

1 NOMENCLATURE

EC number
3.2.1.36

Systematic name
Hyaluronate 3-glycanohydrolase

Recommended name
Hyaluronoglucuronidase

Synonymes
Hyaluronidase
Hyaluronoglucuronidase
Glucuronoglucosaminoglycan hyaluronate lyase
Orgelase [3]

CAS Reg. No.
37288-34-9

---

2 REACTION AND SPECIFICITY

Catalysed reaction
Hyaluronic acid + H_2O →
→ tetrasaccharide ((N-acetylglucosamine)-(glucuronic acid)-(N-acetylglucosamine)-(glucoronic acid))
More (random hydrolysis of 1, 3-linkages between beta-D-glucuronate and N-acetyl-D-glucosamine residues in hyaluronate)

Reaction type
O-Glycosyl bond hydrolysis

Natural substrates
Hyaluronic acid + H_2O

Substrate spectrum
1 Hyaluronic acid + H_2O [1, 2]
2 More (not: chondroitin) [1]

Product spectrum
1 Tetrasaccharide ((N-acetylglucosamine)-(glucuronic acid)-(N-acetylglucosamine)-(glucuronic acid)) [1, 2]
2 ?

Enzyme Handbook © Springer-Verlag Berlin Heidelberg 1991
Duplication, reproduction and storage in data banks are only
allowed with the prior permission of the publishers

Inhibitor(s)
Butane-2, 3-dione [3]; Phenylglyoxal [3]

Cofactor(s)/prostethic group(s)

Metal compounds/salts

Turnover number (min^{-1})

Specific activity (U/mg)

K_m-value (mM)

pH-optimum

pH-range

Temperature optimum (°C)

Temperature range (°C)

---

3 ENZYME STRUCTURE

Molecular weight

Subunits

Glycoprotein/Lipoprotein
–

---

4 ISOLATION/PREPARATION

Source organism
Hirudo medicinalis [1–3]

Source tissue
Salivary glands [1]; Commercial preparation [3]

Localisation in source

Purification

Crystallization
–

Cloned
–

Renaturated

–

---

5 STABILITY

pH

Temperature (°C)

Oxidation

Organic solvent

General stability information

Storage

---

6 CROSSREFERENCES TO STRUCTURE DATABANKS

PIR/MIPS code

Brookhaven code

---

7 LITERATURE REFERENCES

[1] Meyer, K., Hoffman, P., Linker, A. in "The Enzymes", 2nd Ed. (Boyer, P.D., Ed.) 4,
    447–460 (Review) (1960)
[2] Linker, A., Meyer, K., Hoffman, P.: J. Biol. Chem., 235, 924–927 (1960)
[3] Hipkin, J.A.D., Gacesa, P., Olavesen, A.H., Sawyer, R.T.: Biochem. Soc. Trans., 17, 784
    (1989)

1 NOMENCLATURE

EC number
3.2.1.37

Systematic name
1, 4-Beta-D-xylan xylanohydrolase

Recommended name
Xylan 1, 4-beta-xylosidase

Synonymes
Xylosidase, exo-1, 4-.beta.-
Xylobiase
Exo-1, 4-beta-xylosidase
.beta.-D-Xylopyranosidase
.beta.-Xylosidase
Beta-xylosidase
Exo-1, 4-xylosidase
Exo-1, 4-beta-D-xylosidase
Beta-xylosidase/beta-glucosidase (one enzyme with both beta-D-glucosidase glucohydrolase activity (EC 3.2.1.21) and 1,4-beta-D-xylan xylohydrolase activity (EC 3.2.1.37)) [5]

CAS Reg. No.
9025-53-0

---

2 REACTION AND SPECIFICITY

Catalysed reaction
1, 4-Beta-D-xylan $+ H_2O \rightarrow$
$\rightarrow$ D-xylose (hydrolysis of 1, 4-beta-D-xylans, removal of successive
D-xylose residues from non-reducing termini)

Reaction type
O-Glycosyl bond hydrolysis

Natural substrates
Xylan $+ H_2O$

Enzyme Handbook © Springer-Verlag Berlin Heidelberg 1991
Duplication, reproduction and storage in data banks are only
allowed with the prior permission of the publishers

Substrate spectrum

1 1, 4-Beta-D-xylan + H_2O (inactive toward [23]) [7, 21]
2 Xylobiose + H_2O [5, 7, 8, 10, 11, 12, 15, 23, 26]
3 p-Nitrophenyl-beta-D-xylanopyranoside + H_2O [1, 8, 9, 11, 12, 16, 26, 30]
4 o-Nitrophenyl-beta-D-xylanopyranoside + H_2O [7, 11, 12, 30]
5 Phenyl-beta-D-xyloside + H_2O [5, 7, 11, 12]
6 Xylotriose + H_2O [7, 10, 11, 12, 15, 23, 26]
7 Xylotetraose + H_2O [7, 10, 11, 12, 15, 23]
8 Xylopentaose + H_2O [7, 10, 11, 12, 15, 23]
9 Xylooligosaccharides + H_2O [8, 10, 24]
10 Aryl-beta-D-xylosides + H_2O [10]
11 p-Chlorophenyl beta-D-xyloside + H_2O [11]
12 o-Chlorophenyl beta-D-xyloside + H_2O [11, 12]
13 p-Methylphenyl beta-D-xyloside + H_2O [11, 12]
14 o-Methylphenyl beta-D-xyloside + H_2O [11, 12]
15 p-Methoxyphenyl beta-D-xyloside + H_2O [12]
16 o-Methoxyphenyl beta-D-xyloside + H_2O [12]
17 Beta-(1 --> 4)-xylobiose (transxylosylation) [13]
18 p-Nitrophenyl alpha-L-arabinopyranoside + H_2O [14, 17]
19 Benzyl beta-D-xyloside + H_2O [15]
20 Cyclohexyl beta-D-xyloside + H_2O [15]
21 Butyl beta-D-xyloside + H_2O [15]
22 Methyl beta-D-xyloside + H_2O (low [23]) [15]
23 Cellobiose + H_2O (low [21]) [5, 26]
24 More (xylosyltransferase activity [8, 13, 26], glycosyltransferase activity [23], one enzyme with beta-D-glucosidase glucohydrolase activity /EC 3.2.1.21/and 1, 4-beta-D-xylan xylohydrolase activity [19, 5], not: xylan [9]) [7, 8, 9, 13, 15, 16, 23, 26, 30, 31, 19, 5]

Product spectrum

1 D-Xylose
2 D-Xylose (xylooligosaccharides are synthesized from xylobiose and xylotriose [23]) [8, 10]
3 p-Nitrophenol + D-xylose [1]
4 o-Nitrophenol + D-xylose [7]
5 Phenol + D-xylose [7]
6 D-Xylose (xylooligosaccharides are synthesized from xylobiose and xyltriose) [23]
7 D-Xylose
8 D-Xylose
9 D-Xylose
10 D-Xylose + ?
11 D-Xylose + p-chlorophenol
12 D-Xylose + o-chlorophenol

13 D-Xylose + p-methylphenol
14 D-Xylose + o-methylphenol
15 D-Xylose + p-methoxyphenol
16 D-Xylose + o-methoxyphenol
17 Beta-(1 --> 3)-xylobiose + beta-(1 --> 4)-xylotriose +
   O-beta-D-xylanopyranosyl-(1 --> 3)-O-beta-xylanopyranosyl-(1 -->
   4)-D-xylopyranose + O-beta-xylopyranosyl-(1 -->
   4)-[O-beta-D-xylopyranosyl-(1 --> 3)]-O-beta-D-xylopyranosyl-(1 -->
   4)-D-xylopyranose [13]
18 Arabinose + p-nitrophenol [14]
19 D-Xylose + benzylalcohol
20 D-Xylose + cyclohexanol
21 D-Xylose + butanol
22 D-Xylose + methanol
23 ?
24 ?

Inhibitor(s)

D-Xylose [1, 9, 17, 18, 26, 30, 4]; Hg^{2+} [7, 8, 11, 12, 15, 24, 5]; $HgCl_2$ [30];
$AgNO_3$ [30]; Zn^{2+} (no effect [9]) [7, 10]; Cu^{2+} (no effect [9]) [7, 8, 10, 24];
N-Bromosuccinimide [7, 8, 11, 12, 14, 15, 5]; p-Chloromercuribenzoate [7,
10, 11, 24, 26]; SDS [7, 11, 18, 29]; Ag^+ [8]; p-Hydroxymercuribenzoate [8];
Iodoacetamide [8]; Methyl-beta-xylopyranoside [9, 5]; Sodium laurylsulfate
[10]; Methyl 1-thio-beta-D-xylopyranoside [14]; n-Pentyl
1-thio-beta-D-xylanopyranoside [14]; $CuSO_4$ [18]; Fe^{3+} [24]; Methyl-
methanethiosulfonate [29]; Diethylpyrocarbonate [29]; 5,
5-Dithiobis(2-nitrobenzoic acid) [29]; Glucono-(1, 5)-lactone [19, 5];
Xylono-1, 4-lactone [19]; Nojirimycin [5]; Methyl-beta-D-glucoside [5];
1-Thiophenyl-beta-D-xyloside [5]; 1-Thiophenyl-beta-D-glucoside [5]; Conduritol B epoxide [19]

Cofactor(s)/prostethic group(s)

Metal compounds/salts

Fe^{2+} (stimulates) [9]; Mn^{2+} (stimulates) [9]; Ca^{2+} (activates) [10]

Turnover number (min⁻¹)

Specific activity (U/mg)

8.0 [1]; 62.9 (Emericella nidulans) [7, 11]; 10.8 (Trichoderma viride) [7, 12];
2.38 [9]; 10.5 (Malbranchea pulchella) [7]; 40.0 [15]; More [5, 6, 21, 23, 24,
26, 27, 30]

Enzyme Handbook © Springer-Verlag Berlin Heidelberg 1991
Duplication, reproduction and storage in data banks are only
allowed with the prior permission of the publishers

K_m-value (mM)

0.89 (p-nitrophenyl beta-D-xylanopyranoside) [1]; 0.038
(p-nitrophenyl-xylanopyranoside) [8]; 2.86 (xylobiose) [10]; 5
(p-nitrophenyl-xylopyranoside) [9]; 9.8 (o-nitrophenyl beta-D-xyloside) [11];
4.0 (p-chlorophenyl beta-D-xyloside) [11]; 3.4 (o-chlorophenyl
beta-D-xyloside) [11]; 2.6 (p-methylphenyl beta-D-xyloside) [11]; 1.4
(o-methylphenyl beta-D-xyloside) [11]; 2.1 (xylobiose) [12]; 1.3 (xylotriose)
[12]; 1.0 (xylotetraose, xylopentaose) [12]; 2.4 (p-nitrophenyl alpha-L-arabinopyranoside) [14]; 1.4 (methyl beta-D-xylopyranoside) [14]; 0.3
(n-pentyl beta-D-xylopyranoside) [14]; 0.154 (2, 4-dinitrophenyl
beta-D-xylanopyranoside) [14]; 0.20 (2, 4-dinitrophenyl
1-thio-beta-D-xylanopyranoside) [14]; 4.2 (xylobiose, xylosidase 1) [15]; 2.7
(xylotriose, xylosidase 1) [15]; 3.9 (xylotetraose, xylosidase 1) [15]; 3.0
(xylopentaose, xylosidase 1) [15]; More [10, 11, 12, 14, 15, 21, 24, 25, 26, 29,
30, 5]

pH-optimum

5.0–9.0 (high activity, p-nitrophenyl beta-D-xylanopyranoside) [1]; 6.2–6.8
(Malbranchea pulchella) [7]; 4.5–5.0 (Emericella nidulans) [7, 11]; 3.5
(Trichoderma viride) [7, 12]; 3.3–4.0 [14]; 4.5 [8, 9, 12, 24]; 4.0 [17]; 3.8–4.0
[26]; 4.0–4.5 (xylosidase 1) [15]; 3.0–4.0 (xylosidase 2, 3, 4) [15]; 4.5–5.0 [21];
4.85 [30]; 3.8–5.5 [5]; 5.4–6.1 [4]; More (of immobilized enzyme) [25]

pH-range

2.5–6.5 (not active below pH 2.5 and above 6.5) [8]; 3.2–6.2 (3.2: less than
5% of maximal activity, 6.2: about 10% of maximal activity) [30]; 3.5–6.5
(3.5: about 25% of maximal activity, 6.5: less than 10% of maximal activity)
[11]; 3.5–6.0 (3.5: 85% of maximal activity, 6.0: 50% of maximal activity) [17]

Temperature optimum (°C)

40–60 (high activity, p-nitrophenyl beta-D-xylanopyranoside) [1]; 50 (Malbranchea pulchella [7]) [7, 8]; 55 (Trichoderma viride [7, 12], Emericella
nidulans [7, 11], xylosidase 1 [15]) [7, 11, 12, 15, 21]; 65 (xylosidase 2, 4)
[15]; 60–65 (xylosidase 3) [15]; 60 [17]; 75 [24, 30]; 70 [26]; More (of immobilized enzyme) [25]

Temperature range (°C)

40–58 (50% of maximal activity at 40°C and 58°C, Malbranchea pulchella)
[7]; 46–65 (50% of maximal activity at 46°C and 65°C, Trichoderma viride)
[7, 12]; 45–59 (50% of maximal activity at 45°C and 59°C, Emericella
nidulans) [11]; 55–82 (55: about 15% of maximal activity, 82: about 10% of
maximal activity) [30]; 30–75 (about 15% of maximal activity at 30°C and
75°C) [17]

3 ENZYME STRUCTURE

Molecular weight
170000–180000 (gel filtration, SDS-PAGE, Sclerotium rolfsii) [8]
100000 (column chromatography, Trichoderma lignorum [9], about, gel filtration, Penicillium wortmanni [14], SDS-PAGE, Trichoderma reesei [17]) [9, 14, 17]
110000 (gel filtration, Penicillium wortmanni, xylosidase 1) [15]
195000 (gel filtration, Penicillium wortmanni, xylosidase 2) [15]
210000 (gel filtration, Penicillium wortmanni, xylosidase 3) [15]
180000 (gel filtration, Penicillium wortmanni, xylosidase 4) [15]
83000 (SDS-PAGE, Neurospora crassa) [21]
360000 (gel filtration, Aspergillus fumigatus) [24]
253000 (gel filtration, Aspergillus niger) [26]
122000 (SDS-PAGE, Aspergillus niger) [26]
62000 (gel filtration, SDS-PAGE, Saccharum officinarum) [30]
205000 (equilibrium sedimentation, Chaetomium trilaterale) [5]
240000 (gel filtration, Chaetomium trilaterale) [5]
168000 (gel filtration, Thermomonospora fusca) [1]
260000 (gel filtration, Malbranchea pulchella) [7]
101000 (gel filtration, Trichoderma viride) [7, 12]
240000 (gel filtration, Emericella nidulans) [7, 11]

Subunits
Trimer (3 × 56000, SDS-PAGE, Thermomonospora fusca) [1]
Dimer (2 × 116000, SDS-PAGE, Emericella nidulans) [7, 11]
Dimer (Trichoderma lignorum) [9]
Tetramer (4 × 66000, SDS-PAGE, Bacillus subtilis gene cloned in E. coli) [18]
Tetramer (4 × 90000 , SDS-PAGE, Aspergillus fumigatus) [24]
Oligomer (x × 122000, SDS-PAGE, Aspergillus niger) [26]
Dimer (2 × 118000, SDS-PAGE, Chaetomium trilaterale) [5]

Glycoprotein/Lipoprotein
Glycoprotein (no or a few carbohydrate residues, Malbranchea pulchella [7], 4.5% carbohydrate, Trichoderma viride [7, 12], 4%, Emericella nidulans [7, 11], 25%, Trichoderma lignorum [9], 7.5%, Trichoderma reesei [17], 13.5%, arabinose + galactose, Saccharum officinarum [30], 20.7, Chaetomium trilaterale [5], no carbohydrate [26]) [7 , 9, 11, 17, 30, 5]

Enzyme Handbook © Springer-Verlag Berlin Heidelberg 1991
Duplication, reproduction and storage in data banks are only
allowed with the prior permission of the publishers

4 ISOLATION/PREPARATION

Source organism

Thermomonospora fusca [1]; Bacillus pumilus [2, 29]; Caldocellum saccharolyticum [6]; Malbranchea pulchella (var. sulfurea) [7, 10]; Trichoderma viride [7, 12]; Emericella nidulans [7, 11]; Sclerotium rolfsii [8]; Trichoderma lignorum (3 isoenzymes) [9]; Penicillium wortmanni (4 types of beta-xylosidase 1, 2, 3, 4 [15]) [13–15]; Trichoderma reesei (2 electrophoretically different forms [16]) [16, 17]; E. coli (Bacillus subtilis gene cloned in E. coli) [18]; Neurospora crassa [21]; Streptomyces lividans [22]; Aspergillus niger (soluble and immobilized enzyme [25], immobilization [27]) [23, 25, 26, 27]; Aspergillus fumigatus [24]; Cryptococcus albidus [28, 31]; Saccharum officinarum [30]; Cellulomonas (mutant CS1–17) [32]; Mouse (one enzyme with beta-D-glucosidase glucohydrolase activity (EC 3.2.1.21) and 1, 4-beta-D-xylan xylohydrolase activity) [19]; Chaetomium trilaterale [5]; Cellulomonas uda [4]; Butyrivibrio fibrisolvens [3]

Source tissue

Culture filtrate [7, 8, 11, 12, 14, 15, 16, 21, 24]; Cell [28]; Culture supernatant [9]; Commercial preparation [27]; Culture fluid [28]; Stalk [30]; Liver [19]

Localisation in source

Intracellular [1, 22, 31, 32]; Periplasmic space [28]; Extracellular (low [23]) [14, 21, 22, 31, 32]; Cell-wall [18, 26, 31]

Purification

Thermomonospora fusca [1]; Caldocellum saccharolyticum [6]; Malbranchea pulchella [7]; Trichoderma viride [7, 12]; Emericella nidulans [7, 11]; Sclerotium rolfsii [8]; Penicillium wortmanni [14, 15]; Trichoderma reesei [16, 17]; E. coli (Bacillus subtilis gene cloned in E. coli) [18]; Neurospora crassa [21]; Aspergillus niger (immobilization [27]) [23, 27]; Aspergillus fumigatus [24]; Saccharum officinarum [30]; Chaetomium trilaterale [5]

Crystallization

–

Cloned

(Bacillus pumilus gene [2], Bacillus subtilis gene in E. coli [20]) [2, 20]

Renaturated

–

5 STABILITY

pH

6.0–8.0 (room temperature) [1]; 6.3–6.7 (Malbranchea pulchella) [7]; 3–4 (50°C, 30 minutes, Trichoderma viride) [7, 12]; 4.0–5.0 (highest stability) [8]; 3.5–7.5 [9]; 5.0–6.0 (highest stability) [14]; 2–8 (50°C, 20 minutes) [24]; 3–6 (24 hours) [17]; 4–7 (30°C, 24 hours) [23]; 3–8 [26]; 4.0–11.0 (30°C, 3 hours) [5]; More (of immobilized enzyme [25]) [15, 25, 30]

Temperature (°C)

65 (half-life: 8 hours [1], complete loss of activity after 30 minutes [11], stable up to 65°C for 20 minutes [24]) [1, 11, 24]; 70 (half-life: 1.5 hours [1], purification by heat treatment [6], 15 minutes, stable below 70°C [23], 1 hour, 0.1% D-xylose [30]) [1, 6, 23, 30]; 50 (1 hour, stable up to 50°C [26], 15 minutes, 50% loss of activity [8], 30 minutes, pH 4.5–5.0 [11]) [8, 11, 26]; 55 (stable up to [7], pH 5.0, 24 hours, stable up to 55°C [17]) [7, 17]; 75 (30 minutes, without substrate, complete loss of activity, Trichoderma viride [7], 1% D-xylose, stable up to 30 minutes [30]) [7, 30]; 60 (pH 5.5, xylosidase 2, stable up to 60°C [15], pH 4.5, 30 minutes, 20% loss of activity [5]) [15, 5]; 45 (pH 5.5, xylosidase 1, 3, 4, stable up to 45°C) [15]; More (of immobilized enzyme [25], more thermostable in presence of Ca^{2+} [10]) [7, 10, 25]

Oxidation

Photooxidation (20% loss of activity by photooxidation in presence of Methylene Blue, 8 hours) [30]

Organic solvent

General stability information

D-Xylose (especially stable in presence of) [30]

Storage

–15°C, pH 4.5, 6 months [8]; 4°C, pH 6.0, several months [5]

---

6 CROSSREFERENCES TO STRUCTURE DATABANKS

PIR/MIPS code

S00067 (Bacillus pumilus)

Brookhaven code

Enzyme Handbook © Springer-Verlag Berlin Heidelberg 1991
Duplication, reproduction and storage in data banks are only
allowed with the prior permission of the publishers

7 LITERATURE REFERENCES

[1] Bachmann, S.L., McCarthy, A.J.: J. Gen. Microbiol., 135, 293–299 (1989)
[2] Panbanbred, W., Kondo, T., Negoro, S., Shinmyo, A., Okada, H.: Mol. Gen. Genet., 192, 335–341 (1983)
[3] Sewell, G.W., Aldrich, H.C., Williams, D., Mannarelli, B., Wilkie, A., Hespell, R.B., Smith, P.H., Ingram, L.O.: Appl. Environ. Microbiol., 54, 1085–1090 (1988)
[4] Rapp, P., Wagner, F.: Appl. Environ. Microbiol., 51, 746–752 (1986)
[5] Yasui, T., Matsuo, M.: Methods Enzymol., 160, 696–700 (1988)
[6] Patchett, M.L., Neal, T.L., Schofield, L.R., Strange, R.C., Daniel, R.M., Morgan, H.W.: Enzyme Microb. Technol., 11, 113–115 (1989)
[7] Matsuo, M., Yasui, T.: Methods Enzymol., 160, 684–695 (1988)
[8] Lachke, A.H.: Methods Enzymol., 160, 679–684 (1988)
[9] John, M., Schmidt, J.: Methods Enzymol., 160, 662–671 (1988)
[10] Matsuo, M., Yasui, T., Kobayashi, T.: Agric. Biol. Chem., 41, 1601–1606 (1977)
[11] Matsuo, M., Yasui, T.: Agric. Biol. Chem., 48, 1853–1860 (1984)
[12] Matsuo, M., Yasui, T.: Agric. Biol. Chem., 48, 1845–1852 (1984)
[13] Win, M., Kamiyama, Y., Matsuo, M., Yasui, T.: Agric. Biol. Chem., 52, 1151–1158 (1988)
[14] Deleyen, F., Claeyssens, M., Van Beeumen, J., De Bruyne, C.K.: Can. J. Biochem., 56, 43–50 (1978)
[15] Matsuo, M., Fujie, A., Win, M., Yasui, T.: Agric. Biol. Chem., 51, 2367–2379 (1987)
[16] Lappalainen, A.: Biotechnol. Appl. Biochem., 8, 437–448 (1986)
[17] Poutanen, K., Puls, J.: Appl. Microbiol. Biotechnol., 28, 425–432 (1988)
[18] Bernier, R., Desrochers, M., Paice, M.G., Yaguchi, M. : J. Gen. Appl. Microbiol., 33, 409–419 (1987)
[19] Stephens, M.C., Bernatsky, A., Legler, G., Kanfer, J. N.: Biochim. Biophys. Acta, 571, 70–78 (1979)
[20] Bernier, R., Desrochers, M.: J. Gen. Appl. Microbiol. , 31, 513–518 (1985)
[21] Deshpande, V., Lachke, A., Mishra, C., Keskar, S., Rao, M.: Biotechnol. Bioeng., 28, 1832–1837 (1986)
[22] Kluepfel, D., Shareck, F., Mondou, F., Morosoli, R.: Appl. Microbiol. Biotechnol., 24, 230–234 (1986)
[23] Takenishi, S., Tsujisaka, Y., Fukumoto, J.: J. Biochem., 73, 335–343 (1973)
[24] Kitpreechavanich, V., Hayashi, M., Nagai, S.: Agric. Biol. Chem., 50, 1703–1711 (1986)
[25] Ogutimein, G.B., Reilly, P.J.: Biotechnol. Bioeng., 22, 1143–1154 (1980)
[26] Rodionova, N.A., Tavobilov, I.M., Bezborodov, A.M.: J. Appl. Biochem., 5, 300–312 (1983)
[27] Ogumtimein, G.B., Reilly, P.J.: Biotechnol. Bioeng., 22, 1127–1142 (1980)
[28] Notario, V., Villa, T.G., Villanueva, J.R.: Can. J. Microbiol., 22, 312–315 (1976)
[29] Kersters-Hilderson, H., Van Doorslaer, E., Lippens, M., De Bruyne, C.K.: Arch. Biochem. Biophys., 234, 61–72 (1984)
[30] Chinen, I., Oouchi, K., Tamaki, H., Fukuda, N.: J. Biochem., 92, 1873–1881 (1982)
[31] Peciarova, A., Biely, P.: Biochim. Biophys. Acta, 716 , 391–399 (1982)
[32] Peiris, S.P., Rickard, P.A.D., Dunn, N.W.: Eur. J. Appl. Microbiol. Biotechnol., 14, 169–173 (1982)

1 NOMENCLATURE

EC number
3.2.1.38

Systematic name
Beta-D-fucoside fucohydrolase

Recommended name
Beta-D-fucosidase

Synonymes
Beta-fucosidase

CAS Reg. No.
9025-34-7

---

2 REACTION AND SPECIFICITY

Catalysed reaction
Hydrolysis of terminal non-reducing beta-D-fucose residues in
beta-D-fucosides

Reaction type
O-Glycosyl bond hydrolysis

Natural substrates
Beta-D-fucosides + H_2O [1–12]

Substrate spectrum
1 Beta-D-fucosides + H_2O (e.g. p-nitrophenyl beta-D-fucoside [1]) [1–12]
2 Beta-L-fucosides + H_2O (e.g. p-nitrophenyl beta-L-fucopyranoside) [1]
3 Beta-D-glucosides + H_2O (e.g. p-nitrophenyl beta-D-glucoside [3–5, 8])
   [3–5, 8, 10]
4 Beta-D-galactosides + H_2O (e.g. p-nitrophenyl beta-D-galactoside) [3–5,
   8, 10]

Product spectrum
1 Beta-D-fucose + alcohol (e.g. p-nitrophenol) [1–12]
2 Beta-L-fucose + alcohol
3 Beta-D-glucose + alcohol (e.g. p-nitrophenol) [3–5, 8, 10]
4 Beta-D-galactose + alcohol (e.g. p-nitrophenol) [3–5, 8, 10]

Enzyme Handbook © Springer-Verlag Berlin Heidelberg 1991
Duplication, reproduction and storage in data banks are only
allowed with the prior permission of the publishers

Inhibitor(s)

Arsenate [1]; Hg^{2+} [1, 4, 10]; Fucose [1, 4, 5, 11]; Tris buffer [4]; Maltose [4]; Lactones [4, 11, 12]; Glucose [4, 5]; Galactose [4, 5]; Cyanide [4, 10]; Iodine [7]; N-Acetylimidazole [7]; Tetranitromethane [7]; Cu^{2+} [10]

Cofactor(s)/prostethic group(s)

Metal compounds/salts

Turnover number (min^{-1})

Specific activity (U/mg)

430 [1]; 2.5–3.2 [3, 10]

K_m-value (mM)

8.0 (p-nitrophenyl beta-D-fucoside) [1]; 0.04–0.63 (p-nitrophenyl beta-D-fucoside) [3–5, 8, 12]; 0.10–0.60 (p-nitrophenyl beta-D-glucoside) [3–5, 8]; 0.10–5.0 (p-nitrophenyl beta-D-galactoside) [3–5, 8, 10]; 2.5–3.3 (o-nitrophenyl beta-D-galactoside) [8]

pH-optimum

5.5 (p-nitrophenyl beta-D-fucoside) [1, 3, 10]; 6.0–6.5 (p-nitrophenyl beta-D-fucoside) [2]; 4.5–5.5 (p-nitrophenyl beta-D-fucoside) [4, 11]; 4.5–6.5 (p-nitrophenyl beta-D-fucoside) [5]; 5.2–5.6 (p-nitrophenyl beta-D-fucoside) [8]

pH-range

3.5–8.0 (p-nitrophenyl beta-D-fucoside) [1]; 4.5–8.0 (p-nitrophenyl beta-D-fucoside) [2]; 3.5 (not active below, p-nitrophenyl beta-D-fucoside) [5]

Temperature optimum (°C)

55 (p-nitrophenyl beta-D-fucoside) [2]; 51–55 (p-nitrophenyl beta-D-fucoside) [8]

Temperature range (°C)

20–70 (p-nitrophenyl beta-D-fucoside) [2]

3 ENZYME STRUCTURE

Molecular weight

37000 (sedimentation equilibrium, gel filtration, Lactuca sativa) [1]
360000 (gel filtration, Achatina balteata, isoenzyme I) [9]
120000 (gel filtration, Achatina balteata, isoenzyme II) [9]

Subunits
Monomer (SDS-PAGE, Lactuca sativa) [1]
Dimer (190000 + 170000, SDS-PAGE, Achatina balteata, isoenzyme I) [9]
Monomer (SDS-PAGE, Achatina balteata, isoenzyme II) [9]

Glycoprotein/Lipoprotein
Glycoprotein [9]

---

4 ISOLATION/PREPARATION

Source organism
Lactuca sativa [1]; Littorina littorea [3]; Turbo cornutus [10]; Pig [11]; More
(all organisms) [7]

Source tissue
Latex serum [1]; Liver [4, 5]; Lung [6]; Gastropod liver [10]; Kidney [11]

Localisation in source
Extracellular [1, 2, 7]; Lysosomes [6]; Lamellar bodies [6]

Purification
Lactuca sativa [1]; Littorina littorea [3]; Turbo cornutus liver [10]; Pig kidney [11]

Crystallization
–

Cloned
–

Renaturated
–

---

5 STABILITY

pH
7.5 (not stable above) [1]; 4.0–8.0 [2]; 5.5–7.0 [4, 9]; 5.0 (not stable below)
[11]; 4.5–7.5 [12]

Temperature (°C)
50 (not'stable above) [1, 9]; 40 (not stable above) [4, 11]

Oxidation

Organic solvent

Enzyme Handbook © Springer-Verlag Berlin Heidelberg 1991
Duplication, reproduction and storage in data banks are only
allowed with the prior permission of the publishers

General stability information

Storage
Several months, 4°C, pH 5.4–5.5 [1, 9]

---

6 CROSSREFERENCES TO STRUCTURE DATABANKS

PIR/MIPS code

Brookhaven code

---

7 LITERATURE REFERENCES

[1] Giordani, R., Noat, G.: Eur. J. Biochem., 175, 619–625 (1988)
[2] Hébraud, M., Fèvre, M.: J. Gen. Microbiol., 134, 1123–1129 (1988)
[3] Melgar, M.J., Cabezas, J.A., Calvo, P.: Comp. Biochem. Physiol., 80B, 149–156 (1985)
[4] Chinchetru, M.A., Cabezas, J.A., Calvo, P.: Comp. Biochem. Physiol., 75B, 719–728 (1983)
[5] Rodriguez, J.A., Cabezas, J.A., Calvo, P.: Int. J. Biochem., 14, 695–698 (1982)
[6] Hook, G.E.R., Gilmore, L.B.: J. Biol. Chem., 257, 9211–9220 (1982)
[7] Colas, B.: Biochim. Biophys. Acta, 657, 535–538 (1981)
[8] Colas, B.: Biochim. Biophys. Acta, 613, 448–458 (1980)
[9] Colas, B.: Biochim. Biophys. Acta, 527, 150–158 (1978)
[10] Yamada, M., Ikeda, K., Egami, F.: J. Biochem., 73, 69–76 (1973)
[11] Wiederschain, G.Y., Prokopenkov, A.A.: Arch. Biochem. Biophys., 158, 539–543 (1971)
[12] Levvy, G.A., McAllan, A.: Biochem. J., 87, 206–209 (1963)

1 NOMENCLATURE

EC number
3.2.1.39

Systematic name
1, 3-Beta-D-glucan glucanohydrolase

Recommended name
Glucan endo-1, 3-beta-glucosidase

Synonymes
Glucanase, endo-1, 3-.beta.-
Endo-1, 3-.beta.-glucanase
Oligo-1, 3-glucosidase
Endo-1, 3-beta-glucanase
Laminarinase
Laminaranase
Callase
Beta-1, 3-glucanase
Kitalase
1, 3-Beta-D-glucan 3-glucanohydrolase [1]
Endo-(1 , 3)-beta-D-glucanase [3]
Endo-1, 3-beta-glucosidase [3]
Endo-1, 3-beta-D-glucanase [6]
(1 -- > 3)-Beta-glucan 3-glucanohydrolase [5]
Endo-(1 -- > 3)-beta-D-glucanase [10]
(1 -- > 3)-Beta-glucan endohydrolase

CAS Reg. No.
9025-37-0

---

2 REACTION AND SPECIFICITY

Catalysed reaction
Hydrolysis of 1, 3-beta-D-glucosidic linkages in 1, 3-beta-D-glucans
(endoglucanase [4, 5, 10, 12, 17, 20, 23, 25, 31, 32], exoglucanase [8, 9, 33],
action pattern [33]) [4, 5, 8, 9, 10, 12, 17, 20, 23, 25, 31, 32, 33]

Reaction type
O-Glycosyl bond hydrolysis

Natural substrates
1,3-Beta-D-glucans + H_2O (induced in plants infected with pathogens [3, 4])

Enzyme Handbook © Springer-Verlag Berlin Heidelberg 1991
Duplication, reproduction and storage in data banks are only
allowed with the prior permission of the publishers

Substrate spectrum
1 1, 3-Beta-D-glucan + H_2O [1–36]
2 Laminarin + H_2O [4, 5, 6, 8, 9, 16, 18, 20, 21, 22, 23, 24, 27, 28, 30, 31, 32, 33, 35, 36]
3 Paramylon + H_2O
4 Pachyman + H_2O (high activity against insoluble pachyman, low activity against soluble short-chain pachyman [17]) [8, 14, 17, 25]
5 p-Nitrophenyl-beta-D-glucopyranoside + H_2O (not: 4-nitrophenyl-beta-D-glucopyranoside [17]) [9]
6 Laminarioligosaccharides + H_2O (with 4 or more glucose residues) [12]
7 Pendulan + H_2O [12]
8 Laminaritetraose + H_2O (slowly [32]) [12, 23, 30, 35, 36]
9 Laminaripentaose + H_2O (slowly [32]) [12, 16, 35]
10 Laminarihexaose + H_2O [12, 36]
11 Cellwall-glucan + H_2O (yeast [14, 16, 17, 25, 35], no activity on heavily branched yeast cellwall glucan [20]) [14, 16, 17, 25, 35]
12 Laminaritriose + H_2O (not [30, 32, 35]) [16, 23, 36]
13 Carboxymethyl pachyman + H_2O [23, 34]
14 More (not: pustulan, lichenan, cellulose [8], 1, 3-glucosidase and beta(1, 3)-glucanase activity [8], exo-laminarase and beta-D-glucosidase activity [9], only saccharides with beta-1, 3-glucosidic linkage [12, 16, 19, 22, 24], not: oligosaccharides composed of 2–9 glucose units [27], very limited action on mixed-link (1, 3–1, 4)-beta-D-glucans) [8, 9, 12, 14, 16, 19, 22, 23, 27, 32, 35, 36]

Product spectrum
1 ?
2 Laminarisaccharides (of different chain lenths, depending on organism [4, 5, 6, 9, 16, 17, 18, 24, 26, 30, 31]) + glucose (small amounts [24, 26, 30, 31]) [4, 5, 6, 9, 16, 17, 18, 24 , 26, 30, 31]
3 ?
4 ?
5 Glucose + p-nitrophenol
6 Laminaritriose + laminaribiose [12]
7 ?
8 Laminaribiose [12]
9 Laminaribiose + laminaritriose [12]
10 Laminaritriose (main product) + laminaritetraose (small amounts) + laminaribiose (small amounts) [12]
11 ?
12 ?
13 ?
14 ?

Inhibitor(s)

Epoxyalkyl beta-oligoglucosides [1]; $HgCl_2$ [8, 25]; $CuCl_2$ [8]; $CoCl_2$ [8]; 1, 5-D-Gluconolactone [9]; Hg^{2+} [12, 20, 21, 26 , 27, 31, 32]; Cu^{2+} [12, 21, 26, 27, 31, 32]; Pb^{2+} [12, 27]; Glutathione (oxidized) [20]; Iodoacetamide [20]; N-Ethylmaleimide [20]; Ca^{2+} (slight) [21]; Mg^{2+} (slight [21]) [21, 27]; Zn^{2+} [21]; EDTA [21]; SDS [21]; Sodium 7-deoxycholate [21]; Hexadecyltrimethylammonium bromide [21]; $Cr(SO_4)_3 \times 2H_2O$ (slight) [25]; $CuSO_4 \times 5H_2O$ [25]; $NiSO_4 \times 6H_2O$ [25]; $ZnSO_4 \times 7H_2O$ [25]; $AgNO_3$ [25]; $Pb(NO_3)_2$ [25]; $Na_2HAgO_4 \times 7H_2O$ (slight) [25]; $Na_2WO_4 \times 2H_2O$ (slight) [25]; Phenylmercurinitrate [26, 32]; N-Bromosuccinimide [26, 32]; KI [32]; Ag^+ [27, 31]; Mn^{2+} [27]; $KMnO_4$ [27]; o-Phenanthroline [27]; Barbital [27]; Semicarbazide [27]; Iodine [27]; p-Chloromercuribenzoate [21, 27]; Na-Laurylsulfate [27]; Laminarin [27]; Sucrose [27]; Fructose [27]; Mannose [27]; Glucono-delta-lactone [27]; N-Acetylimidazole [32]; Carbodiimide [32]; Lespedeza cuneata cellulase inhibitor [32]; More [27]

Cofactor(s)/prostethic group(s)

More (no metal ion required) [24]

Metal compounds/salts

$CaCl_2$ (activates, Ca^{2+}: no effect [20]) [23]; $MgCl_2$ (activates, Mg^{2+}: no effect [20]) [23]; Na^+ (activates) [23]; Ca^{2+} (activates) [26]; Ba^{2+} (activates) [26]; Co^{2+} (activates) [26]

Turnover number (min⁻¹)

Turnover number (min^{-1})

Specific activity (U/mg)

19.0 [4]; 70.8 [12]; 16 [14]; 1050 [20]; 4.1 [24]; More [4, 18, 19, 21, 22, 25, 28, 31 , 36]

K_m-value (mM)

8 (laminarin) [5]; 0.27 (p-nitrophenol-beta-D-glucopyranoside) [9]; 0.64 (laminarin) [9]; More [8, 16, 17, 18, 22–25, 34, 36]

pH-optimum

5.1 (G1, Pisum sativum) [4]; 5.6 (Hordeum vulgare) [5]; 5. 4 (G2, Pisum sativum) [4]; 5.8 (Arthrobacter) [7]; 4.5 (Neurospora crassa) [8]; 6.5 (Trichoderma viride) [16]; 5.0 (Mexican lime) [17]; 6.0 (Penicillium italicum) [20]; 7.0 (Alternaria tenuissima, Aspergillus vesicolor) [21]; 5.0 (Schizosaccharomyces pombe) [22]; 5.0 (rye, laminarin) [23]; 5.7 (rye, carboxymethylpachyman) [23]; 5.5 (Bacillus circulans, beta(1 --> 3)-glucanase I) [24]; 6.5–7 (Bacillus circulans, beta(1 --> 3)-glucanase II) [24]; 5.0 (Rhizopus chinensis) [25]; 4.8–5.0 (enzyme purified from papain) [26]; 4–6 (strain of fungi imperfecti, glucan depolymerizing activity) [28]; 5.5–8.0 (Bacillus) [30]; 4. 8 (strain of fungi imperfecti, cell-lytic acitvity) [28]; 5.0 (Nicotiana tabacum) [31]; 5.0 (Nicotiana glutinosa) [34]; 5.7 (Streptomyces sp.) [35]

Enzyme Handbook © Springer-Verlag Berlin Heidelberg 1991
Duplication, reproduction and storage in data banks are only
allowed with the prior permission of the publishers

pH-range

3.5–8 (Alternaria tenuissima, Aspergillus vesicolor, 3.5: 20% of maximum activity, 8.0: 40–60% of maximal activity) [21]; 4.0–9.5 [24]; 4.0–7.0 (4: 40% of maximum activity, 7.0: 10% of maximum activity, Rhizopus chinensis) [25]; 4.0–8.0 (Nicotiana tabacum) [31]; 3.2–7.2 (Nicotiana glutinosa) [34];–3.5–7.6 (Arthrobacter) [7]; 3–7 (Neurospora crassa, 3: 1, 3-glucanase I (50% of maximal activity), 1, 3-glucanase II (35% of maximal activity), 7: glucanase I (5% of maximal activity), glucanase II (25% of maximal activity) [8]; 3–9 (20% of maximal activity at pH 3 and 9) [20]; More [30]

Temperature optimum (°C)

30 (Alternaria tenuissima, Aspergillus vesicolor) [21]; 45 (rye) [23]; 55 (Rhizopus chinensis) [25]; 65 (Streptomyces sp.) [35]; 45 (Neurospora crassa) [8]; 45 (Trichoderma viride) [16]; 50 (Porodisculus pendulus) [12]; 50 (short incubation time, Penicillium italicum) [20]

Temperature range (°C)

30–60 (Neurospora crassa, 30°C: glucanase I/II (15% of maximal activity), 60°C: glucanase I (5% of maximal activity), glucanase II (40% of maximal activity)) [8]; 30–80 (Bacillus) [30]

---

3 ENZYME STRUCTURE

Molecular weight

32000 (SDS-PAGE, Hordeum vulgare) [5]
17000 (gel filtration, Artiplex littoralis, EG-1) [6]
26000 (gel filtration, Artiplex littoralis, EG-2) [6]
55000 (SDS-PAGE, Arthrobacter, gene expression in E. coli) [7]
82000 (disc gel electrophoresis, varying polyacrylamide gel concentrations, Neurospora crassa, 1, 3-glucanase I) [8]
12000 (disc gel electrophoresis, varying polyacrylamide gel concentrations, Neurospora crassa, 1, 3-glucanase II) [8]
62000 (SDS-PAGE, Acacia verek) [9]
32000 (sedimentation equilibrium, Nicotiana tabacum) [11]
33000 (SDS-PAGE, Nicotiana tabacum) [11]
43000 (SDS-PAGE, Porodisculus pendulus) [12]
42000 (gel filtration, Porodisculus pendulus) [12]
34700 (gel filtration, Trichoderma viride) [16]
12000 (gel filtration, Mexican lime) [17]
33000 (gel filtration, SDS-PAGE, Glycine max) [19]
65000 (gel filtration, Penicillium italicum) [20]
68000 (SDS-PAGE, Penicillium italicum) [20]
160500 (1 × 78500, 1 × 82000, SDS-PAGE, Schizosaccharomyces pombe, glucanase I) [22]
75000 (SDS-PAGE, Schizosaccharomyces pombe, glucanase II) [22]

24300 (gel filtration, rye) [23]
40000 (SDS-PAGE, Bacillus circulans, glucanase I) [24]
23000 (molecular sieve chromatography, Rhizopus chinensis) [25]
24500 (gel filtration, SDS-PAGE, strain of fungi imperfecti) [28]
16000 (molecular sieve chromatography, Flavobacterium dorminator) [36]
35000 (SDS-PAGE, Lycopersicon esculentum) [2]
34300 (SDS-PAGE, Pisum sativum, G2) [4]
33500 (SDS-PAGE, Pisum sativum, G1) [4]

Subunits
Monomer (1 × 33000, SDS-PAGE, Nicotiana tabacum) [11]
Monomer (1 × 75000, SDS-PAGE, glucanase II, Schizosaccharomyces pombe) [22]
Dimer (1 × 78500, 1 × 82000, Schizosaccharomyces pombe, SDS-PAGE, glucanase I) [22]

Glycoprotein/Lipoprotein
Glycoprotein (1% carbohydrate, only arabinose residues [11], acidic glycoproteins [20]) [11, 20]

---

4 ISOLATION/PREPARATION

Source organism
Acacia verek [9]; Gossypium hirsutum [10]; Nicotiana tabacum [11, 31]; Porodisculus pendulus [12]; Rhizoctonia solani [14]; Streptomyces sp. [15]; Trichoderma viride [16]; Mexican lime [17]; Saccharomyces cerevisiae [18]; Glycine max [19]; Penicillium italicum [20]; Alternaria tenuissima [21]; Aspergillus vesicolor [21]; Schizosaccharomyces pombe (2 forms I/II) [22]; Bacillus circulans (2 forms I/II) [24]; Rhizopus chinensis [25]; Rhizopus sp. [27]; Fungi imperfecti (a strain belonging to) [28]; Flavobacterium [29]; Bacillus (alkalophilic) [30]; Rhizopus arrhizus [32]; Basidiomycete sp. [33]; Streptomyces sp. [35]; Flavobacterium dorminator [36]; Nicotiana glutinosa [1, 32, 24]; Lycopersicon esculentum (inoculated with Chladysporium fulvum) [2]; Trichoderma longibrachiatum [3]; Pisum sativum (2 forms: G1/G2) [4]; Hordeum vulgare [5]; Artiplex littoralis (2 forms: EG-1, EG-2) [6]; Arthrobacter (gene expression in E. coli) [7]; Neurospora crassa [8]; Rye [23]

Source tissue
Mycelium [20]; Papain (commercially available) [26]; Leaves [2, 6]; Pods [4]; Cultured cells [11]; Cotton fibres [10]; Culture filtrate [12, 28]; Cotyledons [19]; More (commercial enzyme preparation from Flavobacterium) [29]

Localisation in source
Cell-wall (associated [8, 10, 18], bound [9]) [8–10, 18, 22]; Cytoplasm [18]; Extracellular [21]

Enzyme Handbook © Springer-Verlag Berlin Heidelberg 1991
Duplication, reproduction and storage in data banks are only
allowed with the prior permission of the publishers

Purification

Pisum sativum (2 forms: G1/G2) [4]; Hordeum vulgare [5]; Neurospora crassa (2 forms: I/II) [8]; Artiplex littoralis (partial) [6]; Acacia verek [9]; Porodisculus pendulus [12]; Rhizoctonia solani [14]; Trichoderma viride [16]; Mexican lime (partial) [17]; Saccharomyces cerevisiae [18]; Glycine max [19]; Penicillium italicum [20]; Alternaria tenuissima [21]; Aspergillus vesicolor [21]; Schizosaccharomyces pombe (2 forms I/II) [22]; Rye [23]; Bacillus circulans [24]; Rhizopus chinensis [25]; Fungi imperfecti (a strain belonging to) [28]; Bacillus [30]; Nicotiana tabacum [31]; Rhizopus arrhizus [32]; Nicotiana glutinosa [34]; Streptomyces sp. [35]; Flavobacterium dorminator [36]

Crystallization

[11, 25, 28]

Cloned

(Arthrobacter gene in E. coli [7]) [3, 7, 13]

Renaturated

–

---

5 STABILITY

pH

4.0–7.0 (12 hours, Neurospora crassa, 1, 3-glucanase I) [8]; 4.0–5.5 (12 hours, Neurospora crassa, 1, 3-glucanase II) [8]; 5.5–7.3 (Porodisculus pendulus) [12]; 4–8 (30 minutes, Alternaria tenuissima, Aspergillus vesicolor) [21]; 7 (highest stability, Bacillus) [30]

Temperature (°C)

45 (EG-2: complete denaturation after 15 minutes, EG-1: 10 % loss of activity after 90 minutes, Artiplex littoralis) [6]; 45 (10 minutes, Neurospora crassa) [8]; 45 (stable below, Porodisculus pendulus) [12]; 60 (10 minutes, 90% loss of activity, Penicillium italicum) [20]; 70 (3 minutes, complete loss of activity, Penicillium italicum) [20]; 45 (stable up to, rye) [23]; 75 (10 minutes, complete inactivation, Rhizopus chinensis) [25]; 50 (pH 5–6, 10 minutes, strains of fungi imperfecti) [28]; 40 (stable below, Nicotiana tabacum) [31]; 65 (10 minutes, inactivated, Nicotiana glutinosa) [34]

Oxidation

Organic solvent

General stability information

More [34]

Storage

6 CROSSREFERENCES TO STRUCTURE DATABANKS

PIR/MIPS code

A32106 (35K, vacuolar, common tobacco, fragment); B32106 (37K, extracellular, common tobacco, fragments); C32106 (36K, extracellular, common tobacco, fragment); D32106 (35K, extracellular, common tobacco, fragment); A30758 (precursor, common tobacco); A31800 (barley); S00396 (barley, fragment); S05510 (II, precursor, barley)

Brookhaven code

7 LITERATURE REFERENCES

[1] Hoj, P.B., Rodriguez, E.B., Stick, R.V., Stone, B.A.: J. Biol. Chem., 264, 4939–4947 (1989)
[2] Joosten, M.H.A.J., De Wit, P.J.G.M.: Plant Physiol., 89, 945–951 (1989)
[3] Shinshi, H., Wenzler, H., Neuhaus, J.-M., Felix, G., Hofsteenge, J., Meins, F.: Proc. Natl. Acad. Sci. USA, 85, 5541–5545 (1988)
[4] Mauch, F., Hadwiger, L.A., Boller, T.: Plant Physiol., 87, 325–333 (1988)
[5] Hoj, P.B., Slade, A.M., Wettenhall, R.E.H., Fincher, G. B.: FEBS Lett., 230, 67–71 (1988)
[6] Boucaud, J., Bigot, J., Devaux, J.: J. Plant Physiol., 128, 337–349 (1987)
[7] Doi, K., Doi, A.: J. Bacteriol., 168, 1272–1276 (1986)
[8] Hiura, N., Kobayashi, M., Nakajima, T., Matsuda, K.: Agric. Biol. Chem., 50, 2461–2467 (1986)
[9] Lienhart, Y., Comtat, J., Barnoud, F.: Biochim. Biophys. Acta, 883, 353–360 (1986)
[10] Bucheli, P., Dürr, M., Buchala, A.J., Meier, H.: Planta, 166, 530–536 (1985)
[11] Shinshi, H., Kato, K.: Agric. Biol. Chem., 47, 1455–1460 (1983)
[12] Iwamuro, Y., Aoki, M., Mikami, Y.: J. Ferment. Technol., 63, 61–66 (1985)
[13] Schwarz, W., Bronnenmeier, K., Staudenbauer, W.L.: Biotechnol. Lett., 7, 859–864 (1985)
[14] Usui, T., Totani, K., Totsuka, A., Oguchi, M.: Biochim. Biophys. Acta, 840, 255–263 (1985)
[15] Beyer, M., Dieckmann, H.: Appl. Microbiol. Biotechnol., 20, 207–212 (1984)
[16] Merc, M., Galas, E.: Acta Biotechnol., 4, 67–74 (1984)
[17] Carrasco, P., Beltran, J.-P., Pereto, J.G., Granell, A.: Phytochemistry, 22, 2699–2701 (1983)
[18] Hien, N.H., Fleet, G.H.: J. Bacteriol., 156, 1204–1213 (1983)
[19] Keen, N.T., Yoshikawa, M.: Plant Physiol., 71, 460–465 (1983)
[20] Sanchez, M., Nombela, C., Villanueva, J.R., Santos, T.: J. Gen. Microbiol., 128, 2047–2053 (1982)
[21] Jirku, V., Kraxnerova, B., Krumphanzl, V.: Folia Microbiol., 25, 24–31 (1980)
[22] Reichelt, B.Y., Fleet, G.H.: J. Bacteriol., 147, 1085–1094 (1981)
[23] Ballance, G.M., Manners, D.J.: Phytochemistry, 17, 1539–1543 (1978)
[24] Rombouts, F.M., Phaff, H.J.: Eur. J. Biochem., 63, 121–130 (1976)
[25] Yamamoto, S., Nagasaki, S.: Agric. Biol. Chem., 39, 2163–2169 (1975)
[26] Wilson, G.: Biochem. Soc. Trans., 2, 550th Meeting, 1115–1116 (1974)

Enzyme Handbook © Springer-Verlag Berlin Heidelberg 1991
Duplication, reproduction and storage in data banks are only
allowed with the prior permission of the publishers

[27] Nagasaki, S., Fukuyama, J., Yamamoto, S., Kobayashi, R.: Agric. Biol. Chem., 38, 349–357 (1974)
[28] Yamamoto, S., Fukuyama, J., Nagasaki, S.: Agric. Biol. Chem., 38, 329–337 (1974)
[29] Manners, D.J., Wilson, G.: Biochem. J., 135, 11–18 (1973)
[30] Horikoshi, K., Atsukawa, Y.: Agric. Biol. Chem., 37, 1449–1456 (1973)
[31] Kato, K., Yamada, A., Noguchi, M.: Agric. Biol. Chem., 37, 1269–1275 (1973)
[32] Moore, A.E., Stone, B.A.: Biochim. Biophys. Acta, 258, 248–264 (1972)
[33] Bochkov, A.F., Sova, V.V., Kirkwood, S.: Biochim. Biophys. Acta, 258, 531–540 (1972)
[34] Moore, A.E., Stone, B.A.: Biochim. Biophys. Acta, 258, 238–247 (1972)
[35] Bielecki, S., Antczak, T., Galas, E. in "Eur. Congr. Biotechnol.", 3rd. Ed., 2, 489–496 (1984)
[36] Nagasaki, S., Noshioka, Y., Mori, H., Yamamoto, S.: Agric. Biol. Chem., 40, 1059–1067 (1976)

1 NOMENCLATURE

EC number
3.2.1.40

Systematic name
Alpha-L-rhamnoside rhamnohydrolase

Recommended name
Alpha-L-rhamnosidase

Synonymes
.alpha.-L-Rhamnosidase
Rhamnosidase, .alpha.-L-
Alpha-L-rhamnosidase T [2]
Alpha-L-rhamnosidase N [2]
More (Narginase: complex of beta-D-glucosidase and al-
pha-L-rhamnosidase) [5–7]

CAS Reg. No.
37288-35-0

---

2 REACTION AND SPECIFICITY

Catalysed reaction
Hydrolysis of terminal non-reducing alpha-L-rhamnose residues in al-
pha-L-rhamnosides

Reaction type
O-Glycosyl bond hydrolysis

Natural substrates
Alpha-L-rhamnoside + H_2O

Substrate spectrum
1 Alpha-L-rhamnoside + H_2O
2 p-Nitrophenyl alpha-L-rhamnoside + H_2O [1, 2]
3 p-Nitrophenyl-alpha-L-rhamnopyranoside + H_2O [6]
4 Naringin + H_2O (Turbo cornutus, not: Aspergillus niger [2]) [2, 5]
5 Quercitrin + H_2O (Aspergillus niger, not: Turbo cornutus [2]) [2, 10]
6 Rutin + H_2O (Aspergillus niger, not: Turbo cornutus [2]) [2, 10]
7 Rutinose + H_2O (6-O-alpha-L-rhamnosyl-D-glucopyranose) [4]
8 Methyl 3-O-alpha-L-rhamnopyranosyl-alpha-D-xylopyranoside + H_2O
[8]

Enzyme Handbook © Springer-Verlag Berlin Heidelberg 1991
Duplication, reproduction and storage in data banks are only
allowed with the prior permission of the publishers

   9 Methyl 3-O-alpha-L-rhamnopyranosyl-alpha-D-mannopyranoside
      + H_2O [8]
  10 Methyl 3-O-alpha-L-rhamnopyranosyl-alpha-L-rhamnopyranoside
      + H_2O [8]
  11 Methyl 4-O-alpha-L-rhamnopyranosyl-alpha-D-galactopyranoside
      + H_2O [8]
  12 Methyl 4-O-alpha-L-rhamnopyranosyl-alpha-D-mannopyranoside
      + H_2O [8]
  13 Methyl 4-O-alpha-L-rhamnopyranosyl-alpha-D-xylopyranoside + H_2O
      [8]
  14 More (not: alpha-L-rhamnoside, rhamnosyl 2-keto-3-deoxyoctonate [2])
      [2, 3]

Product spectrum

   1 Alpha-L-rhamnose + ?
   2 p-Nitrophenol + alpha-L-rhamnose [1]
   3 p-Nitrophenolate anion + alpha-L-rhamnose [6]
   4 4',5,7-Trihydroxyflavanone 7-glucoside + L-rhamnose
   5 Quercetin + alpha-L-rhamnose [10]
   6 ?
   7 L-Rhamnose + D-glucose
   8 L-Rhamnose + methyl alpha-D-xyloside [8]
   9 L-Rhamnose + methyl alpha-D-mannoside [8]
  10 L-Rhamnose + methy alpha-L-rhamnoside [8]
  11 L-Rhamnose + methyl alpha-D-galactoside [8]
  12 L-Rhamnose + methyl alpha-D-mannoside [8]
  13 L-Rhamnose + alpha-D-xyloside [8]
  14 ?

Inhibitor(s)

$CuSO_4$ (slight) [2]; L-Rhamnose (competitive [4]) [4, 6, 7]; L-Lyxose (competitive) [4]; NaN_2 (Turbo cornutus, not: Aspergillus niger) [2];
6-Deoxy-D-glucose (competitive) [4]; Methyl-alpha-D-mannoside (competitive) [4]; Citric acid (no other organic acids, highest inhibition at low pH)
[5]; Glucose [6, 7]

Cofactor(s)/prostethic group(s)

Metal compounds/salts

Turnover number (min⁻¹)

Specific activity (U/mg)

0.324 (Turbo cornutus) [2]; 31.7 (Aspergillus) [2]; More [4]

K_m-value (mM)
2.0 (p-nitrophenyl alpha-L-rhamnoside, Turbo cornutus) [2]; 2.65
(p-nitrophenyl alpha-L-rhamnoside, Aspergillus) [2]; 0.33
(p-nitrophenyl-alpha-L-rhamnoside) [4]; 2.2
(6-O-alpha-L-rhamnosyl-D-glucopyranose) [4]; 1.52
(p-nitrophenyl-alpha-L-rhamnoside) [6]; 7.0 (naringin) [6]; 2.8 (immobilized,
narginase activity) [11]; 7.0 (soluble, narginase activity) [11]

pH-optimum
3.5 (Penicillium sp., no change upon immobilization [7]) [6, 7]; 2.0
(Corticium rolfsii) [1]; 2.8 (Turbo cornutus) [2]; 4.5–5.0 (Aspergillus niger) [2]

pH-range
2.5–5 [6, 7, 11]; 2–5 (Turbo cornutus [2], Corticium rolfsii [1]) [1, 2]; 2–7
(Aspergillus niger) [2]

Temperature optimum (°C)
57 [6, 11]; More (immobilized arginase) [9]

Temperature range (°C)
35–75 [6]

---

3 ENZYME STRUCTURE

Molecular weight
70000 (disc gel electrophoresis, gel filtration, Fagopyrum esculentum) [4]

Subunits

Glycoprotein/Lipoprotein
–

---

4 ISOLATION/PREPARATION

Source organism
Corticium rolfsii [1]; Turbo cornutus [2]; Aspergillus niger [2, 5, 8];
Fagopyrum esculentum (buckwheat) [3, 4]; Penicillium sp. (immobilization
of narginase [7, 9, 11]) [6, 7, 9, 11]; Bacteroides distasonis [10]

Source tissue
Culture filtrate [1]; Liver [2]

Localisation in source

Purification
Aspergillus niger [1]; Turbo cornutus [2]; Fagopyrum esculentum (buckwheat, partial [3], saracen corn [4]) [3, 4]

Crystallization
–

Enzyme Handbook © Springer-Verlag Berlin Heidelberg 1991
Duplication, reproduction and storage in data banks are only
allowed with the prior permission of the publishers

Cloned

−

Renaturated

−

---

5 STABILITY

pH

2.0–7.0 (stable between, 21% loss of activity after 24 hours at 2.0) [1]

Temperature (°C)

57 (immobilized enzyme, 6 months) [7]

Oxidation

Organic solvent

General stability information

Storage

3–5° C, pH 3.5, potassium hydrogen phthalate /HCl buffer, immobilized enzyme, 85–90% of activity retained after 1 year [7]; More (stability of immobilized enzyme) [11]

---

6 CROSSREFERENCES TO STRUCTURE DATABANKS

PIR/MIPS code

Brookhaven code

---

7 LITERATURE REFERENCES

[1] Kaji, A., Ichimi, T.: Agric. Biol. Chem., 37, 431–432 (1973)
[2] Kurosawa, Y., Ikeda, K., Egami, F.: J. Biochem., 73, 31–37 (1973)
[3] Bourbouze, R., Pratviel-Sosa, F., Percheron, F.: Phytochemistry, 14, 1279–1282 (1975)
[4] Bourbouze, R., Percheron, F., Courtois, J.-E.: Eur. J. Biochem., 63, 331–337 (1976)
[5] Ono, M.: J. Ferment. Technol., 58, 387–389 (1980)
[6] Romero, C., Manjon, A., Bastida, J., Iborra, J.L.: Anal. Biochem., 149, 566–571 (1985)
[7] Manjon, A., Bastida, J., Romero, A., Jimeno, A., Iborra, J.L.: Biotechnol. Lett., 7, 477–482 (1985)
[8] Kamiya, S., Esaki, S., Ito-Tanaka, R.: Agric. Biol. Chem., 49, 2351–2358 (1985)
[9] Turecek, P., Pittner, F.: Appl. Biochem. Biotechnol., 13, 1–13 (1986)
[10] Bokkenheuser, V.D., Shackleton, C.H.L., Winter, J.: Biochem. J., 248, 953–956 (1987)
[11] Jimeno, A., Manjon, A., Canovas, M., Iborra, J.L.: Process Biochem., 22, 13–16 (1987)

1 NOMENCLATURE

EC number
3.2.1.41

Systematic name
Alpha-dextrin 6-glucanohydrolase

Recommended name
Alpha-dextrin endo-1, 6-alpha-glucosidase

Synonymes
Limit dextrinase
Debranching enzyme
Amylopectin 6-glucanohydrolase
Pullulanase
R-enzyme
EC 3.2.1.69 (formerly)
Promozyme 200L

CAS Reg. No.
9075-68-7

2 REACTION AND SPECIFICITY

Catalysed reaction
Hydrolysis of 1, 6-alpha-D-glucosidic linkages in pullulan, amylopectin and glycogen, and in the alpha- and beta-amylase limit dextrins of amylopectin and glycogen

Reaction type
O-Glycosyl bond hydrolysis

Natural substrates
Starch + H_2O [1–27]

Substrate spectrum
1 Pullulan + H_2O (r) [1–27]
2 Amylopectin or glycogen or their limit dextrins + H_2O (r) [1–27]

Product spectrum
1 Maltotriose [1–27]
2 Maltooligosaccharides + maltose [1–27]

Enzyme Handbook © Springer-Verlag Berlin Heidelberg 1991
Duplication, reproduction and storage in data banks are only
allowed with the prior permission of the publishers

Inhibitor(s)

EDTA [4, 5, 7, 8, 11, 12, 23]; Zn^{2+} [4, 5, 7, 8, 22, 23]; Cu^{2+} [4, 5, 8, 22, 23]; Hg^{2+} [5, 12, 20, 22, 23]; $KMnO_4$ [5]; Sodium dodecyl sulfate [5, 23]; Pb^{2+} [5, 23]; Mg^{2+} [2]; Ag^{2+} [5, 20]; Ba^{2+} [5]; Bi^{2+} [5]; Fe^{2+} [5, 20, 23]; Sr^{2+} [5]; o-Phenanthroline [5]; Cyclodextrins [7, 8, 10, 21, 25]; N-Bromosuccinimide [7, 23]; Fe^{3+} [7, 23]; Mn^{2+} [12]; p-Chloromercuribenzoate [20, 23]; Al^{3+} [23] Co^{2+} [23]; Ni^{2+} [23]; Cd^{2+} [23]

Cofactor(s)/prostethic group(s)

Metal compounds/salts

Ca^{2+} [4, 8, 9, 11, 12, 23]; Ba^{2+} [4]; Mn^{2+} [4]; Mg^{2+} [4]; Co^{2+} [4, 5, 7]; Fe^{2+} [4]; Sr^{2+} [9]

Turnover number (min⁻¹)

175000–216000 [6]; 16240 [7]; 290000 [9]; 5940–13200 [21]

Specific activity (U/mg)

84.2 [2]; 49.6 [4]; 1352–1790 [5, 9]; 200–220 [6]; 481 [7]; 0.949 [12]; 16.5–29.8 [16–18]; 5.5–6.5 [27]

K_m-value (mM)

2.5 (pullulan, glucose equivalents) [6]; 0.021–0.078 (pullulan, glucose equivalents) [21]

pH-optimum

5.5–6.5 (maltose + cyclomaltooctaose) [1]; 7.0 (pullulan) [2, 5]; 6.5 (pullulan) [3, 12]; 6.0 (pullulan) [4, 20, 23, 24]; 5.0 (pullulan) [6]; 5.5 (pullulan) [7, 9, 15]; 5.0–6.0 (pullulan) [8]; 5.6 (pullulan) [10, 11, 16–18]; 8.5–9.0 (pullulan) [22]

pH-range

4.0 (not active below, maltose + cyclomaltooctaose) [1]; 3.0–10.0 (pullulan) [2]; 4.0–10.0 (pullulan) [4]; 3.0–7.0 (pullulan) [6]; 8.0 (not active above, pullulan) [7, 8]; 4.0–7.5 (pullulan) [9, 11]; 3.0–8.0 (pullulan) [15]; 5.0–10.0 (pullulan) [20]; 4.0–12.0 (pullulan) [22]; 4.0–11.0 (pullulan) [24]

Temperature optimum (°C)

55 (maltose + cyclomaltooctaose) [1]; 50 (pullulan) [2, 20, 24]; 65 (pullulan) [4, 6]; 70 (pullulan) [5]; 90 (pullulan) [7, 8, 11]; 40 (pullulan) [10]; 85 (pullulan) [15]; 55 (pullulan) [19, 22]

Temperature range (°C)

60 (not active above, maltose + cyclomaltooctaose) [1]; 80 (not active above, pullulan) [2]; 75 (not active above, pullulan) [6]; 60 (not active above, pullulan) [20, 22, 24]

3 ENZYME STRUCTURE

Molecular weight
218000–220000 (gel filtration, gel electrophoresis, Bacillus circulans) [2]
77000 (gel filtration, Bacteroides thetaiotaomicron) [3]
83000 (gel electrophoresis, Bacillus stearothermophilus) [4]
79000 (gel filtration, Thermoactinomyces thalpophilus) [5]
115000–116000 (gel electrophoresis, Bacillus acidopullulyticus) [6]
133000 (gel filtration, Clostridium thermohydrosulfuricum) [7]
120000 (HPLC, Thermoanaerobium sp.) [9]
67000 (gel filtration, sugar beet) [10]
80000–83000 (gel filtration, gel electrophoresis) [12]
70000 (gel filtration, rice seeds) [16]
85000 (gel filtration, oat seeds) [17]
110000 (gel filtration, Bacillus cereus var. mycoides) [20]
141000 (gel filtration, Klebsiella aerogenes) [21]
92000 (gel electrophoresis, Bacillus sp.) [22]
80000–90000 (gel filtration, sedimentation equilibrium, Aerobacter aerogenes) [24]
143000 (gel electrophoresis, Aerobacter aerogenes) [26]

Subunits
Monomer (gel electrophoresis) [2–7, 21]

Glycoprotein/Lipoprotein
Glycoprotein [7]; Lipoprotein [13]

---

4 ISOLATION/PREPARATION

Source organism
Bacillus circulans [2]; Bacillus stearothermophilus [4]; Thermoactinomyces thalpophilus [5]; Bacillus acidopullulyticus [6]; Clostridium thermohydrosulfuricum [7]; Thermoanaerobium sp. [9]; Sugar beet [10]; Thermus aquaticus [12]; Rice [16, 18]; Oat [17]; Bacillus cereus var. mycoides [20]; Klebsiella aerogenes [21]; Bacillus sp. [22]; Aerobacter aerogenes [24]; Yeast [22]

Source tissue

Localisation in source
Extracellular (2, 4, 5, 8, 9, 13, 20–23)

Enzyme Handbook © Springer-Verlag Berlin Heidelberg 1991
Duplication, reproduction and storage in data banks are only
allowed with the prior permission of the publishers

Purification

Bacillus circulans [2]; Bacillus stearothermophilus [4]; Thermoactinomyces thalpophilus [5]; Bacillus acidopullulyticus [6]; Clostridium thermohydrosulfuricum [7]; Thermoanaerobium sp. [9]; Sugar beet roots [10]; Thermus aquaticus [12]; Rice seeds [16, 18]; Oat seeds [17]; Bacillus cereus var. mycoides [20]; Klebsiella aerogenes [21] Bacillus sp. [22]; Aerobacter aerogenes [24]

Crystallization
[24]

Cloned
[3, 4, 14]

Renaturated
–

---

5 STABILITY

pH
5.0–8.0 [2]; 6.0–8.5 [4]; 4.0–8.5 [6]; 3.0–5.0 [7]; 4.5–10.5 [11]; 4.5–5.5 [15]; 5.0–9.0 [20]; 6.5–11.0 [22]; 4.5–12.0 [24]

Temperature (°C)
50 (not stable above) [2, 19, 20, 22, 24]; 60 (not stable above) [6]; 90 (not stable above) [7]; 80 (not stable above) [9]; 35 (not stable above) [10]; 85 (not stable above) [11]

Oxidation

Organic solvent
Ethanol (stable in 10%) [15]

General stability information

Storage
4 months, 4 °C [21]

---

6 CROSSREFERENCES TO STRUCTURE DATABANKS

PIR/MIPS code
A32880 (Klebsiella pneumoniae, fragment); A25025 (Klebsiella pneumoniae, fragment); A26879 (precursor, Klebsiella pneumoniae); S02472 (precursor, Klebsiella pneumoniae, fragment)

Brookhaven code

7 LITERATURE REFERENCES

[1] Hizukuri, S., Kawano, S., Abe, J., Koizumi, K., Tanimoto, T.: Biotechnol. Appl. Biochem., 11, 60–73 (1989)
[2] Sata, H., Umeda, M., Kim, C., Taniguchi, H., Maruyama, Y.: Biochim. Biophys. Acta, 991, 388–394 (1989)
[3] Smith, K.A., Salyers, A.A.: J. Bacteriol., 171, 2116–2123 (1989)
[4] Kuruki, T., Park, J., Okada, S., Imanaka, T.: Appl. Environ. Microbiol., 54, 2881–2883 (1988)
[5] Odibo, F.J.C., Obi, S.K.C.: J. Ind. Microbiol., 3, 343–350 (1988)
[6] Kusano, S., Nagahata, N., Takahashi, S., Fujimoto, D., Sakano, Y.: Agric. Biol. Chem., 52, 2293–2298 (1988)
[7] Saha, B.C., Mathupala, S.P., Zeikus, J.G.: Biochem. J., 252, 343–348 (1988)
[8] Koch, R., Zablowski, P., Antranikian, G.: Appl. Microbiol. Biotechnol., 27, 192–198 (1987)
[9] Plant, A.R., Clemens, R.M., Daniel, R.M., Morgan, H.W.: Appl. Microbiol. Biotechnol., 26, 427–433 (1987)
[10] Masuda, H., Takahashi, T., Sugawara, S.: Plant Physiol., 84, 361–365 (1987)
[11] Melasniemi, H.: Biochem. J., 246, 193–197 (1987)
[12] Plant, A.R., Morgan, H.W., Daniel, R.M.: Enzyme Microb. Technol., 8, 668–672 (1986)
[13] Pugsley, A.P., Chapon, C., Schwartz, M.: J. Bacteriol., 166, 1083–1088 (1986)
[14] Michaelis, S., Chapon, C., D'Enfert, C., Pugsley, A.P., Schwartz, M.: J. Bacteriol., 164, 633–638 (1985)
[15] Hyun, H.H., Zeikus, J.G.: Appl. Environ. Microbiol., 49, 1168–1173 (1985)
[16] Yamada, J.: Agric. Biol. Chem., 45, 1269–1270 (1981)
[17] Yamada, J.: Agric. Biol. Chem., 45, 1013–1015 (1981)
[18] Yamada, J., Izawa, M.: Agric. Biol. Chem., 43, 37–44 (1979)
[19] Adams, K.R., Priest, F.G.: FEMS Microbiol. Lett., 1, 269–273 (1977)
[20] Takasaki, Y.: Agric. Biol. Chem., 40, 1523–1530 (1976)
[21] Brandt, C.J., Catley, B.J., Awad Jr., W.M.: J. Bacteriol., 125, 501–508 (1976)
[22] Nakamura, N., Watanabe, K., Horikoshi, K.: Biochim. Biophys. Acta, 397, 188–193 (1975)
[23] Ohba, R., Ueda, S.: Agric. Biol. Chem., 39, 967–972 (1975)
[24] Ohba, R., Ueda, S.: Agric. Biol. Chem., 37, 2821–2826 (1973)
[25] Marshall, J.J.: FEBS Lett., 37, 269–273 (1973)
[26] Eisele, B., Rasched, I.R., Wallenfels, K.: Eur. J. Biochem., 26, 62–67 (1972)
[27] Lee, E.Y.C., Whelan, W.J. in "The Enzymes", 3rd. Ed. (Boyer, P.D., Ed.) 5, 191–234 (1972) (Review)

1 NOMENCLATURE

EC number
3.2.1.42

Systematic name
GDPglucose glucohydrolase

Recommended name
GDPglucosidase

Synonymes
Glucosidase, guanosine diphospho-
Guanosine diphosphoglucosidase
Guanosine diphosphate D-glucose glucohydrolase [1]

CAS Reg. No.
37288-36-1

---

2 REACTION AND SPECIFICITY

Catalysed reaction
GDPglucose + H_2O →
→ GDP + D-glucose

Reaction type
O-Glycosyl bond hydrolysis

Natural substrates
GDPglucose + H_2O (control of intracellular concentration) [1]

Substrate spectrum
1 GDPglucose + H_2O (strictly specific for) [1]

Product spectrum
1 GDP + D-glucose

Inhibitor(s)
Guanosine diphosphate (competitive) [1]

Cofactor(s)/prostethic group(s)

Metal compounds/salts
Mg^{2+} (relieves inhibition by guanosine diphosphate, some activation) [1];
Mn^{2+} (relieves inhibition by guanosine diphosphate, some activation) [1]

Enzyme Handbook © Springer-Verlag Berlin Heidelberg 1991
Duplication, reproduction and storage in data banks are only
allowed with the prior permission of the publishers

Turnover number (min⁻¹)

Specific activity (U/mg)
0.208 [1]

K_m-value (mM)
0.23 (GDPglucose) [1]

pH-optimum
5–7 [1]

pH-range
5–9 [1]

Temperature optimum (°C)
30 (assay at) [1]

Temperature range (°C)

---

3 ENZYME STRUCTURE

Molecular weight

Subunits

Glycoprotein/Lipoprotein
–

---

4 ISOLATION/PREPARATION

Source organism
Baker's yeast [1]

Source tissue
More (starch-free commercial baker's yeast) [1]

Localisation in source
Cytoplasm [1]; Soluble [1]

Purification
Baker's yeast [1]

Crystallization
–

Cloned
–

Renaturated
–

5 STABILITY

pH

Temperature (°C)

Oxidation

Organic solvent

General stability information

Storage
−20° C, 50% loss of activity after 6 months [1]

6 CROSSREFERENCES TO STRUCTURE DATABANKS

PIR/MIPS code

Brookhaven code

7 LITERATURE REFERENCES

[1] Sonnino, S., Carminatti, H., Cabib, E.: Arch. Biochem. Biophys., 116, 26–33 (1966)

Enzyme Handbook © Springer-Verlag Berlin Heidelberg 1991
Duplication, reproduction and storage in data banks are only
allowed with the prior permission of the publishers

1 NOMENCLATURE

EC number
3.2.1.43

Systematic name
Beta-L-rhamnoside rhamnohydrolase

Recommended name
Beta-L-rhamnosidase

Synonymes
Rhamnosidase, .beta.-L-
.beta.-L-Rhamnosidase

CAS Reg. No.
37288-37-2

2 REACTION AND SPECIFICITY

Catalysed reaction
Hydrolysis of terminal, non-reducing beta-L-rhamnose residues in
beta-rhamnosides

Reaction type
O-Glycosyl bond hydrolysis

Natural substrates
Beta-rhamnosides + H_2O

Substrate spectrum
1 Beta-rhamnoside + H_2O [1]

Product spectrum
1 Beta-rhamnose + ? [1]

Inhibitor(s)

Cofactor(s)/prostethic group(s)

Metal compounds/salts

Enzyme Handbook © Springer-Verlag Berlin Heidelberg 1991
Duplication, reproduction and storage in data banks are only
allowed with the prior permission of the publishers

Turnover number (min^{-1})

Specific activity (U/mg)

K_m-value (mM)

pH-optimum

pH-range

Temperature optimum (°C)

Temperature range (°C)

3 ENZYME STRUCTURE

Molecular weight

Subunits

Glycoprotein/Lipoprotein
–

4 ISOLATION/PREPARATION

Source organism
Klebsiella aerogenes (only) [1]

Source tissue
Cell [1]

Localisation in source

Purification

Crystallization
–

Cloned
–

Renaturated
–

5 STABILITY

pH

Temperature (°C)

Oxidation

Organic solvent

General stability information

Storage

6 CROSSREFERENCES TO STRUCTURE DATABANKS

PIR/MIPS code

Brookhaven code

7 LITERATURE REFERENCES

[1] Kamiya, S., Esaki, S., Shiba, N.: Agric. Biol. Chem., 51, 2207–2214 (1987)

Enzyme Handbook © Springer-Verlag Berlin Heidelberg 1991
Duplication, reproduction and storage in data banks are only
allowed with the prior permission of the publishers

1 NOMENCLATURE

EC number
3.2.1.44

Systematic name
Poly(1, 2-alpha-L-fucoside-4-sulfate) glycanohydrolase

Recommended name
Fucoidanase

Synonymes
Alpha-L-fucosidase

CAS Reg. No.
37288-38-3

---

2 REACTION AND SPECIFICITY

Catalysed reaction
Endohydrolysis of 1, 2-alpha-L-fucoside linkages in fucoidan without release of sulfate

Reaction type
O-Glycosyl bond hydrolysis (endoglycosidic)

Natural substrates
Alpha-1, 2-fucoside 4-sulfate + H_2O [1]

Substrate spectrum
1 Alpha-1, 2-fucoside 4-sulfate + H_2O [1]
2 p-Nitrophenyl alpha-L-fucoside + H_2O [2]
3 More (not: alpha-p-nitrophenyl-L-fucoside, blood group substance [1], p-nitrophenyl-L-fucoside [2]) [1, 2]

Product spectrum
1 Alpha-1, 2-fucoside 4-sulfate (polymer, residual) + oligosaccharides (variety of) [1]
2 Alpha-1, 2-fucoside (polymer, residual) + p-nitrophenol [2]
3 ?

Inhibitor(s)
Hg^{2+} [1]; Ag^+ [1]; Mn^{2+} (slight) [1]; EDTA (slight) [1]

Cofactor(s)/prostethic group(s)

Enzyme Handbook © Springer-Verlag Berlin Heidelberg 1991
Duplication, reproduction and storage in data banks are only
allowed with the prior permission of the publishers

Metal compounds/salts

Mg^{2+} (0.01 M, stimulation) [1]; More (metal ion not required for catalysis) [1]

Turnover number (min⁻¹)

Specific activity (U/mg)

K_m-value (mM)

0.22 (p-nitrophenyl-alpha-L-fucoside, rat epididymis) [2]; 0.21 (p-nitrophenyl-alpha-L-fucoside, ox liver) [2]

pH-optimum

5.4 [1]; 5.5 (ox liver) [2]; 6.0 (rat, epididymis) [2]

pH-range

4.5–7.0 [1]; 3–7.0 [2]

Temperature optimum (°C)

38 [1]

Temperature range (°C)

---

3 ENZYME STRUCTURE

Molecular weight

100000–200000 (gel filtration, Haliothus sp.) [1]

Subunits

Glycoprotein/Lipoprotein

–

---

4 ISOLATION/PREPARATION

Source organism

Haliothus rufescens (abalone) [1]; Haliothus corrugata [1]; Ox [2]; Rat [2]; Mouse [2]; Pig [2]; Cattle [2]; Patella vulgata (limpet) [3]

Source tissue

Ovary [2]; Uterus [2]; Liver [2]; Epididymis [2]; Hepatopancreas [1]; Spleen [2]; Kidney [2]; Testis [2]; More (distribution in tissues of: mouse, rat, cattle, pig) [2]

Localisation in source

Purification

Crystallization
–

Cloned
–

Renaturated
–

---

5 STABILITY

pH
5–9 (Haliothus sp., stable between, optimum 7.5) [1]; 7.5 (stability optimum, Haliothus sp.) [1]; 4.5–6.0 (stable between) [1]

Temperature (°C)
50 (purified enzyme resistant to denaturation up to, crude extract: complete loss of activity after 5 minutes) [1]

Oxidation

Organic solvent

General stability information

Storage

---

6 CROSSREFERENCES TO STRUCTURE DATABANKS

PIR/MIPS code

Brookhaven code

---

7 LITERATURE REFERENCES

[1] Thanassi, N.M., Nakada, H.I.: Arch. Biochem. Biophys., 118, 172–177 (1967)
[2] Levvy, G.A., McAllan, A.: Biochem. J., 80, 435–439 (1967)
[3] Conchie, J., Levvy, G.A.: Biochem. J., 65, 389 (1957)

Enzyme Handbook © Springer-Verlag Berlin Heidelberg 1991
Duplication, reproduction and storage in data banks are only
allowed with the prior permission of the publishers

1 NOMENCLATURE

EC number
3.2.1.45

Systematic name
D-Glucosyl-N-acylsphingosine glucohydrolase

Recommended name
Glucosylceramidase

Synonymes
GlcCer-beta-glucosidase [13]
Ceramidase, glucosyl-
. beta.-D-Glucocerebrosidase
Glucosylcerebrosidase
.beta.-Glucosylceramidase
Ceramide glucosidase
Glucocerebrosidase
Glucosylsphingosine.beta.-glucosidase
Glucosylsphingosine.beta.-D-glucosidase
. beta.-Glucocerebrosidase
Glucocerebroside.beta.-glucosidase
Glucosylceramide.beta.-glucosidase
Glucose cerebrosidase
Cerebroside.beta.-glucosidase
Acid beta-glucosidase [13, 15]

CAS Reg. No.
37228-64-1

---

2 REACTION AND SPECIFICITY

Catalysed reaction
D-Glucosyl-N-acylsphingosine + H_2O →
→ D-glucose + N-acylsphingosine

Reaction type
O-Glycosyl bond hydrolysis

Natural substrates
Glucosylceramide + H_2O [5, 17]

Enzyme Handbook © Springer-Verlag Berlin Heidelberg 1991
Duplication, reproduction and storage in data banks are only
allowed with the prior permission of the publishers

Substrate spectrum
1 D-Glucosyl-N-acylsphingosine + H_2O
2 4-Methylumbelliferyl-beta-D-glucopyranoside + H_2O [1, 2, 3, 5, 8, 12, 15]
3 Glucocerebroside + H_2O [5, 16]
4 More (not: galactocerebroside) [16]

Product spectrum
1 D-Glucose + N-acylsphingosine
2 Methylumbelliferone + glucose
3 Glucose + ceramide [5]
4 ?

Inhibitor(s)
Alkyl amines [11]; Alkyl beta-glucosides [11]; Conduritol-beta-epoxide [12, 13, 21, 14]; Bromo-conduritol [13]; N-Hexyl-D-glucosylsphingosine [1, 24]; Conduritol B [6]; 1-Deoxynojirimycin (and N-alkyl derivatives [8]) [8, 13, 24]; Alkyl-1-deoxynojirimycin [11]; p-Hydroxymercuribenzoate (weak) [16]; Gluconolactone [21, 24]; More [11]

Cofactor(s)/prostethic group(s)
More (acidic phospholipids activate [1, 5], phosphatidylserine and heat-stable protein stimulate [2], activated by: sodium taurocholate [6], sodium taurodeoxycholate [6], heat-stable glycoprotein [6], phospholipid or sodium taurocholate required [15]) [1, 2, 5, 6, 15]

Metal compounds/salts

Turnover number (min⁻¹)

Specific activity (U/mg)
26.7 [22]; 2.5 [3]; 3.57 [15]; 16.6 [17]; More [5, 9, 10, 14, 16, 19, 21]

K_m-value (mM)
0.636 (4-methylumbelliferyl glucoside) [2]; 0.08 (glucosylceramide) [19]; 0.055 (beta-D-glucocerebroside) [21]; 3.0 (4-methylumbelliferyl glucoside) [21]; More (effect of phosphatidylserine and heat-stable factor on K_m [1]) [1, 8, 11, 24]

pH-optimum
4.5–5.0 [1, 18]; 5.6 [5]; 6.0 [16, 19]; 4.8–5.8 [6]; 6.2 [21]; 6.0–6.6 [17]; 5.0 (addition of heat-stable factor shifted pH optimum for hydrolysis of methyl-umbelliferyl-beta-D-glucose from 6.5 to 5.0) [17]

pH-range
4–6 [1]; 5.0–8.0 (5.0: about 70% of maximum activity at, 8.0: about 30% of maximum activity at) [16]; 3.5–6.0 (3.5: about 40% of maximum activity at, 6.0: about 20% of maximum activity at) [18]; 4.5–7.5 [19]

Temperature optimum (°C)
37 (assay at) [1, 6]; 25 (assay at) [8]

Temperature range (°C)

3 ENZYME STRUCTURE

Molecular weight
67000 (radiation inactivation, human, normal enzyme) [13]
125000 (radiation inactivation, human, Gaucher disease enzyme) [13]
75000 (SDS-PAGE, human) [5]
67000 (SDS-PAGE, human) [9]
68000 (SDS-PAGE, human) [10]
59000 (SDS-PAGE, human) [12]
63000 (SDS-PAGE, human, Gaucher disease enzyme) [20]
138000 (gel chromatography, bovine) [21]
63000 (SDS-PAGE, bovine 63000 and 56000) [21]
56000 (SDS-PAGE, bovine, 63000 and 57000) [21]
More [14, 15, 18, 23, 25]

Subunits
Dimer (2 × 57000, SDS-PAGE, mouse) [2]
More (subunit coupling in Gaucher disease enzyme [13], in tissues:
monomer or aggregate, interconvertible [23]) [13, 23]
Tetramer (4 × 60000, human) [17]

Glycoprotein/Lipoprotein
Glycoprotein [2]

4 ISOLATION/PREPARATION

Source organism
Rat [1, 16]; Calf [8, 25]; Bovine [21]; Mouse [2]; Human (normal and
Gaucher disease enzyme [11]) [3, 4, 5, 7–20, 22]; More [4]

Source tissue
Mononuclear white blood cells (Gaucher disease) [20]; Placenta [3, 5, 9, 10,
11, 15, 17, 18, 19]; Liver [1, 2, 9]; Spleen [9, 11, 12, 16, 25]; Brain [9, 21];
Fibroblasts (cultured) [14]; Skin (fibroblasts, cultured) [14]; More [4]

Localisation in source
Membrane (bound to acidic lipids in membranes [2], bound [1, 2, 14]) [1, 2,
12, 14, 23]; Lysosomes [1, 2, 6, 12, 13, 14, 23]

Purification
Mouse [2]; Human [3, 5, 9, 10, 14, 15, 16, 19]; Bovine [21]

Enzyme Handbook © Springer-Verlag Berlin Heidelberg 1991
Duplication, reproduction and storage in data banks are only
allowed with the prior permission of the publishers

Crystallization

–

Cloned

(human [7, 19], bovine [21], complex-type) [7, 19, 21]

Renaturated

–

---

5 STABILITY

pH

4.0 (30 minutes, 10% loss of activity [13]) [6, 13]

Temperature (°C)

Oxidation

Organic solvent

General stability information

Dialysis (stable) [16]; Glycerol (stabilizes in latter stages of chromatographic separation) [17]; Dithiothreitol (stabilizes in latter stages of chromatographic separation) [17]; Freezing (destroys activity) [6]; Phospholipid (preincubation stabilizes) [15]

Storage

More [16]; –43°C, 50% ethylene glycerol, several weeks [6]; 0°C, 90% ethylene glycol, several weeks [9]; –20°C, 25% glycerol, several months [15]

---

6 CROSSREFERENCES TO STRUCTURE DATABANKS

PIR/MIPS code

EUHUGC (precursor, human); A25130 (human, fragment); A32931 (precursor, mouse); A30367 (precursor, human); A27306 (precursor, human, fragment)

Brookhaven code

7 LITERATURE REFERENCES

[1] Basu, A., Glew, R.H.: Biochem. J., 224, 515–524 (1984)
[2] Imai, K.: J. Biochem., 98, 1405–1416 (1985)
[3] Dale, G.L., Beutler, E.: Proc. Natl. Acad. Sci. USA, 73, 4672–4674 (1976)
[4] Schliemann, W., Schliemann, B.: Pharmazie, 4, 243–250 (1982) (Review)
[5] Strasberg, P.M., Lowden, J.A., Mahuran, D.: Can. J. Biochem., 60, 1025–1031 (1982)
[6] Daniels, L.B., Glew, R.H. in "Methods Enzym. Anal.", 3 Rd. Ed. (Bergmeyer, H.U.,
    Ed.) 4, 217–226 (1984) (Review)
[7] Ginns, E.I., Choudary, P.V., Martin, B.M., Winfield, S., Stubblefield, B., Mayor, J.,
    Merkle-Lehman, D., Murray, G.J., Bowers, L.A., Barranger, J.A.: Biochem. Biophys.
    Res. Commun., 123, 574–580 (1984)
[8] Legler, G., Liedtke, H.: Biol. Chem. Hoppe-Seyler, 366, 1113–1122 (1985)
[9] Aerts, J.M.F.G., Donker-Koopman, W.E., Murray, G.J., Barranger, J.A., Tager, J.M.,
    Schram, A.W.: Anal. Biochem., 154, 655–663 (1986)
[10] Choy, F.Y.M.: Anal. Biochem., 156, 515–520 (1986)
[11] Osiecki-Newman, K., Fabbro, D., Legler, G., Desnick, R.J., Grabowski, G.A.: Biochim.
     Biophys. Acta, 915, 87–100 (1987)
[12] Schram, A.W., Aerts, J.M.F.G., Van Weely, S., Barranger, J.A., Tager, J.M.: Methodol.
     Surv. Biochem. Anal., 17, 113–126 (1987) (Review)
[13] Maret, A., Salvayre, R., Potier, M., Legler, G., Beauregard, G., Douste-Blazy, L.:
     NATO ASI Ser. Ser. A, 150, 57–61 (1988)
[14] Choy, F.Y.M.: Anal. Biochem., 179, 312–318 (1989)
[15] Shafit-Zagardo, B., Turner, B.M.: Biochim. Biophys. Acta, 659, 7–14 (1981)
[16] Brady, R.O., Kanfer, J., Shapiro, D.: J. Biol. Chem., 240, 39–43 (1965)
[17] Pentchev, P.G., Brady, R.O., Hibbert, S.R., Gal, A.E., Shapiro, D.: J. Biol. Chem., 248,
     5256–5261 (1973)
[18] Braidman, I.P., Gregoriadis, G.: Biochem. J., 164, 439–445 (1977)
[19] Furbish, F.S., Blair, H.E., Shiloach, J., Pentchev, P.G., Brady, R.O.: Methods En-
     zymol., 50, 529–532 (1978) (Review)
[20] Pirruccello, S., Barranger, J.A., Barton, N.W., Brady, R.O., Ginns, E.I.: Biochem. Med.,
     31, 73–79 (1984)
[21] Reddy, P.U.M., Murray, G.J., Barranger, J.A.: Biochem. Med., 33, 200–210 (1985)
[22] Murray, G.J., Youle, R.J., Gandy, S.E., Zirzow, G.C., Barranger, J.A.: Anal. Biochem.,
     147, 301–310 (1985)
[23] Tager, J.M., Aerts, J.M.F.G., Jonsson, L.M.V., Murray, G.J., Van Weeley, S., Strijland,
     A., Ginns, E.I., Reuser, A.J.J., Schram, A.W., Barranger, J.A.: NATO ASI Ser. Ser. A,
     116, 735–745 (1986) (Review)
[24] Legler, G.: NATO ASI Ser. Ser. A, 150, 63–72 (1988)
[25] Liedtke, H., Legler, G.: NATO ASI Ser. Ser. A, 150, 353–358 (1988)

Enzyme Handbook © Springer-Verlag Berlin Heidelberg 1991
Duplication, reproduction and storage in data banks are only
allowed with the prior permission of the publishers

1 NOMENCLATURE

EC number
3.2.1.46

Systematic name
D-Galactosyl-N-acylsphingosine galactohydrolase

Recommended name
Galactoceramidase

Synonymes
Ceramidase, galcatosyl-
Cerebroside galactosidase
Galactocerebroside.beta.-galactosidase
Galactosylcerebrosidase
Galactocerebrosidase
Ceramide galactosidase
Galactocerebroside galactosidase
Galactosylceramide.beta.-galactosidase
Cerebroside. beta.-galactosidase
.beta.-Galactocerebrosidase
Lactosylceramidase
Lactosylceramidase I [1]
Galactocerebroside-.beta.-D-galactosidase
.beta.-Galactosylceramidase
Galactosylceramidase I [3]

CAS Reg. No.
9027-89-8

---

2 REACTION AND SPECIFICITY

Catalysed reaction
D-Galactosyl-N-acylsphingosine + H_2O →
→ D-galactose + N-acylsphingosine

Reaction type
O-Glycosyl bond hydrolysis

Natural substrates
D-Galactosyl-N-acylsphingosine + H_2O

Enzyme Handbook © Springer-Verlag Berlin Heidelberg 1991
Duplication, reproduction and storage in data banks are only
allowed with the prior permission of the publishers

Substrate spectrum

1 D-Galactosyl-N-acylsphingosine + H_2O [1–11]
2 Lactosylsphingosine + H_2O [1]
3 4-Methylumbelliferyl beta-galactoside + H_2O [5]

Product spectrum

1 D-Galactose + N-acylsphingosine
2 Lactose + sphingosine
3 Methylumbelliferone + galactose

Inhibitor(s)

N-Acetyl psychosine [8]; N-Decanoyl psychosine [8]; Lactosyl ceramide [8];
Galactonolactone [8]; N-(n-Hexyl)psychosine [8]; Taurocholate (high con-
centrations, above 0.3% w/v) [9]; NaCl (human, mouse: no effect) [9];
D-Galactonamide [11]; Lactose [9]; Galactose [9]; More [8]

Cofactor(s)/prostethic group(s)

More (bile salts stimulate [1, 6], sodium taurocholate activates [3, 9],
phosphatidylserine from bovine brain activates in absence of sodium
taurocholate [6]) [1, 3, 6, 9]

Metal compounds/salts

Turnover number (min^{-1})

Specific activity (U/mg)

0.12 [7]

K_m-value (mM)

0.100 (galactosylceramide, taurocholate activated) [6]; 0.200 (galactosylceramide, phosphatidylserine activated) [6]; More (K_m of mutant enzyme is
14 times higher than of the normal) [1]

pH-optimum

4.2 (taurocholate activated) [6]; 4.7 (phosphatidylserine activated) [6]; 4.1
(mouse) [9]; 4.4 (human) [9]

pH-range

More [6]; 3.5–6.0 [6]; 3.5–5.3 (3.5: about 10% of maximal activity, 5.3: about
40% of maximal activity) [9]

Temperature optimum (°C)

37 (assay at) [6]

Temperature range (°C)

3 ENZYME STRUCTURE

Molecular weight
121000 (gel filtration, human, monomer) [5]
760000 (gel filtration, human, hexamer) [5]
750000 (gel filtration, human) [7]

Subunits
Hexamer, dimer or monomer (aggregation from monomer to dimer, from
dimer to tetramer or hexamer: 6 × 125000, human, active form in vivo) [5]

Glycoprotein/Lipoprotein
–

---

4 ISOLATION/PREPARATION

Source organism
Human (normal [5] and mutant enzyme /Krabbe disease [5, 7, 10],
Lesch-Nyhan cells [9]) [1, 3, 4, 5, 6, 7, 9, 10, 11]; Mouse (LMTK-cells [9]) [2,
9]; Rat [4, 8]

Source tissue
Spleen [11]; Liver [4, 7]; Brain [1, 4, 6, 8]; Fibroblasts [1, 3, 9]; Kidney [4];
Skin [10]; Leukocytes [4]; Cell culture (human: Lesch-Nyhan cells [9],
mouse: LMTK-cells [9]) [1, 9]

Localisation in source

Purification
Human (Krabbe disease) [7]

Crystallization
–

Cloned
–

Renaturated
–

---

5 STABILITY

pH

Enzyme Handbook © Springer-Verlag Berlin Heidelberg 1991
Duplication, reproduction and storage in data banks are only
allowed with the prior permission of the publishers

Temperature (°C)

54 (mouse: 20 minutes, 45% inactivation) [9]; 52 (4 minutes, 50% loss of activity (mutant enzyme), 35 minutes, 50% loss of activity (normal enzyme)) [7]

Oxidation

Organic solvent

General stability information

Dialysis (mutant enzyme: against 1.3 M urea, 50% loss of activity, normal enzyme: against 5.6 M urea, 50% loss of activity) [7]; Urea (dialysis, mutant enzyme: against 1.3 M urea, 50% loss of activity, normal enzyme: against 5.6 M urea, 50% loss of activity) [7]

Storage

---

6 CROSSREFERENCES TO STRUCTURE DATABANKS

PIR/MIPS code

Brookhaven code

---

7 LITERATURE REFERENCES

[1] Goda, S., Kobayashi, T., Goto, I.: Biochim. Biophys. Acta, 920, 259–265 (1987)
[2] Kobayashi, T., Shinnoh, N., Kuroiwa, Y.: Biochim. Biophys. Acta, 879, 215–220 (1986)
[3] Kobayashi, T., Shinnoh, N., Goto, I., Kuroiwa, Y.: J. Biol. Chem., 260, 14982–14987 (1985)
[4] Kato, T., Suzuki, Y.: Anal. Biochem., 126, 44–51 (1982)
[5] Ben-Yoseph, Y., Hungerford, M., Nadler, H.L.: Biochem. J., 189, 9–15 (1980)
[6] Hanada, E., Suzuki, K.: Biochim. Biophys. Acta, 575, 410–420 (1979)
[7] Ben-Yoseph, Y., Hungerford, M., Nadler, H.L.: Arch. Biochem. Biophys., 196, 93–101 (1979)
[8] Arora, R.C., Lin, Y.-N., Radin, N.S.: Arch. Biochem. Biophys., 156, 77–83 (1973)
[9] Rushton, A.R., Dawson, G.: Biochim. Biophys. Acta, 388, 92–105 (1975)
[10] Besley, G.T.N.: Biochem. Soc. Trans. (554th Meeting) 3, 241–244 (1975)
[11] Chiao, Y.-B., Moffitt, K., Smallwood, Y., Glew, R.H.: Arch. Biochem. Biophys., 192, 1–9 (1979)

1 NOMENCLATURE

EC number
3.2.1.47

Systematic name
D-Galactosyl-D-galactosyl-D-glucosyl-N-acylsphingosine galactohydrolase

Recommended name
Galactosylgalactosylglucosylceramidase

Synonymes
Ceramidase, galactosylgalactosylglucosyl-
Trihexosyl ceramide galactosidase
Ceramide trihexosidase
Ceramidetrihexoside-.alpha.-galactosidase
Trihexosylceramide.alpha.-galactosidase
Trihexosylceramide alpha-galactosidase
Ceramidetrihexosidase [8]

CAS Reg. No.
9023-01-2

---

2 REACTION AND SPECIFICITY

Catalysed reaction
D-Galactosyl-D-galactosyl-D-glucosyl-N-acylsphingosine + H_2O →
→ lactosyl-N-acylsphingosine + D-galactose

Reaction type
O-Glycosyl bond hydrolysis

Natural substrates
D-Galactosyl-D-galactosyl-D-glucosyl-N-acylsphingosine + H_2O

Substrate spectrum
1 D-Galactosyl-D-galactosyl-D-glucosyl-N-acylspingosine + H_2O [1–8]
2 More (methylumbelliferyl alpha-galactosidase and ceramidetrihexoside
alpha-galactosidase activity closely associated or identical [5], enzyme
with ceramide trihexosidase activity has no activity against 4-methylum-
belliferyl alpha-galactoside and p-nitrophenyl alpha-galactoside [7],
some evidence for a single enzyme with ceramide trihexosidase activity
and p-nitrophenyl-alpha-galactosidase and 4-methylumbel-
liferyl-alpha-galactosidase activity [8], not: related sphingolipids [3],
p-nitrophenyl-alpha-galactoside [4]) [3, 4, 5, 7, 8]

Enzyme Handbook © Springer-Verlag Berlin Heidelberg 1991
Duplication, reproduction and storage in data banks are only
allowed with the prior permission of the publishers

Product spectrum
1 Lactosyl-N-acylspingosine + D-galactose
2 ?

Inhibitor(s)
Human serum albumin [5]; NaCl [5]; Glucosylsphingosine (psychosine) [3]; Digalactosylceramide [3]; Ceramide digalactoside [5]; Ceramide trihexoside (above 0.13 mM) [5]; EDTA [9]; Galactose [9]; Lactosyl ceramide [5, 9]

Cofactor(s)/prosthetic group(s)
More (absolutely dependent on a mixture of sodium taurocholate and Triton X-100 [5], no activity in absence of detergent [9]) [5, 9]

Metal compounds/salts

Turnover number (min^{-1})

Specific activity (U/mg)
4.02 [1]; 0.0396 [4]; 0.076 [6]; More [3, 5]

K$_m$-value (mM)
1.55 (4-methylumbelliferyl-alpha-D-galactopyranoside) [1]; 0.05 (ceramide trihexoside) [5]; 0.09–0.150 (ceramide trihexoside) [9]; More (non-hyperbolic substrate-saturation curve) [5]

pH-optimum
4.5 (methylumbelliferyl alpha-galactoside) [5]; 3.5 (ceramide trihexoside) [5]; 4.4 (4-methylumbelliferyl-alpha-D-galactopyranoside) [1]; 4.1 (D-galactosyl-D-galactosyl-D-glucosyl-N-acylsphingosine) [1]; 5.0 [3]; 4.4 [9]; 5.4 (form A) [6]; 7.2 (form B) [6]

pH-range
3.5–6.0 (3.5, 6.0: about 10% of maximal activity) [3]; 3.5–5.5 (3.5: about 60% of maximal activity, 5.5: about 10% of maximal activity) [9]

Temperature optimum (°C)
37 (assay at) [3, 7]

Temperature range (°C)

3 ENZYME STRUCTURE

Molecular weight
103000 (gel filtration, human) [1]
95000 (sucrose density gradient centrifugation, gel filtration, human) [4]

Subunits
Tetramer (SDS-PAGE, presence of sulfhydryl reducing agents, human, 4 × 24000) [4]

Glycoprotein/Lipoprotein
Glycoprotein (contains sialic acid [1, 6], multiple forms differ in sialic acid content [6]) [1, 6]

---

4 ISOLATION/PREPARATION

Source organism
Human (2 forms, non-interconvertible: form A-1, form A-2 [4], multiple forms: A, B [6]) [1, 4, 5, 6, 7, 8, 9]; Rat [3]

Source tissue
Plasma [4, 6, 7]; Placenta [1]; Hepatocytes [2]; Kupffer cells [2]; Spleen [3]; Brain [3]; Liver [3, 5]; Kidney [3]; Small intestine [3]; Urine [8]; Leukocytes [9]

Localisation in source

Purification
Human [1, 5, 6]; Rat [3]

Crystallization
–

Cloned
–

Renaturated
–

---

5 STABILITY

pH

Temperature (°C)
52 (1 hour, 90% loss of activity) [1]; 100 (A-1: 75% loss of activity after boiling, A-2: 100% loss of activity after boiling) [2]

Oxidation

Organic solvent

General stability information
Human serum albumin (stabilizes) [3]

Storage
0°C, 2 weeks [3]

Enzyme Handbook © Springer-Verlag Berlin Heidelberg 1991
Duplication, reproduction and storage in data banks are only
allowed with the prior permission of the publishers

6 CROSSREFERENCES TO STRUCTURE DATABANKS

PIR/MIPS code

Brookhaven code

---

7 LITERATURE REFERENCES

[1] Kusiak, J.W., Quirk, J.M., Brady, R.O.: Methods Enzymol., 50, 533–537 (1978) (Review)
[2] Barranger, J.A., Pentchev, P.G., Furbish, F.S.: Biochem. Biophys. Res. Commun., 83, 1055–1060 (1978)
[3] Brady, R.O., Gal, A.E., Bradley, R.M., Martensson, E.: J. Biol. Chem., 242, 1021–1026 (1967)
[4] Mapes, C.A., Suelter, C.H., Sweeley, C.C.: J. Biol. Chem., 248, 2471–2479 (1973)
[5] Mae Wan Ho: Biochem. J., 133, 1–10 (1973)
[6] Mapes, C.A., Sweeley, C.C.: Arch. Biochem. Biophys., 158, 297–304 (1973)
[7] Mapes, C.A., Sweeley, C.C.: FEBS Lett., 25, 279–281 (1972)
[8] Rietra, P.J.G.M., Tager, J.M., Borst, P.: Biochim. Biophys. Acta, 279, 436–445 (1972)
[9] Poulos, A., Beckman, K.: Clin. Chim. Acta, 89, 35–45 (1978)

1 NOMENCLATURE

EC number
3.2.1.48

Systematic name
Sucrose alpha-D-glucohydrolase

Recommended name
Sucrose alpha-glucosidase

Synonymes
Sucrose alpha-glucohydrolase
Sucrase-isomaltase
Sucrase
Glucosidase, sucrose.alpha.-
Sucrose.alpha.-glucohydrolase
Intestinal sucrase
Sucrase(invertase)
More (enzyme isolated from intestinal mucosa is a single polypeptide chain
also displaying activity towards isomaltose (oligo-1, 6-glucosidase, EC
3.2.1.10))

CAS Reg. No.
37288-39-4

2 REACTION AND SPECIFICITY

Catalysed reaction
Hydrolysis of sucrose and maltose by an alpha-D-glucosidase-type action

Reaction type
O-Glycosyl bond hydrolysis

Natural substrates
Sucrose + H_2O

Substrate spectrum
1 Sucrose + H_2O [13]
2 Maltose + H_2O [3, 13]
3 Isomaltose + H_2O [3]
4 p-Nitrophenyl-alpha-glucoside + H_2O [13]
5 6-Bromo-2-naphthyl-alpha-glucoside + H_2O [13]
6 More [13]

Enzyme Handbook © Springer-Verlag Berlin Heidelberg 1991
Duplication, reproduction and storage in data banks are only
allowed with the prior permission of the publishers

Product spectrum
1 Glucose [13]
2 Glucose [13]
3 Glucose
4 Nitrophenol + glucose
5 6-Bromonaphthol + glucose
6 ?

Inhibitor(s)
p-Nitrophenyl-alpha-glucoside [13]; H$^+$ [4]; Na$^+$ (high concentration) [3, 4]; Li$^+$ (high concentration) [4]; Castanospermine [5]; Acarbose [8]; Nojirimycin [8]; Deoxynojirimycin [8]; Tris [3, 11]; 6-Bromo-2-naphthyl-alpha-glucoside [13]; Maltose [13]; Sucrose (high concentration) [3]

Cofactor(s)/prostethic group(s)

Metal compounds/salts
Na$^+$ (low concentration: activation [3, 4, 6, 11]) [4, 6, 11, 13]; Li$^+$ (low concentration: activation) [4]

Turnover number (min^{-1})

Specific activity (U/mg)
47 [9]; 17.5 [11]; More [12]

K$_m$-value (mM)
48 (sucrose) [1]; 16 (sucrose) [2]; 20 (sucrose) [7, 9]; 8.4 (sucrose) [11]; 8.2 (maltose) [13]; 1.34 (p-nitrophenyl-alpha-glucoside) [13]; 0.12 (6-bromo-2-naphthyl-alpha-glucoside) [13]; 10.5 (sucrose) [3]

pH-optimum
6.5 [1, 3]; 6–6.5 [9]; 6.7 [13]

pH-range
4–7.8 (4: 35% of maximal activity, 7.8: 50% of maximal activity) [9]; 4.1–8.8 (activity almost neglegible at 4.1 and 8.8) [3]

Temperature optimum (°C)
37 (assay at) [3, 9]

Temperature range (°C)

3 ENZYME STRUCTURE

Molecular weight

130000 (gel chromatography, density gradient, equilibrium centrifugation, human, sucrase subunit) [9]

28000 (gel chromatography, density gradient equilibrium centrifugation, human, sucrase-isomaltase complex) [9]

140000 (SDS-PAGE, pig, intact pancreatic duct, subunit I) [12]

150000 (SDS-PAGE, pig, intact pancreatic duct, subunit II) [12]

260000 (SDS-PAGE, pig, disconnected pancreatic duct) [12]

140000 (SDS-PAGE, rabbit, sucrase-isomaltase) [11]

140000 (gel chromatography, rat, luminal) [2]

290000 (gel chromatography, rat, sucrase-isomaltase complex, intestinal mucosa) [2]

270000 (chicken, SDS-PAGE) [3]

Subunits

Dimer (rat) [7]

Dimer (pig, intact pancreatic duct, SDS-PAGE) [12]

Monomer (pig, disconnected pancreatic duct, SDS-PAGE) [12]

More [3]

Glycoprotein/Lipoprotein

Lipoprotein (lipid content: 2–4%) [11]; Glycoprotein (glucosamine and galactosamine) [11]

---

4 ISOLATION/PREPARATION

Source organism

Pig [12]; Rabbit [4, 6, 11]; Human (sucrase/isomaltase molecule is an hybrid consisting of 2 distinct alpha-glucosidases [9]) [5, 9]; Streptococcus salivarius (expression in E. coli) [1]; E. coli (Streptococcus salivarius gene) [1]; Rat (single protein with sucrase and isomaltase activity [7], sucrase-isomaltase synthesized as a one-chain polypeptide, precursor that is split into the subunits after its transfer to the microvillus membrane [10], sucrase-isomaltase [13]) [2, 7, 10, 13]; Chicken [3]

Source tissue

Small intestine (luminal content: sucrase without isomaltase activity [2], intestinal mucosa [2, 3, 9, 10]: sucrase-isomaltase complex [2], brush border [4, 6, 7, 11]) [2, 3, 4, 6, 7, 9, 10, 11, 12, 13]; Cultured cells (enterocyte-like cell line Caco-2) [5]

Localisation in source

Purification

Chicken [3]; Rat [13]; Human [9]; Pig [12]

Enzyme Handbook © Springer-Verlag Berlin Heidelberg 1991
Duplication, reproduction and storage in data banks are only
allowed with the prior permission of the publishers

Crystallization
–

Cloned
(Streptococcus salivarius gene in E. coli) [1]

Renaturated
–

---

5 STABILITY

pH

Temperature (°C)
50 (pH 5.9, inactivation) [13]

Oxidation

Organic solvent

General stability information

Storage
4°C or –20°C, 0.02% sodium azide, Triton X-100, several weeks,
sucrase-isomaltase [11]

---

6 CROSSREFERENCES TO STRUCTURE DATABANKS

PIR/MIPS code
D26967 (rabbit, fragment); JU0091 (Vibrio alginolyticus); A25562 (Bacillus
subtilis); A27326 (oligo-1, 6-glucosidase EC 3.2.1.10, precursor, human,
fragment); A25987 (oligo-1, 6-glucosidase EC 3.2.1.10, pig, fragments);
A23945 (oligo-1, 6-glucosidase EC 3.2.1.10, precursor, rabbit); A29286
(oligo-1, 6-glucosidase EC 3.2.1.10, precursor, rat, fragments)

Brookhaven code

7 LITERATURE REFERENCES

[1] Houck, C.M., Pear, J.R., Elliott, R., Perchorowicz, J. T.: J. Bacteriol., 169, 3679–3684 (1987)

[2] Abe, M., Yamada, K., Hosoya, N., Moriuchi, S.: J. Nutr. Sci. Vitaminol., 31, 243–252 (1985)

[3] Matsushita, S.: Comp. Biochem. Physiol., 76B, 465–470 (1983)

[4] Vasseur, M., Van Melle, G., Frangne, R., Alvarado, F.: Biochem. J., 251, 667–675 (1988)

[5] Trugnan, G., Rousset, M., Zweibaum, A.: FEBS Lett., 195, 28–32 (1986)

[6] Vasseur, M., Tellier, Ch., Alvarado, F.: Arch. Biochem. Biophys., 218, 263–274 (1982)

[7] Montgomery, R.K., Sybicki, M.A., Forcier, A.G., Grand, R.J.: Biochim. Biophys. Acta, 661, 346–349 (1981)

[8] Hanozet, G., Pircher, H.-P., Vanni, P., Oesch, B., Semenza, G.: J. Biol. Chem., 256, 3703–3711 (1981)

[9] Conklin, K.A., Yamashiro, K.M., Gray, G.M.: J. Biol. Chem., 250, 5735–5741 (1975)

[10] Hauri, H.-P., Quaroni, A., Isselbacher, K.J.: Proc. Natl. Acad. Sci. USA, 76, 5183–5186 (1979)

[11] Sigrist, H., Ronner, P., Semenza, G.: Biochim. Biophys. Acta, 406, 433–446 (1975)

[12] Sjöström, H., Noren, O., Christiansen, L., Wacker, H. , Semenza, G.: J. Biol. Chem., 255, 11332–11338 (1980)

[13] Kolinska, J., Kraml, J.: Biochim. Biophys. Acta, 284, 235–247 (1972)

Enzyme Handbook © Springer-Verlag Berlin Heidelberg 1991
Duplication, reproduction and storage in data banks are only
allowed with the prior permission of the publishers

1 NOMENCLATURE

EC number
3.2.1.49

Systematic name
Alpha-N-acetyl-D-galactosaminide N-acetylgalactosaminohydrolase

Recommended name
Alpha-N-acetylgalactosaminidase

Synonymes
Alpha-acetylgalactosaminidase
N-Acetyl-alpha-D-galactosaminidase
N-Acetyl-alpha-galactosaminidase

CAS Reg. No.
9075-63-2

---

2 REACTION AND SPECIFICITY

Catalysed reaction
Hydrolysis of terminal non-reducing N-acetyl-D-galactosamine residues in
N-acetyl-alpha-D-galactosaminides

Reaction type
O-Glycosyl bond hydrolysis

Natural substrates
N-Acetyl-alpha-D-galactosaminides + H_2O [1–20]

Substrate spectrum
1 N-Acetyl-alpha-D-galactosaminides + H_2O [1–20]
2 More (splits N-acetylgalactosaminyl groups from 0–3 of Ser and Thr)

Product spectrum
1 N-Acetyl-D-galactosamine + H_2O [1–20]
2 ?

Inhibitor(s)
Hg^{2+} [1, 5, 8, 14]; Ag^+ [3, 5, 12, 14]; N-Acetylgalactosamine [3, 5, 14, 16, 18];
Galactose [3, 5, 14, 16]; Cu^{2+} [5, 12]; Fe^{2+} [18]

Cofactor(s)/prostethic group(s)

Metal compounds/salts

Enzyme Handbook © Springer-Verlag Berlin Heidelberg 1991
Duplication, reproduction and storage in data banks are only
allowed with the prior permission of the publishers

Turnover number (min⁻¹)

Specific activity (U/mg)
82–121 [1, 2, 5]; 17.3–30.2 [3, 5, 8]; 4.7–6.7 [9, 11]; 52.6 [14]; 1.8 [20]

K_m-value (mM)
0.6–6.4 (p-nitrophenyl-alpha-N-acetylgalactosaminide) [1–3, 5, 7–9, 14–16, 18]; 0.0543–0.44 (ovine submaxillary asialoglycoprotein) [2, 8]; 0.036–0.59 (Forssman hapten) [9, 13, 14]; 5.1–10 (4-methylumbel-liferyl-alpha-galactopyranoside) [9–13]; 14.7 (p-nitrophenyl-2-deoxy-alpha-D-galactopyranoside) [10]; 8.8 (o-nitrophenyl-alpha-D-fucopyranoside) [10]; 1.3 (o-nitrophenyl-alpha-N-acetylgalactosaminide) [10, 11, 13]; 3.7 (globopentaose) [11]; 50.0 (p-nitrophenyl-alpha-N-acetylgalactosaminide) [19]; More [11]

pH-optimum
4.0–4.5 (p-nitrophenyl-alpha-N-acetylgalactosaminide) [1]; 4.0 (p-nitrophenyl-alpha-N-acetylgalactosaminide) [2]; 3.8 (p-nitrophenyl-alpha-N-acetylgalactosaminide) [3, 14]; 3.0–4.2 (p-nitrophenyl-alpha-N-acetylgalactosaminide) [5]; 4.5 (p-nitrophenyl-alpha-N-acetylgalactosaminide) [7]; 6.0 (p-nitrophenyl-alpha-N-acetylgalactosaminide) [8]; 4.3–4.7 (p-nitrophenyl-alpha-N-acetylgalactosaminide) [9]; 4.8 (4-methylumbel-liferyl-alpha-galactopyranoside) [11]; 4.5 (4-methylumbel-liferyl-alpha-galactopyranoside) [12]; More (other substrates) [11]; 4.3 (p-nitrophenyl-alpha-N-acetylgalactosaminide) [16, 18]; 4.4 (p-nitrophenyl-alpha-N-acetylgalactosaminide) [19]

pH-range
2.5–8.0 (p-nitrophenyl-alpha-N-acetylgalactosaminide) [1]; 7.0 (not active above, p-nitrophenyl-alpha-N-acetylgalactosaminide) [2]

Temperature optimum (°C)
50 (p-nitrophenyl-alpha-N-acetylgalactosaminide) [1]

Temperature range (°C)

3 ENZYME STRUCTURE

Molecular weight
55000 (gel filtration, Acremonium, Solanocera melantho) [1, 3]
80000 (molecular sieve chromatography, Katsuwonus pelmis) [2]
500000 (gel filtration, squid liver, isoenzyme I) [5]
230000 (gel filtration, squid liver, isoenzyme II) [5]
48000 (gel filtration, man) [6]
102000 (gel filtration, pig) [9]
200000 (gel filtration, Patella vulgata) [14]
155000 (gel filtratio, bovine) [17]

Subunits
Monomer (SDS-PAGE) [1]
Dimer (2 × 35000, SDS-PAGE) [2]
Dimer (2 × 52000, SDS-PAGE) [9]
Tetramer (4 × 50000, SDS-PAGE) [14]
Tetramer (4 × 30000–42000, gel filtration) [17]

Glycoprotein/Lipoprotein
Glycoprotein (7.0–7.2 % neutral sugars) [1, 6, 9]

---

4 ISOLATION/PREPARATION

Source organism
Acremonium [1]; Katsuwonus pelmis [2]; Solanocera melantho [3]; Rumen bacteria and protozoa [4]; Squid [5]; Mammals [6, 9, 11]; Clostridium perfringens [8]; Snails [14, 16, 20]; Trichomonas fetus [16]; Pig [9, 18]; Human [11, 16]; Ox [19]

Source tissue
Katsuwonus pelmis liver [2]; Solanocera melantho hepatopancreas [3]; Squid liver [5]; Mammalian tissues [16, 18, 19]; Spleen [19]

Localisation in source
Lysosomes [6, 16, 18]

Purification
Acremonium sp. [1]; Katsuwonas pelmis liver [2]; Solanocera melantho hepatopancreas [3]; Squid liver [5]; Clostridium perfringens [8]; Pig liver [9, 18]; Human liver [11, 16]; Patella vulgata [14]; Ox spleen [19]; Helix pomatia [20]

Crystallization
−

Cloned
−

Enzyme Handbook © Springer-Verlag Berlin Heidelberg 1991
Duplication, reproduction and storage in data banks are only
allowed with the prior permission of the publishers

Renaturated

–

---

5 STABILITY

pH

6.0–7.5 [1]; 4.0–6.0 [2]; 3.0–7.5 [3]; 4.5–5.5 [5]; 3.6–6.8 [14]

Temperature (°C)

40 (not stable above) [1–3, 14]

Oxidation

Organic solvent

General stability information

Storage

1 month, 0°C [1]; 2 weeks, 0°C, pH 4.2 [2]; Several years, lyophilized [2, 8]; 6 months, –20°C [9, 20]

---

6 CROSSREFERENCES TO STRUCTURE DATABANKS

PIR/MIPS code

A33265 (precursor, human); A35485 (precursor, human)

Brookhaven code

---

7 LITERATURE REFERENCES

[1] Kadowaki, S., Ueda, T., Yamamoto, K., Kumagai, H., Tochikura, T.: Agric. Biol. Chem., 53, 111–120 (1989)

[2] Nakagawa, H., Asakawa, M., Enomoto, N.: J. Biochem., 101, 855–862 (1987)

[3] Chien, S.F.: J. Chin. Biochem. Soc., 15, 86–96 (1986)

[4] Williams, A.G., Withers, S.E., Coleman, G.S.: Curr. Microbiol., 10, 287–294 (1984)

[5] Itoh, T., Uda, Y.: J. Biochem., 95, 959–970 (1984)

[6] Sweeley, C.C., Ledonne, N.C., Robbins, P.W.: Arch. Biochem. Biophys., 223, 158–165 (1983)

[7] Salvayre, R., Negre, A., Maret, A., Lenoir, G., Douste-Blazy, L.: Biochim. Biophys. Acta, 659, 445–456 (1981)

[8] Levy, G.N., Aminoff, D.: J. Biol. Chem., 255, 11737–11742 (1980)

[9] Sung, S.S.J., Sweely, C.C.: J. Biol. Chem., 255, 6589–6594 (1980)

[10] Dean, K.J., Sweeley, C.C.: J. Biol. Chem., 254, 10006–10010 (1979)

[11] Dean, K.J., Sweeley, C.C.: J. Biol. Chem., 254, 10001–10005 (1979)

[12] Salvayre, R., Maret, A., Negre, A., Douste-Blazy, L.: Eur. J. Biochem., 100, 377–383 (1979)

[13] Dean, K.J., Sung, S.S.J., Sweeley, C.C.: Biochem. Biophys. Res. Commun., 77, 1411–1417 (1977)

[14] Uda, Y., Li, S.C., Li, Y.T.: J. Biol. Chem., 252, 5194–5200 (1977)

[15] Schram, A.W., Hamers, M.N., Tager, J.M.: Biochim. Biophys. Acta, 482, 138–144 (1977)

[16] Callahan, J.W., Lassila, E.L., Den Tandt, W., Philippart, M.: Biochem. Med., 7, 424–431 (1973)

[17] Wang, C.T., Weissmann, B.: Biochemistry, 10, 1067–1072 (1971)

[18] Weissmann, B., Hinrichsen, D.F.: Biochemistry, 8, 2034–2043 (1969)

[19] Werries, E., Wollek, E., Gottschalk, A., Buddecke, E.: Eur. J. Biochem., 10, 445–449 (1969)

[20] Tuppy, H., Staudenbauer, W.L.: Biochemistry, 5, 1742–1747 (1966)

Enzyme Handbook © Springer-Verlag Berlin Heidelberg 1991
Duplication, reproduction and storage in data banks are only
allowed with the prior permission of the publishers

1 NOMENCLATURE

EC number
3.2.1.50

Systematic name
Alpha-N-acetyl-D-glucosaminide N-acetylglucosaminohydrolase

Recommended name
Alpha-N-acetylglucosaminidase

Synonymes
Alpha-acetylglucosaminidase
N-Acetyl-alpha-D-glucosaminidase
N-Acetyl-alpha-glucosaminidase
Alpha-N-acetyl-D-glucosaminidase
Alpha-D-2-acetamido-2-deoxyglucosidase

CAS Reg. No.
37288-40-7

2 REACTION AND SPECIFICITY

Catalysed reaction
Hydrolysis of terminal non-reducing N-acetyl-D-glucosamine residues in
N-acetyl-alpha-D-glucosaminides

Reaction type
O-Glycosyl bond hydrolysis

Natural substrates
N-Acetyl-alpha-D-glucosaminides + H_2O [1–11]
Heparan sulfate + H_2O [4, 8]
UDP-alpha-N-acetylglucosamine + H_2O [10]

Substrate spectrum
1 N-Acetyl-alpha-D-glucosaminides + H_2O (e.g. p-nitrophenyl-alpha-N-
acetylglucosaminide [3, 5], phenyl-alpha-N-acetylglucosaminide [3, 5, 6,
8, 10], UDP-alpha-N-glucosamine [10]) [1–11]

Product spectrum
1 N-Acetyl-D-glucosamine [1–11]

Enzyme Handbook © Springer-Verlag Berlin Heidelberg 1991
Duplication, reproduction and storage in data banks are only
allowed with the prior permission of the publishers

Inhibitor(s)
   Hg^{2+} [5, 8]; Cu^{2+} [5]

Cofactor(s)/prosthetic group(s)

Metal compounds/salts

Turnover number (min^{-1})

Specific activity (U/mg)
   0.115 [1]; 0.99–1.35 [3, 5, 9]; 0.009 [8]

K_m-value (mM)
   0.13–0.14 (p-nitrophenyl-alpha-N-acetylglucosaminide) [3, 5]; 0.043–0.61
   (o-nitrophenyl-alpha-N-acetylglucosaminide) [3, 5, 8]; 0.12–1.6
   (phenyl-alpha-N-acetylglucosaminide) [3, 5, 6, 8, 10]; 0.39
   (UDP-N-acetylglucosamine) [5]

pH-optimum
   4.4 (p-nitrophenyl-alpha-N-acetylglucosaminide) [3]; 4.5
   (o-nitrophenyl-alpha-N-acetylglucosaminide) [3]; 4.8
   (phenyl-alpha-N-acetylglucosaminide) [3]; 4.4 (heparin) [4]; 4.8
   (p-nitrophenyl-alpha-N-acetylglucosaminide) [5]; 4.6
   (phenyl-alpha-N-acetylglucosaminide) [5, 6, 11]; 4.2
   (UDP-N-acetylglucosamine) [5]; 4.7 (phenyl-alpha-N-acetylglucosaminide)
   [8]; 4.5 (phenyl-alpha-N-acetylglucosaminide) [10]; 4.7
   (p-nitrophenyl-alpha-N-acetylglucosaminide) [11]

pH-range
   3.0–9.0 (p-nitrophenyl-alpha-N-acetylglucosaminide) [3]; 3.5–6.5 (heparin)
   [4]

Temperature optimum (°C)

Temperature range (°C)

3 ENZYME STRUCTURE

Molecular weight
   240000 (sucrose gradient centrifugation, man) [1]
   304000–307000 (gel filtration, man) [3, 5]
   127500 (gel filtration, ox) [8]
   217000 (gel electrophoresis, Patella vulgata) [9]

Subunits
   Monomer (SDS-PAGE, Patella vulgata) [9]

Glycoprotein/Lipoprotein
   Glycoprotein (23.4 % carbohydrates) [3, 5]

4 ISOLATION/PREPARATION

Source organism
Human [1, 3, 5, 7]; Bovine [8, 10]; Patella vulgata [9]; Pig [11]

Source tissue
Human urine [1, 5, 7]; Human placenta [3]; Bovine spleen [8, 10]; Pig liver [11]

Localisation in source
Lysosomes [1, 2, 11]; Lamellar bodies [2]

Purification
Human urine [1, 5, 7]; Human placenta [3]; Bovine spleen [8, 10]; Patella vulgata [9]; Pig liver [11]

Crystallization
—

Cloned
—

Renaturated
—

5 STABILITY

pH
5.0–9.0 [3]; 5.0–8.5 [5]; 5.0 (not stable below) [11]

Temperature (°C)
50 (not stable above) [3, 5, 8]

Oxidation

Organic solvent

General stability information

Storage
Several months, 4°C, pH 6 [11]

6 CROSSREFERENCES TO STRUCTURE DATABANKS

PIR/MIPS code

Brookhaven code

Enzyme Handbook © Springer-Verlag Berlin Heidelberg 1991
Duplication, reproduction and storage in data banks are only
allowed with the prior permission of the publishers

7 LITERATURE REFERENCES

[1] Salvatore, D., Bonatti, S., Di Natale, P.: Bull. Mol. Biol. Med., 9, 111–121 (1984)

[2] Hook, G.E.R., Gilmore, L.B.: J. Biol. Chem., 257, 9211–9220 (1982)

[3] Roehrborn, W., Von Figura, K.: Hoppe-Seyler's Z. Physiol. Chem., 359, 1353–1362 (1978)

[4] Von Figura, K.: Eur. J. Biochem., 80, 535–542 (1977)

[5] Von Figura, K.: Eur. J. Biochem., 80, 525–533 (1977)

[6] Hultberg, B., Lindsten, J., Sjöblad, S.: Biochem. J., 155, 599–605 (1976)

[7] Von Figura, K., Kresse, H.: Eur. J. Biochem., 61, 581–588 (1976)

[8] Mersman, G., Von Figura, K., Buddecke, E.: Biochim. Biophys. Acta, 364, 88–96 (1974)

[9] Bannister, J.V., Phizackerley, P.J.R.: FEBS Lett., 29, 313–317 (1973)

[10] Werries, E., Wollek, E., Gottschalk, A., Buddecke, E.: Eur. J. Biochem., 10, 445–449 (1969)

[11] Weissmann, B., Rowin, G., Marshall, J., Friederici, D.: Biochemistry, 6, 207–214 (1967)

1 NOMENCLATURE

EC number
3.2.1.51

Systematic name
Alpha-L-fucoside fucohydrolase

Recommended name
Alpha-L-fucosidase

Synonymes
Alpha-fucosidase

CAS Reg. No.
9037-65-4

2 REACTION AND SPECIFICITY

Catalysed reaction
An alpha-L-fucoside + H_2O →
→ an alcohol + L-fucose

Reaction type
O-Glycosyl bond hydrolysis

Natural substrates
Alpha-L-fucosides + H_2O [1–33]

Substrate spectrum
1 Alpha-L-fucosides + H_2O (e.g. p-nitrophenyl alpha-L-fucoside [2, 5, 9, 11, 20–22, 24–28, 33], 4-methylumbelliferyl alpha-L-fucopyranoside [6, 7, 12, 14, 15, 26, 27], lacto-N-fucopentaitol [8, 12, 23, 30]) [1–33]

Product spectrum
1 L-Fucose + an alcohol (corresponding) [1–33]

Inhibitor(s)
Hg^{2+} [2, 4, 5, 9, 17, 20, 21, 25, 27, 28, 30, 31]; L-Fucose [2, 5, 9–11, 17, 20–22, 24, 25, 27]; Cd^{2+} [4, 31]; Cu^{2+} [4, 20, 21, 28, 31]; Mg^{2+} [4]; Ni^{2+} [4, 20, 31]; Pb^{2+} [4, 31]; Zn^{2+} [4, 31]; Beta-L-fucosylamine [10]; p-Chloromercuribenzoate [11, 21, 28]; 1-Cyclohexyl-3-(2-morpholinoethyl)-carbodiimide [11]; p-Chloromercuriphenylsulfonate [20, 24]; Ag^{2+} [21]; Lactones [25]; Ag^+ [27, 28, 31]; Mn^{2+} [30]; EDTA [30]; Co^{2+} [31]

Enzyme Handbook © Springer-Verlag Berlin Heidelberg 1991
Duplication, reproduction and storage in data banks are only
allowed with the prior permission of the publishers

Cofactor(s)/prostethic group(s)

Metal compounds/salts

Turnover number (min^{-1})

Specific activity (U/mg)
110 [2]; 0.12–0.37 [3, 30]; 0.96–4.76 [5, 17, 18]; 24.2–38.9 [9, 15, 20, 24, 27]; 10.6–15.7 [10, 22, 26, 31]; 7.8 [11]; 1620 [25]

K_m-value (mM)
0.07–0.87 (p-nitrophenyl alpha-L-fucoside) [2, 5, 9, 11, 20–22, 24–28, 33]; 0.049–0.65 (4-methylumbelliferyl alpha-L-fucopyranoside) [6, 7, 12, 14, 15, 26, 27]; 0.67–2.5 (2'-fucosyllactitol) [8, 23, 30]; 0.000025–2.0 (lacto-N-fucopentaitol) [8, 12, 23, 30]; More [10, 30]

pH-optimum
4.5–6.0 (p-nitrophenyl alpha-L-fucoside) [2, 5]; 5.25–5.5 (p-nitrophenyl alpha-L-fucoside) [4, 16]; 4.0–6.8 (4-methylumbelliferyl alpha-L-fucopyranoside, isoenzymes) [6]; 5.5–6.0 (4-methylumbelliferyl alpha-L-fucopyranoside, isoenzymes) [7]; 5.5 (2'-fucosyllactitol) [8, 23]; 5.0–6.5 (lacto-N-fucopentaitol) [8, 23]; 5.3–5.7 (p-nitrophenyl alpha-L-fucoside) [9]; 5.0 (p-nitrophenyl alpha-L-fucoside) [10, 21]; 5.1 (p-nitrophenyl alpha-L-fucoside) [11]; 5.3 (4-methylumbelliferyl alpha-L-fucopyranoside) [12]; 6.1–6.3 (p-nitrophenyl alpha-L-fucoside) [13, 31]; 3.9–4.0 (4-methylumbelliferyl alpha-L-fucopyranoside) [15]; 5.5–5.9 (p-nitrophenyl alpha-L-fucoside) [20, 24, 33]; 5.2 (p-nitrophenyl alpha-L-fucoside) [25]; 5.5 (p-nitrophenyl alpha-L-fucoside) [26, 28]; 6.0 (4-methylumbelliferyl alpha-L-fucopyranoside) [26]; 4.6 (p-nitrophenyl alpha-L-fucoside) [27]; 5.0–5.6 (p-nitrophenyl alpha-L-fucoside) [32]; 2.0 (p-nitrophenyl alpha-L-fucoside) [32]

pH-range
2.0–10.0 (p-nitrophenyl alpha-L-fucoside) [2]; 2.0–9.0 (p-nitrophenyl alpha-L-fucoside) [5]; 4.2 (not active below, 4-methylumbelliferyl alpha-L-fucopyranoside) [7]; 8.0 (not active above, p-nitrophenyl alpha-L-fucoside) [10]

Temperature optimum (°C)
60 (p-nitrophenyl alpha-L-fucoside) [2]; 50 (p-nitrophenyl alpha-L-fucoside) [5]; 70 (p-nitrophenyl alpha-L-fucoside) [9]; 50–65 (p-nitrophenyl alpha-L-fucoside) [10]; 45 (p-nitrophenyl alpha-L-fucoside) [28]; 55 (p-nitrophenyl alpha-L-fucoside) [32]

Temperature range (°C)

3 ENZYME STRUCTURE

Molecular weight
75000–80000 (gel filtration, Fusarium oxysporum) [2, 5]
200000–330000 (gel filtration, bull, isoenzymes) [4]
50000–390000 (gel filtration, man, isoenzymes) [7, 10, 12, 18, 26, 27]
70000–75000 (gel filtration, Octopus vulgaris) [9]
192000 (density gradient centrifugation, pig) [11]
255000 (gel filtration, pig) [11]
68000–83000 (gel filtration, Bacteroides fragilis, isoenzymes) [13]
200000–220000 (gel filtration, marine molluscs) [17]
217000 (sedimentation equilibrium, rat) [20, 24, 31]
285000 (gel filtration, monkey) [21]
200000 (gel filtration, Venus gallina) [25]
70000–80000 (gel filtration, Bacillus fulminans) [30]

Subunits
Monomer (SDS-PAGE) [2, 9]
Tetramer (4 × 55000, SDS-PAGE) [11, 20, 24]
Monomer to hexamer (1–6 × 50000, SDS-PAGE) [12, 26, 27]
Tetramer (4 × 73500, SDS-PAGE) [21]
Tetramer (2 × 60000 + 2 × 47000, SDS-PAGE) [31]

Glycoprotein/Lipoprotein
Glycoprotein [2, 11, 12, 15, 17, 18, 20, 24, 27, 31]

4 ISOLATION/PREPARATION

Source organism
Mouse [1, 6]; Fusarium oxysporum [2, 5]; Bull [3, 4]; Human [7, 10, 12, 14, 15, 18, 26, 27, 29]; Almond [8, 19, 23]; Octopus vulgaris [9]; Pig [11]; Bacteroides fragilis [13]; Venus mercenaria [16]; Rat [20, 22, 24, 31]; Monkey [21]; Clostridium perfringens [22]; Limulus polyphemus [22]; Venus gallina [25]; Bacillus fulminans [30]; Abalone [32]; More (most organisms) [7]

Source tissue
Liver [1, 6]; Seminal plasma [3, 4]; Reproductive organs [4]; Brain [7]; Emulsin [8, 19, 23]; Hepatopancreas [9]; Human serum [10]; Pig thyroid glands [11]; Human spleen [12]; Human liver [14, 15, 27]; Venus mercenaria hepatopancreas [16]; Human placenta [18, 26, 29]; Rat liver [20, 24]; Monkey brain [21]; Rat epididymis [22, 31]; Venus gallina hepatopancreas [25]; Abalone liver [32]

Localisation in source
Lysosomes [1, 6, 7, 10, 20, 24, 28]

Enzyme Handbook © Springer-Verlag Berlin Heidelberg 1991
Duplication, reproduction and storage in data banks are only
allowed with the prior permission of the publishers

Purification

Mouse liver [1, 6]; Fusarium oxysporum [2, 5]; Bull seminal plasma [3, 4];
Bull reproductive organs [4]; Human brain [7]; Almond emulsin [8, 19, 23];
Octopus vulgaris hepatopancreas [9]; Human serum [10]; Pig thyroid
glands [11]; Human spleen [12]; Bacteroides fragilis [13]; Human liver [14,
15, 27]; Venus mercenaria hepatopancreas [16]; Human placenta [18, 26,
29]; Rat liver [20, 24]; Monkey brain [21]; Rat epididymis [22, 31];
Clostridium perfringens [22]; Limulus polyphemus [22]; Venus gallina
hepatopancreas [25]; Bacillus fulminans [30]; Abalone liver [32]

Crystallization

–

Cloned

–

Renaturated

–

---

5 STABILITY

pH

5.5–10.0 [2]; 4.0–8.0 [5, 20]; 4.0–6.0 [25, 33]; 3.0–5.0 [32]

Temperature (°C)

45 (not stable above) [2, 4, 5]; 55 (not stable above) [9]; 60 (not stable
above) [17, 30]

Oxidation

Organic solvent

General stability information

Storage

29 days, 2–4°C [7]; 3–6 months, 0°C [8, 21, 27]; Several months, –20°C, pH 6,
20 % glycerol [20]

---

6 CROSSREFERENCES TO STRUCTURE DATABANKS

PIR/MIPS code

HWHUFA (tissue, human, fragment); A33427 (precursor, human); S03891
(slime mold, Dictyostelium discoideum); A30364 (precursor, Dictyostelium
discoideum)

Brookhaven code

7 LITERATURE REFERENCES

[1] Laury-Kleintop, L.D., Damjanov, I., Alhadeff, J.A.: Biochem. J., 245, 589–593 (1987)
[2] Yamamoto, K., Tsuji, Y., Kumagai, H., Tochikura, T.: Agric. Biol. Chem., 50, 1689–1695 (1986)
[3] Srivastava, P.N., Arbtan, K., Takei, G.H., Huang, T.T.F., Yanagimachi, R.: Biochem. Biophys. Res. Commun., 137, 1061–1068 (1986)
[4] Jauhiainen, A., Vanha-Perttula, T.: Biochim. Biophys. Acta, 880, 91–95 (1986)
[5] Yano, T., Yamamoto, K., Kumagai, H., Tochikura, T., Yokoyama, T., Seno, T., Yamaguchi, H.: Agric. Biol. Chem., 49, 3179–3187 (1985)
[6] Laury-Kleintop, L.D., Damjanov, I., Alhadeff, J.A.: Biochem. J., 230, 75–82 (1985)
[7] Hopfer, R.L., Alhadeff, J.A.: Biochem. J., 229, 679–685 (1985)
[8] Kobata, A.: Methods Enzymol., 83, 625–631 (1982)
[9] D'Aniello, A., Hakimi, J., Cacace, G.M., Ceccarini, C.: J. Biochem., 91, 1073–1080 (1982)
[10] DiCioccio, R.A., Barlow, J.J., Matta, K.L.: J. Biol. Chem., 257, 714–718 (1982)
[11] Grove, D.S., Serif, G.S.: Biochim. Biophys. Acta, 662, 246–255 (1981)
[12] Chien, S.F., Dawson, G.: Biochim. Biophys. Acta, 614, 476–488 (1980)
[13] Berg, J.O., Lindqvist, L., Nord, C.E.: Appl. Environ. Microbiol., 40, 40–47 (1980)
[14] Alhadeff, J.A., Andrews-Smith, G.L.: Biochem. J., 187, 45–51 (1980)
[15] Kress, B.C., Freeze, H.H., Herd, J.K., Alhadeff, J.A., Miller, A.L.: J. Biol. Chem., 255, 955–961 (1980)
[16] Concha-Slebe, I., Presper, K.A., Basu, S. in "Glycoconjugate Research" (Gregory, J.D., Jeanloz, R.W., Eds.) 2, 873–875 (1979)
[17] Cabezas, J.A., Reglero, A., Calvo, P., Angeles De Pedro, M. in "Glycoconjugate Research" (Gregory, J.D., Jeanloz, R.W., Eds.) 2, 867–871 (1979)
[18] Turner, B.M.: Biochim. Biophys. Acta, 578, 325–336 (1979)
[19] Yoshima, H., Takasaki, S., Ito-Mega, S., Kobata, A.: Arch. Biochem. Biophys., 194, 394–398 (1979)
[20] Opheim, D.J., Touster, O.: Methods Enzymol., 50, 505–510 (1978)
[21] Alam, T., Balasubramanian, A.S.: Biochim. Biophys. Acta, 524, 373–384 (1978)
[22] Jain, R.S., Binder, R.L., Levy-Benshimol, A., Buck, C.A., Warren, L.: J. Chromatogr., 139, 283–290 (1977)
[23] Ogata-Arakawa, M., Muramatsu, T., Kobata, A.: Arch. Biochem. Biophys., 181, 353–358 (1977)
[24] Opheim, D.J., Touster, O.: J. Biol. Chem., 252, 739–743 (1977)
[25] Reglero, A., Cabezas, J.A.: Eur. J. Biochem., 66, 379–387 (1976)
[26] Di Matteo, G., Orfeo, M.A., Romeo, G.: Biochim. Biophys. Acta, 429, 527–537 (1976)
[27] Alhadeff, J.A., Miller, A.L., Wenaas, H., Vedvick, T., O'Brien, J.S.: J. Biol. Chem., 250, 7106–7113 (1975)
[28] Avila, J.L., Convit, J.: Biochim. Biophys. Acta, 358, 308–318 (1974)
[29] Aldaheff, J.A., Miller, A.L., O'Brien, J.S.: Anal. Biochem., 60, 424–430 (1974)
[30] Kochibe, N.: J. Biochem., 74, 1141–1149 (1973)
[31] Carlsen, R.B., Pierce, J.G.: J. Biol. Chem., 247, 23–32 (1972)
[32] Tanaka, K., Nakano, T., Noguchi, S., Pigman, W.: Arch. Biochem. Biophys., 126, 624–633 (1968)
[33] Levvy, G.A., McAllan, A.: Biochem. J., 80, 435–439 (1961)

1 NOMENCLATURE

EC number
3.2.1.52

Systematic name
Beta-N-acetyl-D-hexosaminide N-acetylhexosaminohydrolase

Recommended name
Beta-N-acetylhexosaminidase

Synonymes
Hexosaminidase
Beta-acetylaminodeoxyhexosidase
N-Acetyl-beta-D-hexosaminidase
N-Acetyl-beta-hexosaminidase
Beta-acetylhexosaminidase
Beta-hexosaminidase
Beta-acetylhexosaminidinase
Beta-D-N-acetylhexosaminidase
Beta-N-acetyl-D-hexosaminidase
Beta-D-hexosaminidase
N-Acetylhexosaminidase
Hexosaminidase A
Beta-N-acetylglucosaminidase (some of the enzymes cited in EC 3.2.1.30
probably are N-acetylhexosaminidase, discussion of IUB Nomenclature
[31])

CAS Reg. No.
9027-52-5

2 REACTION AND SPECIFICITY

Catalysed reaction
Hydrolysis of terminal non-reducing N-acetyl-D-hexosamine residues in
N-acetyl-beta-D-hexosaminides

Reaction type
O-Glycosyl bond hydrolysis

Natural substrates
N-Acetyl-beta-D-hexosaminides + H_2O [1–30]

Enzyme Handbook © Springer-Verlag Berlin Heidelberg 1991
Duplication, reproduction and storage in data banks are only
allowed with the prior permission of the publishers

Substrate spectrum

1 N-Acetyl-beta-D-hexosaminides + H_2O (e.g.
p-nitrophenyl-N-acetyl-beta-D-glucosaminide [1–3, 6–8, 10, 11, 14, 17, 18,
20–22, 24–26, 29, 30], p-nitrophenyl-N-acetyl-beta-D-galactosaminide
[1–3, 7, 8, 10, 11, 14, 17, 20–22, 24–26, 29, 30], 4-methylumbel-
liferyl-beta-D-glucosaminide [9, 13, 19]) [1–30]

Product spectrum

1 N-Acetyl-D-hexosamines [1–30]

Inhibitor(s)

N-Acetyl-D-glucosamine [1–3, 13, 17, 19–22, 26, 29, 30];
N-Acetyl-D-galactosamine [1, 3, 10, 19–22, 26, 29, 30];
2-Acetamido-2-deoxygluconolactone [1, 21, 22, 25, 29]; Hg^{2+} [1, 7, 8, 10, 11,
14, 16, 18, 20–22, 24, 29]; Ag^+ [1, 3, 14, 16, 18, 21, 24, 29, 30]; Cu^{2+} [1, 7, 14];
Glucosamine [3, 19, 30]; Galactosamine [3, 19, 30]; Zn^{2+} [3]; Fe^{2+} [7, 10, 16,
24]; p-Chloromercuriphenylsulphonate [10, 19]; p-Chloromercuribenzoate
[16, 18, 24]; Monoiodoacetate [16]; Acetate [19, 22, 23, 29, 30]; Cd^{2+} [20];
Beta-mercaptoethanol [20]; N-Acetylgalactosaminolactone [21, 22, 25];
Thimerasol [21]; p-Hydroxymercuribenzoate [21, 30]; Fe^{3+} [22, 24, 29]; Mannose [22]; EDTA [23]; Acetamide [29]

Cofactor(s)/prostethic group(s)

Metal compounds/salts

Turnover number (min⁻¹)

Specific activity (U/mg)

56.6–101.9 [2, 6, 12, 15–18, 29]; 1.2–3.5 [3, 13, 21]; 21.3–30.9 [5, 22, 23];
186–214 [11, 14]; 0.12–0.56 [19, 20, 24]; 731 [25]

K_m-value (mM)

0.04–1.9 (p-nitrophenyl-N-acetyl-beta-D-glucosaminide) [1–3, 6–8, 10, 11,
14, 17, 18, 20–22, 24–26, 29, 30]; 0.064–1.71
(p-nitrophenyl-N-acetyl-beta-D-galactosaminide) [1–3, 7, 8, 10, 11, 14, 17,
20–22, 24–26, 29, 30]; 0.45–1.0 (4-methylumbelliferyl-beta-D-glucosaminide)
[9, 13, 19]; 0.3–0.4 (4-methylumbelliferyl-beta-D-galactosaminide) [19]

pH-optimum

5.2 (p-nitrophenyl-N-acetyl-beta-D-glucosaminide) [1]; 4.4
(p-nitrophenyl-N-acetyl-beta-D-galactosaminide) [1]; 2.2
(p-nitrophenyl-N-acetyl-beta-D-glucosaminide) [2, 17]; 3.7
(p-nitrophenyl-N-acetyl-beta-D-galactosaminide) [2, 17, 30]; 2.5 (N,
N-diacetylchitobiose) [2]; 5.0
(p-nitrophenyl-N-acetyl-beta-D-glucosaminide) [3]; 4.6
(p-nitrophenyl-N-acetyl-beta-D-galactosaminide) [3]; 4.0
(p-nitrophenyl-N-acetyl-beta-D-glucosaminide) [5, 16, 18]; 5.0–5.5
(p-nitrophenyl-N-acetyl-beta-D-glucosaminide) [6, 29]; 4.5–5.0
(p-nitrophenyl-N-acetyl-beta-D-glucosaminide) [7]; 4.0–5.0
(p-nitrophenyl-N-acetyl-beta-D-glucosaminide) [8]; 5.5–6.5
(p-nitrophenyl-N-acetyl-beta-D-galactosaminide) [8]; 4.5
(p-nitrophenyl-N-acetyl-beta-D-glucosaminide) [10, 22]; 4.0
(p-nitrophenyl-N-acetyl-beta-D-galactosaminide) [10, 14]; 3.0–4.5
(p-nitrophenyl-N-acetyl-beta-D-glucosaminide,
p-nitrophenyl-N-acetyl-beta-D-galactosaminide) [11]; 5.2 (4-methylumbel-
liferyl-beta-D-glucosaminide) [13]; 4.2
(p-nitrophenyl-N-acetyl-beta-D-glucosaminide) [14, 21, 30]; 5.0
(4-methylumbelliferyl-beta-D-glucosaminide) [19]; 4.6
(p-nitrophenyl-N-acetyl-beta-D-glucosaminide) [20]; 5.8
(p-nitrophenyl-N-acetyl-beta-D-galactosaminide) [20]; 4.2
(p-nitrophenyl-N-acetyl-beta-D-galactosaminide) [21]; 4.5
(p-nitrophenyl-N-acetyl-beta-D-galactosaminide) [22]; 4.8–6.0
(p-nitrophenyl-N-acetyl-beta-D-glucosaminide, isoenzymes) [24]; 4.4–5.5
(p-nitrophenyl-N-acetyl-beta-D-galactosaminide, isoenzymes) [24]; 6.2
(p-nitrophenyl-N-acetyl-beta-D-glucosaminide,
p-nitrophenyl-N-acetyl-beta-D-galactosaminide) [26]; 3.5–4.0
(p-nitrophenyl-N-acetyl-beta-D-galactosaminide) [29]

pH-range

2.4–8.0 (p-nitrophenyl-N-acetyl-beta-D-glucosaminide,
p-nitrophenyl-N-acetyl-beta-D-galactosaminide) [1, 11]; 7.0 (not active
above, p-nitrophenyl-N-acetyl-beta-D-glucosaminide) [16]; 1.5 (not active
below, p-nitrophenyl-N-acetyl-beta-D-glucosaminide) [17]; 3.4 (not active
below, 4-methylumbelliferyl-beta-D-glucosaminide) [19]; 2.0–10.0
(p-nitrophenyl-N-acetyl-beta-D-glucosaminide,
p-nitrophenyl-N-acetyl-beta-D-galactosaminide) [20]

Temperature optimum (°C)

55 (p-nitrophenyl-N-acetyl-beta-D-glucosaminide,
p-nitrophenyl-N-acetyl-beta-D-galactosaminide) [11]; 35–37
(p-nitrophenyl-N-acetyl-beta-D-glucosaminide,
p-nitrophenyl-N-acetyl-beta-D-galactosaminide) [20]; 40–45
(p-nitrophenyl-N-acetyl-beta-D-glucosaminide,
p-nitrophenyl-N-acetyl-beta-D-galactosaminide) [24]

Enzyme Handbook © Springer-Verlag Berlin Heidelberg 1991
Duplication, reproduction and storage in data banks are only
allowed with the prior permission of the publishers

Temperature range (°C)

3 ENZYME STRUCTURE

Molecular weight
  120000 (gel filtration, Pycnoporus cinnabarinus) [2, 17]
  170000 (gel filtration, Mugil cephalus) [5]
  150000 (gel filtration, watermelon) [6]
  125000 (gel filtration, Mucor fragilis) [8]
  100000–110000 (gel electrophoresis, man, isoenzymes) [9]
  141000–143000 (gel filtration, sedimentation equilibrium, Penicillium oxalicum) [11]
  525000 (gel filtration, rat brain) [13]
  330000 (gel filtration, Halocynthia roretzi) [14]
  132000 (gel filtration, Styela plicata) [16]
  152000–154000 (gel filtration, man, isoenzyme S) [19]
  100000 (gel filtration, Paecilomyces persicinus) [20]
  120000–145000 (gel filtration, Helicella ericetorum) [22]
  62500 (gel filtration, ox) [23]
  72000–150000 (gel filtration, fenugreek, isoenzymes) [24]
  150000 (gel filtration, Trichomonas foetus) [26]
  100000 (gel filtration, jack bean) [29]
  21000 (gel filtration, pea) [7]

Subunits
  Dimer (2 × 65000–70000, SDS-PAGE) [2, 8, 11, 16]
  Tetramer (4 × 74000, SDS-PAGE, Halocynthia roretzi) [14]
  Monomer (SDS-PAGE) [20, 23]
  Trimer (3 × 24000–28000, fenugreek isoenzymes I and II, SDS-PAGE) [24]
  Hexamer (6 × 30500, fenugreek isoenzyme III, SDS-PAGE) [24]
  Pentamer (5 × 30000, fenugreek isoenzyme IV, SDS-PAGE) [24]

Glycoprotein/Lipoprotein
  Glycoprotein [7–9, 11, 15, 24]

4 ISOLATION/PREPARATION

Source organism
  Wheat [1]; Pycnoporus cinnabarinus [2, 17]; Soybean [3]; Mugil cephalus [5]; Watermelon [6]; Pea [7]; Mucor fragilis [8]; Human [4, 9, 15, 19, 28]; Black cherry [10]; Penicillium oxalicum [11]; Hen [12]; Halocynthia roretzi [14]; Styela plicata [16]; Rabbit [18]; Paecilomyces persicinus [20]; Helicella ericetorum [22]; Fenugreek [24, 25]; Trichomonas foetus [26]; Jack bean [29]; Calf [30]; Rat [13, 21]; Bovine [23, 27]

Source tissue

Leaves [1]; Soybean seeds [3]; Human placenta [4]; Mugil cephalus roe [5]; Watermelon fruits [6]; Pea seeds [7]; Human liver [9, 19]; Black cherry seeds [10]; Egg white [12]; Hen oviduct [12]; Rat brain [13]; Human urine [15]; Rabbit semen [18]; Human brain [19]; Rat colon [21]; Bovine testes [23]; Fenugreek seeds [24, 25]; Trichomonas foetus [26]; Bull epididymis [27]; Human skin fibroblasts [28]; Calf brain [30]

Localisation in source

Lysosomes [4, 12]; Microsomes [12]; Cytoplasm [13]; Extracellular [20]

Purification

Wheat leaves [1]; Pycnoporus cinnabarinus [2, 17]; Soybean seeds [3]; Human placenta [4]; Mugil cephalus roe [5]; Watermelon fruits [6]; Pea seeds [7]; Mucor fragilis [8]; Human liver [9, 19]; Black cherry seeds [10]; Penicillium oxalicum [11]; Egg white [12]; Hen oviduct [12]; Rat brain [13]; Halocynthia roretzi [14]; Human urine [15]; Styela plicata [16]; Rabbit semen [18]; Human brain [19]; Paecilomyces persicinus [20]; Rat colon [21]; Helicella ericetorum [22]; Bovine testes [23]; Fenugreek seeds [24, 25]; Trichomonas foetus [26]; Bull epididymis [27]; Human skin fibroblasts [28]; Jack bean [29]; Calf brain [30]

Crystallization

[29]

Cloned

–

Renaturated

–

5 STABILITY

pH

2.0–4.0 [2, 17]; 6.5–8.0 [6, 29]; 4.5–8.5 [8]; 7.0–9.0 [11]; 4.0–7.5 [14]; 5.0–6.0 [16] 3.8–4.6 [22]

Temperature (°C)

50 (not stable above) [2, 19, 21, 22, 24]; 52 (not stable above) [4]; 45 (not stable above) [8, 11, 17]; 40 (not stable above) [14]; 60 (not stable above) [20, 29]

Oxidation

Organic solvent

General stability information

Storage

4 months, 0°C, 10 % glycerol [16]; 1 year, –20°C [23]; 2 months, 4°C [24, 26]

Enzyme Handbook © Springer-Verlag Berlin Heidelberg 1991
Duplication, reproduction and storage in data banks are only
allowed with the prior permission of the publishers

6 CROSSREFERENCES TO STRUCTURE DATABANKS

PIR/MIPS code

AOHUBA (alpha chain, precursor, human); A34204 (human, fragment); A30766 (A, precursor, Dictyostelium discoideum); A31778 (A, precursor, slime mold, Dictyostelium discoideum); S02609 (human, fragment); A28139 (alpha chain, human, fragment); A23842 (alpha chain, precursor, human, fragment); A26727 (alpha chain, precursor, human, fragment); A28948 (B, precursor, human, fragment); A31250 (beta chain, human); A22081 (beta chain, human, fragment); A30153 (beta chain, precursor, human); B28139 (beta-a chain, human, fragment); C28139 (beta-b chain, human, fragment); S01328 (beta chain, mouse)

Brookhaven code

---

7 LITERATURE REFERENCES

[1] Barber, M.S., Ride, J.P.: Plant Sci., 60, 163–172 (1989)
[2] Ohtakara, A.: Methods Enzymol., 161, 462–470 (1988)
[3] Gers-Barlag, H., Bartz, I., Ruediger, H.: Phytochemistry, 27, 3739–3741 (1988)
[4] Kinoshita, K., Taniguchi, N., Makita, A., Narita, M., Oikawa, K.: J. Biochem., 104, 827–831 (1988)
[5] DeGaspari, R., Li, S.C., Li, Y.T.: J. Biol. Chem., 263, 1325–1328 (1988)
[6] Nakagawa, H., Enomoto, N., Asakawa, M., Uda, Y.: Agric. Biol. Chem., 52, 2223–2230 (1988)
[7] Harley, S.M., Beevers, L.: Plant Physiol., 85, 1118–1122 (1987)
[8] Yamamoto, K., Tsuji, Y., Matsushita, S., Kumagai, H., Tochikura, T.: Appl. Environ. Microbiol., 51, 1019–1023 (1986)
[9] Dewji, N.N., De-Keyser, D.R., Stirling, J.L.: Biochem. J., 234, 157–162 (1986)
[10] Poulton, J.E., Thomas, M.A., Ottwell, K.K., McCormick, S.J.: Plant Sci., 42, 107–114 (1985)
[11] Yamamoto, K., Lee, K.M., Kumagai, H., Tochikura, T.: Agric. Biol. Chem., 49, 611–619 (1985)
[12] Ogawa, Y., Nakamura, R., Sato, Y.: Agric. Biol. Chem., 47, 2085–2089 (1983)
[13] Izumi, T., Suzuki, K.: J. Biol. Chem., 258, 6991–6999 (1983)
[14] Uda, Y., Itoh, T.: J. Biochem., 93, 847–855 (1983)
[15] Kress, B.C., Hirani, S., Freeze, H.H., Little, L., Miller, A.L.: Biochem. J., 207, 421–428 (1982)
[16] Shigeta, S., Matsuda, A., Oka, S.: J. Biochem., 92, 163–172 (1982)
[17] Ohtakara, A., Yoshida, M., Murakami, M., Izumi, T.: Agric. Biol. Chem., 45, 239–247 (1981)
[18] Farooqui, A.A., Srivastava, P.N.: Biochem. J., 191, 827–834 (1980)
[19] Potier, M., Teitelbaum, J., Melancon, S.B., Dallaire, L.: Biochim. Biophys. Acta, 566, 80–87 (1979)
[20] Eriquez, L.A., Pisano, M.A.: J. Bacteriol., 137, 620–626 (1979)
[21] Mian, N., Herries, D.G., Cowen, D.M., Batte, E.A.: Biochem. J., 177, 319–330 (1979)
[22] Calvo, P., Reglero, A., Cabezas, J.A.: Biochem. J., 175, 743–750 (1978)
[23] Sarber, R.L., Distler, J.J., Jourdian, G.W.: Methods Enzymol., 50, 520–523 (1978)

[24] Bouquelet, S., Spik, G.: Eur. J. Biochem., 84, 551–559 (1978)
[25] Bouquelet, S., Spik, G.: FEBS Lett., 63, 95–101 (1976)
[26] Edwards, R.G., Thomas, P., Westwood, J.H.: Biochem. J., 151, 145–148 (1975)
[27] Pokorny, M., Glaudemans, C.P.J.: FEBS Lett., 50, 66–69 (1975)
[28] Dawson, G., Propper, R.L., Dorfman, A.: Biochem. Biophys. Res. Commun., 54, 1102–1110 (1973)
[29] Li, S.C., Li, Y.T.: J. Biol. Chem., 245, 5153–5160 (1970)
[30] Frohwein, Y.Z., Gatt, S.: Biochemistry, 6, 2774–2782 (1967)
[31] Cabezas, J.A.: Biochem. J., 261, 1059–1061 ((1989))

1 NOMENCLATURE

EC number
3.2.1.53

Systematic name
Beta-N-acetyl-D-galactosaminide N-acetylgalactosaminohydrolase

Recommended name
Beta-N-acetylgalactosaminidase

Synonymes
N-Acetyl-beta-galactosaminidase
N-Acetyl-beta-D-galactosaminidase
Beta-acetylgalactosaminidase
Beta-D-N-acetylgalactosaminidase
Beta-N-acetylgalactosaminidase
N-Acetylgalactosaminidase

CAS Reg. No.
9054-43-7

---

2 REACTION AND SPECIFICITY

Catalysed reaction
Hydrolysis of terminal non-reducing N-acetyl-D-galactosamine residues in
N-acetyl-beta-D-galactosaminides

Reaction type
O-Glycosyl bond hydrolysis

Natural substrates
N-Acetyl-beta-D-galactosaminides + H_2O [1, 2]

Substrate spectrum
1 N-Acetyl-beta-D-galactosaminides + H_2O (e.g. 4-methylumbelliferyl
  beta-N-acetylgalactosaminide) [1, 2]

Product spectrum
1 N-Acetyl-D-galactosamine + an alcohol (e.g. 4-methylumbelliferone) [1, 2]

Inhibitor(s)
N-Acetylgalactosamine [1, 2]; p-Hydroxymercuribenzoate [2]; $AgNO_3$ [2]

Cofactor(s)/prostethic group(s)

Metal compounds/salts

Enzyme Handbook © Springer-Verlag Berlin Heidelberg 1991
Duplication, reproduction and storage in data banks are only
allowed with the prior permission of the publishers

Turnover number (min^{-1})

Specific activity (U/mg)
9.8 [1]

K_m-value (mM)
0.42 (4-methylumbelliferyl beta-N-acetylgalactosaminide) [1]; 0.35
(p-nitrophenyl beta-N-acetylgalactosaminide) [2]

pH-optimum
6.0 (4-methylumbelliferyl beta-N-acetylgalactosaminide) [1]; 5.5
(p-nitrophenyl beta-N-acetylgalactosaminide) [2]

pH-range
4.0 (not active below, p-nitrophenyl beta-N-acetylgalactosaminide) [2]

Temperature optimum (°C)

Temperature range (°C)

---

3 ENZYME STRUCTURE

Molecular weight
210000 (gel filtration, rat) [1]
82000 (sucrose density centrifugation, rat) [1]

Subunits

Glycoprotein/Lipoprotein
–

---

4 ISOLATION/PREPARATION

Source organism
Rat [1]; Calf [2]

Source tissue
Brain [1]

Localisation in source
Lysosomes [1]; Soluble [2]

Purification
Rat brain [1]; Calf brain [2]

Crystallization
–

Cloned
–

Renaturated
–

---

5 STABILITY

pH
4.0 (unstable below) [2]

Temperature (°C)

Oxidation

Organic solvent
Acetone (inactivation) [2]

General stability information
Dithiothreitol (stabilization) [2]; Cysteine (stabilization) [2]

Storage
–20°C, inactivaton after several weeks [2]

---

6 CROSSREFERENCES TO STRUCTURE DATABANKS

PIR/MIPS code

Brookhaven code

---

7 LITERATURE REFERENCES

[1] Izumi, T., Suzuki, K.: J. Biol. Chem., 258, 6991–6999 (1983)
[2] Frohwein, Y.Z., Gatt, S.: Biochemistry, 6, 2775–2782 (1967)

Enzyme Handbook © Springer-Verlag Berlin Heidelberg 1991
Duplication, reproduction and storage in data banks are only
allowed with the prior permission of the publishers

1 NOMENCLATURE

EC number
3.2.1.54

Systematic name
Cyclomaltodextrin dextrin-hydrolase (decyclizing)

Recommended name
Cyclomaltodextrinase

Synonymes
Glucanase, cyclohepta-
Glucanase, cyclohexa-
Cycloheptaglucanase
Cyclohexaglucanase
E.C. 3.2.1.12 (formerly)
E.C. 3.2.1.13 (formerly)
Cyclodextrinase [1]

CAS Reg. No.
37288-41-8

---

2 REACTION AND SPECIFICITY

Catalysed reaction
Cyclomaltodextrin + H_2O →
→ linear maltodextrin

Reaction type
O-Glycosyl bond hydrolysis

Natural substrates
Cyclomaltodextrins + H_2O

Substrate spectrum
1 Cyclomaltodextrin + H_2O (alpha-, beta-and gamma [1]) [1–5]
2 Maltodextrins + H_2O (linear)
3 Maltotetraose + H_2O [1, 3, 5]
4 Maltopentaose + H_2O [1, 3, 5]
5 Maltohexaose + H_2O [1, 3, 5]
6 Schardinger dextrins + H_2O [3]
7 Maltoheptaose + H_2O [3]
8 Maltotriose + H_2O (slowly [1]) [1, 5]

Enzyme Handbook © Springer-Verlag Berlin Heidelberg 1991
Duplication, reproduction and storage in data banks are only
allowed with the prior permission of the publishers

9 Cyclomaltohexaose + H_2O [5]
10 Cyclomaltoheptaose + H_2O [5]
11 More (not: maltose [1], slowly: short chain amylose [1], hydrolyses by detaching one maltose unit, recognizes maltose residues at non-reducing end of various maltosides and hydrolyzes (1 --> 4)-, (1 --> 2)-, (1 --> 3)- and (1 --> 6)-alpha-D-glucosidic linkages and the glucosidic linkages between D-glucose and phenol, D-glucose and D-glucitol, and D-glucose and D-fructose adjacent to the maltose molecule [5]) [1, 5]

Product spectrum
1 Maltodextrin (linear, Bacteroides ovatus: glucose only [4], Bacteroides distasonis: series of maltooligomers [4])
2 ?
3 Maltose [3, 5]
4 Maltose + maltotriose [3, 5]
5 Maltotetraose + maltose [3, 5]
6 ?
7 Maltooligosaccharides (lower, with the possible exception of maltohexaose) [3]
8 ?
9 ?
10 ?
11 ?

Inhibitor(s)
Hg^{2+} [1]; Fe^{2+} [1]; Fe^{3+} [1]

Cofactor(s)/prostethic group(s)

Metal compounds/salts
Mg^{2+} (increases activity) [1]

Turnover number (min^{-1})

Specific activity (U/mg)
More [1, 2]

K_m-value (mM)
5.3 (maltotriose) [5]; 5.5 (phenyl alpha-maltoside) [5]; 17 (4G-O-alpha-D-glucosylsucrose) [5]; 21 (2-O-alpha-maltosylglucose) [5]; 28 (isopanose) [5]; 110 (maltotriitol) [5]; 4.5 (maltotriose) [1]; 4.0 (maltotetraose) [1]; 2.3 (maltopentaose) [1]; 1.5 (maltohexaose, short chain amylose) [1]; 10 (alpha-cyclodextrin) [1]; 2.8 (beta-cyclodextrin) [1]; 0.47 (gamma-cyclodextrin) [1]; 2.62 (alpha-cyclodextrin) [2]; 2.65 (beta-cyclodextrin) [2]

pH-optimum
  6.2 [1]; 7.0 (crude enzyme) [4]; 6.2–6.4 [2]

pH-range
  5–8 (5: about 40% of maximal activity, 8: about 20% of maximal activity)
  [2]; More (crude enzyme) [4]

Temperature optimum (°C)
  50 [1]; 55 (crude enzyme, Bacteroides distasonis) [4]; 42 (crude enzyme,
  Bacteroides ovatus) [4]

Temperature range (°C)

---

3 ENZYME STRUCTURE

Molecular weight
  62000 (SDS-PAGE, Bacillus coagulans) [1]

Subunits

Glycoprotein/Lipoprotein
  –

---

4 ISOLATION/PREPARATION

Source organism
  Bacillus coagulans [1, 5]; Bacillus macerans [2, 3]; Bacteroides distasonis
  [4]; Bacteroides ovatus [4]

Source tissue
  Cell [1]

Localisation in source
  Intracellular [1, 4]; Cell-bound [4]

Purification
  Bacillus coagulans [1]; Bacilllus macerans [2]

Crystallization
  –

Cloned
  –

Renaturated
  –

Enzyme Handbook © Springer-Verlag Berlin Heidelberg 1991
Duplication, reproduction and storage in data banks are only
allowed with the prior permission of the publishers

5 STABILITY

pH
6.0–7.3 (40°C, 2 hours stable) [1]

Temperature (°C)
45 (pH 7, stable up to 45°C) [1]; 4–55 (48 hours, little loss of activity, Bacteroides distasonis) [4]; 37 (24 hours, significant loss of activity, Bacteroides ovatus) [4]

Oxidation

Organic solvent

General stability information

Storage

6 CROSSREFERENCES TO STRUCTURE DATABANKS

PIR/MIPS code

Brookhaven code

7 LITERATURE REFERENCES

[1] Kitahata, S., Taniguchi, M., Beltran, S.D., Sugimoto, T., Okada, S.: Agric. Biol. Chem., 47, 1441–1447 (1983)
[2] DePinto, J.A., Campbell, L.L.: Biochemistry, 7, 121–125 (1968)
[3] DePinto, J.A., Campbell, L.L.: Arch. Biochem. Biophys., 125, 253–258 (1968)
[4] Antenucci, R.N., Palmer, J.K.: J. Agric. Food Chem., 32, 1316–1321 (1984)
[5] Kitahata, S., Okada, S.: Carbohydr. Res., 137, 217–225 (1985)

1 NOMENCLATURE

EC number
3.2.1.55

Systematic name
Alpha-L-arabinofuranoside arabinofuranohydrolase

Recommended name
Alpha-L-arabinofuranosidase

Synonymes
Arabinosidase
Alpha-arabinosidase
Alpha-L-arabinosidase
Alpha-arabinofuranosidase
Polysaccharide alpha-L-arabinofuranosidase
Alpha-L-arabinofuranoside hydrolase
L-Arabinosidase
Alpha-L-arabinanase
EC 3.2.1.79 (formerly)

CAS Reg. No.
9067-74-7

2 REACTION AND SPECIFICITY

Catalysed reaction
Hydrolysis of terminal non reducing alpha-L-arabinofuranoside residues in alpha-L-arabinosides

Reaction type
O-Glycosyl bond hydrolysis

Natural substrates
Alpha-L-arabinosides + H_2O [1–15]

Substrate spectrum
1 Alpha-L-arabinosides + H_2O (e.g. p-nitrophenyl-alpha-L-arabinofuranoside [1–5, 7, 9, 10, 12, 13]) [1–15]
2 L-Arabinans + H_2O [1–15]
3 Arabinoxylans + H_2O [1–4, 12]
4 Arabinans + H_2O [8]

Enzyme Handbook © Springer-Verlag Berlin Heidelberg 1991
Duplication, reproduction and storage in data banks are only
allowed with the prior permission of the publishers

Product spectrum

1 Alpha-L-arabinofuranosides [1–15]
2 L-Arabinose [1–15]
3 L-Arabinose [1–4, 12]
4 Arabinotriose [8]

Inhibitor(s)

Zn^{2+} [1, 13]; Cu^{2+} [1, 12, 13]; Arabinose [1, 2, 13]; L-Arabono-1, 4-lactone
[1, 9, 13]; Hg^{2+} [2, 8–10, 12]; Ag^+ [2, 9, 10, 12]; Iodoacetamide [7];
p-Chloromercuribenzoate [7, 10]; Ca^{2+} [10]; Sn^{2+} [12]

Cofactor(s)/prostethic group(s)

Metal compounds/salts

Turnover number (min⁻¹)

Specific activity (U/mg)

0.611 [1, 13]; 388.6–396.6 [2, 15]; 19.2 [4]; 655.5 [7]; 32.5–44.0 [8, 12, 14]; 82.9
[9]; 3.4 [10]

K_m-value (mM)

0.082–9.1 (p-nitrophenyl-alpha-L-arabinofuranoside) [1–5, 7, 9, 10, 12, 13];
9.22 (alpha-arabinofuranosyl-1, 3-arabinose) [1]; 2.0–8.06 (al-
pha-arabinofuranosyl-1, 5-arabinose) [1, 6]; 2.86–5.0
(phenyl-alpha-L-arabinofuranoside) [2, 14]; 0.48 (4-methylumbelliferone alpha-L-arabinofuranosides) [5]; 4.6 (phenyl-alpha-L-arabinofuranoside) [5,
6]; 0.017 (L-arabinan) [6]; More (aryl alpha-L-arabinofuranosides) [5]

pH-optimum

4.8 (p-nitrophenyl-alpha-L-arabinofuranoside) [1, 13]; 3.8
(p-nitrophenyl-alpha-L-arabinofuranoside) [2, 15]; 4.0
(p-nitrophenyl-alpha-L-arabinofuranoside) [3]; 5.0–5.5
(p-nitrophenyl-alpha-L-arabinofuranoside) [4]; 3.8–4.0 (arabinan) [6, 15];
6.9 (p-nitrophenyl-alpha-L-arabinofuranoside) [7]; 6.0 (arabinan) [8, 10] 6.5
(p-nitrophenyl-alpha-L-arabinofuranoside) [9]; 2.0 (arabinan) [12]; 2.5
(phenyl-alpha-L-arabinofuranoside, arabinan) [14]

pH-range

6.5 (not active above, p-nitrophenyl-alpha-L-arabinofuranoside) [6];
3.5–11.0 (p-nitrophenyl-alpha-L-arabinofuranoside) [9]; 3.0–11.0 (arabinan)
[10]

Temperature optimum (°C)

Temperature range (°C)

3 ENZYME STRUCTURE

Molecular weight
62000 (gel filtration, Scopolia japonica) [1, 13]
53000 (gel filtration, Aspergillus niger) [2, 15]
53000 (SDS-PAGE, Trichoderma reesei) [3]
90000 (gel filtration, Clostridium acetobutylicum) [4]
40000 (gel filtration, Minilinia fructigena) [5]
310000 (gel filtration, Ruminococcus albus) [7]
495000 (gel filtration, Streptomyces purpurascens) [9]
92000 (gel filtration, Streptomyces sp.) [10]

Subunits
Monomer (SDS-PAGE, Streptomyces sp. [10], Trichoderma reesei [3]) [3–5, 10]
Tetramer (4 × 75000, SDS-PAGE, Ruminococcus albus) [7]
Octamer (8 × 62000, SDS-PAGE, Streptomyces purpurascens) [9]

Glycoprotein/Lipoprotein
Glycoprotein [6, 7]

---

4 ISOLATION/PREPARATION

Source organism
Scopolia japonica [1, 13]; Aspergillus niger [2, 11, 15]; Trichoderma reesei [3]; Clostridium acetobutylicum [4]; Monilinia fructigena [5]; Ruminococcus albus [7]; Erwinia carotovora [8]; Streptomyces purpurascens [9]; Streptomyces sp. [10]; Rhodotorula flava [12]; Corticium rolfsii [14]; Plant tissue degrading microorganisms [3]

Source tissue

Localisation in source
Extracellular [3–5, 7, 8, 12, 13, 15]

Purification
Scopolia japonica [1, 13]; Aspergillus niger [2, 11, 15]; Trichoderma reesei [3]; Clostridium acetobutylicum [4]; Monilinia fructigena [5]; Ruminococcus albus [7]; Erwinia carotovora [8]; Streptomyces purpurascens [9]; Streptomyces sp. [10]; Rhodotorula flava [12]; Corticium rolfsii [14]

Crystallization
[2, 15]

Cloned
–

Renaturated
–

Enzyme Handbook © Springer-Verlag Berlin Heidelberg 1991
Duplication, reproduction and storage in data banks are only
allowed with the prior permission of the publishers

5 STABILITY

pH
3.5 (not stable below) [1, 13]; 2.5–8.5 [2]; 4.0–8.0 [3]; 5.5–8.0 [4]; 3.0–8.0 [6];
5.0–11.0 [8]; 4.0–9.0 [10]; 2.0–5.0 [12]; 1.5–10.0 [14]

Temperature (°C)
50 (not stable above) [3]

Oxidation

Organic solvent

General stability information

Storage
1 month, 4°C, pH 5.5 [1, 13]; 3 months, 5°C [2]

6 CROSSREFERENCES TO STRUCTURE DATABANKS

PIR/MIPS code

Brookhaven code

7 LITERATURE REFERENCES

[1] Uchida, T., Tanaka, M.: Methods Enzymol., 160, 712–719 (1988)
[2] Tagawa, K., Kaji, A.: Methods Enzymol., 160, 707–712 (1988)
[3] Poutanen, K.: J. Biotechnol., 7, 271–282 (1988)
[4] Lee, S.F., Forsberg, C.W.: Can. J. Microbiol., 33, 1011–1016 (1987)
[5] Kelly, M.A., Sinnott, M.L., Herrchen, M.: Biochem. J., 245, 843–849 (1987)
[6] Whitacker, J.R.: Enzyme Microb. Technol., 6, 341–349 (1984) (Review)
[7] Greve, L.C., Labavitch, J.M., Hungate, R.E.: Appl. Environ. Microbiol., 47, 1135–1140 (1984)
[8] Kaji, A., Shimokawa, K.: Agric. Biol. Chem., 48, 67–72 (1984)
[9] Komae, K., Kaji, A., Sato, M.: Agric. Biol. Chem., 46, 1899–1905 (1982)
[10] Kaji, A., Sato, M., Tsutsui, Y.: Agric. Biol. Chem., 45, 925–931 (1981)
[11] Waibel, R., Amadò, R., Neukom, H.: J. Chromatogr., 197, 86–91 (1980)
[12] Uesaka, E., Sato, M., Raiju, M., Kaji, A.: J. Bacteriol., 133, 1073–1077 (1978)
[13] Tanaka, M., Uchida, T.: Biochim. Biophys. Acta, 522, 531–540 (1978)
[14] Kaji, A., Yoshihara, O.: Biochim. Biophys. Acta, 250, 367–371 (1971)
[15] Kaji, A., Tagawa, K.: Biochim. Biophys. Acta, 207, 456–464 (1970)

1 NOMENCLATURE

EC number
   3.2.1.56

Systematic name
   3-D-Glucuronosyl-N^2, 6-disulfo-beta-D-glucosamine glucuronohydrolase

Recommended name
   Glucuronosyl-disulfoglucosamine glucuronidase

Synonymes
   Glucuronidase, glucuronosyldisulfoglucosamine
   Glycuronidase [1–3]

CAS Reg. No.
   37288-42-9

---

2 REACTION AND SPECIFICITY

Catalysed reaction
   3-D-Glucuronosyl-N^2, 6-disulfo-beta-D-glucosamine + H_2O →
   → D-glucuronate + N^2, 6-disulfo-D-glucosamine

Reaction type
   O-Glycosyl bond hydrolysis

Natural substrates
   Disaccharide + H_2O (sulfated, degradation of heparin) [1, 2]

Substrate spectrum
   1 3-D-Glucuronosyl-N^2, 6-disulfo-beta-D-glucosamine + H_2O
   2 Disaccharide + H_2O (disulfated disaccharides, overview) [1]
   3 More (not: trisulfated disaccharide [2]) [2, 3]

Product spectrum
   1 N^2, 6-Disulfo-D-glucosamine + D-glucuronate
   2 Glucosamine 2 , 6-disulfate + uronic acid [1]
   3 ?

Inhibitor(s)

Cofactor(s)/prostethic group(s)

Metal compounds/salts

Enzyme Handbook © Springer-Verlag Berlin Heidelberg 1991
Duplication, reproduction and storage in data banks are only
allowed with the prior permission of the publishers

Turnover number (min^{-1})

Specific activity (U/mg)
  More (6.470 mg/h) [2]

K$_m$-value (mM)

pH-optimum
  6.5 [1]

pH-range
  5.5–8.0 [1]

Temperature optimum (°C)
  30 [1]

Temperature range (°C)
  5–45 [1]

---

3 ENZYME STRUCTURE

Molecular weight

Subunits

Glycoprotein/Lipoprotein
  –

---

4 ISOLATION/PREPARATION

Source organism
  Flavobacterium heparinum [1, 2]; Flavobacterium [3]

Source tissue
  Cell [1–3]

Localisation in source

Purification
  Flavobacterium heparinum [2]

Crystallization
  –

Cloned
  –

Renaturated
  –

5 STABILITY

pH

Temperature (°C)

Oxidation

Organic solvent

General stability information

Storage

6 CROSSREFERENCES TO STRUCTURE DATABANKS

PIR/MIPS code

Brookhaven code

7 LITERATURE REFERENCES

[1] Dietrich, C.P.: Biochemistry, 8, 2089–2094 (1969)
[2] Dietrich, C.P., Silva, M.E., Michelacci, Y.M.: J. Biol. Chem., 248, 6408–6415 (1973)
[3] Hovingh, P., Linker, A.: Biochem. J., 165, 287–293 (1977)

Enzyme Handbook © Springer-Verlag Berlin Heidelberg 1991
Duplication, reproduction and storage in data banks are only
allowed with the prior permission of the publishers

1 NOMENCLATURE

EC number
3.2.1.57

Systematic name
Pullulan 4-glucanohydrolase

Recommended name
Isopullulanase

Synonymes
Pullulanase, iso-
More (Arthrobacter globiformis T6, single enzyme with G2-dextranase, EC 3.2.1.94 and isopullulanase activity) [3]

CAS Reg. No.
37288-43-0

---

2 REACTION AND SPECIFICITY

Catalysed reaction
Hydrolysis of pullulan to isopanose (6-alpha-maltosylglucose)

Reaction type
O-Glycosyl bond hydrolysis

Natural substrates
Pullulan + H_2O

Substrate spectrum
1 4-Alpha-isomaltosylglucose + H_2O [1–4]
2 Pullulan + H_2O [1–4]
3 6^3-Alpha-glucosylmaltotriose + H_2O [1]
4 6^2-Alpha-maltosylmaltose + H_2O [1]
5 More (no action on starch [4], attacks reducing end of alpha-1, 4-glucosidic linkages adjacent to alpha-1, 6-glucosidic linkages in pullulan, 6^3-alpha-glucosylmaltotriose, 6^2-alpha-maltosylmaltose, panose [1]) [1, 4]

Product spectrum
1 Isomaltose + glucose
2 Isopanose + tetrasaccharide (small amount) [1–4]
3 Isomaltose + maltose [1]
4 Isopanose + glucose [1]
5 ?

Enzyme Handbook © Springer-Verlag Berlin Heidelberg 1991
Duplication, reproduction and storage in data banks are only
allowed with the prior permission of the publishers

Inhibitor(s)
   Ag^+ (partial, 5 mM) [3]; Hg^{2+} [3]; Fe^{3+} [3]; $KMnO_4$ [3]; N-Bromosuccinimide
   [3]; More (not: EDTA, p-chloromercuribenzoate) [2]

Cofactor(s)/prostethic group(s)

Metal compounds/salts

Turnover number (min^{-1})

Specific activity (U/mg)
   0.034 [3]

K_m-value (mM)

pH-optimum
   3.0–3.5 [1]; 3.5–4.0 [4]; 4.8 [2]

pH-range

Temperature optimum (°C)
   40 (at pH 3.5) [1]; 60 [2]; 30–40 [4]

Temperature range (°C)

---

3 ENZYME STRUCTURE

Molecular weight
   74000 (Aspergillus niger, gel filtration) [1]

Subunits

Glycoprotein/Lipoprotein
   –

---

4 ISOLATION/PREPARATION

Source organism
   Aspergillus niger (ATCC 9642 [4]) [1, 4]; Arthrobacter globiformis T6 (single
   enzyme with G2-dextranase EC 3.2.1.94 and isopullulanase activity [3]) [2,
   3]

Source tissue
   Cell (of wheat bran culture) [1]

Localisation in source

Purification
Aspergillus niger [1]; Arthrobacter [1]; Arthrobacter globiformis T6 [3]; More (single enzyme with G2-dextranase EC 3.2.1.94 and isopullulanase activity) [3]

Crystallization
–

Cloned
–

Renaturated
–

5 STABILITY

pH
4.0–7.0 (stable between) [4]; 2.0–8.0 (5° C, 24 hours) [1]; 2.3–6.0 (30° C, 24 hours: stable, loss of activity below 2.4 and above 6.0) [1]; 2.8–8.0 (4° C) [3]; 4.1–8.0 (45° C) [3]

Temperature (°C)
50 (30 minutes, pH 3.7–4.5: activity remains intact) [1]; 60 (30 minutes, stable up to) [2]; 50 (10 minutes, completely stable below) [3]; 70 (10 minutes, 80% loss of activity) [3]; 80 (10 minutes, completely inactivated) [3]

Oxidation
Photooxidation in presence of Rose Bengal [3]

Organic solvent

General stability information

Storage

6 CROSSREFERENCES TO STRUCTURE DATABANKS

PIR/MIPS code

Brookhaven code

Enzyme Handbook © Springer-Verlag Berlin Heidelberg 1991
Duplication, reproduction and storage in data banks are only
allowed with the prior permission of the publishers

7 LITERATURE REFERENCES

[1] Sakano, Y., Higuchi, M., Kobayashi, T.: Arch. Biochem. Biophys., 153, 180–187 (1972)

[2] Tago, M., Aoji, M., Sakano, Y., Kobayashi, T., Sawai, T.: Agric. Biol. Chem., 41, 909–910 (1977)

[3] Okada, G., Takayanagi, T., Miyahara, S., Sawai, T.: Agric. Biol. Chem., 52, 829–836 (1988)

[4] Sakano, Y., Masuda, N., Kobayashi, T.: Agric. Biol. Chem., 35, 971–973 (1971)

1 NOMENCLATURE

EC number
3.2.1.58

Systematic name
1, 3-Beta-D-glucan glucohydrolase

Recommended name
Glucan 1, 3-beta-glucosidase

Synonymes
Beta-1, 3-glucan exo-hydrolase [1]
Exo (1 --> 3)-beta-glucanase [6]
Exo-1, 3-beta-glucosidase [7]
Glucosidase, exo-1, 3-.beta.-
Exo-.beta.-(1.fwdarw.3)-D-glucanase
Exo-. beta.-1, 3-glucanase
Exo-1, 3-beta-glucosidase
Exo-1, 3-. beta.-glucanase
Exo-1, 3-.beta.-glucosidase
Exo-.beta-1, 3-glucanase
Exo-1, 3-.beta.-D-glucanase
Exo-beta-1, 3-glucanase [2]
Exo-beta-1, 3-D-glucanase [9]
Exo-beta-(1–3)-D-glucanase [18]
Exo-beta-(1 --> 3)-glucanohydrolase [28]

CAS Reg. No.
9073-49-8

---

2 REACTION AND SPECIFICITY

Catalysed reaction
Successive hydrolysis of beta-D-glucose units from the non-reducing end
of 1, 3-beta-D-glucans, releasing alpha-glucose

Reaction type
O-Glycosyl bond hydrolysis

Natural substrates
Beta-1, 3-glucan + H_2O
Laminarin + H_2O [3]
More (modification of the cell wall at the side of budding [3], involvement in
cell extension growth [14]) [3, 14]

Enzyme Handbook © Springer-Verlag Berlin Heidelberg 1991
Duplication, reproduction and storage in data banks are only
allowed with the prior permission of the publishers

Substrate spectrum

1 Beta-1, 3-glucan + H_2O [1–34]
2 Laminarin + H_2O [1, 3, 6, 8, 13, 17, 25, 26, 29, 32, 33, 34]
3 Laminarin + H_2O (periodate oxidized and reduced, unbranched) [7]
4 Laminaran + H_2O [19, 22]
5 Paramylon + H_2O [1]
6 Pachyman + H_2O [1, 19, 30, 33]
7 Oligoglucosides + H_2O (oligoglucosides from pachyman) [1]
8 Laminaritriose + H_2O [1, 6, 12]
9 Laminaritetraose + H_2O [1, 12, 33]
10 Laminaripentaose + H_2O [1, 12, 33]
11 Claviceps glucan + H_2O [1]
12 Yeast glucan + H_2O [1, 13, 33, 34]
13 Lichenin + H_2O (weak [1]) [1, 6]
14 Barley beta-glucan + H_2O (weak) [1]
15 Oat beta-glucan + H_2O (weak) [1]
16 p-Nitrophenyl-beta-glucoside + H_2O (weak [19]) [3, 10, 17, 19, 23, 32]
17 Glucan + H_2O (cell wall glucan, Sclerotinia sclerotiorum) [4]
18 Laminaribiose + H_2O (not [1, 12]) [6]
19 Cellodextrin + H_2O [6]
20 Luteose + H_2O [6]
21 Gentiobiose + H_2O [6]
22 Pustulan + H_2O [6, 17, 30, 32]
23 Curdlan + H_2O [22]
24 Laminarioligosaccharides + H_2O [12]
25 Salicin + H_2O [17]
26 Cellobiose + H_2O [17]
27 Schizophyllan + H_2O [22]
28 Laminaran + H_2O [24]
29 p-Nitrophenyl-beta-D-glucopyranoside + H_2O [25, 34]
30 More (intermediates: 3^2-O-beta-D-glucosyl-gentiobiose, 3^2-O-beta-gentiobiosyl-gentiobiose, 3^3-O-beta-D-glucosyl-gentiotriose, 3^4-O-beta-D-glucosyl-3^2-O-gentiobiosyl-gentiobiose, 3^3-O-beta-D-gentiobiosylgentiotriose [24], no transglycosylation from laminarin to glucose or reversion reaction with glucose alone observed [1], mixed linkage (1 --> 3, 1--> 4)-beta-glucans extensively degraded, beta-(1 --> 6)- and beta-(1 --> 4)-linked glucose polymers slowly, broad substrate specificity [6], single protein hydrolyses beta-(1 --> 6)-glucosidic and beta-(1 --> 3)-glucosidic linkages [8], beta beta-glucosidase activity (EC 3.2.1.21) [21]) [1, 6, 8, 21, 24, 26, 30]

Product spectrum

1 Alpha-glucose
2 More (different combinations and amounts of following products: glucose (main product), laminaribiose, gentiobiose, oligosaccharides, laminaritriose) [1, 3, 6, 8, 13, 17, 25, 26, 29, 32, 33, 34]
3 Glycerol + glycolaldehyde + D-glucose [7]
4 Glucose [19]
5 Glucose [1]
6 Glucose [1]
7 Glucose [1]
8 Glucose (only [6]) + laminaribiose [1, 6, 12]
9 Glucose + laminaribiose + laminaritriose (different combinations and amount of this products, depending on organism) [1, 6, 12]
10 Glucose + laminaribiose + laminaritriose + laminaritetraose (different combinations and amount of this products, depending on organism) [1, 5, 6]
11 Glucose + gentiobiose (+ laminaribiose [13]) [1, 13]
12 Glucose [1]
13 Glucose [1, 6]
14 Glucose [1, 6]
15 Glucose [1, 6]
16 Glucose + nitrophenol + glucan (incomplete degraded) [10]
17 Glucose + gentiobiose [4]
18 Glucose [6]
19 Glucose [6]
20 Glucose [6]
21 Glucose [6]
22 Glucose [8]
23 ?
24 Glucose + oligosaccharides (shorter chain) [12]
25 ?
26 Glucose
27 Glucose [22]
28 Glucose + gentiobiose + gentiotriose [24]
29 Nitrophenol + glucose
30 ?

Enzyme Handbook © Springer-Verlag Berlin Heidelberg 1991
Duplication, reproduction and storage in data banks are only
allowed with the prior permission of the publishers

Inhibitor(s)

$MnCl_2$ [31, 33]; $HgCl_2$ [6, 19, 22, 31]; $AgNO_3$ [6]; $FeCl_2$ [6]; $CuSO_4$ [6, 31, 33];
Ammonium molybdate [6]; Phenyl mercuric nitrate [6]; $CoCl_2$ [19]; $CuCl_2$
[19]; Pb^{2+} [22]; $FeSO_4 \times 7H_2O$ [22]; Hg^{2+} [29, 33]; N-Bromosuccinimide [6,
33]; Gluconic acid delta-lactone [6]; Methiolate [8]; Zn^{2+} [8, 33]; Hg^{2+} [8,
11, 12, 34]; Auxin (slight) [8]; Glucono-delta-lactone [8, 11, 28]; Ag^+ [12, 12,
34]; 5-Amino-1H-tetrazole (diazotized) [18]; Tetranitromethane [18];
D-Erythritol [28]; p-Chloromercuribenzoate [33]; Ethanol [33]; $CoSO_4$ [33];
$LiSO_4$ [33]; $Al_2(SO_4)_3$ [33]; $NiSO_4$ [33]; Na_2HAsO_4 [33]; Ag_2SO_4 [33]; Lactanes
[34]

Cofactor(s)/prostethic group(s)

Metal compounds/salts

$CaCl_2$ (enhances activity) [31]; $MgSO_4$ (enhances activity) [31]; $ZnSO_4$ (enhances activity) [31]; $CoCl_2$ (enhances activity) [31]

Turnover number (min^{-1})

Specific activity (U/mg)

24.2 [6]; 180 [8]; 0.73 (enzyme I) [10]; 12.622 [11]; 27.7 [22]; 2.5 (enzyme II)
[10]; 0.845 [19]; More [1, 12, 25, 31, 32, 33 , 34]

K_m-value (mM)

0.111 (laminaripentaose) [1]; 1.49 (laminaritriose) [12]; 0.06
(laminaritetraose) [12]; 0.28 (laminaripentaose) [12]; 0.36 (laminarihexaose)
[12]; 0. 85 (laminarin) [12]; 4.04 (p-nitrophenyl-beta-D-glucopyranoside)
[25]; 0.37 (pustulan) [32]; 5.0 (p-nitrophenyl-beta-D-glucoside) [32]; More
(laminaran: 0.080 mg/ml [22], laminarihexaose: 0.086 mg/ml [22],
laminaripentaose: 0.107 mg/ml [22], laminaritriose: 0.172 mg/ml [22],
laminaribiose: 4.06 mg/ml [22], no Michaelis-Menten kinetics, Hill coefficient: 1.6 [28], pustulan: 166 mg/ml [8], insoluble laminarin: 0.008 % [1],
lichenin: 2.22 mg/ml [6], laminarin: 1.22 mg/ml [6], 6.25 mg/ml [8], 1. 7
mg/ml [11]) [1, 6, 8, 11, 12, 14, 19, 22, 28, 30, 32, 33]

pH-optimum

5.5–6.0 [11]; 4.6 [12]; 5.7 [14]; 4.5 [19]; 4.5–5.0 [22, 26]; 4.7–5.2 [1]; 4.4–4.5
[6]; 5.0 [8, 34]; 4.75 (3 optima: 4. 75, 5.5, 6.5, p-nitrophenyl beta-glucoside)
[3]; 5.5 (3 optima: 4.75, 5.5, 6.5 , p-nitrophenyl beta-glucoside) [3]; 6.5 (3 optima: 4.75, 5.5, 6.5 , p-nitrophenyl beta-glucoside) [3]; 5.4 [28]; 6.0 [31, 33]

pH-range

3–7 (inactive below pH 3 and above pH 7) [1]; 3.0–6.5 (low activity at pH 3.0
and 6.5) [6]; 5.0–7.0 (below pH 5.0 and above pH 7.0: at least 70% reduction in activity) [11]; 3.5–7.5 (3.5: 40% of maximal activity, 7.5: less than 10%
of maximal activity) [22]; 4.0–7.5 (low activity at pH 4.0 and 7.5) [26]; 3.8–8.2
(3.8: 20% of maximal activity, 8.2 : 45% of maximal activity) [28]; 3.0–8.0
(40% of maximal activity at 3.0 and 8.0) [31]; More [33]

Temperature optimum (°C)
   53 [6]; 46 [12]; 45 [22]; 50 [11, 31, 33]; 45 [19]; 60 [28]; 37 [34]

Temperature range (°C)
   10–70 (10°C: less than 10% of maximal activity, 70 C: about 20 % of maximal activity) [25]

---

3 ENZYME STRUCTURE

Molecular weight
   100000 (about, Candida albicans, non-dissociating polyacrylamide gel
   electrophoresis) [11]
   315000 (gel filtration, Trichoderma harzianum) [12]
   82000 (disc gel electrophoresis, using varying polyacrylamide gel concentrations) [19]
   55000 (SDS-PAGE, Porodisculus pendulus) [22]
   59500 (gel filtration, Porodisculus pendulus) [22]
   58800 (Saccharomyces cerevisiae, native form) [25]
   43000 (Saccharomyces cerevisiae, underglycosylated form) [25]
   6000000 (self associated beta-glucanases from different yeasts) [27]
   61000 (SDS-PAGE, from different yeast) [27]
   68000 (SDS-PAGE, Strongylocentrotus, sea urchin) [28]
   60000 (Zea mays, gel filtration) [29]
   70000 (gel filtration, Trichoderma reesei) [30]
   82000 (polyacrylamide gel electrophoresis, denaturing conditions, Helix
   pomatia) [6]
   43000 (Schizosaccharomyces japonicus [8], Kluyveromyces aestuarii, gel
   filtration [32], fungi imperfecti, molecular sieve method [33]) [8, 32, 33]
   47000 (Saccharomyces cerevisiae, deglycosylated form, SDS-PAGE, 2
   polypeptides 47000 and 48000) [10]
   More (exo-beta-glucanases from yeast are aggregates of many functional
   units) [27]

Subunits
   Monomer (1 × 82000, Helix pomatia) [6]
   Monomer (1 × 43000 , gel filtration under denaturing conditions,
   Kluyveromyces aestuarii) [32]
   Dimer (1 × 63000, 1 × 44000, SDS-PAGE, Candida albicans) [11]
   More (exo-beta-glucanases from yeast are aggregates of many functional
   units) [27]

Glycoprotein/Lipoprotein
   Glycoprotein (Kluyveromyces aestuarrii: 24% carbohydrate) [32]

Enzyme Handbook © Springer-Verlag Berlin Heidelberg 1991
Duplication, reproduction and storage in data banks are only
allowed with the prior permission of the publishers

4 ISOLATION/PREPARATION

Source organism

Basidiomycete sp. QM 806 [7, 13, 18, 24]; Schizosaccharomyces japonicus [8]; Candida albicans [11, 16]; Trichoderma harzianum [12]; Artiplex littoralis (bifunctional enzyme, exo-1, 3-beta-D-glucanase and beta-glucosidase) [14]; Pinus pinaster [15]; Neurospora crassa [19]; Streptomyces sp. 1228 [29]; Gossypium hirsutum (enzyme has also beta-glucosidase activity) [21]; Porodisculus pendulus [22]; Valerianella olitoria [23]; Zea mays [26, 29]; Candida utilis [27, 34]; Geotrichum lactis [27]; Kluyveromyces phaeseolosporus [27]; Strongylocentrotus purpuratus (sea urchin) [28]; Trichoderma reesei [30]; Streptomyces murinus [31]; Kluyveromyces aestuarii [32]; Fungi imperfecti [33]; Euglena gracilis [1, 2]; Saccharomyces cerevisiae (several enzymes [3], 2 enzymes: I /II [10], underglycosylated form produced in presence of tunicamycin [25]) [3, 10, 17, 25, 27]; Coniothyrium minitans [4]; Trichoderma viride [5]; Helix pomatia [6]

Source tissue

Cell [2, 8, 17]; Culture fluid [4, 8, 34]; Culture medium [19, 17, 20]; Digestive juice [6]; Leaf [14]; Seedlings [26, 29]; Eggs [28]; Cotton fibres [21]; More (commercial cellulase preparation) [30]

Localisation in source

Extracellular [3, 10, 11, 20, 25]; Intracellular [3, 11, 32]; Cell wall [14, 15, 19, 23, 26]; Periplasm [11, 16, 17, 32]; Protoplast [17]; More [1, 5]

Purification

Euglena gracilis [1]; Helix pomatia [6]; Schizosaccharomyces japonicus [8]; Candida albicans [11]; Trichoderma harzianum [12]; Neurospora crassa [19]; Porodisculus pendulus [22]; Strongylocentrotus purpuratus (sea urchin) [28]; Zea mays (partial) [29]; Trichoderma reesei [30]; Streptomyces murinus [31]; Kluyveromyces aestuarii [32]; Fungi imperfecti [33]; Candida utilis [34]

Crystallization

–

Cloned

–

Renaturated

–

5 STABILITY

pH
 4.0–7.0 [19]; 5.5–10.5 [22]; 3.1 (75 minutes, glycosylated enzyme: 44% loss of activity, underglycosylated enzyme: 93% loss of activity) [25]; 4.0–6.0 [31]

Temperature (°C)
 45 (45 minutes, glycosylated enzyme: no loss of activity, underglycosylated enzyme: 35% loss of activity [25], 10 minutes, stable up to 45°C [19]) [19, 25]; 37 [3]; 50 (30 minutes, stable up to 50°C, rapidly inactivated at higher temperatures [6], 15 minutes, 82% loss of activity [29], stable up to 30 minutes [8], stable below [22]) [6, 8, 22, 29]; 60 (15 minutes, 100% loss of activity) [29]; 30 (stable below) [31]; More [33]

Oxidation

Organic solvent

General stability information
 Polyethylene glycol (stabilizes) [28]; Trypsin (90 minutes, glycosylated form, 48% loss of activity) [25]; Freeze-drying (no loss of activity [6], 50% loss of activity [8]) [6, 8]; Freezing and thawing (stable) [6, 8]; Dilution (in presence of human serum albumin, stable) [6]

Storage
 –20°C, 5 months (25% loss of activity) [1]; 4°C, 1 week (no loss of activity) [3]; –20°C, several months [3]; –15°C, 10 days (21% loss of activity) [30]; 1–4°C, pH 5.0–5.5, 0.05 M sodium succinate buffer (stable for at least 3 months) [8]; More [30]

---

6 CROSSREFERENCES TO STRUCTURE DATABANKS

PIR/MIPS code

Brookhaven code

Enzyme Handbook © Springer-Verlag Berlin Heidelberg 1991
Duplication, reproduction and storage in data banks are only
allowed with the prior permission of the publishers

7 LITERATURE REFERENCES

[1] Barras, D.R., Stone, B.A.: Biochim. Biophys. Acta, 191 , 342–353 (1969)
[2] Barras, D.R., Stone, B.A.: Biochim. Biophys. Acta, 191 , 329–341 (1969)
[3] Cortat, M., Matile, P., Wiemken, A.: Arch. Mikrobiol., 82, 189–205 (1972)
[4] Jones, D., Gordon, A.H., Bacon, J.S.D.: Biochem. J., 140, 47–55 (1974)
[5] Dabbagh, R., Conant, N.F., Burns, R.O.: J. Gen. Microbiol., 85, 190–202 (1974)
[6] Marshall, J.J., Grand, R.J.A.: Arch. Biochem. Biophys. , 167, 165–175 (1975)
[7] Nelson, T.E.: Biochim. Biophys. Acta, 377, 139–145 (1975)
[8] Fleet, G.H., Phaff, H.J.: Biochim. Biophys. Acta, 401, 318–332 (1975)
[9] Nebreda, A.R., Villa, T.G., Villanueva, J.R., Del Rey, F.: Gene, 47, 245–259 (1986)
[10] Ramirez, M., Hernandez, L.M., Larriba, G.: Arch. Microbiol., 151, 391–389 (1989)
[11] Molina, M., Cenamor, R., Sanchez, M., Nobela, C.: J. Gen. Microbiol., 135, 309–314
     (1989)
[12] Kitamoto, Y., Kono, R., Shimotori, A., Mori, N., Ichikawa, Y.: Agric. Biol. Chem., 51
     (12) , 3385–3386 (1987)
[13] Ryan, E.M., Ward, O.P.: Biotechnol. Lett., 9, 405–410 (1987)
[14] Boucaud, J., Bigot, J., Devaux, J.: J. Plant Physiol., 128, 337–349 (1987)
[15] Llamazares, J., Acebes, J.L., Zarra, I.: J. Plant Physiol., 127, 11–22 (1987)
[16] Molina, M., Cenamor, R., Nobela, C.: J. Gen. Microbiol., 133, 609–617 (1987)
[17] Cenamor, R., Molina, M., Galdona, J., Sanchez, M., Nombela, C.: J. Gen. Microbiol.,
     133, 619–628 (1987)
[18] Jeffcoat, R., Kirkwood, S.: J. Biol. Chem., 262, 1088–1091 (1987)
[19] Hiura, N., Kobayashi, M., Nakajima, T., Matsuda, K.: Agric. Biol. Chem., 50,
     2461–2467 (1986)
[20] Bielecki, S., Antczak, T., Galas, E. in "Eur. Congr. Biotechnol.", 3rd Ed., 2, 489–496
     (1984)
[21] Bucheli, P., Dürr, M., Buchala, A.J., Meier, H.: Planta, 166, 530–536 (1985)
[22] Iwamuro, Y., Aoki, M., Mikami, Y.: J. Ferment. Technol., 63, 405–409 (1985)
[23] Lienart, Y., Barnoud, F.: Planta, 165, 68–75 (1985)
[24] Nanjo, F., Usui, T., Suzuki, T.: Agric. Biol. Chem., 48, 1523–1532 (1984)
[25] Sanchez, A., Villanueva, J.R., Villa, T.G.: J. Gen. Microbiol., 128, 3051–3060 (1982)
[26] Huber, D.J., Nevins, D.J.: Planta, 155, 467–472 (1982)
[27] Sanchez, A., Nebreda, A.R., Villa, T.G.: FEBS Lett., 145, 213–216 (1982)
[28] Talbot, C.F., Vacquier, V.D.: J. Biol. Chem., 257, 742–746 (1982)
[29] Huber, D.J., Nevins, D.J.: Planta, 151, 206–214 (1981)
[30] Bamforth, C.W.: Biochem. J., 191, 863–866 (1980)
[31] Yokotsuka, K., Kakehashi, Y., Oyama, S., Kato, A., Akatsuka, S., Yoshitake, S.,
     Kushida, T.: J. Ferment. Technol., 56, 599–605 (1978)
[32] Lachance, M.-A., Villa, T.G., Phaff, H.J.: Can. J. Biochem., 55, 1001–1006 (1977)
[33] Nagasaki, S., Saito, K., Yamamoto, S.: Agric. Biol. Chem., 41, 493–502 (1977)
[34] Villa, T.G., Notario, V., Benitez, T., Villanueva, J. R.: Can. J. Biochem., 54, 927–934
     (1976)

1 NOMENCLATURE

EC number
3.2.1.59

Systematic name
1, 3(1, 3;1, 4)-Alpha-D-glucan 3-glucanohydrolase

Recommended name
Glucan endo-1, 3-alpha-glucosidase

Synonymes
Glucanase, endo-1, 3-.alpha.-
Mutanase
Endo-(1 --> 3)-alpha-glucanase [5]
Endo-1, 3-alpha-glucanase
Cariogenase
Cariogenanase
Endo-1, 3-.alpha.-glucanase
Endo-1, 3-alpha-D-glucanase [2]
Mutanase (Novo) [3]

CAS Reg. No.
9075-84-7

2 REACTION AND SPECIFICITY

Catalysed reaction
Endohydrolysis of 1, 3-alpha-D-glucosidic linkages in isolichenin, pseudo-nigeran and nigeran

Reaction type
O-Glycosyl bond hydrolysis

Natural substrates
Pseudonigeran + H_2O
Nigeran + H_2O
Isolichenin + H_2O
Glucan + H_2O

Enzyme Handbook © Springer-Verlag Berlin Heidelberg 1991
Duplication, reproduction and storage in data banks are only
allowed with the prior permission of the publishers

Substrate spectrum
1 Pseudonigeran + H_2O (not [5]) [7]
2 Nigeran + H_2O
3 Isolichenin + H_2O
4 Limit glucan + H_2O [2]
5 Glucan + H_2O (glucan produced from sucrose by Streptococcus mutans [4], (1 --> 3)-alpha-glucan [5, 7, 9]) [4, 5, 7, 9]
6 Glucan + H_2O (insoluble, sticky glucan of Streptococcus mutans) [8]
7 Lentinus alpha-1, 3-glucan + H_2O [7]
8 Nigerotriose + H_2O [7]
9 Nigerotetraose + H_2O [7]
10 Nigeropentaose + H_2O [7]
11 Mutan + H_2O [7]
12 More (not: nigerose [7]) [5, 7, 8, 9]

Product spectrum
1 Nigerose + alpha-D-glucose
2 ?
3 ?
4 ?
5 Nigerose + glucose [5 , 7]
6 Isomaltose + nigerose + nigerotriose + oligosaccharides [8]
7 ?
8 Glucose + nigerose [7]
9 Glucose + nigerose [7]
10 Glucose + nigerose [7]
11 ?
12 ?

Inhibitor(s)
Ag^+ [2, 5, 8, 9]; Fe^{3+} [2]; Fe^{2+} (no effect [4]) [3, 9]; Hg^{2+} [2, 4, 5, 7, 8]; Pb^{2+} [2, 5, 8]; Zn^{2+} [2, 5, 9]; EDTA (no effect [4], stimulates [5]) [2]; p-Chloromercuribenzoic acid [2]; Mn^{2+} [4, 5, 7]; Cu^{2+} [4, 5]; Ag^{2+} [4]; Merthiolate [4]; Iodoacetamide [5]

Cofactor(s)/prostethic group(s)

Metal compounds/salts
Co^{2+} (increases activity [2], no effect [5]) [2]; Ca^{2+} (increases activity [2], no effect [4, 5]) [2]; More (no requirement of divalent metal) [5]

Turnover number (min⁻¹)

Specific activity (U/mg)
78.1 [2]; 0.968 [4]; 11.2 [5]; 7.52 [5]; 0.831 [8]; More [9]

K$_m$-value (mM)
80.0 (limit glucan) [2]; 0.46 (pseudonigeran) [9]; 0.89 (glucan) [4]; More
[5, 7]

pH-optimum
5.0 [2]; 5.5–6.0 [4]; 7.5–8.5 [5]; 5.4 [7]; 6.3 [8]; 4.5 [9]

pH-range
4.5–8.5 (4.5: 40% of optimal activity, 8.5: 70% of optimal activity) [8]; 4.5–8.0
(sharp decrease below 4.5, gradual decrease up to 8.0) [2]; 3–7 (3: 30% of
optimum activity, 7: 20% of optimum activity) [7]; 3–8 [9]

Temperature optimum (°C)
56 [2]; 55 [4]; 60 [7]; 42 [8]; 50 [9]

Temperature range (°C)
30–80 (30°C: 18% of optimum activity, 80°C: 25% of optimum activity) [7]

---

3 ENZYME STRUCTURE

Molecular weight
67400 (gel filtration, Pseudomonas, subunit ?) [2]
279000 (gel filtration, Pseudomonas, aggregated form) [2]
68000 (SDS-PAGE, Streptomyces chartreusis) [4]
134000 (SDS-PAGE, Bacillus circulans) [5]
78000 (SDS-PAGE, Streptomyces sp.) [6]
68000 (SDS-PAGE, Flavobacterium) [8]
47000 (gel filtration, Trichoderma viride) [9]

Subunits
Oligomer (possibility that native enzyme exists in aggregated form or is on
oligomer, molecular weight larger than 300000, x × 68000, SDS-PAGE,
Streptomyces chartreusis) [4]

Glycoprotein/Lipoprotein
More (carbohydrate not detected) [5]

---

4 ISOLATION/PREPARATION

Source organism
Bacteroides oralis [1]; Flavobacterium [8]; Trichoderma viride [9]; Pseudomonas NRRL-B-12324 [2]; Streptomyces chartreusis [4]; Bacillus circulans
[5]; Streptomyces sp. [6, 7]

Source tissue
Cell [1]; Culture supernatant [2, 8]; Culture filtrate [5, 9]; Culture medium [6]

Enzyme Handbook © Springer-Verlag Berlin Heidelberg 1991
Duplication, reproduction and storage in data banks are only
allowed with the prior permission of the publishers

Localisation in source
Extracellular [5]; Cytoplasm (only) [1]

Purification
Pseudomonas NRRL-B-12324 [2]; Streptomyces chartreusis [4]; Bacillus circulans [5]; Streptomyces sp. [6, 7]; Trichoderma viride [9]; Flavobacterium [8]

Crystallization
–

Cloned
–

Renaturated
–

---

5 STABILITY

pH
5.25–10.5 (30°C, stable for at least 1.5 hours) [5]; 3.8 (50% loss of activity after 10 minutes) [5]; 4.2 (50% loss of activity after 30 minutes) [5]

Temperature (°C)
60 (complete loss of activity after 30 minutes) [8]; 50 (rapid inactivation) [9]; 60 (loss of activity at or above) [2]; 37 (pH 5.5, 3 days, 3% loss of activity) [2]; 65 (pH 6.0, 10 minutes, complete inactivation) [4]; 40 (5% loss of activity after 90 minutes) [5]; 45 (25% loss of activity after 90 minutes) [5]; 50 (complete loss of activity after 20 minutes) [5]; 50 (stable up to 10 minutes) [7]

Oxidation

Organic solvent

General stability information
Freezing (–15°C, appreciable loss in activity) [5]

Storage
4°C, 18 months (17% loss of activity) [2]; 4°C, 2 years, pH 6 (stable) [5]

---

6 CROSSREFERENCES TO STRUCTURE DATABANKS

PIR/MIPS code

Brookhaven code

7 LITERATURE REFERENCES

[1] Takahshi, N., Satoh, Y., Takamori, K.: J. Gen. Microbiol., 131, 1077–1082 (1985)

[2] Simonson, L.G., Gaugler, R.W., Lamberts, B.L., Reiher, D.A.: Biochim. Biophys. Acta, 715, 189–195 (1982)

[3] Stephen, E.R., Nasim, A.: Can. J. Microbiol., 27, 550–553 (1981)

[4] Takehara, T., Inoue, M., Morioka, T., Yokogawa, K.: J. Bacteriol., 145, 729–735 (1981)

[5] Meyer, M.T., Phaff, H.J.: J. Gen. Microbiol., 118, 197–208 (1980)

[6] Imai, K., Kikuta, T., Kobayashi, M., Matsuda, K.: Agric. Biol. Chem., 41, 1339–1346 (1977)

[7] Imai, K., Kobayashi, M., Matsuda, K.: Agric. Biol. Chem. , 41, 1889–1895 (1977)

[8] Ebisu, S., Kato, K., Kotani, S., Misaki, A.: J. Bacteriol., 124, 1489–1501 (1975)

[9] Hasegawa, S., Nordin, J.H.: J. Biol. Chem., 244, 5460–5470 (1969)

Enzyme Handbook © Springer-Verlag Berlin Heidelberg 1991
Duplication, reproduction and storage in data banks are only
allowed with the prior permission of the publishers

1 NOMENCLATURE

EC number
3.2.1.60

Systematic name
1, 4-Alpha-D-glucan maltotetraohydrolase

Recommended name
Glucan 1, 4-alpha-maltotetraohydrolase

Synonymes
Exo-malto tetraohydrolase
Maltotetraohydrolase, exo-
Exomaltotetraohydrolase
Exo-maltotetraohydrolase [2]

CAS Reg. No.
37288-44-1

2 REACTION AND SPECIFICITY

Catalysed reaction
Hydrolysis of 1, 4-alpha-D-glucosidic linkages in amylaceous polysaccharides so as to remove succesive maltotetraose residues from the non-reducing chain ends

Reaction type
O-Glycosyl bond hydrolysis

Natural substrates
Polysaccharide + H_2O (amylaceous polysaccharides)

Substrate spectrum
1 Starch + H_2O [1, 3, 4, 5]
2 Amylose + H_2O [4, 5]
3 Amylopectin + H_2O [6]
4 Glycogen + H_2O [4]
5 Maltohexaose + H_2O [6]
6 Maltoheptaose + H_2O [6]
7 More (not: cyclomaltodextrins [4]) [4, 6]

Enzyme Handbook © Springer-Verlag Berlin Heidelberg 1991
Duplication, reproduction and storage in data banks are only
allowed with the prior permission of the publishers

Product spectrum
1 Maltotetraose [1]
2 Maltotetraose [4]
3 Maltotetraose [6]
4 Maltotetraose [4, 6]
5 Maltotetraose + maltose dimer [6]
6 Maltotetraose + maltotriose [6]
7 ?

Inhibitor(s)
N-Bromosuccinimide [4]; Cyclomaltodextrins [4]; Zn^{2+} [4]; Ni^{2+} [4]; Cu^{2+} [4]; Hg^{2+} [4]; Fe^{3+} [4]; Fe^{2+} [4]; Co^{2+} [4]; More (slight inhibition by several other metals) [4]

Cofactor(s)/prostethic group(s)

Metal compounds/salts

Turnover number (min^{-1})

Specific activity (U/mg)
102 [4]; 2500 [6]; 287 [5]

K_m-value (mM)
0.32 (amylose, F-1/F-2) [4]; 5.9 (soluble starch, F-1) [4]; 5.7 (soluble starch, F-2) [4]; 5.9 (shellfish glycogen, F-1/F-2) [4]; 7.1 (oyster glycogen, F-1) [4]; 6.7 (oyster glycogen, F-2) [4]

pH-optimum
7.8–8.2 [4]; 8 [6]

pH-range
4.5–10.5 (more than 95% of maximum activity at 7 and 9.5, 4.5: 40% of maximal activity, 10.5: 82% of maximal activity) [6]

Temperature optimum (°C)
55 [3]; 45 [4]; 47 [6]

Temperature range (°C)
25–65 [4]; 15–60 (15°C: 20% of maximal activity, 60°C: 30% of maximal activity) [6]

3 ENZYME STRUCTURE

Molecular weight
55000 (estimated by SDS-PAGE, Pseudomonas stutzeri) [4]
12500 (subunit, gel filtration, Pseudomonas stutzeri) [6]

Subunits
Oligomer (dimers, tetramers, hexamers, octamers or decamers, concentration dependent) [6]

Glycoprotein/Lipoprotein
–

4 ISOLATION/PREPARATION

Source organism
Pseudomonas stutzeri (2 forms, F-1/F-2 [4]) [1, 3, 4, 5, 6]; Pseudomonas saccharophila (gene expression in E. coli) [2]

Source tissue
Culture supernatant [4, 6]

Localisation in source
Extracellular [5, 6]

Purification
Pseudomonas stutzeri (affinity chromatography [5]) [4, 5, 6]

Crystallization
–

Cloned
(Pseudomonas saccharophila gene in E. coli) [2]

Renaturated
–

5 STABILITY

pH
5.5–11.0 (30°C, 60 minutes, presence of $CaCl_2$, stable) [4]; 6.5–8.5 (absence of Ca^{2+}, 30°C, 60 minutes stable) [4]

Enzyme Handbook © Springer-Verlag Berlin Heidelberg 1991
Duplication, reproduction and storage in data banks are only
allowed with the prior permission of the publishers

Temperature (°C)
55 (pH 8.0, 60 minutes, absence of Ca^{2+}, complete loss of activity) [4];More (Ca^{2+} improves thermostability) [1]; 20–40 (1 hour, stable, rapid loss above 40) [6]; 55 (1 hour, complete loss of activity) [6]; 40 (stable below, pH 8.0, 60 minutes, absence of Ca^{2+}) [4]; 60 (pH 8.0, 60 minutes, presence of Ca^{2+}, complete loss of activity) [4]

Oxidation

Organic solvent

General stability information
Substrate (20–30%, stabilizes immobilized enzyme) [1]; Ca^{2+} (stabilizes immobilized enzyme [1], stabilizes [4]) [1, 4]; Dialysis (4°C, prolonged periods, complete loss of activity) [6]

Storage

---

6 CROSSREFERENCES TO STRUCTURE DATABANKS

PIR/MIPS code
A32803 (precursor, Pseudomonas stutzeri); A35138 (Pseudomonas stutzeri, fragment); S05667 (precursor, Pseudomonas saccharophila)

Brookhaven code

---

7 LITERATURE REFERENCES

[1] Kimura, T., Ogata, M., Yoshida, M., Nakakuki, T.: Biotechnol. Bioeng., 33, 845–855 (1989)
[2] Zhou, J., Takano, T., Kobayashi, S.: Agric. Biol. Chem., 53, 301–302 (1989)
[3] Kimura, T., Ogata, M., Yoshida, M., Nakakuki, T.: Biotechnol. Bioeng., 32, 669–676 (1988)
[4] Sakano, Y., Kashiyama, E., Kobayashi, T.: Agric. Biol. Chem., 47, 1761–1768 (1983)
[5] Dellweg, H., John, M., Schmidt, J.: Eur. J. Appl. Microbiol. Biotechnol., 1, 191–198 (1975)
[6] Robyt, J.F., Ackerman, R.J.: Arch. Biochem. Biophys., 145, 105–114 (1971)

1 NOMENCLATURE

EC number
3.2.1.61

Systematic name
1, 3–1, 4-Alpha-D-glucan 4-glucanohydrolase

Recommended name
Mycodextranase

Synonymes
Dextranase, myco-

CAS Reg. No.
9047-04-5

---

2 REACTION AND SPECIFICITY

Catalysed reaction
Endohydrolysis of 1, 4-alpha-D-glucosidic linkages in alpha-D-glucans containing both 1, 3- and 1, 4-bonds

Reaction type
O-Glycosyl bond hydrolysis
More (endo-multichain mechanism) [1]

Natural substrates
Nigeran + H_2O [1, 3]

Substrate spectrum
1 Alpha-D-glucan + H_2O (containing both 1, 3- and 1, 4-bonds)
2 Nigeran + H_2O (or oligosaccharides of nigeran) [1, 5]
3 More (no hydrolysis of alpha-D-glucans containing only 1, 3- or 1,4-bonds)

Product spectrum
1 Nigerose + 4-alpha-D-nigerosylglucose
2 Nigerose + 4-alpha-D-glucopyranosyl-(1 --> 3)-D-glucopyranosyl-(1 --> 4)-O-alpha-D-gluco-pyranosyl-(1 --> 3)-D-glucose [1, 5]
3 More (no production of glucose) [1]

Enzyme Handbook © Springer-Verlag Berlin Heidelberg 1991
Duplication, reproduction and storage in data banks are only
allowed with the prior permission of the publishers

Inhibitor(s)
Ag$^+$ [1]; Hg^{2+} [1]; N-Chlorosuccinimide [1]; N-Bromosuccinimide [1];
Nigeran trisaccharides [1]; Urea [1]; Nigerose [1]; Maltose [1]; Maltotriose
[1]; Nigeran tetrasaccharides [1]

Cofactor(s)/prostethic group(s)

Metal compounds/salts

Turnover number (min^{-1})

Specific activity (U/mg)
27 [1]

K$_m$-value (mM)

pH-optimum
4–5 [1]

pH-range
2–9 [1]

Temperature optimum (°C)
70 [1]

Temperature range (°C)
23–80 [1]

---

3 ENZYME STRUCTURE

Molecular weight
40000 (sedimentation equilibrium, Penicillium melinii QM 1931, both forms)
[1]

Subunits
Monomer (guanidine hydrochloride treatment) [1]

Glycoprotein/Lipoprotein
Glycoprotein (fast form: 16% carbohydrate, slow form: 15% carbohydrate,
only mannose and glucose [1, 2]) [1, 2, 4]; More (chemical constitution of
carbohydrate portion) [2]

---

4 ISOLATION/PREPARATION

Source organism
Penicillium melinii (QM 1931, 2 forms: fast and slow electrophoretic form) [1,
2, 4, 5]; Aspergillus niger [3]; Aspergillus awamori [3]

Source tissue
Culture medium [1]

Localisation in source
Extracellular [1, 4]

Purification
Penicillium melinii (QM 1031, 2 forms fast and slow electrophoretic form) [1]

Crystallization
–

Cloned
–

Renaturated
–

---

5 STABILITY

pH

Temperature (°C)

Oxidation

Organic solvent

General stability information

Storage
4° C, 2 months (no loss of activity) [1]

---

6 CROSSREFERENCES TO STRUCTURE DATABANKS

PIR/MIPS code

Brookhaven code

---

7 LITERATURE REFERENCES

[1] Tung, K.K., Rosenthal, A., Nordin, J.H.: J. Biol. Chem., 246, 6722–6736 (1971)
[2] Rosenthal, A.L., Nordin, J.H.: J. Biol. Chem., 256, 5295–5303 (1975)
[3] Marchessault, R.H., Revol, J.-F., Bobbitt, F., Nordin, J.H.: Biopolymers, 19, 1069–1080 (1980)
[4] Samuel, O., Nordin, J.H.: Biochem. Biophys. Res. Commun. , 45, 1376–1383 (1971)
[5] Tung, K.K., Nordin, J.H.: Biochim. Biophys. Acta, 158, 154–156 (1968)

Enzyme Handbook © Springer-Verlag Berlin Heidelberg 1991
Duplication, reproduction and storage in data banks are only
allowed with the prior permission of the publishers

1 NOMENCLATURE

EC number
3.2.1.62

Systematic name
Glycosyl-N-acylsphingosine glycohydrolase

Recommended name
Glycosylceramidase

Synonymes
Ceramidase, glycosyl-
Glycosyl ceramide glycosylhydrolase
Phlorizin hydrolase
Glucosidase, phloretin
Phloretin-glucosidase
Phlorizin hydrolase
Cerebrosidase
Phloridzin glucosidase
Phloridzin.beta.-glucosidase
Glycosylceramidase
Phloretin glucosidase
Lactase-phlorizin hydrolase [1–5, 9–13]
Lactase/phlorizin hydrolase [6, 7]

CAS Reg. No.
9033-10-7

2 REACTION AND SPECIFICITY

Catalysed reaction
Glycosyl-N-acylsphingosine + $H_2O \rightarrow$
$\rightarrow$ a sugar + N-acylsphingosine

Reaction type
O-Glycosyl bond hydrolysis

Natural substrates
Glucosylceramides + H_2O (component of fat globules in milk) [12]
Lactosylceramides + H_2O (component of fat globules in milk) [12]

Enzyme Handbook © Springer-Verlag Berlin Heidelberg 1991
Duplication, reproduction and storage in data banks are only
allowed with the prior permission of the publishers

Substrate spectrum

1 Glycosyl-N-acylsphingosine + H_2O
2 Phlorizin + H_2O
3 Methylumbelliferyl beta-D-galactoside + H_2O [8]
4 Methylumbelliferyl beta-D-glucoside + H_2O [8]
5 Methylumbelliferyl beta-D-xyloside + H_2O [8]
6 Galactosylceramide + H_2O [8]
7 Glucosylceramide + H_2O [8]
8 Lactosylceramide + H_2O [8]
9 Galactosylsphingosine + H_2O [8]
10 Glucosylsphingosine + H_2O [8]
11 More (phlorizin is the only substrate hydrolyzed by the phlorizin hydrolase active site [11], lactase-phlorizin hydrolase EC 3.2.1.23/3.2.1.62 contains both the lactase and the phlorizin hydrolase activity in a single polypeptide chain [1], lactase/phlorizin hydrolase is composed of 2 monomers [6], broad specificity, the intestinal enzyme is a complex which also catalyzed the reaction of E.C. 3.2.1.108) [1–6, 10, 11, 15]

Product spectrum

1 Sugar + N-acylsphingosine
2 Phloretin + glucose
3 Methylumbelliferone + galactose
4 Methylumbelliferone + glucose
5 Methylumbelliferone + xylose
6 Galactose + ceramide
7 Glucose + ceramide
8 Lactose + ceramide
9 Galactose + sphingosine
10 Glucose + sphingosine
11 ?

Inhibitor(s)

Pyruvic acid [13]; Ag^+ [15]; Hg^{2+} [15]; More (tartaric acid stimulated phlorizin hydrolase activity at pH 3.3 is inhibited by a number of organic acids and by anions like SO_4^{2-} and Cl^- [13]) [13, 14]

Cofactor(s)/prostethic group(s)

More (active towards glycolipid substrates in presence of sodium taurodeoxycholate) [8]

Metal compounds/salts

Turnover number (min^{-1})

2

Specific activity (U/mg)
   More [6, 7, 8, 10]

K_m-value (mM)
   More [8, 10, 14]; 0.44 (phlorizin) [6]; 0.24 (galactosylsphingosine) [8]; 0.37
   (glucosylsphingosine) [8]; 0.065 (galactosylceramide) [8]; 0.010 (glucosylceramide) [8]; 0.27 (lactosylceramide) [8]; 28 (lactose) [8]; 67 (lactulose) [8];
   0.73 (N-palmitoyl dihydrocerebroside) [9]

pH-optimum
   5.5–5.7 (disaccharides or artificial substrates) [8]; 6.0 [9]; 5.2–5.6 (absence
   of tartaric acid) [13]; 3.3–3.6 (presence of tartaric acid) [13]; 5–6 [14]; 3.5
   (brush-border [15]) [14, 15]; 5.5 (lysosomal) [15]

pH-range
   More [12]; 2.5–7.5 [15]

Temperature optimum (°C)
   37 (assay at) [10, 15]

Temperature range (°C)

---

3 ENZYME STRUCTURE

Molecular weight
   22000 (SDS-PAGE, rat, lactase-phlorizin hydrolase) [2]
   160000 (pig, SDS-PAGE, human) [5]
   320000 (human, amphiphilic form, gel filtration [6], gel filtration, pig,
   lactase-phlorizin hydrolase [7]) [6, 7]
   130000 (SDS-PAGE, mouse) [8]
   300000 (gel filtration, mouse) [8]

Subunits
   Dimer (lactase/phlorizin hydrolase: composed of 2 monomers, 2 × 160000,
   human, SDS-PAGE [6], pig, SDS-PAGE, 2 × 160000 [7]) [6, 7]

Glycoprotein/Lipoprotein
   Glycoprotein [1]

---

4 ISOLATION/PREPARATION

Source organism
   Human [1, 3, 5, 6]; Rabbit [1]; Rat (2 forms [14]) [2, 9, 11, 12 , 14]; Pig [4, 7];
   Mouse [8]; Monkey [10, 13]; Hamster (2 forms: brush-border and lysosomal)
   [15]

Enzyme Handbook © Springer-Verlag Berlin Heidelberg 1991
Duplication, reproduction and storage in data banks are only
allowed with the prior permission of the publishers

Source tissue
Small intestine [11–15]

Localisation in source
Brush-border membrane [1, 6, 7, 11]; Microvillus membrane [2, 4]; More (ER-Golgi fractions) [2]; Cytoplasm [15]; Lysosomes [15]

Purification
Human (lactase-phlorizin hydrolase) [6]; Pig (lactase-phlorizin hydrolase) [7]; Mouse [8]; Rat (lactase-phlorizin hydrolase) [9]; Monkey (lactase-phlorizin hydrolase) [10]

Crystallization
—

Cloned
—

Renaturated
—

---

5 STABILITY

pH

Temperature (°C)
49 (40% loss of activity after 60 minutes) [7]; 45 (60 minutes, stable) [9]; More [12]

Oxidation

Organic solvent

General stability information

Storage

---

6 CROSSREFERENCES TO STRUCTURE DATABANKS

PIR/MIPS code

Brookhaven code

7 LITERATURE REFERENCES

[1] Mantei, N., Villa, M., Enzler, T., Wacker, H., Boll, W., James, P., Hunziker, W., Semenza, G.: EMBO J., 7, 2075–2713 (1988)

[2] Büller, H.A., Montgomery, R.K., Sasak, W.V., Grand, R. J.: J. Biol. Chem., 262, 17206–17211 (1987)

[3] Naim, H.Y., Sterchi, E.E., Lentze, M.J.: Biochem. J., 241, 427–434 (1987)

[4] Danielsen, E.M., Skovbjerg, H., Noren, O., Sjöström, H.: Biochem. Biophys. Res. Commun., 122, 82–90 (1984)

[5] Skovbjerg, H., Danielsen, E.M., Noren, O., Sjöström, H.: Biochim. Biophys. Acta, 798, 247–251 (1984)

[6] Skovbjerg, H., Sjöström, H., Noren, O.: Eur. J. Biochem., 114, 653–661 (1981)

[7] Skovbjerg, H., Noren, O., Sjöström, H., Danielsen, E. M., Enevoldsen, B.S.: Biochim. Biophys. Acta, 707, 89–97 (1982)

[8] Kobayashi, T., Suzuki, K.: J. Biol. Chem., 256, 7768–7773 (1981)

[9] Cousineau, J., Green, J.R.: Biochim. Biophys. Acta, 615, 147–157 (1980)

[10] Ramaswamy, S., Radhakrishnan, A.N.: Biochim. Biophys. Acta, 403, 446–455 (1975)

[11] Birkenmeier, E., Alpers, D.H.: Biochim. Biophys. Acta, 350, 100–112 (1974)

[12] Leese, H.J., Semenza, G.: J. Biol. Chem., 248, 8170–8173 (1973)

[13] Ramaswamy, S., Radhakrishnan, A.N.: Biochem. Biophys. Res. Commun., 54, 197–204 (1973)

[14] Kraml, J., Kolinska, J., Ellederova, D., Hirsova, D.: Biochim. Biophys. Acta, 258, 520–530 (1972)

[15] Malathi, P., Crane, R.K.: Biochim. Biophys. Acta, 173, 245–256 (1969)

Enzyme Handbook © Springer-Verlag Berlin Heidelberg 1991
Duplication, reproduction and storage in data banks are only
allowed with the prior permission of the publishers

1 NOMENCLATURE

EC number
3.2.1.63

Systematic name
2-Alpha-L-fucopyranosyl-beta-D-galactoside fucohydrolase

Recommended name
1, 2-Alpha-L-fucosidase

Synonymes
Fucosidase, 1, 2-.alpha.-L-
1, 2-.alpha.-L-Fucosidase
Alpha-(1 -- > 2)-L-fucosidase

CAS Reg. No.
37288-45-2

---

2 REACTION AND SPECIFICITY

Catalysed reaction
Methyl-2-alpha-L-fucopyranosyl-beta-D-galactoside + H_2O →
→ L-fucose + methyl beta-D-galactoside (high specificity for non-reducing terminal L-fucose residues linked to D-galactose residues by 1,2-alpha-linkage)

Reaction type
O-Glycosyl bond hydrolysis

Natural substrates
Methyl-2-alpha-L-fucopyranosyl-beta-D-galactoside + H_2O

Substrate spectrum
1 2-O-Alpha-L-fucopyranosyl-D-galactose + H_2O [1]
2 2-O-Alpha-L-fucosyllactose + H_2O [1]
3 Lacto-N-fucopentaose I + H_2O [1]
4 Mucins + H_2O (porcine and canine, submaxillary [1], porcine sub-maxillary and glycopeptide of desialyzed mucin [5]) [1, 5]
5 2-Fucosyllactitol + H_2O [2]
6 2-Fucosyllactose + H_2O [2]
7 Glycoprotein + H_2O (hog, submaxillary glycoprotein) [4]

Enzyme Handbook © Springer-Verlag Berlin Heidelberg 1991
Duplication, reproduction and storage in data banks are only
allowed with the prior permission of the publishers

8 More (not: p-nitrophenyl-alpha-L-fucoside, 2-O-, 3-O- and
   4-O-alpha-L-fucopyranosylfucoses, 3-O- and
   4-O-alpha-L-fucopyranosyl-D-galactoses, action on oligosaccharides
   and glycoproteins, not: simple methyl nitrophenyl fucosides [4], highly
   specific for non-reducing terminal L-fucose residues linked to
   D-galactose residues by 1, 2-alpha linkage [1, 2]) [1, 2, 4, 5]
9 Lacto-N-fucopentaose I + H_2O [5]
10 2'-Fucosyllactose + H_2O [5]

Product spectrum

1 L-Fucose + galactose [1]
2 L-Fucose + lactose [1]
3 L-Fucose + ?
4 Fucose [1]
5 Fucose + ?
6 Fucose + ?
7 ?
8 ?
9 ?
10 ?

Inhibitor(s)

$MgCl_2$ [2]; Mn^{2+} [5]; Fe^{2+} [4]; Fe^{3+} [4]; Hg^{2+} [4, 5]; Zn^{2+} [4]; Cu^{2+} [4];
p-Chloromercuribenzoate [4]; Iodoacetamide [4]; EDTA [4, 5]; SDS [5];
Deoxycholate [5]

Cofactor(s)/prostethic group(s)

Metal compounds/salts

Turnover number (min⁻¹)

Specific activity (U/mg)

6.7 [1]; 0.00186 [2]; 70 [4]; 0.12 [5]

K_m-value (mM)

0.083 (methyl 2-O-alpha-L-fucopyranosyl-beta-D-galactoside) [1]; 0.675
(2-fucosyllactitol) [2]; 0.33 (porcine submaxillary mucin) [5]; 1.6 (glycopeptide desialyzed porcine submaxillary mucin) [5]; 2.0 (lacto-N-fucopentaose
I) [5]; 2.5 (2'-fucosyllactose) [5]; 0.175 (glycoprotein, hog, submaxillary) [5]

pH-optimum

3.8 (Aspergillus niger) [1]; 5–6.5 (almond) [2]; 6.0 (Clostridium perfringens)
[4]

pH-range

2.4–7.5 [1]; 4.5–8.0 [4]

Temperature optimum (°C)
  37 (assay at) [1]

Temperature range (°C)

---

3 ENZYME STRUCTURE

Molecular weight
  200000 (greater than, multiple isozymic forms, Clostridium perfringens) [4]
  70000–80000 (gel filtration, Bacillus fulminans) [5]

Subunits

Glycoprotein/Lipoprotein
  –

---

4 ISOLATION/PREPARATION

Source organism
  Aspergillus niger [1]; Almond [2]; Bifidobacterium bifidum (strain: VIII-210)
  [3]; Bifidobacterium infantis (strain: VIII-240) [3]; Ruminococcus torquens
  (strains: VIII-239 and IX-70) [3]; Ruminococcus AB (strain: VI-268) [3];
  Clostridium perfringens [4]; Bacillus fulminans [5]

Source tissue
  Emulsin [2]; More (commercial preparation) [1]

Localisation in source

Purification
  Aspergillus niger [1]; Almond (partial) [2]; Bacillus fulminans [5];
  Clostridium perfringens [4]

Crystallization
  –

Cloned
  –

Renaturated
  –

Enzyme Handbook © Springer-Verlag Berlin Heidelberg 1991
Duplication, reproduction and storage in data banks are only
allowed with the prior permission of the publishers

5 STABILITY

pH

Temperature (°C)
  50 (5 minutes, no loss of activity) [5]; 60 (5 minutes, 7% loss of activity) [5];
  70 (5 minutes, 77% loss of activity) [5]; 80 (5 minutes, complete loss of activity) [5]

Oxidation

Organic solvent

General stability information
  Repeated freezing and thawing (inactivation) [4]; Dialysis (water: rapid loss
  of activity, buffers: precipitation with considerable inactivation below pH
  5.5) [4]; Lyophilization (inactivation) [4]

Storage

6 CROSSREFERENCES TO STRUCTURE DATABANKS

PIR/MIPS code

Brookhaven code

7 LITERATURE REFERENCES

[1] Bahl, O.P.: J. Biol. Chem., 245 (2) , 299–304 (1970)
[2] Ogata-Arakawa, M., Muramatsu, T., Kobata, A.: Arch. Biochem. Biophys., 181,
    353–358 (1977)
[3] Larson, G., Falk, P., Hoskin, L.C.: J. Biol. Chem., 263, 10790–10798 (1988)
[4] Aminoff, D., Frurukawa, K.: J. Biol. Chem., 245, 1659–1669 (1970)
[5] Kochibe, N.: J. Biochem., 74, 1141–1149 (1973)

1 NOMENCLATURE

EC number
3.2.1.64

Systematic name
2, 6-Beta-D-fructan 6-beta-D-fructofuranosyl fructohydrolase

Recommended name
2, 6-Beta-fructan 6-levanbiohydrolase

Synonymes
Levanbiohydrolase, 2, 6-.beta.-fructan 6-
2, 6-.beta.-Fructan 6-levanbiohydrolase
Beta 2, 6-fructan-6-levanbiohydrolase [1]

CAS Reg. No.
37288-46-3

2 REACTION AND SPECIFICITY

Catalysed reaction
Hydrolysis of 2, 6-beta-D-fructan so as to remove successive levanbiose residues from the end of the chain

Reaction type
O-Glycosyl bond hydrolysis

Natural substrates
2, 6-Beta-D-fructan + H_2O

Substrate spectrum
1 2, 6-Beta-D-fructan + H_2O [1]

Product spectrum
1 Levanbiose (6-beta-fructosylfructose) [1]

Inhibitor(s)

Cofactor(s)/prostethic group(s)

Metal compounds/salts

Turnover number (min^{-1})

Enzyme Handbook © Springer-Verlag Berlin Heidelberg 1991
Duplication, reproduction and storage in data banks are only
allowed with the prior permission of the publishers

Specific activity (U/mg)

K_m-value (mM)

pH-optimum

pH-range

Temperature optimum (°C)

Temperature range (°C)

---

3 ENZYME STRUCTURE

Molecular weight

Subunits

Glycoprotein/Lipoprotein

--

---

4 ISOLATION/PREPARATION

Source organism
  Pseudomonas sp. (strain 11) [1]

Source tissue
  Culture supernatant [1]

Localisation in source
  Extracellular [1]

Purification

Crystallization
  --

Cloned
  --

Renaturated
  --

---

5 STABILITY

pH

Temperature (°C)

Oxidation

Organic solvent

General stability information

Storage

---

6 CROSSREFERENCES TO STRUCTURE DATABANKS

PIR/MIPS code

Brookhaven code

---

7 LITERATURE REFERENCES

[1] Avigad, G., Zelikson, R.: Bull. Res. Counc. Isr., Vol.11 A4, 253–257 (1963)

1 NOMENCLATURE

EC number
3.2.1.65

Systematic name
2, 6-Beta-D-fructan fructanohydrolase

Recommended name
Levanase

Synonymes
Levan hydrolase

CAS Reg. No.
9041-11-6

---

2 REACTION AND SPECIFICITY

Catalysed reaction
Random hydrolysis of 2, 6-beta-D-fructofuranosidic linkages in
2,6-beta-D-fructans (levans) containing more than 3 fructose units

Reaction type
O-Glycosyl bond hydrolysis

Natural substrates
Levans + H_2O [1–8]

Substrate spectrum
1 Levans + H_2O [6, 8]
2 Levans + H_2O [7]

Product spectrum
1 Oligofructosides [6, 8]
2 Fructose [7]

Inhibitor(s)
Mn^{2+} [3]; Ba^{2+} [3]; Fe^{3+} [3]; Zn^{2+} [3]; Hg^{2+} [3, 7, 8]; EDTA [3]; Ag^{2+} [7];
Cu^{2+} [7, 8]; p-Chloromercuribenzoate [7]; Pb^{2+} [8]; Ag^+ [8]; Tris buffer [8]

Cofactor(s)/prostethic group(s)

Metal compounds/salts

Enzyme Handbook © Springer-Verlag Berlin Heidelberg 1991
Duplication, reproduction and storage in data banks are only
allowed with the prior permission of the publishers

Turnover number (min⁻¹)

Specific activity (U/mg)
851 [3]; 810 [7]; 0.5 [8]

K_m-value (mM)
0.5 (levan) [8]; 10 (tetraose) [8]

pH-optimum
6.0 (levan) [3]; 6.5 (levan) [7] 5.5–6.5 (levan) [8]

pH-range
4.0 (not active below, levan) [3]

Temperature optimum (°C)
45 (levan) [3]; 40–50 (levan) [7]

Temperature range (°C)
55 (not active above, levan) [3, 7]; 60 (not active above, levan) [8]

---

3 ENZYME STRUCTURE

Molecular weight
73000–74000 (SDS-PAGE, Bacillus subtilis) [1, 4]
89000 (SDS-PAGE, Actinomyces viscosus) [3]
75866 (DNA sequence analysis, Bacillus subtilis) [4]
100000 (SDS-PAGE, Streptococcus salivarius) [7]

Subunits
Monomer (SDS-PAGE, Actinomyces viscosus [3], Streptococcus salivarius [7]) [3, 7]

Glycoprotein/Lipoprotein
–

---

4 ISOLATION/PREPARATION

Source organism
Actinomyces viscosus [3]; Streptococcus salivarius [7]; Arthrobacter sp. [8]; Yeasts [6]

Source tissue

Localisation in source
Extracellular [1, 3–5]

Purification
Actinomyces viscosus [3]; Streptococcus salivarius [7]; Arthrobacter sp. [8]

Crystallization
−

Cloned
[1, 2, 4, 5]

Renaturated
−

---

5 STABILITY

pH
4.6–7.5 [8]

Temperature (°C)

Oxidation

Organic solvent
Ethanol (90 % ethanol, 1 day)

General stability information

Storage
Several months, −15 °C [8]

---

6 CROSSREFERENCES TO STRUCTURE DATABANKS

PIR/MIPS code
A27286 (Bacillus subtilis); S06353 (precursor, Bacillus subtilis)

Brookhaven code

---

7 LITERATURE REFERENCES

[1] Schoergendorfer, K., Schwab, H., Lafferty, R.M.: J. Biotechnol., 7, 247–258 (1988)
[2] Schoergendorfer, K., Schwab, H., Lafferty, R.M.: Nucleic Acids Res., 15, 9606 (1987)
[3] Igarashi, T., Takahashi, M., Yamamoto, A., Etoh, Y., Takamori, K.: Infect. Immun., 55, 3001–3005 (1987)
[4] Martin, I., Débarbouillé, M., Ferrari, E., Klier, A., Rapoport, G.: Mol. Gen. Genet., 208, 177–184 (1987)
[5] Friehs, K., Schoergendorfer, K., Schwab, H., Lafferty, R.M.: J. Biotechnol., 3, 333–341 (1986)
[6] Fuchs, A., De Bruijn, J.M., Niedeveld, C.J.: Antonie Leeuwenhoek, 51, 333–351 (1985)
[7] Takahashi, N., Mizuno, F., Takamori, K.: Infect. Immun., 42, 231–236 (1983)
[8] Avigad, G., Bauer, S.: Methods Enzymol., 8, 621–628 (1966)

---

1 NOMENCLATURE

EC number
3.2.1.66

Systematic name
Quercitrin 3-rhamnohydrolase

Recommended name
Quercitrinase

Synonymes

CAS Reg. No.
37288-47-4

---

2 REACTION AND SPECIFICITY

Catalysed reaction
Quercitrin + $H_2O \rightarrow$
$\rightarrow$ rhamnose + quercetin

Reaction type
O-Glycosyl bond hydrolysis

Natural substrates
Quercitrin + H_2O (carbon source) [1]

Substrate spectrum
1 Quercitrin + H_2O (quercitrin is identical with quercetin 3-rhamnoside) [1]
2 Myricitrin + H_2O [1]
3 Robinin + H_2O [1]

Product spectrum
1 Rhamnose + quercetin [1]
2 Rhamnose + quercetin [1]
3 Rhamnose + quercetin [1]

Inhibitor(s)

Cofactor(s)/prostethic group(s)

Metal compounds/salts

Enzyme Handbook © Springer-Verlag Berlin Heidelberg 1991
Duplication, reproduction and storage in data banks are only
allowed with the prior permission of the publishers

Turnover number (min^{-1})

Specific activity (U/mg)

K$_m$-value (mM)

pH-optimum
  6.4 [1]

pH-range
  4.5–8.0 [1]

Temperature optimum (°C)

Temperature range (°C)

---

3 ENZYME STRUCTURE

Molecular weight

Subunits

Glycoprotein/Lipoprotein
  −

---

4 ISOLATION/PREPARATION

Source organism
  Aspergillus flavus (synthesized by only a few members of the Aspergillus flavus group) [1]

Source tissue
  Mycelium [1]; Culture medium [1]

Localisation in source
  Extracellular (78%) [1]

Purification

Crystallization
  −

Cloned
  −

Renaturated
  −

5 STABILITY

pH

Temperature (°C)
35 (stable above) [1]

Oxidation

Organic solvent

General stability information

Storage
−17°C, several weeks (little loss of activity) [1]

6 CROSSREFERENCES TO STRUCTURE DATABANKS

PIR/MIPS code

Brookhaven code

7 LITERATURE REFERENCES

[1] Westlake, D.W.S.: Can. J. Microbiol., 9, 211–220 (1963)

Enzyme Handbook © Springer-Verlag Berlin Heidelberg 1991
Duplication, reproduction and storage in data banks are only
allowed with the prior permission of the publishers

1 NOMENCLATURE

EC number
3.2.1.67

Systematic name
Poly(1, 4-alpha-D-galacturonide)galacturonohydrolase

Recommended name
Galacturan 1, 4-alpha-galacturonidase

Synonymes
Exopolygalacturonase
Poly(galacturonate)hydrolase
Exo-Polygalacturonase
Exo-D-galacturonase
Exo-D-galacturonanase
Exopoly-D-galacturonase

CAS Reg. No.
9045-35-6

---

2 REACTION AND SPECIFICITY

Catalysed reaction
$(1, 4\text{-Alpha-D-galacturonide})_n + H_2O \rightarrow$
$\rightarrow (1, 4\text{-alpha-D-galacturonide})_{n-1} + \text{D-galacturonate}$

Reaction type
O-Glycosyl bond hydrolysis

Natural substrates
Pectic acid $+ H_2O$ [1–17]

Substrate spectrum
1 D-Galacturonans $+ H_2O$ [1–17]

Product spectrum
1 D-Galacturonate [1–17]

Inhibitor(s)
Citrate [1, 4, 11, 15]; EDTA [1, 4, 11, 15]; Hg^{2+} [2, 4, 9]; Cu^{2+} [2, 9]; Ba^{2+} [2, 9, 15]; Mn^{2+} [2, 9]; Cd^{2+} [4]; Na^+ [4]; NaCl [14]

Cofactor(s)/prostethic group(s)

Enzyme Handbook © Springer-Verlag Berlin Heidelberg 1991
Duplication, reproduction and storage in data banks are only
allowed with the prior permission of the publishers

Metal compounds/salts
  Ca^{2+} [1, 4, 11, 14, 15]; Cd^{2+} [1]; Sr^{2+} [4, 11, 14, 15]

Turnover number (min^{-1})

Specific activity (U/mg)
  16.0 [1]; 0.08–0.35 [2, 9]; 0.59 [4]; 3.05 [16]; 28.6 [17]

K_m-value (mM)
  0.016–0.120 (polygalacturonate) [2, 14]; 0.012–0.015 (pectate) [14];
  0.074–0.159 (octagalacturonate, heptagalacturonate) [14]; 0.63–0.97
  (trigalacturonate) [14]; 2.2–2.8 (digalacturonate) [14]

pH-optimum
  5.3 (polygalacturonate) [1]; 3.6–4.8 (polygalacturonate) [2, 9]; 4.5–5.5
  (polygalacturonate) [4, 14]; 5.1 (pectate) [5, 12]; 5.6 (pectate) [6]; 4.4
  (digalacturonate, tetragalacturonate) [10]; 4.6 (hexagalacturonate, heptagalacturonate) [10]; 4.9 (polygalacturonate) [10]; 4.5–5.0 (polygalacturonate) [11]; 5.2 (digalacturonate) [13]; 5.5 (polygalacturonate) [15];
  6.0–6.5 (digalacturonate) [17]

pH-range
  4.0–7.5 (polygalacturonate) [1]

Temperature optimum (°C)
  50 (pectate) [5]; 50 (digalacturonate) [10]; 60–70 (pectate) [12]; 45 (pectate)
  [13]

Temperature range (°C)
  60 (not active above, digalacturonate) [10]

---

3 ENZYME STRUCTURE

Molecular weight
  51000 (gel filtration, corn) [1]
  48000 (gel filtration, carrot) [2]
  76000 (gel filtration, liverwort) [2, 9]
  47000 (gel filtration, tomato) [4]
  58000 (gel filtration, apple) [11]
  68000 (gel filtration, peach) [15]

Subunits

Glycoprotein/Lipoprotein
  –

2

4 ISOLATION/PREPARATION

Source organism
Corn [1]; Carrot [2, 12, 14]; Tomato [4]; Butyrivibrio fibrisolvens [6];
Trichoderma reesei [7]; Penicillium oxalicum [8]; Banana [10]; Apple [11];
Aspergillus niger [13]; Peach [15]; Collectotrichum lindemuthianum [16];
Bacillus sp. [17]; Liverwort [2, 9]

Source tissue
Pollen [1]; Carrot root [2, 12, 14]; Tomato fruit [4]; Banana peel [10]; Apple
cortical tissue (partially) [11]

Localisation in source
Intracellular [6]; Extracellular [13, 16]

Purification
Corn pollen (partially) [1]; Carrot root [2, 12, 14]; Liverwort [2, 9]; Tomato
fruit [4]; Butyrivibrio fibrisolvens [6]; Trichoderma reesei [7]; Penicillium
oxalicum [8]; Banana peel [10]; Apple cortical tissue (partially) [11];
Aspergillus niger (partially) [13]; Peach [15]; Collectotrichum
lindemuthianum [16]; Bacillus sp. [17]

Crystallization
–

Cloned
–

Renaturated
–

---

5 STABILITY

pH
3.5–6.5 [2, 9]; 5.2 [13]

Temperature (°C)
55 (not stable above) [2, 9]; 30 (not stable above) [5]; 50 (not stable above)
[10]; 40 (not stable above) [13]

Oxidation

Organic solvent

General stability information

Storage
2 months, –20°C [2, 9]

Enzyme Handbook © Springer-Verlag Berlin Heidelberg 1991
Duplication, reproduction and storage in data banks are only
allowed with the prior permission of the publishers

6 CROSSREFERENCES TO STRUCTURE DATABANKS

PIR/MIPS code

Brookhaven code

7 LITERATURE REFERENCES

[1] Pressey, R., Reger, B.J.: Plant Sci., 59, 57–62 (1989)
[2] Konno, H.: Methods Enzymol., 161, 373–380 (1988)
[3] Ikotun, T., Balogun, O.: J. Basic Microbiol., 27, 347–354 (1987)
[4] Pressey, R.: Phytochemistry, 26, 1867–1870 (1987)
[5] Heinrichova, K., Lehoczki, M., Zliechovcova, D.: Collect. Czech. Chem. Commun., 51, 2291–2298 (1986)
[6] Heinrichova, K., Wojciechowicz, M., Ziolecki, A.: J. Gen. Microbiol., 131, 2053–2058 (1985)
[7] Markovic, O., Slezarik, A., Labudova, I.: FEMS Microbiol. Lett., 27, 267–271 (1985)
[8] Ikotun, T.: Z. Allg. Mikrobiol., 24, 247–252 (1984)
[9] Konno, H., Yamasaki, Y., Katoh, K.: Plant Physiol., 73, 216–222 (1983)
[10] Heinrichova, K.: Collect. Czech. Chem. Commun., 47, 2433–2439 (1982)
[11] Bartley, I.M.: Phytochemistry, 17, 213–216 (1978)
[12] Heinrichova, K.: Collect. Czech. Chem. Commun., 42, 3214–3221 (1977)
[13] Heinrichova, K., Rexova-Benkova, L.: Biochim. Biophys. Acta, 422, 349–356 (1976)
[14] Pressey, R., Avants, J.K.: Phytochemistry, 14, 957–961 (1975)
[15] Pressey, R., Avants, J.K.: Plant Physiol., 52, 252–256 (1973)
[16] Keegstra, K., English, P.D., Albersheim, P.: Phytochemistry, 11, 1873–1880 (1972)
[17] Hasegawa, S., Nagel, C.W.: Arch. Biochem. Biophys., 124, 513–520 (1968)

1 NOMENCLATURE

EC number
3.2.1.68

Systematic name
Glycogen 6-glucanohydrolase

Recommended name
Isoamylase

Synonymes
Debranching enzyme

CAS Reg. No.
9067-73-6

2 REACTION AND SPECIFICITY

Catalysed reaction
Hydrolysis of 1,6-alpha-glucosidic linkages in glycogen, amylopectin and their beta-limit dextrins

Reaction type
O-Glycosyl bond hydrolysis

Natural substrates
Starch + H_2O [1–8]

Substrate spectrum
1 Glycogen, amylopectin or their beta-limit dextrins + H_2O (r) [1–8]
2 More (distinguished from alpha-dextrin endo-1,6-alpha-glucosidase (EC 3.2.1.41) by the inability of isoamylase to attack pullulan, and by limited action on alpha-limit dextrins, action on glycogen however is complete in contrast to limited action by alpha-dextrin glucanohydrolase. 1,6-linkage hydrolysed only at branch point)

Product spectrum
1 Maltose + maltooligosaccharides [1–8]
2 ?

Enzyme Handbook © Springer-Verlag Berlin Heidelberg 1991
Duplication, reproduction and storage in data banks are only
allowed with the prior permission of the publishers

Inhibitor(s)
p-Chloromercuribenzoate [2, 8]; N-Bromosuccinimide [6];
2-Hydroxy-5-nitrobenzyl bromide [6]; Iodine [6]; Iodoacetate [6, 8]; Phenyl-
mercuriacetate [6]; 2, 4-Dinitrofluorobenzene [6]; Hg^{2+} [6]; Ag^{2+} [6];
Oligosaccharides with alpha-1, 4-glucosidic linkages [6]

Cofactor(s)/prostethic group(s)

Metal compounds/salts

Turnover number (min^{-1})

Specific activity (U/mg)
8.0 [2]; 190 [5]

K$_m$-value (mM)

pH-optimum
5.5–6.0 (amylopectin) [2]; 5.6 (starch) [3]; 3.0–4.0 (starch) [8]

pH-range
3.5 (not active below, amylopectin) [2]; 1.0–8.0 (starch) [8]

Temperature optimum (°C)
50 (amylopectin) [2]; 30 (starch) [3]; 52 (starch) [8]

Temperature range (°C)
60 (not active above, amylopectin) [2]; 70 (not active above, starch) [8]

---

3 ENZYME STRUCTURE

Molecular weight
80807 (DNA sequence analysis, Pseudomonas amyloderamosa) [1]
520000 (gel filtration, potato tubers) [2]
65000 (gel filtration, Lipomyces kononenkoae) [3]
86000–94000 (gel filtration, gel electrophoresis, Pseudomonas
amyloderamosa) [4]
94000 (sedimentation equilibrium, Pseudomonas sp.) [6]
95000 (gel electrophoresis, Pseudomonas sp.) [8]

Subunits
Monomer (gel electrophoresis) [4]
Dimer (2 × 46000–52000, sedimentation equilibrium, gel filtration) [6]

Glycoprotein/Lipoprotein
–

2

4 ISOLATION/PREPARATION

Source organism
Potato [2]; Lipomyces kononenkoae [3]; Pseudomonas amyloderamosa [5]; Pseudomonas sp. [8]; Plants [2, 7]; Bacteria [2, 3, 7]; Yeasts [3, 7]

Source tissue
Tuber [2]

Localisation in source
Extracellular [3]

Purification
Potato tubers [2]; Lipomyces kononenkoae (partially) [3]; Pseudomonas amyloderamosa [5]; Pseudomonas sp. [8]

Crystallization
[4, 6]

Cloned
[1]

Renaturated
–

5 STABILITY

pH
2.5–7.5 [8]

Temperature (°C)
45 (not stable above) [2, 8]

Oxidation

Organic solvent

General stability information

Storage
Several months, 4°C [8]

6 CROSSREFERENCES TO STRUCTURE DATABANKS

PIR/MIPS code
A28109 (precursor, Pseudomonas sp.)

Brookhaven code

Enzyme Handbook © Springer-Verlag Berlin Heidelberg 1991
Duplication, reproduction and storage in data banks are only
allowed with the prior permission of the publishers

7 LITERATURE REFERENCES

[1] Amemura, A., Chakraborty, R., Fujita, M., Noumi, T., Masamitsu, F.: J. Biol. Chem., 263, 9271–9275 (1988)

[2] Ishizaki, Y., Taniguchi, H., Maruyama, Y., Nakamura, Y.: Agric. Biol. Chem., 47, 771–779 (1983)

[3] Spencer-Martins, I.: Appl. Environ. Microbiol., 44, 1253–1257 (1982)

[4] Amemura, A., Konishi, Y., Harada, T.: Biochim. Biophys. Acta, 611, 390–393 (1980)

[5] Kato, K., Konishi, Y., Amemura, A., Harada, T.: Agric. Biol. Chem., 41, 2077–2080 (1977)

[6] Kitagawa, H., Amemura, A., Harada, T.: Agric. Biol. Chem., 39, 989–994 (1975)

[7] Lee, E.Y.C., Whelan, W.J. in "The Enzymes", 3rd. Ed. (Boyer, P. D., Ed.) 5, 191–234 (1972) (Review)

[8] Yokobayashi, K., Misaki, A., Harada, T.: Biochim. Biophys. Acta, 212, 458–469 (1970)

1 NOMENCLATURE

EC number
3.2.1.70

Systematic name
1, 6-Alpha-D-glucan glucohydrolase

Recommended name
Glucan 1, 6-alpha-glucosidase

Synonymes
Exo-1, 6-alpha-glucosidase
Glucodextranase
Dextran glucosidase
Alpha-1, 6-glucosidase
Exodextranase

CAS Reg. No.
37288-48-5

---

2 REACTION AND SPECIFICITY

Catalysed reaction
Hydrolysis of successive glucose residues from 1, 6-alpha-D-glucans and derived oligosaccharides

Reaction type
O-Glycosyl bond hydrolysis

Natural substrates
Dextrans + H_2O [1–11]

Substrate spectrum
1 Dextrans + H_2O (ir) [1–11]
2 Isomaltodextrins + H_2O (ir) [1–11]
3 Isomaltose + H_2O (very slowly)
4 More (beta-D-glucose formed by inversion)

Product spectrum
1 Beta-D-glucose [1–11]
2 Beta-D-glucose [1–11]
3 ?
4 ?

Enzyme Handbook © Springer-Verlag Berlin Heidelberg 1991
Duplication, reproduction and storage in data banks are only
allowed with the prior permission of the publishers

Inhibitor(s)
Pb^{2+} [1, 2, 7]; Zn^{2+} [1, 2, 7]; Cu^{2+} [1, 2, 7, 10]; Hg^{2+} [1, 2, 10]; Fe^{3+} [1, 2];
$KMnO_4$ [1, 2]; N-Bromosuccinimide [1, 2, 10]; Phenyl alpha-D-glucoside [2];
Methyl-alpha-D-glucoside [2]; Glucose [4]; EDTA [7]; Iodine [10];
p-Chloromercuribenzoate [10]

Cofactor(s)/prostethic group(s)

Metal compounds/salts

Turnover number (min^{-1})
5714 [2]

Specific activity (U/mg)
25.6 [1, 2]; 50.0 [4]; 34.3–38.4 [8, 10]

K_m-value (mM)
0.012 (dextran) [4]

pH-optimum
6.0 [1, 2, 9]; 4.5–5.0 [4]; 7.0–7.5; 4.0–4.4 [11]

pH-range
3.0–8.0 [4, 11]

Temperature optimum (°C)
45 [1, 2, 4]

Temperature range (°C)
60 (not active above) [4]

---

3 ENZYME STRUCTURE

Molecular weight
120000 (gel electrophoresis, Arthrobacter globiformis) [2]
29000 (gel filtration, Lipomyces lipofer) [4]
70000 (gel filtration, Arthrobacter globiformis) [6]

Subunits
Monomer (gel electrophoresis, also active as glucoamylase) [1, 4, 6]

Glycoprotein/Lipoprotein
–

4 ISOLATION/PREPARATION

Source organism
Fungi [4]; Arthrobacter globiformis [1, 2, 6, 7, 9]; Lipomyces lipofer [4];
Streptococcus mitis [8]; Brevibacterium fuscum var. dextranlyticum [10];
Achromobacter sp. [11]

Source tissue

Localisation in source
Extracellular [1, 2, 4, 6, 11]; Cytoplasm [3, 5]

Purification
Arthrobacter globiformis [1, 2, 6, 7, 9]; Lipomyces lipofer [4]; Streptococcus
mitis [8]; Brevibacterium fuscum var. dextranlyticum [10]; Achromobacter
sp. (partially) [11]

Crystallization
[6]

Cloned
–

Renaturated
–

---

5 STABILITY

pH
5.5–7.5 [2]; 4.5–6.0 [4]; 5.0–11.0 [10]

Temperature (°C)
55 (not stable above) [2, 4, 10]

Oxidation

Organic solvent

General stability information

Storage

---

6 CROSSREFERENCES TO STRUCTURE DATABANKS

PIR/MIPS code

Brookhaven code

Enzyme Handbook © Springer-Verlag Berlin Heidelberg 1991
Duplication, reproduction and storage in data banks are only
allowed with the prior permission of the publishers

7 LITERATURE REFERENCES

[1] Okada, G., Unno, T.: Agric. Biol. Chem., 53, 223–228 (1989)
[2] Okada, G., Unno, T., Sawai, T.: Agric. Biol. Chem., 52, 2169–2176 (1988)
[3] Takahashi, N., Satoh, Y., Takamori, K.: J. Gen. Microbiol., 131, 1077–1082 (1985)
[4] Ramos, A., Spencer-Martins, I.: Antonie Leeuwenhoek, 49, 183–190 (1983)
[5] Linder, L., Andersson, C., Sund, M.L., Shockman, G.D.: Infect. Immun., 40, 1146–1154 (1983)
[6] Ohya, T., Suzuki, H., Sawai, T.: Agric. Biol. Chem., 44, 203–204 (1980)
[7] Ohya, T., Sawai, T., Uemura, S., Abe, K.: Agric. Biol. Chem., 42, 571–577 (1978)
[8] Wheatley, M.A., Moo-Young, M.: Biotechnol. Bioeng., 19, 219–233 (1977)
[9] Sawai, T., Yamaki, T., Ohya, T.: Agric. Biol. Chem., 40, 1293–1299 (1976)
[10] Sugiura, M., Ito, A., Yamaguchi, T.: Biochim. Biophys. Acta, 350, 61–70 (1974)
[11] Sawai, T., Toriyama, K., Yano, K.: J. Biochem., 75, 105–112 (1974)

1 NOMENCLATURE

EC number
3.2.1.71

Systematic name
1, 2-Beta-D-glucan glucanohydrolase

Recommended name
Glucan endo-1, 2-beta-glucosidase

Synonymes
Glucanase, endo-1, 2-.beta.
Endo-1, 2-.beta.-glucanase
Endo-1, 2-beta-glucanase
.beta.-D-1, 2-Glucanase
Endo-(1 -> 2)-beta-D-glucanase [2, 3]

CAS Reg. No.
37288-49-6

---

2 REACTION AND SPECIFICITY

Catalysed reaction
Random hydrolysis of 1, 2-glucosidic linkages in 1, 2-beta-D-glucans

Reaction type
O-Glycosyl bond hydrolysis

Natural substrates
Beta-D-1, 2-glucan + H_2O

Substrate spectrum
1 Beta-D-1, 2-glucan + H_2O [1]
2 Cyclic (1 --> 2)-beta-D-glucan + H_2O [2]
3 Cyclic (1 --> 2)-beta-D-glucan + H_2O [3]
4 More (not: laminaran, curdlan, carboxymethyl-cellulose) [2]

Product spectrum
1 Oligosaccharides (beta-1, 2-linked, including sophorose) [1]
2 Sophorose (mainly) [2]
3 Oligosaccharides (beta (1 --> 2) linked) + glucose (small amount) [3]
4 ?

Enzyme Handbook © Springer-Verlag Berlin Heidelberg 1991
Duplication, reproduction and storage in data banks are only
allowed with the prior permission of the publishers

Inhibitor(s)

Cofactor(s)/prostethic group(s)

Metal compounds/salts

Turnover number (min^{-1})

Specific activity (U/mg)
More (1400 U/OD) [2]; 0.149 [3]

K$_m$-value (mM)

pH-optimum
4.0 [1]; 4.0–4.5 [2]

pH-range
2.5–7.0 [2]; 2.8–7.0 (at 7.0: activity very low, at 2.8: 25% of optimum) [1]

Temperature optimum (°C)
60 [1]; 35 (assay at) [3]

Temperature range (°C)
40–60 (activity rises over range) [1]

---

3 ENZYME STRUCTURE

Molecular weight
36000 (SDS-PAGE, Acremonium sp.) [2]

Subunits

Glycoprotein/Lipoprotein
–

---

4 ISOLATION/PREPARATION

Source organism
Fungi (adaptive enzyme) [1]; Acremonium sp. 15 [2]; Cytophaga arvensicola [3]; Aspergillus auratus [1]; Aspergillus aureolus [1]; Aspergillus quadricinctus [1]; Aspergillus sydowi [1]; Aspergillus unguis [1]; Beauveria bassiana [1]; Fusarium oxysporum [1]; Penicillium brefeldianum [1]; Penicillium funiculosum [1]; Penicillium melinii [1]; Penicillium parvum [1]; Penicillium quadrilineatum [1]; Penicillium verruculosum [1]

Source tissue
Cell [3]

Localisation in source
Cell-bound [3]; Intracellular [3]

Purification
Acremonium sp. 15 [2]; Cytophaga arvensicola (partial) [3]

Crystallization
−

Cloned
−

Renaturated
−

---

5 STABILITY

pH
4.5–5.5 (40°C, 2 hours) [2]

Temperature (°C)
40 (stable below, 2 hours, 20 mM acetate buffer) [2]

Oxidation

Organic solvent

General stability information

Storage

---

6 CROSSREFERENCES TO STRUCTURE DATABANKS

PIR/MIPS code

Brookhaven code

---

7 LITERATURE REFERENCES

[1] Reese, E.T., Parrish, F.W., Mandels, M.: Can. J. Microbiol., 7, 309–317 (1961)
[2] Kitahata, S., Edagawa, S.: Agric. Biol. Chem., 51, 2701–2708 (1987)
[3] Mendoza, N.S., Amemura, A.: J. Ferment. Technol., 61, 473–481 (1983)

Enzyme Handbook © Springer-Verlag Berlin Heidelberg 1991
Duplication, reproduction and storage in data banks are only
allowed with the prior permission of the publishers

1 NOMENCLATURE

EC number
3.2.1.72

Systematic name
1, 3-Beta-D-xylan xylohydrolase

Recommended name
Xylan 1, 3-beta-xylosidase

Synonymes
Exo-1, 3-beta-xylosidase
Exo-1, 3-.beta.-xylosidase
Beta-1 , 3'-xylanase [1]
Exo-beta-1, 3'-xylanase [1]

CAS Reg. No.
37288-50-9

---

2 REACTION AND SPECIFICITY

Catalysed reaction
Hydrolysis of successive xylose residues from the non-reducing termini of
1,3-beta-D-xylans

Reaction type
O-Glycosyl bond hydrolysis

Natural substrates
1, 3-Beta-D-xylans + H_2O

Substrate spectrum
1 Beta-1, 3-xylan + H_2O [1]
2 Beta-1, 3-xylo-oligosaccharides + H_2O [1]

Product spectrum
1 Xylose + glucose (small amount) [1]
2 Xylose + xylobiose [1]

Inhibitor(s)

Cofactor(s)/prostethic group(s)

Metal compounds/salts

Enzyme Handbook © Springer-Verlag Berlin Heidelberg 1991
Duplication, reproduction and storage in data banks are only
allowed with the prior permission of the publishers

Turnover number (min^{-1})

Specific activity (U/mg)
    More [1]

K$_m$-value (mM)

pH-optimum

pH-range

Temperature optimum (°C)

Temperature range (°C)

---

3 ENZYME STRUCTURE

Molecular weight

Subunits

Glycoprotein/Lipoprotein
    –

---

4 ISOLATION/PREPARATION

Source organism
    Chaetonium globosum A2 [1]

Source tissue

Localisation in source

Purification
    Chaetonium globosum A2 [1]

Crystallization
    –

Cloned
    –

Renaturated
    –

---

5 STABILITY

pH

Temperature (°C)

Oxidation

Organic solvent

General stability information

Storage

---

6 CROSSREFERENCES TO STRUCTURE DATABANKS

PIR/MIPS code

Brookhaven code

---

7 LITERATURE REFERENCES

[1] Fukui, S., Suzuki, T., Kitahara, K., Miwa, K., Miwa, T.: J. Gen. Appl. Microbiol., 6, 270–282 (1960)

1 NOMENCLATURE

EC number
3.2.1.73

Systematic name
1, 3–1, 4-Beta-D-glucan glucanohydrolase

Recommended name
Lichenase

Synonymes
1, 3–1, 4-Beta-D-glucan 4-glucanohydrolase [1]
Beta-(1 --> 3), (1 --> 4)-D-glucan 4-glucanohydrolase [6]
1, 3; 1, 4-Beta-glucan endohydrolase [7]
1, 3; 1, 4-Beta-glucan 4-glucanohydrolase [7]

CAS Reg. No.
37288-51-0

---

2 REACTION AND SPECIFICITY

Catalysed reaction
Hydrolysis of 1, 4-beta-D-glucosidic linkages in beta-D-glucans containing 1, 3- and 1, 4-bonds

Reaction type
O-Glycosyl bond hydrolysis

Natural substrates
Lichenin + H_2O
Beta-D-glucan + H_2O (depolymerization of endosperm glucan in germination) [5, 7]

Substrate spectrum
1 Lichenin + H_2O [3, 6, 9, 10, 11]
2 Beta-D-glucan + H_2O [3, 4, 6, 8, 9, 11]
3 Lichenin + H_2O [8]
4 Barley glucan + H_2O [11]
5 More (does not act on beta-D-glucans containing only 1, 3- or 1,4-bonds, not: starch, laminarin, xylan, carboxymethyl-pachyman, carboxymethyl-cellulose [3]) [3, 8]

Enzyme Handbook © Springer-Verlag Berlin Heidelberg 1991
Duplication, reproduction and storage in data banks are only
allowed with the prior permission of the publishers

Product spectrum

1 Trisaccharides + pentasaccharides (main products) [3]
2 Trisaccharides + pentasaccharides (main products) [4]
3 3-O-Beta-D-cello-biosyl-D-glucose [8]
4 3-O-Beta-cellobiosyl-D-glucose + 3-O-beta-D-cellotriosyl-D-glucose [11]
5 ?

Inhibitor(s)

D-Glucone-1, 5-lacton [8, 10]; EDTA (no effect [11]) [8];
p-Chloromercuribenzoate (no effect [11]) [8]; Epoxyalkyl
beta-oligoglucosides [2]

Cofactor(s)/prostethic group(s)

Metal compounds/salts

Turnover number (min^{-1})

Specific activity (U/mg)

1372.3 [3]; 1329 [8]; More [4, 11]

K_m-value (mM)

More (lichenin: 0.35 mg/ml [3], lichenan: 1.43 mg/ml [8], oat beta-D-glucan:
0.71 mg/ml [3], barley glucan: 0.3 mg/ml [4], 1.15 mg/ml [8]) [3, 4, 8]

pH-optimum

5.75–6.25 (broad) [3]; 7.0 [11]; 6.5–7.0 [4]; 6–8 [8]; 5.0 [6]; More (pH-optima
of different Bacillus species) [10]

pH-range

5.0–8.0 (5.0: oat beta-D-glucan, 95% of maximal activity, lichenin, 20% of
maximal activity, 8.0: oat beta-D-glucan, 70% of maximal activity, lichenin,
90% of maximal activity) [3]; 4–10 (4: 10% of maximal activity, 10: 30% of
maximal activity) [8]; 5.2–8.0 (half-maximal activities at pH 5.2 and 8.0) [4];
4–10 [11]; More (pH-range of different Bacillus species) [10]

Temperature optimum (°C)

50 [3]; 60 [4, 8]; 50–60 [11]

Temperature range (°C)

3 ENZYME STRUCTURE

Molecular weight
27000 (Bacillus pumilus) [11]
26000 (SDS-PAGE, Bacillus) [8]
37200 (gel filtration, Bacteroides succinogenes, cloned in E. coli) [3]
35200 (SDS-PAGE, Bacillus succinogenes, cloned in E. coli) [3]
33000 (SDS-PAGE, Hordeum vulgare) [6]
28000 (SDS-PAGE, ultrafiltration, amino acid composition, Hordeum vulgare, enzyme I) [7]
33000 (SDS-PAGE, ultrafiltration, amino acid analysis, Hordeum vulgare, enzyme II) [7]

Subunits
Monomer (Hordeum vulgare) [7]

Glycoprotein/Lipoprotein
Glycoprotein (Hordeum vulgare, enzyme I: traces of associated carbohydrate, enzyme II: 3.6% carbohydrate, 3 residues N-acetylglucosamine) [7]

---

4 ISOLATION/PREPARATION

Source organism
Bacteroides succinogenes (cloned in E. coli) [3]; E. coli (Bacteroides succinogenes gene) [3]; Hordeum vulgare (2 forms I/II [7]) [5–7]; Bacillus [8]; Bacillus subtilis [2, 4, 9, 10]; Bacillus pumilus [9, 10, 11]; Bacillus polymyxa [9, 10]; Bacillus amyloliquefaciens [9, 10]; Bacillus circulans [9, 10]; Bacillus laterosporus [9, 10]

Source tissue
Cell [3]; Scutella [5]; Aleurone layer [5]; Malt [6]; Culture medium [8]

Localisation in source

Purification
Bacteroides succinogenes (cloned in E. coli) [3]; E. coli (Bacteroides succinogenes gene) [3]; Bacillus subtilis [4]; Hordeum vulgare [6, 7]; Bacillus [8]; Bacillus pumilus [11]

Crystallization
–

Cloned
(Bacteroides succinogenes gene in E. coli) [1]

Renaturated
–

Enzyme Handbook © Springer-Verlag Berlin Heidelberg 1991
Duplication, reproduction and storage in data banks are only
allowed with the prior permission of the publishers

5 STABILITY

pH
4.0–5.0 [6]; 5–7 (highest stability) [8]; 5.0–7.0 [11]

Temperature (°C)
60 (30 minutes, 20% loss of activity) [3]; 60 (pH 6.5, 15 minutes, stable up to) [4]; 60 (complete loss of activity after 10 minutes) [6]; 60 (pH 6, presence of $CaCl_2$ and serum albumin, stable for 120 minutes) [8]; 60 (35% loss of activity after 10 minutes) [11]

Oxidation

Organic solvent

General stability information
Repeated freeze-thaw cycles (stable) [4]; Ca^{2+} (stabilizes) [10]; Bovine serum albumin (stabilizes) [11]

Storage
4°C, or 20°C, citrate/phosphate (pH 2.5–7.5) or phosphate/Tris/HCl (pH 5.5–9.0) buffer, 20 hours [4]; –15°C, 12 months [7]

---

6 CROSSREFERENCES TO STRUCTURE DATABANKS

PIR/MIPS code
LXBS (precursor, Bacillus subtilis); A29091 (beta, Bacillus amyloli-quefaciens); JU0110 (precursor, Bacillus subtilis); A18891 (I, barley, fragment); B18891 (II, barley, fragment); A25455 (precursor, barley)

Brookhaven code

---

7 LITERATURE REFERENCES

[1] Irvin, J.E., Teather, R.M.: Appl. Environ. Microbiol., 54, 2672–2676 (1988)
[2] Hoj, P.H., Rodriguez, E.B., Stick, R.V., Stone, B.A.: J. Biol. Chem., 264, 4939–4947 (1989)
[3] Erfle, J.D., Teather, R.M., Wood, P.J., Irvin, J.E.: Biochem. J., 255, 833–841 (1988)
[4] McCleary, B.V.: Methods Enzymol., 160, 572–575 (1988)
[5] Stuart, I.M., Loi, L., Fincher, G.B.: Plant Physiol., 80, 310–314 (1986)
[6] Yamashita, H., Hayase, F., Kato, H.: Agric. Biol. Chem., 49, 1313–1320 (1985)
[7] Woodward, J.R., Fincher, G.B.: Eur. J. Biochem., 121, 663–669 (1982)
[8] Borriss, R.: Z. Allg. Mikrobiol., 21, 7–17 (1981)
[9] Borriss, R., Zemek, J.: Zentralbl. Bakteriol. Parasitenkd. Infektionskr. Hyg., Abt. II, 135, 696–703 (1980)
[10] Borriss, R., Zemek, J.: Zentralbl. Bakteriol. Parasitenkd. Infektionskr. Hyg., Abt. II, 136, 63–69 (1981)
[11] Suzuki, H., Kaneko, T.: Agric. Biol. Chem., 40, 577–586 (1976)

1 NOMENCLATURE

EC number
   3.2.1.74

Systematic name
   1, 4-Beta-D-glucan glucohydrolase

Recommended name
   Glucan 1, 4-beta-glucosidase

Synonymes
   Glucosidase, exo-1, 4-.beta.-
   Exocellulase
   Exo-1, 4-.beta.-glucosidase
   Exo-.beta.-1, 4-glucosidase
   Exo-1, 4-beta-glucosidase
   Exo-.beta.-1, 4-glucanase
   .beta.-1, 4-.beta.-Glucanase
   Exo-.beta.-1, 4-glucosidase
   Cellodextrin glucohydrolase
   Beta-glucosidase [2]
   Exo-1, 4-beta-glucanase [6]
   Exo-beta-1, 4-glucanase [5]
   Cellodextrin glucohydrolase [4]
   (1 --> 4)-Beta-D-glucan glucanohydrolase [7]
   .beta.-1, 4-Glucosidase
   Exo-1, 4-.beta.-D-glucosidase

CAS Reg. No.
   37288-52-1

---

2 REACTION AND SPECIFICITY

Catalysed reaction
   Hydrolysis of 1,4-linkages in 1,4-beta-D-glucans so as to remove successive
   glucose units

Reaction type
   O-Glycosyl bond hydrolysis

Natural substrates
   1, 4-Beta-D-glucan + H_2O

Enzyme Handbook © Springer-Verlag Berlin Heidelberg 1991
Duplication, reproduction and storage in data banks are only
allowed with the prior permission of the publishers

Substrate spectrum
  1 1, 4-Beta-D-glucans + H_2O
  2 p-Nitrophenyl-beta-D-glucopyranoside + H_2O [1, 2]
  3 Cellodextrins + H_2O [1]
  4 Cellotetraose + H_2O [1]
  5 Cellobiose + H_2O (slowly [1], transglycosidase activity: production of higher oligosaccharides from cellobiose) [1, 4, 7]
  6 Aryl-beta-D-glycosides + H_2O [1]
  7 Cellulose + H_2O (low activity [1], not: high molecular-weight cellulose [2]) [1, 3, 7]
  8 Beta-glucan + H_2O (low activity) [1]
  9 Sophorose + H_2O [1, 7]
  10 Maltose + H_2O (not [2]) [1]
  11 Cellooligomers + H_2O (e.g. salicin) [2, 3, 4, 7]
  12 p-Nitrophenylcellobiose + H_2O [2]
  13 o-Nitrophenylcellobiose + H_2O [7]
  14 Laminaribiose + H_2O [7]
  15 More (transglycosidase activity: production of higher oligosaccharides from cellobiose, transfer activity) [4]

Product spectrum
  1 D-Glucose
  2 Glucose + nitrophenol [2]
  3 Beta-D-glucose [1]
  4 Beta-D-glucose [1]
  5 Beta-D-glucose [1]
  6 Glucose + ?
  7 Glucose [1]
  8 Glucose
  9 ?
  10 Glucose
  11 Glucose (+ cellooligomers [2]) [2, 4]
  12 Glucose + nitrophenol [2]
  13 Glucose + p-nitrophenol
  14 Glucose + p-nitrophenol
  15 ?

Inhibitor(s)
  D-Glucose [1, 4, 7]; D-Cellobiose [1]; Sulfhydryl reagents [2]; D-Glucono-1,5-lactone [1, 3, 7]; Zn^{2+} [2]; Hg^{2+} [2]; Cu^{2+} [2]; p-Chloromercuribenzoate [2]; Iodoacetamide [2]

Cofactor(s)/prostethic group(s)

Metal compounds/salts
  More (no essential metal ion) [1]

Turnover number (min⁻¹)

Specific activity (U/mg)
276.7 [1]; 13.0 [2]; 7.0 [3]; More [6]

K_m-value (mM)
2.8 (p-nitrophenyl-beta-D-glucopyranoside) [1]; 2.2
(p-nitrophenyl-beta-D-glucopyranoside) [2]; 26 (cellobiose) [2]; 0.54 (cellobiose) [4]; 0.08 (cellotriose) [4]; 0.15 (cellotetraose) [4]; 0.16 (cellopentaose) [4]; 0.18 (cellohexaose) [4]; 0.77 (o-nitrophenyl
beta-D-glucopyranoside) [7]; 10.0 (cellobiose) [7]; 0.44 (cellotriose) [7]; 0.77
(cellotetraose) [7]; 0.37 (cellopentaose) [7]; 172 (H_3PO_4-swollen cellulose)
[7]; More (cellulose: 10 mg/ml, cellopentaose: 0.83 mg/ml, cellotriose: 0.55
mg/ml) [3]

pH-optimum
4.25 (p-nitrophenyl-beta-D-glucopyranoside) [1]; 5.0 (Walseth cellulose [3],
o-nitrophenyl-beta-D-glucopyranoside and H_3PO_4-swollen cellulose [7]) [3,
7]; 6. 5 [2]

pH-range
4.25–6.0 (4.25: optimum, 6.0: 20% of maximal activity) [1]

Temperature optimum (°C)
50 (p-nitrophenyl-beta-D-glucopyranoside [1], Walseth cellulose [3]) [1, 3];
30–35 [2]

Temperature range (°C)
50–55 (50°C: optimum, 55°C: 10% of maximal activity) [1]

---

3 ENZYME STRUCTURE

Molecular weight
151000 (high performance size exclusion chromatography, Torulopsis wickerhamii) [1]
116000 (SDS-PAGE, Ruminococcus albus) [2]
20000 (SDS-PAGE, Penicillium funiculosum) [3]
48600 (calculation from partial specific volume, ultracentrifugation data
and amino acid composition, Sporotrichum pulverulentum) [6]
65000 (Penicillium funiculosum, gel filtration) [7]
143600 (sedimentation equilibrium, Torulopsis wickerhamii) [1]
82000 (gradient slab polyacrylamide gel electrophoresis, gel filtration,
Ruminococcus albus) [2]

Subunits
Dimer (2 × 83500, SDS-PAGE, Torulopsis wickerhamii) [1]

Enzyme Handbook © Springer-Verlag Berlin Heidelberg 1991
Duplication, reproduction and storage in data banks are only
allowed with the prior permission of the publishers

Glycoprotein/Lipoprotein
Glycoprotein (12% carbohydrate, mostly mannose [1], no carbohydrate
[6]) [1, 2]

---

4 ISOLATION/PREPARATION

Source organism
Botrytis cinerea [5]; Chrysosporium pannorum [5]; Cladosporium
chladosporicides [5]; Chladosporium resinae [5]; Fusarium culmorum [5];
Fusarium equiseti [5]; Fusarium pore [5]; Fusarium solani [5]; Gilmaniella
humicola [5]; Myrothecium striatisporum [5]; Phoma sp. [5]; Rhizopus
arrhizus [5]; Stachybotrys chartarum [5]; Stemphylium botrysum [5]; Tham-
nidium elegans [5]; Thielavia basicola [5]; Trichoderma koningii [5];
Trichoderma polysporum [5]; Trichoderma viride [5]; Verticillium nigrescens
[5]; Torulopsis wickerhamii [1]; Ruminococcus albus [2]; Penicillium
funiculosum [3, 7]; Trichoderma reesei [4]; Sporotrichum pulverulentum (i.e.
Chrysosporium lignorum) [6]; More [5]

Source tissue
Growth supernatant [1]; Cell [2]; Culture filtrate [3, 4]

Localisation in source

Purification
Torulopsis wickerhamii [1]; Ruminococcus albus [2]; Penicillium
funiculosum [3, 7]; Trichoderma reesei [4]; Sporotrichum pulverulentum (i.e.
Chrysosporium linorum) [6]

Crystallization
–

Cloned
–

Renaturated
–

---

5 STABILITY

pH

Temperature (°C)
50 (pH 5.0, 10 minutes) [7]; 60 (pH 5.0, 10 minutes, significant loss of activity) [7]; 70 (pH 5.0, 10 minutes, complete inactivation) [7]; 30 (10 minutes,
stable below 30°C) [2]; 37 (10 minutes, 20% loss of activity) [2]

4

Oxidation

Organic solvent

General stability information

Storage

6 CROSSREFERENCES TO STRUCTURE DATABANKS

PIR/MIPS code

Brookhaven code

7 LITERATURE REFERENCES

[1] Himmel, M.E., Tucker, M.P., Lastick, S.M., Oh, K.K., Fox, J.W., Spindler, D.D., Groh-
    mann, K.: J. Biol. Chem., 261, 12948–12955 (1986)
[2] Ohmiya, K., Shimizu, S.: Methods Enzymol., 160, 408–414 (1988)
[3] Rao, M., Mishra, C.: Appl. Microbiol. Biotechnol., 30, 130–134 (1989)
[4] Schmid, G., Wandrey, C.: Biotechnol. Bioeng., 33, 1445–1460 (1989)
[5] Augustin, J., Zemek, J., Kuniak, B., Fassatiova, O.: Folia Microbiol., 26, 14–18 (1981)
[6] Eriksson, K.-E., Petterson, B.: Eur. J. Biochem., 51, 213–218 (1975)
[7] Wood, T.M., McCrae, S.I.: Carbohydr. Res., 110, 291–303 (1982)

Enzyme Handbook © Springer-Verlag Berlin Heidelberg 1991
Duplication, reproduction and storage in data banks are only
allowed with the prior permission of the publishers

1 NOMENCLATURE

EC number
  3.2.1.75

Systematic name
  1, 6-Beta-D-glucan glucanohydrolase

Recommended name
  Glucan endo-1, 6-beta-glucosidase

Synonymes
  Glucanase, endo-1, 6-.beta.-
  Endo-1, 6-.beta.-glucanase
  Endo-1, 6-.beta. glucanase
  Endo-1, 6-beta-glucanase
  .beta.-1.fwarw.6-Glucan hydrolase
  Endo-beta-1, 6-glucanase [1]
  Endo-(1 -- > 6)-beta-D-glucanase [2, 5]
  Beta-1, 6-glucanase-pustulanase [3]
  Beta-1, 6-glucan 6-glucanohydrolase [12, 13]

CAS Reg. No.
  37278-39-0

---

2 REACTION AND SPECIFICITY

Catalysed reaction
  Random hydrolysis of 1, 6-linkages in 1, 6-beta-D-glucans

Reaction type
  O-Glycosyl bond hydrolysis

Natural substrates
  1, 6-Beta-D-glucan + H_2O

Substrate spectrum
  1 Lutean + H_2O [6, 13, 15]
  2 Luteose + H_2O [5, 11]
  3 Pustulan + H_2O [1, 3, 4, 6, 9, 14]
  4 1, 6-Oligo-beta-D-glucosides + H_2O
  5 Laminaran + H_2O (low activity [1, 3, 6, 9]) [1, 3, 6, 9, 11, 12]
  6 Gentiooligosaccharides + H_2O [3, 13]
  7 3^2-Beta-gentiobiosyl-gentiobiose + H_2O [13]

Enzyme Handbook © Springer-Verlag Berlin Heidelberg 1991
Duplication, reproduction and storage in data banks are only
allowed with the prior permission of the publishers

8 3^2-Gentiotriosyl-gentiobiose + H_2O [12]
9 Gyrophoran + H_2O [13]
10 Gentiobiose + H_2O (not [3]) [5]
11 p-Nitrophenyl-beta-D-glucoside + H_2O (low activity) [5]
12 Octasaccharide + H_2O (desuccinylated, repeating unit of suc-
    cinoglycan, composed of D-glucose, D-galactose, pyruvic acid, succinic
    acid, 7: 1: 1: 1 /molar ratio) [5]
13 Gentiotriose + H_2O [6]
14 Gentiotetraose + H_2O [6]
15 Glucan + H_2O (baker's yeast glucan) [6, 9, 11]
16 Gentiopentaose + H_2O [6]
17 Islandican + H_2O [8]
18 Pachyman + H_2O (low activity) [9]
19 Beta-1, 6-glucan + H_2O [11]
20 Beta-1, 6-glucan + H_2O [14]
21 More (partially hydrolyses (beta-1, 3-beta-1 , 6-mixed linked) yeast
    glucan) [3, 6, 8, 9, 11, 12]

Product spectrum
1 Gentiobiose + gentiotriose + glucose [13]
2 Gentiobiose + gentiotriose + gentiotetraose [11]
3 Gentio-oligosaccharides + gentiobiose + glucose + gentiotriose [1 ,
    3, 6, 9]
4 Monosaccharides + disaccharides + oligosaccharides
5 Gentiobiose [9, 12] + glucose [12] + laminaribiose [12]
6 Glucose + gentiobiose + gentiotriose
7 Gentiobiose [12]
8 Gentiobiose + gentiotriose + 6-beta-laminaribiosyl glucose [12]
9 Glucose + gentiobiose + gentiotriose [13]
10 ?
11 p-Nitrophenol + glucose
12 Tetrasaccharide (composed of D-glucose and D-galactose, 3: 1 molar
    ratio) + tetrasaccharide (composed of D-glucose, succinic acid and
    pyruvic acid, 4: 1: 1 molar ratio) [5]
13 Glucose + gentiobiose [6]
14 Glucose + gentiobiose + gentiotriose [6]
15 Oligosaccharides + glucose [6]
16 Gentiotetraose + glucose + gentiotriose + gentiobiose [6]
17 (1 --> 6)-Beta-D-oligosaccharides (not glucose) [8]
18 Gentiobiose + glucose [9]
19 Gentiobiose + gentiotriose [11]
20 Oligosaccharides (beta-1, 6-linked) + gentiobiose [14]
21 ?

Inhibitor(s)

Co^{2+} (no effect [10], slight [1]) [1]; Hg^{2+} (slight [1]) [1, 8, 10, 11]; Cu^{2+} (slight [1]) [1, 3, 8, 11]; Fe^{2+} (slight [1]) [1, 3, 11]; Fe^{3+} (slight [1]) [1, 8]; Zn^{2+} (no effect [10]) [3, 8, 11]; Mn^{2+} [3]; Cd^{2+} [8]; Glycono-delta-lactone (not [11]) [10]; Ni^{2+} [11]

Cofactor(s)/prostethic group(s)

Metal compounds/salts

More (no metal ion requirement) [9, 11]

Turnover number (min⁻¹)

Specific activity (U/mg)

13.4 [1]; 624 [4]; 0.22 [5]; 6.2 [9]; More [3, 6, 8, 11]

K_m-value (mM)

0.0123 (pustulan) [3]; 0.00012 (lutean) [13]; 0.014 (gentiotetraose) [13]; More (luteose: 0.169% [11], pustulan: 0.8 mg/ml [1], 2.78 mg/ml [4], 0.173% [8], 0.29 mg/ml [9]) [1, 4, 8, 9, 11]

pH-optimum

5.0 [1, 13]; 4.2–4.6 [3]; 4.2 [4]; 4.0 [14]; 4.5 [8]; 6.0 [9]; 5.5 [10]; 5.5–6.0 [11]; 5–5.5 [15]

pH-range

2.0–6.0 (2.0: about 20% of maximal activity, 6.0: 55% of maximal activity) [4]; 3.0–7.0 (3.0: 60% of maximal activity, 7.0: 15% of maximal activity) [8]; 2–7 (2: 50% of maximal activity, 7: less than 10% of maximal activity) [14]

Temperature optimum (°C)

50 [1, 4]; 60 [11]; 55 [8]; 50–60 [14]

Temperature range (°C)

20–80 (20°C: about 20% of maximal activity, 80°C: less than 10 % of maximal activity) [4]; 10–70 [11]; 20–70 (about 30% of maximal activity at 20°C and 70°C) [14]

3 ENZYME STRUCTURE

Molecular weight

30000 (gel filtration, Rhizopus chinensis) [1]
47000 (SDS-PAGE, Neurospora crassa) [1]
20000 (gel filtration, Chlamys abbidus) [3]
44000 (SDS-PAGE, Penicillium brefeldianum) [4]
54000 (SDS-electrophoresis in slab gels, Bacillus circulans) [9]

Enzyme Handbook © Springer-Verlag Berlin Heidelberg 1991
Duplication, reproduction and storage in data banks are only
allowed with the prior permission of the publishers

Subunits
Monomer (Bacillus circulans) [9]

Glycoprotein/Lipoprotein
Glycoprotein (26.8 mole glucose per mole of enzyme) [11]

---

4 ISOLATION/PREPARATION

Source organism
Helminthosporium [15]; Gibberella [15]; Aspergillus [15]; Macrosporium [15]; Fusarium [15]; Aspergillus usamii [15]; Gibberella fujikuroi [12, 13]; Fungi [14]; Penicillium [14]; Neurospora crassa [1]; Trichoderma harzianum [7]; Saccharomyces cerevisiae [7]; Mucor hiemalis [8]; Bacillus circulans [9, 10]; Rhizopus chinensis [11] Trichoderma viride [7]; Chlamys abbidus [3]; Penicillium brefeldianum [4]; Flavobacterium M64 [5]; Acinetobacter sp. [6]; More [15]

Source tissue
Culture filtrate [1, 4, 6, 11, 13, 15]; Cell [5]; Culture liquid [8, 10]; Commercial enzyme preparation [2, 15]; Crystalline style [3]

Localisation in source

Purification
Neurospora crassa [1]; Chlamys abbidus [3]; Penicillium brefeldianum [3]; Flavobacterium M64 [5]; Acinetobacter sp. [6]; Mucor hiemalis [8]; Bacillus circulans [9, 10]; Rhizopus chinensis [11]; Gibberella fujikuroi [13]; Aspergillus usamii [15]

Crystallization
–

Cloned
–

Renaturated
–

---

5 STABILITY

pH
5.0–8.0 (4°C, 12 hours) [1]; 4.9–5.5 (stable up to 5 hours) [3]; 5.0–8.0 (low stability below 5.0 and above 8.0) [4]; 3–7 [8]; 4.0–6.0 (unstable below 4.0 and above 6.0) [11]; 4.0–8.0 (30°C) [13]; More [4, 13, 14]

Temperature (°C)

50 (pH 5.5, 10 minutes, stable up to) [1]; 30–60 [8]; 65 (10 minutes) [11];
More [4, 13]

Oxidation

Organic solvent

General stability information

Freezing and thawing (10% loss of activity after 12 cycles [4], little effect
[9]) [4, 9]; Bovine serum albumin (protects) [11]

Storage

–20°C, pH 6.0 (stable for at least 2 months) [4]; 2°C, pH 5–7, buffers of
0.002–1.0 M (20% loss of activity after 1 week) [9]; –25°C (10% loss of activity after 1 months) [9]; 4°C, half a year (stable) [3]

---

6 CROSSREFERENCES TO STRUCTURE DATABANKS

PIR/MIPS code

Brookhaven code

---

7 LITERATURE REFERENCES

[1] Hiura, N., Nakajima, T., Matsuda, K.: Agric. Biol. Chem., 51, 3315–3321 (1987)
[2] Dubourdieu, D., Desplanques, C., Villettaz, J.-C., Ribereau-Gayon, P.: Carbohydr.
    Res., 144, 277–287 (1985)
[3] Rudakova, V.Y., Shevchenko, N.M., Elyakova, L.A.: Comp. Biochem. Physiol., 81B,
    677–682 (1985)
[4] Schep, G.P., Shepherd, M.G., Sullivan, P.: Biochem. J., 233, 707–714 (1984)
[5] Abe, J., Amemura, A., Harada, T.: Agric. Biol. Chem., 44, 1877–1884 (1980)
[6] Katohda, S., Suzuki, F., Katsuki, S., Sato, T.: Agric. Biol. Chem., 43, 2029–2034
    (1979)
[7] Del Rey, F., Garcia-Acha, I., Nombela, C.: J. Gen. Microbiol., 110, 83–89 (1979)
[8] Miyazaki, T., Oikawa, N.: Carbohydr. Res., 48, 209–216 (1976)
[9] Rombouts, F.M., Phaff, H.J.: Eur. J. Biochem., 63, 109–120 (1976)
[10] Fleet, G.H., Phaff, H.J.: J. Bacteriol., 119, 207–219 (1974)
[11] Yamamoto, S., Kobayashi, R., Nagasaki, S.: Agric. Biol. Chem., 38, 1493–1500 (1974)
[12] Shibata, Y.: J. Biochem., 75, 85–92 (1974)
[13] Shibata, Y., Fukimbara, T.: J. Ferment. Technol., 51, 216–226 (1973)
[14] Reese, E.T., Parrish, F.W., Mandels, M.: Can. J. Microbiol., 8, 327–329 (1962)
[15] Shibata, Y., Fukimbara, T.: J. Ferment. Technol., 50, 388–396 (1972)

Enzyme Handbook © Springer-Verlag Berlin Heidelberg 1991
Duplication, reproduction and storage in data banks are only
allowed with the prior permission of the publishers

1 NOMENCLATURE

EC number
3.2.1.76

Systematic name
Glycosaminoglycan alpha-L-iduronohydrolase

Recommended name
L-Iduronidase

Synonymes
Iduronidase, .alpha.-L-
.alpha.-L-Iduronidase
Alpha-L-iduronidase [2–5]
L-Iduronidase [1]

CAS Reg. No.
9073-56-7

---

2 REACTION AND SPECIFICITY

Catalysed reaction
Hydrolysis of alpha-L-iduronosidic linkages in desulfated dermatan

Reaction type
O-Glycosyl bond hydrolysis

Natural substrates
Dermatan + H_2O (degradation) [1]

Enzyme Handbook © Springer-Verlag Berlin Heidelberg 1991
Duplication, reproduction and storage in data banks are only
allowed with the prior permission of the publishers

Substrate spectrum
1 Heparan (desulfated) + H_2O [1]
2 Dermatan (desulfated) + H_2O [1]
3 Phenyl-alpha-L-iduronide + H_2O [2, 3, 4, 5, 14]
4 Anhydromannitol iduronide + H_2O [5]
5 4-Methylumbelliferyl alpha-L-iduronide + H_2O [11, 13–15]
6 Alpha-L-iduronosyl-(alpha 1 --> 4)-anhydromannitol 6-sulfate + H_2O [6, 11]
7 Alpha-L-iduronosyl-(alpha 1 --> 3)2, 5-anhydrotalitol 4-sulfate + H_2O [6, 11]
8 Alpha-L-iduronosyl-(1 --> 4)2, 5-anhydro-D-mannitol + H_2O [11]
9 More (bovine, 2 distinct enzyme species, MW 290000: phenyl alpha-L-iduronidase activity, no 4-methylumbelliferyl beta-D-glucuronidase activity, MW 78000: 4-methylumbelliferyl beta-D-glucuronidase activity, no phenyl alpha-L-iduronidase activity [15]) [11, 14, 15]

Product spectrum
1 Iduronolactone [1]
2 L-Iduronic acid [1]
3 Phenol + iduronic acid [2]
4 Anhydromannitol + iduronic acid
5 4-Methylumbelliferone + iduronic acid + iduronolactone [15]
6 Iduronic acid + anhydromannitol 6-sulfate
7 Iduronic acid + anhydromannitol 4-sulfate
8 Iduronic acid + anhydromannitol
9 ?

Inhibitor(s)
NaCl [11]; $NaSO_4$ [11]; NaH_2PO_4 [11]; $CuCl_2$ [11]; Alpha-L-idosyl(1 --> 4)2, 5-anhydro-D-mannitol 6-sulfate [11]; D-Saccharolactone [2]; p-Chloromercuribenzoate [5, 14]; Cu^{2+} [5]; Fe^{3+} (slight) [5]; 2(S)-Carboxy-3(R), 4(R), 5(S)-trihydroxypiperidine (specific inhibitor of beta-D-glucuronidase and alpha-L-iduronidase) [8]

Cofactor(s)/prostethic group(s)

Metal compounds/salts

Turnover number (min^{-1})
960 (phenyl iduronide) [5]; 16200 (anhydromannitol iduronide) [5]

Specific activity (U/mg)
34.9 [7]; 2.08 [10]; 4.73 [13]; 4.16 [14]; More [5, 6]

Km-value (mM)
   0.20 (alpha-L-iduronide) [2]; 0.45 (phenyl alpha-L-iduronide, form I) [4];
   0.31 (phenyl alpha-L-iduronide, form II) [4]; 1.5 (phenyl iduronide) [5, 14]; 9
   (anhydromannitol iduronide) [5, 14]; 0.065 (4-methylumbel-
   liferyl-alpha-L-iduronide) [7]; 0.09 (4-methylumbelliferyl-alpha-L-iduronide)
   [14]; More [15]

pH-optimum
   3.0–3.5 [4, 7]; 4.5 [11]; 4.0 (phenyl iduronide [5, 14], 4-methylumbelliferyl
   iduronide [14]) [5, 14]; 4.5 (anhydromannitol) [5, 14]

pH-range
   2.5–4.5 (2.5: about 20–30% of maximal activity, 4.5: about 5% of maximal
   activity) [4]; More [5]

Temperature optimum (°C)

Temperature range (°C)

---

3 ENZYME STRUCTURE

Molecular weight
   87000 (gel filtration, human, corrective form) [4]
   67000 (gel filtration, human, noncorrective form) [4]
   60000 (sedimentation equilibrium, human) [5, 14]
   65000 (gel permeation chromatography, human, native form I/II) [6]
   70000 (gel filtration, pig) [7]
   65000 (SDS-PAGE, human) [10]
   85000 (gel filtration, human, membrane-bound) [13]
   68000 (gel filtration, human, soluble) [13]
   290000 (gel filtration, bovine, 2 distinct enzyme species with different MW
   (290000 and 78000) and specificity) [15]
   78000 (gel filtration, bovine, 2 distinct enzyme species with different MW
   (290000 and 78000) and specificity) [15]

Subunits
   Dimer (2 × 31000, SDS-PAGE in presence of dithiothreitol, human) [5]

Glycoprotein/Lipoprotein
   Glycoprotein [14]

4 ISOLATION/PREPARATION

Source organism

Human (Hurler corrective form I [4, 5], noncorrective form II [4], form I/II [6], 2 forms: soluble/low-uptake form and membrane-bound/high uptake form [13]) [1, 3, 4, 5, 6, 9, 10, 11, 13, 14]; Rat [2, 15]; Pig [7]; Bovine (2 distinct enzyme species, with different MW and specificity) [15]

Source tissue

Liver (fibroblast, cultured cells [1, 6], lysosomes [2], 2 forms [15]) [1, 2, 6, 7, 10, 15]; Preputial gland [15]; Skin (fibroblast, cultured cells [1], fibroblast [3]) [1, 3]; Amniotic fluid cells [3]; Kidney [5, 6, 14]; Lung [6, 13]; Fibroblast [9]

Localisation in source

Soluble [5, 13]; Lysosomes (membrane [7]) [7, 9]; Membrane [7, 13]

Purification

Human (immunopurification with monoclonal antibodies [6]) [5, 6, 10, 13, 14]; Pig [7]

Crystallization

–

Cloned

(human) [12]

Renaturated

–

---

5 STABILITY

pH

3.5–6.0 [7]; More [7]

Temperature (°C)

55 (30 minutes) [4]; 65 (form I: about 40% loss of activity after 15 minutes, form II: about 80% loss of activity after 15 minutes [4], 50% loss of activity after 8.5 minutes [5, 14]) [4, 5, 14]; More [15]

Oxidation

Organic solvent

General stability information

NaCl (stabilizes at pH 7) [7]; Bovine serum albumin (stabilizes diluted enzyme) [14]; Repeated freezing and thawing (stable) [14]

Storage

4°C, several months [14]

6 CROSSREFERENCES TO STRUCTURE DATABANKS

PIR/MIPS code

Brookhaven code

7 LITERATURE REFERENCES

[1] Matalon, R., Cifonelli, J.A., Dorfman, A.: Biochem. Biophys. Res. Commun., 42, 340–345 (1971)

[2] Weissmann, B., Santiago, R.: Biochem. Biophys. Res. Commun., 46, 1430–1433 (1972)

[3] Hall, C.W., Neufeld, E.F.: Arch. Biochem. Biophys., 158, 817–821 (1973)

[4] Shapiro, L.J., Hall, C.W., Leder, I.G., Neufeld, F.: Arch. Biochem. Biophys., 172, 156–161 (1976)

[5] Rome, L.H., Garvin, A.J., Neufeld, E.F.: Arch. Biochem. Biophys., 189, 344–353 (1978)

[6] Clements, P.R., Brooks, D.A., McCourt, P.A.G., Hopwood, J.J.: Biochem. J., 259, 199–208 (1989)

[7] Ohshita, T., Sakuda, H., Nakasone, S., Iwamasa, T.: Eur. J. Biochem., 179, 201–207 (1989)

[8] Cenci Di Bello, I., Dorling, P., Fellows, L., Wichester, B.: FEBS Lett., 176, 60–64 (1984)

[9] Myerowitz, R., Neufeld, E.F.: J. Biol. Chem., 256, 3044–3048 (1981)

[10] Clements, P.R., Brooks, D.A., Saccone, G.T.P., Hopwood, J.J.: Eur. J. Biochem., 152, 21–28 (1985)

[11] Clements, P.R., Muller, V., Hopwood, J.J.: Eur. J. Biochem., 152, 29–34 (1985)

[12] Schuchman, E.H., Astrin, K.H., Aula, P., Desnick, R. J.: Proc. Natl. Acad. Sci. USA, 81, 1169–1173 (1984)

[13] Schuchman, E.H., Guzman, N.A., Desnick, R.J.: J. Biol. Chem., 259, 3132–3140 (1984)

[14] Rome, L.H.: Methods Enzymol., 83, 578–582 (1982)

[15] Kosaka, H., Isemura, M., Ono, T., Nishimura, Y., Kato, K.: J. Biochem., 88, 69–75 (1980)

Enzyme Handbook © Springer-Verlag Berlin Heidelberg 1991
Duplication, reproduction and storage in data banks are only
allowed with the prior permission of the publishers

1 NOMENCLATURE

EC number
3.2.1.77

Systematic name
1, 2–1, 3-Alpha-D-mannan mannohydrolase

Recommended name
Mannan 1, 2-(1, 3)-alpha-mannosidase

Synonymes
Exo-1, 2–1, 3-alpha-mannosidase

CAS Reg. No.
37288-53-2

2 REACTION AND SPECIFICITY

Catalysed reaction
Hydrolysis of 1, 2- and 1, 3-linkages in yeast mannan, releasing mannose

Reaction type
O-Glycosyl bond hydrolysis

Natural substrates
Mannans + H_2O [1–3]

Substrate spectrum
1 Mannans + H_2O (ir, a 1,6-alpha-D-mannan backbone remains after action on yeast mannan, this is further attacked, but slowly) [1–3]

Product spectrum
1 Mannose + 1, 6-alpha-D-mannan-backbone [1–3]

Inhibitor(s)
Mg^{2+} [3]; Cu^{2+} [3]; Zn^{2+} [3]; EDTA [3]

Cofactor(s)/prostethic group(s)

Metal compounds/salts
Ca^{2+} [3]

Turnover number (min^{-1})

Specific activity (U/mg)
6.0 [3]

Enzyme Handbook © Springer-Verlag Berlin Heidelberg 1991
Duplication, reproduction and storage in data banks are only
allowed with the prior permission of the publishers

K_m-value (mM)
   0.01 (mannan) [3]; 0.5–2.0 (oligosaccharides from mannan) [3]

pH-optimum
   6.5–7.0 (mannan) [3]

pH-range
   5.0–9.0 (mannan) [3]

Temperature optimum (°C)

Temperature range (°C)

---

3 ENZYME STRUCTURE

Molecular weight

Subunits

Glycoprotein/Lipoprotein
   –

---

4 ISOLATION/PREPARATION

Source organism
   Penicillium charlesii [1]; Arthrobacter sp. [2, 3]

Source tissue

Localisation in source
   Extracellular [1–3]

Purification
   Arthrobacter sp. (partially) [3]

Crystallization
   –

Cloned
   –

Renaturated
   –

---

5 STABILITY

pH

Temperature (°C)
    70 (not stable above) [3]

Oxidation

Organic solvent

General stability information

Storage
    2 days, 4°C [3]; Indefinitely, –10°C [3]

---

6 CROSSREFERENCES TO STRUCTURE DATABANKS

PIR/MIPS code

Brookhaven code

---

7 LITERATURE REFERENCES

[1] Preston, J.F., Lapis, E., Gander, J.E.: Arch. Microbiol., 88, 71–76 (1973)
[2] Jones, G.H., Ballou, C.E.: J. Biol. Chem., 244, 1052–1059 (1969)
[3] Jones, G.H., Ballou, C.E.: J. Biol. Chem., 244, 1043–1051 (1969)

Enzyme Handbook © Springer-Verlag Berlin Heidelberg 1991
Duplication, reproduction and storage in data banks are only
allowed with the prior permission of the publishers

1 NOMENCLATURE

EC number
3.2.1.78

Systematic name
1, 4-Beta-D-mannan mannanohydrolase

Recommended name
Mannan endo-1, 4-beta-mannosidase

Synonymes
Endo-1, 4-beta-mannanase
Endo-.beta.-1, 4-mannase
Mannanase, endo-1, 4-.beta.-
Beta-mannanase B [5]
Beta-1, 4-mannan 4-mannanohydrolase [1]
Endo-beta-mannanase [3, 6]
Beta-D-mannanase [7]

CAS Reg. No.
37288-54-3

---

2 REACTION AND SPECIFICITY

Catalysed reaction
Random hydrolysis of 1, 4-beta-D-mannosidic linkages in mannans, galactomannans and glucomannans

Reaction type
O-Glycosyl bond hydrolysis

Natural substrates
Mannan + H_2O (break down of mannan-rich cell walls [6], break down of
endosperm cell wall [3]) [3, 6, 13]

Enzyme Handbook © Springer-Verlag Berlin Heidelberg 1991
Duplication, reproduction and storage in data banks are only
allowed with the prior permission of the publishers

Substrate spectrum

1 Mannans (codium, coffee [9]) + H_2O [1, 9, 15, 17]
2 Mannotetraose + H_2O [4, 7, 9, 15]
3 Mannotriose + H_2O (not [9]) [4, 15]
4 Galactomannans + H_2O [7, 9, 14]
5 Glucomannans + H_2O [9, 10, 17]
6 Mannopentaose + H_2O [17]
7 Galactoglucomannan + H_2O [17]
8 More (transglycosylation reaction, Streptomyces: transfer of one mannose unit from the oligosaccharides, fenugreek: transfer of oligoman-
   nose residues [8], not: mannobiose [4, 9, 15]) [1, 4, 9, 15]

Product spectrum

1 Oligosaccharides (dimers, trimers and other) [1]
2 Mannobiose + mannose (low amount) + mannotriose (small amount)
   [4, 15]
3 Mannobiose + mannose [4, 15]
4 Hydrolyzed galactomannans [7]
5 Oligosaccharides (of various size) [9, 10]
6 Mannose + mannobiose + mannotriose + mannotetraose
7 ?
8 ?

Inhibitor(s)
Ag^+ [9, 15]; Hg^{2+} [9, 14]; Pb^{2+} [9]; Zn^{2+} [9]; Fe^{3+} [9];
p-Chloromercuribenzoate [9]; N-Bromosuccinimide [15]; Cycloheximide
[18]; Cu^{2+} [9]

Cofactor(s)/prostethic group(s)

Metal compounds/salts
Ca^{2+} (stimulates) [9]

Turnover number (min^{-1})

Specific activity (U/mg)
201 [9]; 21 [14]; 470 [15]; More [4, 5, 11, 17]

K_m-value (mM)
More (copra beta-mannan: 2.0 mg/ml, locust bean beta-mannan: 3.8
mg/ml, konjak beta-mannan: 7.7 mg/ml [15]) [7, 15, 17]; 1.9 (soluble mannan) [5]

pH-optimum
6 [1]; 3.6 [4]; 5.5 [9]; 5 [14]; 9 (M-I, M-II) [15]; 8.5 (M-III) [15]; 4.5 (lucerne) [7];
3.0 (Basidiomycetes sp., Aspergillus niger) [7]; 4.5–5.5 (snail) [7]; 5.0–6.0
(Bacillus subtilis) [7]; More [7, 14, 17]

pH-range
6–8 [1]; 3–6 [1]; 6–10 [15]; 4–7 [1]; 3–8 (3: 15% of maximal activity) [14];
More [1, 17]

Temperature optimum (°C)
60–70 [1]; 65 (M-III [15]) [15]; 80 [4]; 70 [14]; 60 (M-I, M-II) [15]; More [17, 14]

Temperature range (°C)
50–70 [15]; 30–70 [1]; 30–80 (30°C: about 25% of maximal activity, 80°C:
about 30% of maximal activity) [14]; More [17]

---

3 ENZYME STRUCTURE

Molecular weight
41000 (lucerne, SDS-PAGE) [7]
53000 (Basidiomycetes sp., SDS-PAGE) [7]
37000 (snail, Bacillus subtilis, SDS-PAGE) [7]
24000 (Aeromonas, gel filtration) [9]
27000 (Chlorella fusca, gel filtration, beta-1,4-mannanase activity separated
in two peaks (MW 27000 and 70000) [11]
70000 (Chlorella fusca, gel filtration, beta-1,4-mannanase activity separated
in two peaks (MW 27000 and 70000)) [11]
57000 (Penicillium purpurogenum, SDS-PAGE) [14]
42000 (Bacillus subtilis, M-III, SDS-PAGE) [15]
58000 (Bacillus subtilis, M-I, SDS-PAGE) [15]
59000 (Bacillus subtilis, M-II, SDS-PAGE) [15]
46000 (Lactuca sativa, SDS-PAGE) [18]
44000 (Chlorella fusca, SDS-PAGE) [13]
25000 (gel filtration, Chlorella fusca) [13]

Subunits
Monomer (SDS-gel electrophoresis) [7, 14]

Glycoprotein/Lipoprotein
–

---

4 ISOLATION/PREPARATION

Source organism
Fungi [1]; Lactuca sativa L. [3, 6, 18, 17]; Aspergillus niger (commercial
preparations [7]) [4, 7, 17]; Lucerne [17, 5, 7]; Streptomyces sp. [8];
Trigonella foenum-graecum [8, 16]; Aeromonas hydrolphila [9]; Tyromyces
palustris [10]; Chlorella fusca [11, 13]; Hyphomycetes [12]; Penicillium purpurogenum [14]; Bacillus sp. (M-I, M-II, M-III) [17, 15]; Helix pomatia [17]; Irpex lateus [17]; Lycopersicon esculentum [2]; Basidiomycetes sp. (commercial preparation) [7]; Snail (commercial preparation) [7]; Bacillus subtilis
(commercial preparation [7]) [2, 7]; Cyamopsis tetragonolobus [17]

Enzyme Handbook © Springer-Verlag Berlin Heidelberg 1991
Duplication, reproduction and storage in data banks are only
allowed with the prior permission of the publishers

Source tissue
Endosperm [6, 18, 17]; Seeds [3, 5, 6]; Culture filtrate [1, 14]; Gut [7]; Culture fluid [9]; Commercial preparation [7]

Localisation in source
Extracellular [15]

Purification
Aspergillus niger [17, 4]; Lucerne [5, 6]; More (commercial preparation) [7]; Aeromonas hydrophila [9]; Chlorella fusca [11, 13]; Penicillium purpurogenum [14]; Helix pomatia [17]; Bacillus subtilis [17]; Irpex lateus [17] Lactuca sativa [6, 18]; Bacillus sp. (M-I, M-II, M-III) [15]; Cyamopsis tetragonolobus [17]

Crystallization
[2, 4]

Cloned
–

Renaturated
–

---

5 STABILITY

pH
More [7, 14, 17]; 5–9 [9]; 4.5–8 [14]; 4–8 [17]; 4–8 (40°C, 1 hour, lucerne) [7]

Temperature (°C)
70 (stable up to) [17]; 40 (stable up to) [9]; 65 (stable up to) [14]; 39 (24 hours, stable between pH 5 and 7) [4]; 50 (stable below, lucerne, snail, Bacillus subtilis) [7]; 60 (stable below, Basidiomycetes sp.) [7]; 70 (stable below, Aspergillus niger) [7]; More [14, 17]

Oxidation

Organic solvent
Acetone (may be used in preparations) [1]

General stability information

Storage
Cold [1]; Dialyzed [9]; –80°C [17]

---

6 CROSSREFERENCES TO STRUCTURE DATABANKS

PIR/MIPS code
PL0111 (Bacillus sp., fragment)

Brookhaven code

7 LITERATURE REFERENCES

[1] Reese, E.T., Shibata, Y.: Can. J. Microbiol., 11, 167–18 (1965)
[2] Groot, S.P.C., Kieliszewska-Rokicka, B., Vermeer, E., Karsson, C.M.: Planta, 174,
    500–504 (1988)
[3] Halmer, P., Bewley, J.D., Thorpe, T.A.: Nature, 258, 716–718 (1975)
[4] Yamazaki, N., Sinner, M., Dietrichs, H.H.: Holzforschung, 30, 101–109 (1976)
[5] McCleary, B.V.: Phytochemistry, 17, 651–653 (1978)
[6] Halmer, P., Bewley, J.D.: Planta, 144, 333–340 (1979)
[7] McCeary, B.V.: Phytochemistry, 18, 757–763 (1979)
[8] Coulombel, C., Clermont, S., Foglietti, M.J., Percheron, F.: Biochem. J., 195, 333–335
    (1981)
[9] Araki, T.: J. Fac. Agric. Kyushu Univ., 27, 89–98 (1983)
[10] Shimizu, K., Ishihara, M.: Agric. Biol. Chem., 47, 949–955 (1983)
[11] Loos, E., Meindl, D.: Planta, 160, 357–362 (1984)
[12] Zemek, J., Marvanova, L.4, Kuniak, L., Kadlecikova, B.: Folia Microbiol., 30, 363–372
    (1985)
[13] Loos, E., Meindl, D.: Planta, 166, 557–562 (1985)
[14] Park, G.G., Kusakabe, I., Komatsu, Y., Kobayashi, H., Yasui, T., Murakami, K: Agric.
    Biol. Chem., 51, 2709–2716 (1987)
[15] Akino, T., Nakamura, N., Horikoshi, K.: Agric. Biol. Chem., 52, 773–779 (1988)
[16] Spyropoulos, C.G., Reid, J.S.G.: Planta, 174, 473–478 (1988)
[17] McCleary, B.: Methods Enzymol., 160, 596–610 (1988)
[18] Dulson, J., Bewley, J.D.: Phytochemistry, 28, 363–369 (1989)

Enzyme Handbook © Springer-Verlag Berlin Heidelberg 1991
Duplication, reproduction and storage in data banks are only
allowed with the prior permission of the publishers

1 NOMENCLATURE

EC number
3.2.1.80

Systematic name
Beta-D-fructan fructohydrolase

Recommended name
Fructan beta-fructosidase

Synonymes
Exo-beta-D-fructosidase
Fructofuranosidase, polysaccharide.beta.
Exo-.beta.-fructosidase
Polysaccharide.beta.-fructofuranosidase
Fructan exohydrolase

CAS Reg. No.
37288-56-5

---

2 REACTION AND SPECIFICITY

Catalysed reaction
Hydrolysis of terminal, non-reducing 2, 1- and 2, 6-linked
beta-D-fructofuranose residues in fructans

Reaction type
O-Glycosyl bond hydrolysis

Natural substrates
Fructans + H_2O (degradation) [3]

Substrate spectrum
1 Inulin + H_2O [1, 2]
2 Raffinose + H_2O [1, 2]
3 Levan + H_2O [1, 2]
4 Sucrose + H_2O [1, 2]
5 More (not: melizitose, dextran, pseudonigeran) [2]

Enzyme Handbook © Springer-Verlag Berlin Heidelberg 1991
Duplication, reproduction and storage in data banks are only
allowed with the prior permission of the publishers

Product spectrum
1 D-Fructose [1, 2]
2 D-Fructose [2]
3 D-Fructose [1, 2]
4 D-Fructose [2]
5 ?

Inhibitor(s)
Hg^{2+} [1, 2]; Fe^{3+} [2]; Ag^{2+} [1, 2]; Cu^{2+} (partial [1]) [1, 2];
p-Chloromercuribenzoate [2]

Cofactor(s)/prostethic group(s)

Metal compounds/salts
Mn^{2+} (slight stimulation) [1, 2]; Co^{2+} (slight stimulation) [1]; KCN (slight stimulation) [2]

Turnover number (min^{-1})

Specific activity (U/mg)
204.5 [1]; 433 [2]

K_m-value (mM)

pH-optimum
5.5 (fructan polymers) [1]; 4.5 (sucrose) [1]; 7.0 [2]

pH-range
3.5–7.0 (little fructanase activity at or above 7.0) [1]; 4.5–8 [2]

Temperature optimum (°C)
50 [2]

Temperature range (°C)

---

3 ENZYME STRUCTURE

Molecular weight
140000 (SDS-PAGE, Streptococcus mutans) [1]
83000–85000 (gel filtration, SDS-PAGE, Streptococcus salivarius) [2]

Subunits

Glycoprotein/Lipoprotein
−

4 ISOLATION/PREPARATION

Source organism
Streptococcus mutans [1]; Streptococcus salivarius [2]; Helianthus tuberosus L. [3]

Source tissue
Tuber [3]

Localisation in source
Vacuole [3]

Purification
Streptococcus mutans [1]; Streptococcus salivarius [2]; Helianthus tuberosus [3]

Crystallization
–

Cloned
(Streptococcus mutans gene in E. coli) [1]

Renaturated
–

---

5 STABILITY

pH
7.0 (37°C, 30 minutes) [2]; 4.5 (almost complete loss of activity at) [2]

Temperature (°C)
37 (pH 7.0, 30 minutes) [2]; 55 (pH 7.0, 15 minutes: 20% loss of activity) [2]; 60 (pH 7.0, 15 minutes: rapid inactivation) [2]

Oxidation

Organic solvent

General stability information
Freezing and thawing (accelerates inactivation) [2]

Storage
4° C, 0.02 mg/ml, 20 mM phosphate buffer, pH 7.0 (stable at least for several days, 90% loss of activity after 2 weeks) [2]; Higher stability at –20° C than at 4° C [2]

Enzyme Handbook © Springer-Verlag Berlin Heidelberg 1991
Duplication, reproduction and storage in data banks are only
allowed with the prior permission of the publishers

6 CROSSREFERENCES TO STRUCTURE DATABANKS

PIR/MIPS code

Brookhaven code

7 LITERATURE REFERENCES

[1] Burne, R.A., Schilling, K., Bowen, W.H., Yasbin, R.E.: J. Bacteriol., 169, 4507–4517 (1987)

[2] Takahashi, N., Mizuno, F., Takamori, K.: Infect. Immun., 47, 271–276 (1985)

[3] Frehner, M., Keller, F., Wiemken, A.: J. Plant Physiol., 116, 197–208 (1984)

1 NOMENCLATURE

EC number
3.2.1.81

Systematic name
Agarose 3-glycanohydrolase

Recommended name
Agarase

Synonymes
Beta-agarase

CAS Reg. No.
37288-57-6

---

2 REACTION AND SPECIFICITY

Catalysed reaction
Hydrolysis of 1, 3-beta-D-galactosidic linkages in agarose, giving the tetramer as the predominant product

Reaction type
O-Glycosyl bond hydrolysis

Natural substrates
Agarose + H_2O [1–13]

Substrate spectrum
1 Agarose + H_2O (ir) [1–13]
2 Porphyran + H_2O (ir) [6, 10–13]

Product spectrum
1 Neoagarotetraose [1–13]
2 Neoagaro-oligosaccharides (methylated, sulphated or unsubstituted) [6, 10–13]

Inhibitor(s)
Ca^{2+} [3]; Mn^{2+} [3]

Cofactor(s)/prostethic group(s)

Metal compounds/salts

Enzyme Handbook © Springer-Verlag Berlin Heidelberg 1991
Duplication, reproduction and storage in data banks are only
allowed with the prior permission of the publishers

Turnover number (min^{-1})

Specific activity (U/mg)
22000 [4]

K$_m$-value (mM)

pH-optimum
6.0 (agarose) [3, 4]; 7.0 (agarose) [7]; 6.7 (agarose) [8]; 6.3 (agarose) [9];
7.2 (porphyran) [12, 13]

pH-range
4.0–11.0 (agarose) [4]; 3.0–9.0 (agarose) [7]

Temperature optimum (°C)
43 (agarose) [8]; 35 (agarose) [9]; 30 (neoagarohexaose) [9]; 40–41 (porphyran) [12]; 34 (porphyran) [13]

Temperature range (°C)
50 (acive up to, agarose) [9]

---

3 ENZYME STRUCTURE

Molecular weight
35132 (DNA sequence analysis, Streptomyces coelicolor) [2]
28000–29000 (gel electrophoresis, extracellular, Streptomyces coelicolor) [3,
5]
34000 (gel electrophoresis, intracellular, Streptomyces coelicolor) [3]
49000–50000 (gel filtration, gel electrophoresis, Vibrio harveyi) [4]
32000 (gel electrophoresis, Pseudomonas atlantica) [7]
62000–64000 (gel filtration, gel electrophoresis, Pseudomonas sp.) [8]
26500 (gel filtration, Cytophaga flevensis) [9]

Subunits
Monomer (gel electrophoresis) [4, 8]

Glycoprotein/Lipoprotein
–

---

4 ISOLATION/PREPARATION

Source organism
Streptomyces coelicolor [3]; Vibrio harveyi [4]; Pseudomonas atlantica [7];
Pseudomonas sp. [8]; Cytophaga flevensis [9]; Cytophaga sp. [12, 13]

Source tissue

Localisation in source
 Extracellular [1–13]; Intracellular [3]

Purification
 Streptomyces coelicolor [3]; Vibrio harveyi [4]; Pseudomonas atlantica [7];
 Pseudomonas sp. [8]; Cytophaga flevensis (partially) [9]; Cytophaga sp.
 (partially) [12, 13]

Crystallization
 —

Cloned
 [1–3, 5]

Renaturated
 —

---

5 STABILITY

pH
 6.0–9.0 [9]

Temperature (°C)
 40 (not stable above) [7, 9]

Oxidation

Organic solvent

General stability information

Storage
 2 months, 20°C [7]; 6 months, –20°C [8]; 2 years, –20°C [9]

---

6 CROSSREFERENCES TO STRUCTURE DATABANKS

PIR/MIPS code
 S07322 (precursor, Streptomyces coelicolor)

Brookhaven code

Enzyme Handbook © Springer-Verlag Berlin Heidelberg 1991
Duplication, reproduction and storage in data banks are only
allowed with the prior permission of the publishers

7 LITERATURE REFERENCES

[1] Belas, R., Bartlett, D., Silverman, M.: Appl. Environ. Microbiol., 54, 30–37 (1988)
[2] Buttner, M.J., Fearnley, I.M., Bibb, M.J.: Mol. Gen. Genet., 209, 101–109 (1987)
[3] Bibb, M.J., Jones, G.H., Joseph, R., Buttner, M.J., Ward, J.M.: J. Gen. Microbiol., 133, 2089–2096 (1987)
[4] Fukasawa, S., Kobayashi, H.: Agric. Biol. Chem., 51, 269–270 (1987)
[5] Kendall, K., Cullum, J.: Gene, 29, 315–321 (1984)
[6] Morrice, L.M., McLean, M.W., Long, W.F., Williamson, F.B.: Eur. J. Biochem., 137, 149–154 (1983)
[7] Morrice, L.M., McLean, M.W., Williamson, F.B., Long, W.F.: Eur. J. Biochem., 135, 553–558 (1983)
[8] Malmqvist, M.: Biochim. Biophys. Acta, 537, 31–43 (1978)
[9] Van Der Meulen, H.J., Harder, W.: Antonie Leeuwenhoek, 41, 431–447 (1975)
[10] Duckworth, M., Turvey, J.R.: Biochem. J., 113, 693–696 (1969)
[11] Duckworth, M., Turvey, J.R.: Biochem. J., 113, 687–692 (1969)
[12] Duckworth, M., Turvey, J.R.: Biochem. J., 113, 139–142 (1969)
[13] Turvey, J.R., Christison, J.: Biochem. J., 105, 311–316 (1967)

1 NOMENCLATURE

EC number
3.2.1.82

Systematic name
Poly(1, 4-alpha-D-galactosiduronate) digalacturonohydrolase

Recommended name
Exo-poly-alpha-galacturonosidase

Synonymes
Exopolygalacturonosidase
Exopolygalacturanosidase

CAS Reg. No.
37288-58-7

---

2 REACTION AND SPECIFICITY

Catalysed reaction
Hydrolysis of pectic acid from non-reducing end, releasing digalacturonate

Reaction type
O-Glycosyl bond hydrolysis

Natural substrates
Pectic acid + H_2O [1–5]

Substrate spectrum
1 Polygalacturonates + H_2O [1–5]

Product spectrum
1 Digalacturonate [1–5]

Inhibitor(s)

Cofactor(s)/prostethic group(s)

Metal compounds/salts

Turnover number (min^{-1})

Specific activity (U/mg)
21.36 [1]; 591 [4]

Enzyme Handbook © Springer-Verlag Berlin Heidelberg 1991
Duplication, reproduction and storage in data banks are only
allowed with the prior permission of the publishers

K$_m$-value (mM)
0.05 (D-galacturonan) [4]

pH-optimum
7.0 (pectate) [1]; 6.6 (tetragalacturonate) [1]; 6.0 (D-galacturonan) [4]

pH-range
3.0–9.0 (D-galacturonan) [4]

Temperature optimum (°C)
40 (pectate) [1]

Temperature range (°C)

---

3 ENZYME STRUCTURE

Molecular weight
67000–68000 (SDS-PAGE, Erwinia chrysanthemi) [2, 4]
47000 (gel filtration, Erwinia chrysanthemi) [4]
58000–65000 (sedimentation equilibrium, Erwinia chrysanthemi) [4]

Subunits

Glycoprotein/Lipoprotein
–

---

4 ISOLATION/PREPARATION

Source organism
Selenomonas ruminantium [1]; Erwinia chrysanthemi [4]; Pseudomonas sp.
[5]

Source tissue

Localisation in source
Extracellular [3–5]

Purification
Selenomonas ruminantium [1]; Erwinia chrysanthemi [4]; Pseudomonas sp.
[5]

Crystallization
–

Cloned
–

Renaturated

–

---

5 STABILITY

pH
  7 [1]

Temperature (°C)
  40 (not stable above) [1]

Oxidation

Organic solvent

General stability information

Storage

---

6 CROSSREFERENCES TO STRUCTURE DATABANKS

PIR/MIPS code

Brookhaven code

---

7 LITERATURE REFERENCES

[1] Heinrichova, K., Wojciechowicz, M., Ziolecki, A.: J. Appl. Bacteriol., 66, 169–174 (1989)
[2] Ried, J.L., Collmer, A.: Appl. Environ. Microbiol., 50, 615–622 (1985)
[3] Kegoya, Y., Masuda, H., Hatanaka, C.: Agric. Biol. Chem., 48, 1911–1912 (1984)
[4] Collmer, A., Whalen, C.H., Beer, S.V., Bateman, D.F.: J. Bacteriol., 149, 626–634 (1982)
[5] Hatanaka, C., Imamura, T.: Agric. Biol. Chem., 38, 2267–2268 (1974)

Enzyme Handbook © Springer-Verlag Berlin Heidelberg 1991
Duplication, reproduction and storage in data banks are only
allowed with the prior permission of the publishers

1 NOMENCLATURE

EC number
3.2.1.83

Systematic name
Kappa-carrageenan 4-beta-D-glycanohydrolase

Recommended name
Kappa-carrageenase

Synonymes
Carrageenase, .kappa.-
.kappa.-Carrageenanase

CAS Reg. No.
37288-59-8

---

2 REACTION AND SPECIFICITY

Catalysed reaction
Hydrolysis of 1, 4-beta-D-linkages between D-galactose 4-sulfate and
3,6-anhydro-D-galactose in various carrageenans

Reaction type
O-Glycosyl bond hydrolysis

Natural substrates
Carrageenans + H_2O

Substrate spectrum
1 Carrageenans + H_2O (hydrolysis of 1, 4-beta-D-linkage)
2 More [6]

Product spectrum
1 Oligomers (O^3-(3, 6-anhydro-alpha-D-galactopyranosyl)-D-galactose
O^4-sulfate) [1]
2 More [6]

Inhibitor(s)
$HgCl_2$ [7]; $AgNO_3$ [7]; NaCl [7]; $ZnCl_2$ [7]; Cu^{2+} [7]; K^+ [7]; Mg^{2+} [7]; Mn^{2+}
[7]

Cofactor(s)/prostethic group(s)

Enzyme Handbook © Springer-Verlag Berlin Heidelberg 1991
Duplication, reproduction and storage in data banks are only
allowed with the prior permission of the publishers

Metal compounds/salts

Turnover number (min⁻¹)

Specific activity (U/mg)
5.0 [10]

K_m-value (mM)

pH-optimum
7.5 [1]; 7.6 [3, 4, 10]; 7.0 [7]

pH-range
5–9 [1]; 4–10 [2]; 5.0–9.5 [3, 4, 7]; 7.2–8.3 [10]

Temperature optimum (°C)
30–45 [7]; 25 [10]

Temperature range (°C)
20–55 [2]; 25–50 [7]; 20–30 [10]

---

3 ENZYME STRUCTURE

Molecular weight
35000 (Pseudomonas carrageenovora, SDS-gel electrophoresis) [2]
100000 (Cytophaga, SDS-gel electrophoresis) [10]

Subunits
Monomer (Pseudomonas carrageenovora, Cytophaga, SDS-gel electrophoresis) [2, 10]

Glycoprotein/Lipoprotein
–

---

4 ISOLATION/PREPARATION

Source organism
Pseudomonas carrageenovora [1, 2, 5, 6]; Diadema setosum [3, 4];
Cytophaga 1k-C783 [7, 8, 10]; Hyphomycetes [9]

Source tissue

Localisation in source
Extracellular [6, 7]

Purification
Pseudomonas carrageenovora [1, 2]; Diadema setosum [3, 4]; Cytophaga [10]

Crystallization
–

Cloned
–

Renaturated
–

5 STABILITY

pH

Temperature (°C)
40 (3.5 hours) [1]; 50 (unstable) [7]

Oxidation

Organic solvent

General stability information

Storage
–20°C [1, 3]; –70°C [2]

6 CROSSREFERENCES TO STRUCTURE DATABANKS

PIR/MIPS code

Brookhaven code

7 LITERATURE REFERENCES

[1] Weigl, J., Yaphe, W.: Can. J. Microbiol., 12, 939–947 (1966)
[2] McLean, M.M., Williamson, F.B.: Eur. J. Biochem., 93, 553–558 (1979)
[3] Benites, L.V., Macaranas, J.M.: Proc. Int. Seaweed Symp., 9, 353–359 (1977)
[4] Benites, L.V., Macaranas, J.M.: Inf. Ser. -N. Z. Dep. Sci. Ind. Res., 137 (Proc. Int. Symposium Mar. Biogeogr. Evol. South. Hemishere) , 2, 371–376 (1978)
[5] McLean, M.W., Williamson, F.B.: Proc. Int. Seaweed Symp., 10, 479–484 (1981)
[6] Bellion, C., Hamer, G.K., Yaphe, W.: Can. J. Microbiol., 28, 874–880 (1982)
[7] Sarwar, G.: Nippon Suisan Gakkaishi, 49, 1689–1694 (1983)
[8] Sarwar, G., Oda, H., Sakawa, T., Kakimoto, D.: Microbiol. Immunol., 29, 405–411 (1985)
[9] Zemek, J., Marvanova, L., Kuniak, L., Kadlecikova, B.: Folia Microbiol., 30, 363–372 (1985)
[10] Sarwar, G., Matayoshi, S., Oda, H.: Microbiol. Immunol., 31, 869–877 (1987)

Enzyme Handbook © Springer-Verlag Berlin Heidelberg 1991
Duplication, reproduction and storage in data banks are only
allowed with the prior permission of the publishers

1 NOMENCLATURE

EC number
3.2.1.84

Systematic name
1, 3-Alpha-D-glucan 3-glucohydrolase

Recommended name
Glucan 1, 3-alpha-glucosidase

Synonymes
Exo-1, 3-alpha-glucanase
Glucosidase II

CAS Reg. No.
9073-99-8

---

2 REACTION AND SPECIFICITY

Catalysed reaction
$Glc_2 Man_9 GlcNAc_2 + H_2O \rightarrow$
$\rightarrow Man_9 GlcNAc_2 + 2$ glucose [2, 7, 8, 9]
More (hydrolysis of terminal 1,3-alpha-D-glucosidic links in 1,3-alpha-D-glucans)

Reaction type
O-Glycosyl bond hydrolysis

Natural substrates
Glucose, alpha-1, 3-linked to high-mannose oligosaccharides N-linked to asparagine residues of glycoproteins [2, 3]

Substrate spectrum
1 $Glc_2 Man_9 GlcNAc_2 + H_2O$ (specific for alpha-(1–3) and alpha-(1–4) linkages of glucose [6], substrate specificity [9]) [2, 5, 7, 8]
2 4-Methylumbelliferyl-alpha-D-glucopyranoside + H_2O [1, 6]
3 p-Nitrophenyl-alpha-glucoside + H_2O [8]
4 Pseudonigeran + H_2O [10, 11]
5 Maltose + H_2O [8]
6 Alpha-1, 3-glucans + H_2O (exo-wise hydrolysis from non-reducing ends) [10]
7 More (does not action nigeran)

Enzyme Handbook © Springer-Verlag Berlin Heidelberg 1991
Duplication, reproduction and storage in data banks are only
allowed with the prior permission of the publishers

Product spectrum

1 $Man_9 GlcNAc_2$ + glucose
2 4-Methylumbelliferone + alpha-D-glucose
3 p-Nitrophenol + alpha-glucose
4 Alpha-glucose
5 Alpha-glucose
6 Alpha-glucose
7 ?

Inhibitor(s)

2, 6-Dideoxy-2, 6-imino-
7-O-(beta-D-glucopyranosyl)-D-glycero-L-glucoheptitol (MDL) [1];
Castanospermine [1]; Maltose [2, 6, 8]; Turanose [2, 6]; Ag^+ [2, 10]; Hg^{2+}
[2, 10]; Tris [6, 8]; p-Nitrophenyl-alpha-D-glucoside [7];
p-Nitrophenyl-beta-D-glucoside [7]; p-Nitrophenyl-alpha-D-mannoside [7];
p-Nitrophenyl-beta-D-mannoside [7]; p-Chloromercuribenzenesulfonate
[8]; Glucose [8]; D-Glucono-1 , 5-lactone [8]; Glycerol [8]; Fe^{3+} [10]; Fe^{2+}
[10]

Cofactor(s)/prostethic group(s)

Mannose (activation hydrolysis of p-nitrophenyl-alpha-glucoside) [8];
2-Deoxy-D-glucose (activation substrate p-nitrophenyl-alpha-glucoside) [8]

Metal compounds/salts

Turnover number (min^{-1})

Specific activity (U/mg)

0.096 [2]; 0.151 [6]; 11 [11]; More [5, 7, 8, 10]

K_m-value (mM)

0.013 (4-methylumbelliferyl-alpha-D-glucopyranoside, similar value [6]) [2];
0.85 (p-nitrophenyl-alpha-glucoside) [8]; 4.8 (maltose) [8]; 13 (pseudonigeran) [10]; 7.1 (pseudonigeran) [11]; More [10, 12]

pH-optimum

4.5 [10, 11]; 5.5 [12]; 6.0–7.5 [6]; 6.5–7.0 [2]; 6.6–7.0 (depending on substrate) [8]; 6.7 [7]

pH-range

3–8 [12]

Temperature optimum (°C)

47 [10]; 50 [12]; 55 [11]

Temperature range (°C)

30–60 [12]

3 ENZYME STRUCTURE

Molecular weight

400000 (pig, non denaturing gel electrophoresis) [6]
260000–290000 (cow, gel filtration [2], rat, gel electrophoresis, gel filtration
[8]) [2, 8]
100000–123000 (rat, immunoprecipitation [3], rat, SDS-PAGE with mercaptoethanol [5]) [3, 5]
88000 (Cladosporium resinae, gel filtration, SDS-PAGE) [10]
47000 (Trichoderma viride, gel filtration) [11]
36000 (Aspergillus nidulans, gel filtration) [12]

Subunits

Tetramer (4 × 63000–65000, cow, SDS-PAGE, comparison with other values
[2], rat, SDS-PAGE [8], 4 × 100000, pig, SDS-PAGE [6]) [2, 6, 8]

Glycoprotein/Lipoprotein

Glycoprotein [3, 5, 8]

---

4 ISOLATION/PREPARATION

Source organism

Mung bean [1, 13]; Cow [2]; Rat [3, 5, 8, 9]; Pig [4, 6]; Calf [7];
Cladosporium resinae [10]; Trichoderma [11]; Aspergillus nidulans [12]

Source tissue

Mammary gland [2]; Hepatoma cell line [3]; Kidney [4, 6]; Liver [5, 7–9]; Culture medium [10]; Culture homogenate [10]

Localisation in source

Microsomes (membranes) [2, 5, 6, 8]; Soluble part of cell [3]; Endoplasmic
reticulum (rough and smooth [4, 7], lumen [6]) [4, 6, 7]; Extracellular [10];
More (subcellular localisation) [4]

Purification

Cow [2]; Rat [5, 8]; Pig [6]; Calf [7]; Cladosporium resinae [10];
Trichoderma viride (from commercial cellulase preparation) [11]

Crystallization

–

Cloned

–

Renaturated

–

Enzyme Handbook © Springer-Verlag Berlin Heidelberg 1991
Duplication, reproduction and storage in data banks are only
allowed with the prior permission of the publishers

5 STABILITY

pH
3.5–7.0 (at 4°C) [10, 11]; 4.0–7.0 (at 30°C) [10]; 3.5–6.0 (at 30°C) [11]; 6.0–8.0 (with 5 mM mercaptoethanol) [8]; 6.0 (inactivation below) [6]

Temperature (°C)
36 [8]; 38 (at pH 7.0) [10]; 40 (inactivation) [8]; 43 (up to at pH 3.5) [10]; 47 (up to at pH 4.5) [10]; 50 (up to) [11]

Oxidation

Organic solvent

General stability information
Glycerol (stabilization) [6]

Storage
Liquid nitrogen, 100 mM phosphate buffer pH 7.0, 100 mM maltose, 5 mM mercaptoethanol, several weeks [8]; –70°C, 0.9% NaCl, several months [6]; 4°C, 0.1 M sodium phosphate buffer pH7, 0.5 mM DTT, 1% Triton [7]; 4°C, with Streptomyces pepsin inhibitor, at least 6 months [10]

---

6 CROSSREFERENCES TO STRUCTURE DATABANKS

PIR/MIPS code

Brookhaven code

---

7 LITERATURE REFERENCES

[1] Kaushal, G.P., Pan, Y.T., Tropea, J.E., Mitchell, M., Liu, P., Elbein, A.D.: J. Biol. Chem., 263, 17278–17283 (1988)
[2] Saxena, S., Shailubhai, K., Dong-Yu, B., Vijay, I.K.: Biochem. J., 247, 563–570 (1987)
[3] Strous, G.J., Van Kerkhof, P., Brok, R., Roth, J., Brada, D.: J. Biol. Chem., 262, 3620–3625 (1987)
[4] Lucocq, J.M., Brada, D., Roth J.: J. Cell Biol., 102, 2137–2146 (1986)
[5] Hino, Y., Rothman, J.E.: Biochemistry, 24, 800–805 (1985)
[6] Brada, D., Dubach, U.C.: Eur. J. Biochem., 141, 149–156 (1984)
[7] Tabas, I., Kornfeld, S.: Methods Enzymol., 83, 416–429 (1982)
[8] Burns, D.M., Touster, O.: J. Biol. Chem., 257, 9991–10000 (1982)
[9] Grinna, L.S., Robbins, P.W.: J. Biol. Chem., 255, 2255–2258 (1980)
[10] Tsunoda, A., Sakano, Y., Kobayashi, T.: Agric. Biol. Chem., 42, 1045–1053 (1978)
[11] Tsunoda, A., Nagaki, T., Sakano, Y., Kobayashi, T.: Agric. Biol. Chem., 41, 939–943 (1977)
[12] Zonneveld, B.J.M.: Biochim. Biophys. Acta, 258, 541–547 (1972)
[13] Szumilo, T., Kaushal, P., Elbein, A.D.: Arch. Biochem. Biophys., 247, 261–271 (1986)

1 NOMENCLATURE

EC number
3.2.1.85

Systematic name
6-Phospho-beta-D-galactoside 6-phosphogalactohydrolase

Recommended name
6-Phospho-beta-galactosidase

Synonymes
Phospho-beta-galactosidase
Beta-D-phosphogalactoside galactohydrolase
Phospho-beta-D-galactosidase
6-Phospho-beta-D-galactosidase

CAS Reg. No.
37237-42-6

---

2 REACTION AND SPECIFICITY

Catalysed reaction
A 6-phospho-beta-D-galactoside + H_2O →
→ an alcohol + 6-phospho-D-galactose

Reaction type
O-Glycosyl bond hydrolysis

Natural substrates
Lactose 6-phosphate + H_2O [1–12]

Substrate spectrum
1 6-Phospho-beta-D-galactosides + H_2O (r) [1–12]

Product spectrum
1 6-Phospho-D-galactose + alcohols (corresponding) [1–12]

Inhibitor(s)
Galactose-6-phosphate [7, 8, 10]; Co^{2+} [8]; Zn^{2+} [8]; Cu^{2+} [8, 9, 12];
Phosphoenolpyruvate [8]; ATP [8]; Glucose-6-phosphate [8]; Hg^{2+} [9, 12];
Mn^{2+} [9, 12]; Mg^{2+} [10]; Li^+ [10]; p-Hydroxymercuribenzoate [10]

Enzyme Handbook © Springer-Verlag Berlin Heidelberg 1991
Duplication, reproduction and storage in data banks are only
allowed with the prior permission of the publishers

Cofactor(s)/prostethic group(s)

Metal compounds/salts

Turnover number (min^{-1})

Specific activity (U/mg)
38–47 [9, 12]; 18.71 [10]; More [11]

K_m-value (mM)
1.59–3.0 (o-nitrophenyl-beta-D-galactopyranoside-6-phosphate) [7, 9, 11, 12]; 0.19–0.58 (o-nitrophenyl-beta-D-galactopyranoside-6-phosphate) [8, 10]

pH-optimum
4.3 [7]; 6.5 [8]; 7.0 [9, 12]; 5.0–8.0 [10]; 5.0 [11]

pH-range

Temperature optimum (°C)
45–52 [10]; 37 [11]

Temperature range (°C)
58 (active below, o-nitrophenyl-beta-D-galactopyranoside-6-phosphate) [11]

---

3 ENZYME STRUCTURE

Molecular weight
53989 (DNA sequence analysis, Lactobacillus casei) [1]
55096–56500 (DNA sequence analysis, gel electrophoresis, Streptococcus lactis) [2]
35000–56000 (gel filtration, gel electrophoresis, Lactobacillus casei) [5, 6]
40000 (gel filtration, Streptococcus mutans) [8]
46000–50000 (gel filtration, gel electrophoresis, Staphylococcus aureus) [9, 12]
67600 (gel filtration, Streptococcus cremoris) [10]
130000 (gel filtration, Lactobacillus casei) [11]

Subunits
Monomer (gel electrophoresis, Staphylococcus aureus) [9, 12]

Glycoprotein/Lipoprotein
–

4 ISOLATION/PREPARATION

Source organism
Streptococcus mutans [8]; Staphylococcus aureus [9, 12]; Streptococcus cremoris [10]; Lactobacillus casei [11]

Source tissue

Localisation in source
Cytoplasm [8]

Purification
Streptococcus mutans [8]; Staphylococcus aureus [9, 12]; Streptococcus cremoris [10]; Lactobacillus casei [11]

Crystallization
–

Cloned
[1–4, 6]

Renaturated
–

5 STABILITY

pH

Temperature (°C)
70 (up to) [10]

Oxidation

Organic solvent

General stability information

Storage
6 months, –70°C [8]

6 CROSSREFERENCES TO STRUCTURE DATABANKS

PIR/MIPS code
A29896 (Lactococcus lactis); A29897 (Lactobacillus casei)

Brookhaven code

Enzyme Handbook © Springer-Verlag Berlin Heidelberg 1991
Duplication, reproduction and storage in data banks are only
allowed with the prior permission of the publishers

7 LITERATURE REFERENCES

[1] Porter, E.V., Chassy, B.M.: Gene, 62, 263–276 (1988)
[2] Boizet, B., Villeval, D., Slos, P., Novel, M., Novel, G., Mercenier, A.: Gene, 62, 249–261 (1988)
[3] Breidt Jr., F., Stewart, G.C.: J. Bacteriol., 166, 1061–1066 (1986)
[4] Maeda, S., Gasson, M.J.: J. Gen. Microbiol., 132, 331–340 (1986)
[5] Jimeno, J., Casey, M., Hofer, F.: FEMS Microbiol. Lett., 25, 275–278 (1984)
[6] Lee, L.J., Hansen, J.B., Jaygusztyn-Krynicka, E.K., Chassy, B.M.: J. Bacteriol., 152, 1138–1146 (1982)
[7] Hall, B.G.: J. Bacteriol., 138, 691–698 (1979)
[8] Calmes, R., Brown, A.T.: Infect. Immun., 23, 68–79 (1979)
[9] Hengstenberg, W., Morse, M.L.: Methods Enzymol., 42, 491–494 (1975)
[10] Johnson, K.G., McDonald, I.J.: J. Bacteriol., 117, 667–674 (1974)
[11] Premi, L., Sandine, W.E., Elliker, P.R.: Appl. Microbiol., 24, 51–57 (1972)
[12] Hengstenberg, W., Penberthy, W.K., Morse, M.L.: Eur. J. Biochem., 14, 27–32 (1970)

1 NOMENCLATURE

EC number
3.2.1.86

Systematic name
6-Phospho-beta-D-glucosyl-(1, 4)-D-glucose glucohydrolase

Recommended name
6-Phospho-beta-glucosidase

Synonymes
Phospho-beta-glucosidase A
Phospho-beta-glucosidase
Phosphocellobiase

CAS Reg. No.
37205-51-9

---

2 REACTION AND SPECIFICITY

Catalysed reaction
6-Phospho-beta-D-glucosyl-(1, 4)-D-glucose + H_2O →
→ D-glucose 6-phosphate + D-glucose

Reaction type
O-Glycosyl bond hydrolysis

Natural substrates
6-Phospho-beta-D-glucosyl-(1, 4)-D-glucose + H_2O [1–4]

Substrate spectrum
1 6-Phospho-beta-D-glucosyl-(1, 4)-D-glucose + H_2O (ir) [1–4]
2 Phospho-beta glucosides + H_2O (ir) [1–4]

Product spectrum
1 D-Glucose 6-phosphate + D-glucose [1–4]
2 Glucose 6-phosphate + alcohols (corresponding) [1–4]

Inhibitor(s)
AMP [3]; Phosphoenolpyruvate [3]

Cofactor(s)/prostethic group(s)

Metal compounds/salts

Enzyme Handbook © Springer-Verlag Berlin Heidelberg 1991
Duplication, reproduction and storage in data banks are only
allowed with the prior permission of the publishers

Turnover number (min^{-1})

Specific activity (U/mg)
0.54 [2, 4]; 85–287 (isoenzymes) [3]

K_m-value (mM)
0.23–0.50 (cellobiose phosphate) [2, 4]; 0.18–0.33
(p-nitrophenyl-beta-D-glucopyranoside-6-phosphate) [3]; 0.46–0.60 (phenyl
beta-glucoside-6-phosphate) [3, 4]; 0.19–0.28 (methyl
beta-glucoside-6-phosphate) [3, 4]; 0.50 (salicin phosphate) [3, 4]; 0.24
(gentiobiose phosphate) [4]; More [3, 4]

pH-optimum
7.0–8.0 (cellobiose phosphate) [2, 4] 6.3
(p-nitrophenyl-beta-D-glucopyranoside-6-phosphate, isoenzyme A) [3];
7.0–9.0 (p-nitrophenyl-beta-D-glucopyranoside-6-phosphate, isoenzyme B)
[3]

pH-range
4.0–10.5 (p-nitrophenyl-beta-D-glucopyranoside-6-phosphate, isoenzyme
A) [3]

Temperature optimum (°C)

Temperature range (°C)

---

3 ENZYME STRUCTURE

Molecular weight
52000 (sucrose density gradient centrifugation, Aerobacter aerogenes) [2,
4]
132000–142000 (molecular sieve chromatography, Escherichia coli) [3]

Subunits
Dimer (identical, 2 × 65000–68000, gel electrophoresis) [3]

Glycoprotein/Lipoprotein
–

---

4 ISOLATION/PREPARATION

Source organism
Aerobacter aerogenes [2, 4]; Escherichia coli [3]

Source tissue

Localisation in source

Purification
   Aerobacter aerogenes [2, 4]; Escherichia coli [3]

Crystallization
   –

Cloned
   [1]

Renaturated
   –

---

5 STABILITY

pH

Temperature (°C)
   28 (isoenzyme B) [3]

Oxidation

Organic solvent

General stability information

Storage
   2 months, 4°C [2, 4]; 1 year, 4°C (isoenzyme A) [3]; 2 months, –20°C (isoenzyme B) [3]

---

6 CROSSREFERENCES TO STRUCTURE DATABANKS

PIR/MIPS code
   A25977 (Escherichia coli)

Brookhaven code

---

7 LITERATURE REFERENCES

[1] Schnetz, K., Toloczyki, C., Rak, B.: J. Bacteriol., 169, 2579–2590 (1987)
[2] Anderson, R.L., Palmer, R.E.: Methods Enzymol., 42 C, 494–497 (1975)
[3] Wilson, G., Fox, C.F.: J. Biol. Chem., 249, 5586–5598 (1974)
[4] Palmer, R.E., Anderson, R.L.: J. Biol. Chem., 247, 3420–3423 (1972)

Enzyme Handbook © Springer-Verlag Berlin Heidelberg 1991
Duplication, reproduction and storage in data banks are only
allowed with the prior permission of the publishers

1 NOMENCLATURE

EC number
3.2.1.87

Systematic name
Aerobacter-capsular-polysaccharide galactohydrolase

Recommended name
Capsular-polysaccharide endo-1, 3-alpha-galactosidase

Synonymes
Polysaccharide depolymerase
Galactohydrolase, capsular polysaccharide

CAS Reg. No.
62213-16-5

2 REACTION AND SPECIFICITY

Catalysed reaction
Random hydrolysis of 1, 3-alpha-D-galactosidic linkages in Aerobacter
aerogenes capsular polysaccharide

Reaction type
O-Glycosyl bond hydrolysis

Natural substrates
Polysaccharide + H_2O (capsular, produced by Aerobacter aerogenes) [2]

Substrate spectrum
1 Polysaccharide + H_2O (capsular, produced by Aerobacter aerogenes,
consists of galactose, mannose and glucuronic acid) [1, 2]
2 More (hydrolyses the galactosyl-alpha-1, 3-D-galactose linkages only in
the complex substrate, bringing about depolymerization)

Product spectrum
1 Oligosaccharides [1, 2]
2 ?

Inhibitor(s)
$FeCl_3$ [2]

Cofactor(s)/prostethic group(s)

Enzyme Handbook © Springer-Verlag Berlin Heidelberg 1991
Duplication, reproduction and storage in data banks are only
allowed with the prior permission of the publishers

Metal compounds/salts
More (addition of metal ion not required) [2]

Turnover number (min^{-1})

Specific activity (U/mg)
235 [2]

K$_m$-value (mM)

pH-optimum
5.2 [2]

pH-range
3.2–8.0 [2]

Temperature optimum (°C)
55 [2]

Temperature range (°C)

3 ENZYME STRUCTURE

Molecular weight
379000 (sedimentation coefficient, diffusion coefficient, Aerobacter
aerogenes, phage induced) [2]

Subunits
Polymer (octamer: 4 × 63200, 4 × 36400 or heptamer: 4 × 63200, 3 × 36400,
Aerobacter aerogenes) [2]

Glycoprotein/Lipoprotein
—

4 ISOLATION/PREPARATION

Source organism
Aerobacter aerogenes (phage induced) [1, 2]

Source tissue
Soluble [2]

Localisation in source

Purification
Aerobacter aerogenes [2]

Crystallization
–

Cloned
–

Renaturated
–

---

5 STABILITY

pH

Temperature (°C)
  68 (52% inactivation after 10 minutes) [2]

Oxidation

Organic solvent

General stability information

Storage

---

6 CROSSREFERENCES TO STRUCTURE DATABANKS

PIR/MIPS code

Brookhaven code

---

7 LITERATURE REFERENCES

[1] Yurewicz, E.C., Ghalambor, M.A., Heath, E.C.: J. Biol. Chem., 246, 5596–5606 (1971)
[2] Yurewicz, E.C., Ghalambor, M.A., Duckworth, D.H., Heath, E.C.: J. Biol. Chem., 246, 5607–5616 (1971)

Enzyme Handbook © Springer-Verlag Berlin Heidelberg 1991
Duplication, reproduction and storage in data banks are only
allowed with the prior permission of the publishers

1 NOMENCLATURE

EC number
3.2.1.88

Systematic name
Beta-L-arabinoside arabinohydrolase

Recommended name
Beta-L-arabinosidase

Synonymes
Arabinosidase, .beta.-L-
.beta.-L-Arabinosidase
Vicianosidase

CAS Reg. No.
39361-63-2

---

2 REACTION AND SPECIFICITY

Catalysed reaction
A beta-L-arabinoside + H_2O →
→ an alcohol + L-arabinose

Reaction type
O-Glycosyl bond hydrolysis

Natural substrates
Beta-L-arabinoside + H_2O

Substrate spectrum
1 Beta-L-arabinoside + H_2O
2 p-Nitrophenyl beta-L-arabinoside + H_2O [1]
3 More (not: alpha-D-galactoside, alpha-D-fucoside [2], p-nitrophenyl alpha-D-galactoside, p-nitrophenyl alpha-D-fucoside, p-nitrophenyl beta-D-galactoside) [2]

Product spectrum
1 Alcohol + L-arabinose
2 p-Nitrophenol + L-arabinose
3 ?

Enzyme Handbook © Springer-Verlag Berlin Heidelberg 1991
Duplication, reproduction and storage in data banks are only
allowed with the prior permission of the publishers

Inhibitor(s)

Cofactor(s)/prostethic group(s)

Metal compounds/salts

Turnover number (min^{-1})

Specific activity (U/mg)

K$_m$-value (mM)
  0.8 (p-nitrophenyl beta-L-arabinoside) [1]; 0.83
  (p-nitrophenyl-beta-L-arabinoside) [2]; More (very small effect of pH in the
  range 2.5–7.2) [1]

pH-optimum
  3–4.6 [2]

pH-range
  2–6.5 [2]

Temperature optimum (°C)
  30 (assay at) [1, 2]

Temperature range (°C)

---

3 ENZYME STRUCTURE

Molecular weight
  259000 (gel filtration, Cajanus indicus) [2]

Subunits

Glycoprotein/Lipoprotein
  –

---

4 ISOLATION/PREPARATION

Source organism
  Cajanus indicus [1, 2]

Source tissue
  Seeds [1, 2]

Localisation in source

Purification
  Cajanus indicus [2]

Crystallization
–

Cloned
–

Renaturated
–

---

5 STABILITY

pH
2–6.5 (30° C, 30 minutes) [2]

Temperature (°C)
4–40 [1]

Oxidation
Photooxidation in presence of Methylene Blue [1]

Organic solvent

General stability information

Storage

---

6 CROSSREFERENCES TO STRUCTURE DATABANKS

PIR/MIPS code

Brookhaven code

---

7 LITERATURE REFERENCES

[1] Dey, P.M.: Biochim. Biophys. Acta, 746, 8–13 (1983)
[2] Dey, P.M.: Biochim. Biophys. Acta, 302, 393–398 (1973)

Enzyme Handbook © Springer-Verlag Berlin Heidelberg 1991
Duplication, reproduction and storage in data banks are only
allowed with the prior permission of the publishers

1 NOMENCLATURE

EC number
3.2.1.89

Systematic name
Arabinogalactan 4-beta-D-galactanohydrolase

Recommended name
Arabinogalactan endo-1, 4-beta-galactosidase

Synonymes
Galactanase, endo-1, 4-.beta.-
Endo-1, 4-beta-galactanase
Arabinogalactanase
Endo-1, 4-beta-D-galactanase [1]
Endo-beta-1, 4-galactanase [6]

CAS Reg. No.
58182-40-4

2 REACTION AND SPECIFICITY

Catalysed reaction
Endohydrolysis of 1, 4-beta-D-galactosidic linkages in arabinogalactans

Reaction type
O-Glycosyl bond hydrolysis

Natural substrates
Galactan + H_2O (involved in degradation of host cell walls in development of Botrytis rot in apple) [5]

Substrate spectrum
1 Arabinogalactan + H_2O (soybean arabinogalactan) [9]
2 o-Nitrophenyl-beta-D-galactopyranoside + H_2O [2, 3]
3 Galactan + H_2O (soybean and lupin galactan) [6]
4 Galactan + H_2O (galactan from citrus pectin) [7]
5 Galactan + H_2O (soybean galactan) [10]
6 More (transfer from soybean arabinogalactan and glycerol [1], apple cell wall [4], attacks substrate in both an exo- and endo-manner [7], not: coffee bean arabinogalactan [10]) [1, 4, 7, 10]

Enzyme Handbook © Springer-Verlag Berlin Heidelberg 1991
Duplication, reproduction and storage in data banks are only
allowed with the prior permission of the publishers

Product spectrum
1 Galactose + galactobiose + galactotriose (not arabinose) [9]
2 Galactooligosaccharides (beta-1, 4-linked) +
  o-nitrophenyl-galactooligosaccharides (beta-1, 4-linked) +
  o-nitrophenol [2]
3 Galactose + arabinose + xylose (small amount) + rhamnose (small
  amount) [6]
4 Tetragalactose (4-linked, main product) [7]
5 Galactobiose (main product, no arabinose) [10]
6 ?

Inhibitor(s)
Hg^{2+} [3, 10]; Fe^{3+} [3, 10]; Ag^+ [10]; Lactose [3]; Galactose [3]; Arabinose
[3]; Glucose [3]; Xylose [3]

Cofactor(s)/prostethic group(s)

Metal compounds/salts
More (no metal ion required) [10]

Turnover number (min^{-1})

Specific activity (U/mg)
45.6 [4]; 0.0597 [7]

K_m-value (mM)
More (soybean arabinogalactan: I /0.043%, II /0.107%, III /0.113% [9]) [3, 9]

pH-optimum
4.5 [3]; 4.5–5.0 [6]; 6.0 [7, 10]; 6.0–7.0 [9]; More [8]

pH-range
4–8 (inactive below 4 and above 8) [7, 9]

Temperature optimum (°C)
55–60 [3, 9]; 50 [7]

Temperature range (°C)
20–60 [9]

---

3 ENZYME STRUCTURE

Molecular weight
67000 (gel filtration, Fusarium roseum) [8]
32000 (Penicillium citrinum, gel filtration) [3]
40000 (SDS-PAGE, Bacillus subtilis) [7, 11]
22000–24000 (gel filtration, Sclerotinia sclerotium) [6]
37000 (sedimentation equilibrium method, Bacillus subtilis) [10]

Subunits
Monomer (Penicillium citrinum, SDS-PAGE) [3]

Glycoprotein/Lipoprotein
Glycoprotein (Penicillium citrinum, about 3%) [3]

---

4 ISOLATION/PREPARATION

Source organism
Penicillium citrinum (2 forms: I, II [3]) [1–3]; Bacillus subtilis (3 forms: I, II, III [9]) [4, 7, 9, 10, 11]; Botrytis cinerea (2 forms: Gal I, Gal II) [5]; Sclerotinia sclerotium [6]; Fusarium roseum [8]

Source tissue
Cell [3]; Culture fluid [4, 9]; Culture filtrate [6, 7, 8]

Localisation in source

Purification
Penicillium citrinum [3]; Bacillus subtilis [4, 7, 9, 10]; Sclerotinia sclerotium [6]

Crystallization
[9, 10]

Cloned
–

Renaturated
–

---

5 STABILITY

pH
4–10 [3]; 5.5–10.5 (stable in presence of Ca^{2+}) [9]; 5.0–9.5 (30°C) [10]

Temperature (°C)
55 (10 minutes, stable below) [3]; 50 (stable below, in presence of Ca^{2+}) [9]; More (in presence of EDTA: enzyme heat-sensitive) [9]

Oxidation

Organic solvent

General stability information
Ca^{2+} (stabilizes) [9]; EDTA (unstable) [9]

Enzyme Handbook © Springer-Verlag Berlin Heidelberg 1991
Duplication, reproduction and storage in data banks are only
allowed with the prior permission of the publishers

Storage

4°C, pH 5.0, 50 mM sodium acetate, several weeks [6]; −20°C (little activity remains in frozen preparations) [6]

---

6 CROSSREFERENCES TO STRUCTURE DATABANKS

PIR/MIPS code

Brookhaven code

---

7 LITERATURE REFERENCES

[1] Nakano, H., Takenishi, S., Watanabe, Y.: Agric. Biol. Chem., 52, 1913–1921 (1988)
[2] Nakano, H., Takenishi, S., Watanabe, Y.: Agric. Biol. Chem., 50, 3005–3012 (1986)
[3] Nakano, H., Takenishi, S., Watanabe, Y.: Agric. Biol. Chem., 49, 3445–3454 (1985)
[4] Voragen, A.-G.-J., Geerst, F., Pilnik, W. in "Util. Enzymes Technol. Aliment. (Symp. Int.)" (Dupuy, P., Ed.) , 497–502 (1982)
[5] Urbanek, H., Zalewska-Sobczak, J.: Biochem. Physiol. Pflanz., 181, 321–329 (1986)
[6] Bauer, W.D., Bateman, D.F., Whalen, C.H.: Phytopathology, 67, 862–868 (1977)
[7] Labavitch, J.M., Freeman, L.E., Albersheim, P.: J. Biol. Chem., 251, 5904–5910 (1976)
[8] Mullen, J.M., Bateman, D.F.: Physiol. Plant Pathol., 6 , 233–246 (1975)
[9] Emi, S., Yamamoto, T.: Agric. Biol. Chem., 36, 1945–1954 (1972)
[10] Emi, S., Fukumoto, J., Yamamoto, T.: Agric. Biol. Chem. , 35, 1891–1898 (1971)
[11] Dekker, R.F.H. in "Biosynthesis And Biodegradation Of Wood Comp." (Higuchi, T., Ed.) , 505–533 (1985) (Review)

1 NOMENCLATURE

EC number
3.2.1.90

Systematic name
Arabinogalactan 3-beta-D-galactanohydrolase

Recommended name
Arabinogalactan endo-1, 3-beta-galactosidase

Synonymes
Endo-1, 3-beta-galactanase
Galactanase
Arabinogalactanase
Galactanase, endo-1, 3-.beta.
Endo-1, 3-beta-D-galactanase [1]

CAS Reg. No.
62213-17-6

2 REACTION AND SPECIFICITY

Catalysed reaction
Endohydrolysis of 1, 3-beta-D-galactosidic linkages in arabinogalactans

Reaction type
O-Glycosyl bond hydrolysis (endoglycosidic)

Natural substrates
Arabinogalactan + H_2O

Substrate spectrum
1 Arabinogalactan + H_2O (coffee arabinogalactan) [2]

Product spectrum
1 D-Galactose [1, 2] + arabinose [2] + 6-D-galactosyl-galactose + D-galactose oligosaccharides (1, 3-and 1, 6-linked) [1, 2]

Inhibitor(s)

Cofactor(s)/prostethic group(s)

Metal compounds/salts

Enzyme Handbook © Springer-Verlag Berlin Heidelberg 1991
Duplication, reproduction and storage in data banks are only
allowed with the prior permission of the publishers

Turnover number (min^{-1})

Specific activity (U/mg)

K_m-value (mM)

pH-optimum

pH-range

Temperature optimum (°C)

Temperature range (°C)

---

3 ENZYME STRUCTURE

Molecular weight

Subunits

Glycoprotein/Lipoprotein
  –

---

4 ISOLATION/PREPARATION

Source organism
  Rhizopus niveus [2]; More (overview) [1]

Source tissue

Localisation in source

Purification

Crystallization
  –

Cloned
  –

Renaturated
  –

---

5 STABILITY

pH

Temperature (°C)

Oxidation

Organic solvent

General stability information

Storage

---

6 CROSSREFERENCES TO STRUCTURE DATABANKS

PIR/MIPS code

Brookhaven code

---

7 LITERATURE REFERENCES

[1] Dekker, R.F.H. in "Biosynthesis And Biodegradation Of Wood Components" (Higuchi, T., Ed.) , 505–533 (1985) (Review)
[2] Hashimoto, Y.: J. Agric. Chem. Soc. Jpn., 45, 147–150 (1971)

Enzyme Handbook © Springer-Verlag Berlin Heidelberg 1991
Duplication, reproduction and storage in data banks are only
allowed with the prior permission of the publishers

1 NOMENCLATURE

EC number
3.2.1.91

Systematic name
1, 4-D-Glucan cellobiohydrolase

Recommended name
Cellulose 1, 4-beta-cellobiosidase

Synonymes

Cellobiohydrolase, exo-
.beta.-1, 4-Glucan cellobiohydrolase
.beta.-1, 4-Glucan cellobiosyl-
  hydrolase
Cellobiosidase, 1, 4-.beta.-glucan
Exo-cellobiohydrolase
1, 4-.beta.-Glucan cellobiosidase
1, 4-.beta.-Glucan cellobiohydrolase
1, 4-.beta.-D-Glucan
  cellobiohydrolase
Exocellobiohydrolase
Cellulase, C_1

Cellobiohydrolase
Exo-.beta.-1, 4-glucan
  cellobiohydrolase
C_1 cellulase
Cellobiohydrolase I
Exocellobiohydrolase (EC 3.2.1.91)
Cellobiosidase
CBH 1 [3]
Avicelase [13]
Exoglucanase [16]
Beta-1, 4-glucan cellobiohydrolase
[30]

CAS Reg. No.
37329-65-0

---

2 REACTION AND SPECIFICITY

Catalysed reaction
Hydrolysis of 1, 4-beta-D-glucosidic linkages in cellulose and cellotetrose, releasing cellobiose from the non-reducing ends of the chain

Reaction type
O-Glycosyl bond hydrolysis

Natural substrates
Cellulose + H_2O

Enzyme Handbook © Springer-Verlag Berlin Heidelberg 1991
Duplication, reproduction and storage in data banks are only
allowed with the prior permission of the publishers

Substrate spectrum

1 p-Nitrophenyl-lactoside + H_2O [1, 8, 37]
2 Cellulose + H_2O [1–37]
3 Cellodextrins + H_2O [1]
4 Cellotetraose + H_2O [1, 4, 5, 13, 32]
5 Cellotriose + H_2O [1, 6, 37]
6 Cellotriitol + H_2O [1]
7 Carboxymethylcellulose + H_2O (not significantly [4], not [31, 37], little
activity [32]) [2, 14]
8 Avicel + H_2O (not [37]) [2, 4, 5, 13, 14, 21]
9 Cellopentaose + H_2O [6, 13]
10 Cellohexaose + H_2O [6, 32]
11 p-Nitrophenyl cellobioside + H_2O [8, 17, 25, 35, 37, 38]
12 Celloligosaccharides + H_2O [37]
13 More (not: cellobiose [1, 6, 37], cellobiose slightly [24], hydrolyses internal glucosidic linkages of beta-1, 3, 1, 4-D-glucan [6]) [1, 6, 10, 28, 30,
33, 35, 37]

Product spectrum

1 p-Nitrophenol + lactose
2 More (different combinations and proportions of following substrates
(depending on organism): cellobiose, glucose, cellotriose) [1–37]
3 Cellobiose + glucose + cellotriose [1]
4 More (different combinations and proportions of following substrates:
cellobiose, glucose, cellotriose) [1, 5]
5 Cellobiose + glucose [1]
6 Cellobiose + sorbitol [1]
7 ?
8 Cellobiose (+ cellotriose, traces [6]) [4, 6, 14]
9 Cellobiose + glucose + cellotriose [6]
10 Cellobiose
11 p-Nitrophenol + cellobiose [8]
12 Cellobiose [37]
13 ?

Inhibitor(s)

o-Phenantroline [35]; Ag^+ (slight) [1]; Pb^+ (slight) [1]; Cu^{2+} (slight) [1];
Hg^{2+} [1]; Fe^{3+} [1]; N-Bromosuccinimide [1, 7, 10]; Cellobiose [1, 3, 21, 35];
Lactose [1]; D-Xylose [21]; Urea [21]; EDTA (slight) [35]; EGTA [35]

Cofactor(s)/prostethic group(s)

Metal compounds/salts

Co^{2+} (activates) [5]; Mn^{2+} (activates) [5]; Ca^{2+} (required for maximal activity) [35]

Turnover number (min^{-1})

Specific activity (U/mg)
0.320 [1]; 0.0571 [5]; 11.93 [8]; 0.151 (CBH I) [10]; 0.114 (CBH II) [10]; 6.67
[12]; More [6, 14, 21, 38]

K$_m$-value (mM)
3 (p-nitrophenyl-lactoside) [1]; 44.1 (Avicel, exo II) [2]; 12.0 (Avicel, exo III)
[2]; 0.1 (p-nitrophenyl cellobioside) [8]; 3.08
(p-nitrophenyl-beta-D-cellobioside) [35]; 0.40 (p-nitrophenyl cellobioside)
[37]; 0.27 (p-nitrophenyl cellobioside, periplasmic) [38]; 0.21 (p-nitrophenyl
cellobioside, extracellular) [38]; More (cellulose: 1.0 mg/ml [21],
H$_3$PO$_4$-swollen cellulose: 0.0067 [27]) [12, 21, 27]

pH-optimum
4–4.5 (II) [13]; 4–5 (I) [13]; 2.5 (I) [23]; 4–6 (thermophilic fungi) [36]; 5.0
(p-nitrophenyl-lactoside [1]) [1, 6]; 5.5–5.7 (exo I) [2]; 4.5–7 (exo III) 3.5–4.0
(amorphous cellulose) [1]; 4.0–5.0 (crystalline cellulose) [1]; 4.5–5.5 [21]; 4.5
(cellulose [7], II [23]) [7, 12, 23]; 4.2 (Avicel) [7]; 4.0 [14]; 4.8 [30]; 5.5–6.5
(Avicel) [4]; 6.2 [8]; 6.8 [25]; More [27]

pH-range
5.9–6.2 [37]; 5.0–7.5 (5.0: 50% of maximal activity, 7. 5: 25% of maximal activity at) [37]; More [8]

Temperature optimum (°C)
40–50 [4]; 39–45 [35]; 55–70 (thermophilic fungi) [30]; 50 (cellulose [7]) [5, 7,
12]; 37 (Avicel [7]) [7, 25]; 55 (I [13]) [13, 21]; 60 (II [13]) [13, 14, 27]; 45–50
[37, 38]

Temperature range (°C)
60–70 (60°C: optimum, 70°C: appreciable amount of activity) [27]; 23–60
(10% of maximal activity at) [37]

---

3 ENZYME STRUCTURE

Molecular weight
62000 (gel filtration, Trichoderma reesei) [4]
81000 (Cellulomonas uda, SDS-PAGE) [5]
65000 (gel filtration, Irpex lacteus) [6]
42000 (gel filtration, Sclerotium rolfsii) [7]
75000 (Bacteroides succinoges, SDS-PAGE) [8]
128000 (Aspergillus ficum, CBH I, SDS-PAGE) [10]
50000 (Aspergillus ficum, CBH I, SDS-PAGE) [10]
64000–68000 (gel filtration, Trichoderma reesei) [11]
84000 (gel filtration, Penicillium purpurogenum, I) [13]

Enzyme Handbook © Springer-Verlag Berlin Heidelberg 1991
Duplication, reproduction and storage in data banks are only
allowed with the prior permission of the publishers

64000 (gel filtration, Penicillium purpurogenum, II) [13]
64000 (gel filtration, Penicillium sp.) [14]
64000 (Trichoderma reesei, SDS-PAGE, CBH I) [15]
53000 (Trichoderma reesei, SDS-PAGE, CBH II) [15]
50300 (light scattering, Trichoderma reesei) [16]
58000 (Cellulomonas fini, SDS-PAGE) [17]
57000 (gel filtration, Fusarium lini, I) [21]
50000 (gel filtration, Fusarium lini, II) [21]
72000 (SDS-gel electrophoresis, Humicola insolens) [22]
46300 (Penicillium pinophilum, I) [23]
50700 (Penicillium pinophilum, II) [23]
65000 (Trichoderma reesei, SDS-PAGE) [24]
200000 (gel filtration, Ruminococcus albus) [25]
46300 (molecular-sieve chromatography, Penicillium funiculosum) [27]
41800 (sedimentation equilibrium analysis, Trichoderma viride) [30]
46000 (chromatography of reduced and alkylated enzyme, Trichoderma viride) [31]
230000 (gel filtration, Ruminococcus flavefaciens) [35]
40000 (gel filtration, Bacteroides succinogenes) [37]
50000 (SDS-PAGE, Bacteroides succinogenes) [38]
More [19]

Subunits

Monomer (carboxymethylated derivative, Sclerotium rolfsii) [7]
Dimer (2 × 100000, SDS-PAGE, Ruminococcus albus) [25]
Dimer (2 × 118000, SDS-PAGE, Ruminococcus flavefaciens)

Glycoprotein/Lipoprotein

Glycoprotein (Sporotrichum thermophile: 8% reducing sugar [1],
Trichoderma koningii: CBH 1 (9%), CBH 1a (33%) [4], Bacteroides succinogenes: (8–16%) [8], Aspergillus ficum: CBH I: (10.7%), CBH II (8.2%)
[10], Trichoderma viride: attached carbohydrate in addition to glycoprotein
constituents [18]) [1, 2, 4, 8, 10, 13–15, 18, 21–13, 29, 30, 31]

---

4 ISOLATION/PREPARATION

Source organism

Chrysosporium thermophile [1]; Trichoderma viride (Exo I /II/III [2]) [2, 16,
18, 29, 30, 31]; Trichoderma reesei (CBH I, CBH II [15]) [3, 11, 15, 16, 19, 24,
26]; Trichoderma koningii (CBH 1, CBH 1a) [4, 28, 32]; Cellulomonas uda
[5]; Irpex lacteus [6, 33]; Trichoderma harzianum [12]; Cellulomonas fini
(gene cloned in: E. coli [7], Saccharomyces cerevisiae [34]) [7, 34];
Sclerotium rolfsii [7]; Bacteroides succinogenes [8, 37, 38]; Aspergillus
ficum (CBH I, CBH II) [10]; Penicillium purpurogenum (2 forms: I/II) [13];

Penicillium sp. [14]; Clostridium stercorarium [20]; Fusarium lini (2 forms: I/II) [21]; Humicola insolens [22]; Humicola grisea [22]; Penicillium pinophilum (2 forms: I /II) [23]; Ruminococcus albus [25]; Penicillium funiculosum [27]; Ruminococcus flavefaciens [35]; Fungi (thermophilic) [36]

Source tissue
Culture filtrate [6, 8, 11, 12, 14, 16, 21, 24, 25, 29, 30, 35]; More (commercial product) [17]

Localisation in source
Extracellular [1, 38, 37]; Periplasm [37, 38]; Cytoplasm [37]

Purification
Chrysosporium thermophile [1]; Trichoderma viride [2, 18, 29 , 30, 31]; Trichoderma reesei [3, 11, 15, 16, 19, 24, 26]; Cellulomonas uda [5]; Cellulomonas fimi (gene cloned in E. coli) [17]; Irpex lacteus [6]; Sclerotium rolfsii [7]; Bacteroides succinogenes [8]; Aspergillus ficum [10]; Trichoderma harzianum [12]; Penicillium purpurogenum [13]; Penicillium sp. [14]; Fusarium lini [21]; Humicola insolens [22]; Humicola grisea [22]; Penicillium pinophilum [23]; Ruminococcus albus [25]; Penicillium funiculosum [27]; Ruminococcus flavefaciens [35]; Bacteroides succinogenes [38]

Crystallization
–

Cloned
(Trichoderma reesei [9], Cellulomonas fimi gene in E. coli [17], Cellulomonas fimi gene in Saccharomyces cerevisiae [34]) [9, 17, 34]

Renaturated
–

---

5 STABILITY

pH
3.0–10.0 (6°C, 48 hours) [14]; 3.5–9.5 [22]; 5.5–8.0 (highest stability) [25]; 5.5–8.0 (25°C, 24 hours) [5]; 4.5–5.0 (highest stability) [7]; 2.5–6.0 (4°C, 24 hours, CBH I) [10]; 2.0–7.0 (4°C, 24 hours, CBH II) [10]

Temperature (°C)
40 (3 hours) [1]; 80 (50% loss of activity after 3 hours) [1]; 60 (pH 5.5, 10 minutes, complete loss of activity [6], stable for 10 minutes [14]) [6, 14]; 100 (95% loss of activity after 10 minutes) [6]; 65 (60% loss of activity after 10 minutes /CBH I [10], stable for 5 minutes [22]) [10, 22]; 70 (CHB II: 40% loss of activity after 10 minutes) [10]; 78 (complete denaturation after 3 minutes) [30]

Enzyme Handbook © Springer-Verlag Berlin Heidelberg 1991
Duplication, reproduction and storage in data banks are only
allowed with the prior permission of the publishers

Oxidation

Organic solvent

General stability information
Freezing and thawing (stable) [7, 27]; Bovine serum albumin (stabilizes) [8, 38]; Dilute solutions (unstable) [21]; Ribonuclease (stabilizes) [38]

Storage
−15°C, pH 4.5, 50 mM citrate buffer [7]; −20°C, concentrated solutions, several months [21]; 0–4°C, 0.05 M acetate buffer, pH 5.0, at least 2 weeks [22]

---

6 CROSSREFERENCES TO STRUCTURE DATABANKS

PIR/MIPS code
EUTQI (I, precursor, Trichoderma reesei); A26160 (II, precursor, Trichoderma reesei); JS0083 (I Phanerochaete chrysosporium); S08674 (Ruminococcus flavefaciens); A24994 (Cellulomonas fimi); S08240 (Imperfect fungus, Humicola grisea); A26472 (II, precursor, Trichoderma reesei)

Brookhaven code

---

7 LITERATURE REFERENCES

[1] Fracheboud, D., Canevascini, G.: Enzyme Microb. Technol., 11, 220–229 (1989)
[2] Voragen, A.G.J., Beldman, G., Rombouts, F.M.: Methods Enzymol., 160, 243–251 (1988)
[3] Schülein, M.: Methods Enzymol., 160, 234–242 (1988)
[4] Wood, T.M.: Methods Enzymol., 160, 221–233 (1988)
[5] Nakamura, K., Kitamura, K.: Methods Enzymol., 160, 211–216 (1988)
[6] Kanda, T., Nisizawa, K.: Methods Enzymol., 160, 403–408 (1988)
[7] Sadana, J.C., Patil, R.V.: Methods Enzymol., 160, 307–314 (1988)
[8] Huang, L., Forsberg, C.W., Thomas, D.Y.: J. Bacteriol., 170, 2923–2932 (1988)
[9] Shoemaker, S., Schweickart, V., Ladner, M., Gelfand, M., Kwok, S., Myambo, K., Innis, M.: Bio/Technology, I, 691–696 (1983)
[10] Hayashida, S., Mo, K., Hosoda, A.: Appl. Environ. Microbiol., 54, 1523–1529 (1988)
[11] Riske, F.J., Eveleigh, D.E., Macmillan, J.D.: J. Ind. Microbiol., 1, 259–264 (1986)
[12] Sidhu, M.S., Kalra, M.K., Sandhu, D.K.: Folia Microbiol., 31, 293–302 (1986)
[13] Kamagata, Y., Sasaki, H., Takao, S.: J. Ferment. Technol., 64, 211–217 (1986)
[14] Funaguma, T., Tsuji, H., Hara, A.: J. Ferment. Technol., 64, 77–80 (1986)
[15] Bhikhabhai, R., Johansson, G., Pettersson, G.: J. Appl. Biochem., 6, 336–345 (1984)
[16] Odegaard, B.H., Anderson, P.C., Lovrien, R.E.: J. Appl. Biochem., 6, 156–183 (1984)
[17] Gilkes, N.R., Langsford, M.L., Kilburn, D.G., Miller, R.C., Warren, R.A.J.: J. Biol. Chem., 259, 10455–10459 (1984)
[18] Alurralde, J.L., Ellenrieder, G.: Enzyme Microb. Technol., 6, 467–690 (1984)

[19] Shoemaker, S., Watt, K., Tsitovsky, G., Cox, R.: Bio/Technology, 1, October, 687–690 (1983)

[20] Creuzet, N., Berenger, J.-F., Frixon, C.: FEMS Microbiol. Lett., 20, 347–350 (1983)

[21] Mishra, C., Vaidya, M., Rao, M., Deshpande, V.: Enzyme Microb. Technol., 5, 430–434 (1983)

[22] Hayashida, S., Ohta, K., Mo, K.: Methods Enzymol., 160, 323–332 (1988)

[23] Wood, T.M.: Methods Enzymol., 160, 398–403 (1988)

[24] Nummi, M., Niku-Paavola, M.-L., Lappalainen, A., Enari, T.-M., Raunio, V.: Biochem. J., 215, 677–683 (1983)

[25] Ohmiya, K., Shimizu, M., Taya, M., Shimizu, S.: J. Bacteriol., 150, 407–409 (1982)

[26] Fägerstam, L.G., Pettersson, L.G.: FEBS Lett., 119, 97–100 (1906)

[27] Wood, T.M., McCrae, S.I., MacFarlane, C.C.: Biochem. J., 189, 51–56 (1980)

[28] Wood, T.M., McCrae, S.I.: Biochem. J., 171, 61–72 (1978)

[29] Gum, E.K., Brown, R.D.: Biochim. Biophys. Acta, 446, 371–386 (1976)

[30] Berghem, L.E.R., Pettersson, L.G., Axiö-Fredriksson, U.-B.: Eur. J. Biochem., 53, 55–62 (1975)

[31] Berghem, L.E.R., Pettersson, L.G.: Eur. J. Biochem., 37, 21–30 (1973)

[32] Wood, T.M., McCrae, S.I.: Biochem. J., 128, 1183–1192 (1972)

[33] Kanda, T., Yatomi, H., Makishima, S., Amano, Y., Nisizawa, K.: J. Biochem., 105, 127–132 (1989)

[34] Wong, W.K.R., Curry, C., Parekh, R.S., Parekh, S.R., Wayman, M., Davies, R.W., Kilburn, D.G., Skipper, N.: Bio/Technology, 6, 713–719 (1988)

[35] Gardner, R.M., Doerner, K.C., White, B.A.: J. Bacteriol., 169, 4581–4588 (1987)

[36] Grajek, W.: Biotechnol. Lett., 8, 587–590 (1986)

[37] Li Huang, Forsberg, C.W.: Appl. Environ. Microbiol., 53, 1034–1041 (1987)

[38] Li Huang, Forsberg, C.W.: Appl. Environ. Microbiol., 54, 1488–1493 (1988)

Enzyme Handbook © Springer-Verlag Berlin Heidelberg 1991
Duplication, reproduction and storage in data banks are only
allowed with the prior permission of the publishers

1 NOMENCLATURE

EC number
3.2.1.92

Systematic name
Peptidoglycan beta-N-acetylmuramoylexohydrolase

Recommended name
Peptidoglycan beta-N-acetylmuramidase

Synonymes
Exo-beta-N-acetyl muramidase
Acetylmuramidase, exo-.beta.-
Exo-beta-N-acetylmuramidase
Beta-2-acetamido-3-O-(D-1-carboxyethyl)-2-deoxy-D-glucoside
acetamidodeoxyglucohydrolase [1]

CAS Reg. No.
52219-03-1

---

2 REACTION AND SPECIFICITY

Catalysed reaction
Hydrolysis of terminal, non-reducing N-acetylmuramic residues

Reaction type
O-Glycosyl bond hydrolysis

Natural substrates
More (hydrolysis of terminal, non-reducing N-acetylmuramic residues)

Substrate spectrum
1 4-Methylumbelliferyl-2-acetamido-3-O-(D-1-carboxyethyl)-2-deoxy-beta-D
-glucose + H_2O [1]
2 O-[2-Acetamido-3-O-(D-1-carboxyethyl)-2-deoxy-beta-D-glucopyranosyl]
-(1 --> 4)-2-acetamido-2-deoxy-D-glucose + H_2O [1]
3 Cell wall (of Micrococcus lysodeikticus) + H_2O [1]
4 O-2-Acetamido-2-deoxy-beta-D-glucopyranosyl-(1 -->
4)-2-acetamido-3-O-(D-1-carboxyethyl)-2-deoxy-D-glucose + H_2O [1]
5 4-Methylumbelliferyl N-acetylmuramide (4-methyl-2-oxo-1, 2-ben-
zopyran-7-yl-2-acetamido-4,
6-O-benzylidene-2-deoxy-beta-D-glucopyranoside) + H_2O [2]

Enzyme Handbook © Springer-Verlag Berlin Heidelberg 1991
Duplication, reproduction and storage in data banks are only
allowed with the prior permission of the publishers

Product spectrum

1 4-Methylumbelliferone + 2-acetamido-3-O-(D-1-carboxyethyl)-2-deoxy-beta-D-glucose

2 2-Acetamido-3-O-(D-1-carboxyethyl)-2-deoxy-beta-D-glucose + 2-acetamido-2-deoxy-D-glucose

3 ?

4 2-Acetamido-2-deoxy-beta-D-glucopyranose + 2-acetamido-3-O-(D-1-carboxyethyl)-2-deoxy-D-glucose

5 ?

Inhibitor(s)

2-Acetamido-3-O-(D-1-carboxyethyl)-2-deoxy-D-glucose [1, 3];
2-Acetamido-2-deoxy-D-glucose [1, 3]; Lactone [1, 3];
2-Acetamido-2-deoxy-D-galactose [1, 3]; 2-Acetamido-2-deoxy-D-mannose [1, 3]

Cofactor(s)/prostethic group(s)

Metal compounds/salts

Turnover number (min^{-1})

Specific activity (U/mg)
24.25 [1]

K_m-value (mM)
0.19 (4-methylumbelliferyl-2-acetamido-3-O-(D-1-carboxyethyl)-2-deoxy-beta-D-glucose) [1]; 0.65 (O-2-acetamido-3-O-(D-1-carboxyethyl)-2-deoxy-beta-D-glucopyranosyl-(1 --> 4)-2-acetamido-2-deoxy-D-glucose) [1]

pH-optimum
5.9–6.0 [3]

pH-range
4.5–10 [3]

Temperature optimum (°C)
37 (assay at) [1]

Temperature range (°C)

3 ENZYME STRUCTURE

Molecular weight
90000 (SDS-PAGE, Bacillus subtilis) [1]

Subunits

Glycoprotein/Lipoprotein
–

4 ISOLATION/PREPARATION

Source organism
Bacillus subtilis B [1, 3]

Source tissue
Culture medium [1]

Localisation in source
Extracellular [1]; More (partly in the medium, partly bound to the cell) [1]

Purification
Bacillus subtilis B [1]

Crystallization
–

Cloned
–

Renaturated
–

5 STABILITY

pH
8.0 (highest stability) [1]

Temperature (°C)

Oxidation

Organic solvent

General stability information

Storage

Enzyme Handbook © Springer-Verlag Berlin Heidelberg 1991
Duplication, reproduction and storage in data banks are only
allowed with the prior permission of the publishers

6 CROSSREFERENCES TO STRUCTURE DATABANKS

PIR/MIPS code

Brookhaven code

7 LITERATURE REFERENCES

[1] Del Rio, L.A., Berkeley, R.C.W.: Eur. J. Biochem., 65, 3–12 (1976)
[2] Del Rio, L.A., Berkeley, R.C.W.: Anal. Biochem., 66, 405–411 (1975)
[3] Del Rio, L.A., Berkeley, R.C.W., Brewer, S.J., Roberts, S.E.: FEBS Lett., 37, 7–9 (1973)

1 NOMENCLATURE

EC number
3.2.1.93

Systematic name
Alpha, alpha-trehalose-6-phosphate phosphoglucohydrolase

Recommended name
Alpha, alpha-phosphotrehalase

Synonymes
Phosphotrehalase

CAS Reg. No.
54576-93-1

---

2 REACTION AND SPECIFICITY

Catalysed reaction
Alpha, alpha-trehalose 6-phosphate + H_2O →
→ D-glucose + D-glucose 6-phosphate

Reaction type
O-Glycosyl bond hydrolysis

Natural substrates
Alpha, alpha-trehalose 6-phosphate + H_2O (trehalose metabolism) [1, 2]

Substrate spectrum
1 Alpha, alpha-trehalose 6-phosphate + H_2O [1, 2]

Product spectrum
1 D-Glucose + D-glucose 6-phosphate [1, 2]

Inhibitor(s)

Cofactor(s)/prostethic group(s)

Metal compounds/salts

Turnover number (min^{-1})

Specific activity (U/mg)
2.83 [2]

Enzyme Handbook © Springer-Verlag Berlin Heidelberg 1991
Duplication, reproduction and storage in data banks are only
allowed with the prior permission of the publishers

K_m-value (mM)
  1.8 (trehalose 6-phosphate) [2]

pH-optimum
  6.5–7.0 [2]

pH-range

Temperature optimum (°C)
  30 (assay at) [2]

Temperature range (°C)

---

3 ENZYME STRUCTURE

Molecular weight

Subunits

Glycoprotein/Lipoprotein
  –

---

4 ISOLATION/PREPARATION

Source organism
  Glycine max (Bradyrhizobium japonicum, bacteroids) [1]; Bacillus popilliae
  [2]

Source tissue
  Nodules (bacteroides) [1]

Localisation in source
  Soluble [1]; Cytoplasm [1]

Purification
  Bacillus popilliae [2]

Crystallization
  –

Cloned
  –

Renaturated
  –

5 STABILITY

pH

Temperature (°C)

Oxidation

Organic solvent

General stability information

Storage

6 CROSSREFERENCES TO STRUCTURE DATABANKS

PIR/MIPS code

Brookhaven code

7 LITERATURE REFERENCES

[1] Salminen, S.O., Streeter, J.G.: Plant Physiol., 81, 538–541 (1986)
[2] Bhumiratana, A., Anderson, R.L., Costilow, R.N.: J. Bacteriol., 119, 484–493 (1974)

Enzyme Handbook © Springer-Verlag Berlin Heidelberg 1991
Duplication, reproduction and storage in data banks are only
allowed with the prior permission of the publishers

1 NOMENCLATURE

EC number
3.2.1.94

Systematic name
1, 6-Alpha-D-glucan isomaltohydrolase

Recommended name
Glucan 1, 6-alpha-isomaltosidase

Synonymes
Exo-isomalto-hydrolase
Isomalto-dextranase
Isomaltohydrolase, exo
Isomaltodextranase
Dextranase, isomalto-
G2-dextranase [1]

CAS Reg. No.
56467-68-6

---

2 REACTION AND SPECIFICITY

Catalysed reaction
Hydrolysis of 1, 6-alpha-D-glucosidic linkages in polysaccharides so as to remove successive isomaltose units from the non-reducing ends of the chains

Reaction type
O-Glycosyl bond hydrolysis
Condensation [8]
More (splitting of (1 --> 2)-alpha-, (1 --> 3)-alpha-, (1 --> 4)-alpha- and (1 --> 6)-alpha-linkages) [7]

Natural substrates
Polysaccharides + H_2O (hydrolysis of 1, 6-alpha-D-glucosidic linkages as to remove successive isomaltose units from the non-reducing ends of the chain)

Enzyme Handbook © Springer-Verlag Berlin Heidelberg 1991
Duplication, reproduction and storage in data banks are only
allowed with the prior permission of the publishers

Substrate spectrum

1 Dextran + H_2O [1, 4, 5, 9]
2 Panose + H_2O [3]
3 Oligosaccharides + H_2O [1]
4 Panose + H_2O [9]
5 Isomaltotriitol + H_2O [9]
6 More (optimum activity is on those 1, 6-alpha-D-glucans containing 6, 7 and 8 glucose units, those containing 3, 4 and 5 glucose units are hydrolysed at slower rates)

Product spectrum

1 Isomaltose [1, 4, 5, 9]
2 Alpha-isomaltose [3]
3 Isomaltose [1]
4 Isomaltose + glucose [9]
5 Isomaltose + sorbitol [9]
6 ?

Inhibitor(s)

Ag^+ (partial) [1, 2]; N-Bromosuccinimide [1, 2]; Hg^{2+} [1, 2]; Fe^{3+} [1, 2]; $KMnO_4$ [1, 2]

Cofactor(s)/prostethic group(s)

Metal compounds/salts

Turnover number (min⁻¹)

Turnover number (min^{-1})

Specific activity (U/mg)

18.8 [1, 2]; 15.3 [9]

K_m-value (mM)

pH-optimum

5.3 [1, 2]; 5.0 [4]; 4.0–4.4 [9]

pH-range

3–7.5 [4]; 2.5–7 [9]

Temperature optimum (°C)

65 [1, 2]

Temperature range (°C)

30–70 [2]

3 ENZYME STRUCTURE

Molecular weight
69000 (SDS-PAGE, Arthrobacter globiformis) [2]

Subunits

Glycoprotein/Lipoprotein
–

4 ISOLATION/PREPARATION

Source organism
Arthrobacter globiformis T6 (dextranase and isopullulanase activity [1])
[1–7]; Actinomadura (strain R10) [4]; Achromobacter sp. [9]

Source tissue
Culture medium [1, 2, 9]

Localisation in source
Extracellular [4]

Purification
Arthrobacter globiformis [1, 2]; Actinomadura (strain R 10) [4]

Crystallization
–

Cloned
–

Renaturated
–

5 STABILITY

pH
2.8–8.0 (4° C, 24 hours) [1, 2]; 4.1–8.0 (45° C, 2 hours) [1]

Temperature (°C)
50 (10 minutes, stable below) [1]; 70 (10 minutes, 80% loss of activity) [1];
80 (10 minutes, complete inactivation) [1]

Oxidation
Photooxidation with Rose Bengal [1]

Organic solvent

Enzyme Handbook © Springer-Verlag Berlin Heidelberg 1991
Duplication, reproduction and storage in data banks are only
allowed with the prior permission of the publishers

General stability information

Storage

6 CROSSREFERENCES TO STRUCTURE DATABANKS

PIR/MIPS code

Brookhaven code

7 LITERATURE REFERENCES

[1] Okada, G., Takayanagi, T., Miyahara, S., Sawai, T.: Agric. Biol. Chem., 52, 829–836 (1988)

[2] Okada, G., Takayanagi, T., Sawai, T.: Agric. Biol. Chem., 52, 495–501 (1988)

[3] Takayanagi, T., Okada, G., Chiba, S.: Agric. Biol. Chem., 51, 2337–2341 (1987)

[4] Sawai, T., Ohara, S., Ichimi, Y., Okaji, S., Hisada, K., Fukaya, N.: Carbohydr. Res., 89, 289–299 (1981)

[5] Sawai, T., Tohyama, T., Natsumi, T.: Carbohydr. Res., 66, 195–205 (1978)

[6] Sawai, T., Ukigai, Y., Nawa, A.: Agric. Biol. Chem., 40, 1249–1250 (1976)

[7] Torii, M., Sakakibara, K., Misaki, A., Sawai, T.: Biochem. Biophys. Res. Commun., 70, 459–464 (1976)

[8] Sawai, T., Niwa, Y.: Agric. Biol. Chem., 39, 1077–1083 (1975)

[9] Sawai, T., Toriyama, K., Yano, K.: J. Biochem., 75, 105–112 (1974)

1 NOMENCLATURE

EC number
3.2.1.95

Systematic name
1, 6-Alpha-D-glucan isomaltotriohydrolase

Recommended name
Dextran 1, 6-alpha-isomaltotriosidase

Synonymes
Exo-isomaltotriohydrolase
Isomaltotriohydrolase, exo-
Exoisomaltotriohydrolase

CAS Reg. No.
72561-11-6

---

2 REACTION AND SPECIFICITY

Catalysed reaction
Hydrolysis of 1, 6-alpha-D-glucosidic linkages in dextrans, so as to remove
successive isomaltotriose units from the non-reducing ends of the chains

Reaction type
O-Glycosyl bond hydrolysis

Natural substrates
Dextran + H_2O

Substrate spectrum
1 Dextran + H_2O [1]
2 Isomaltodextrins (reduced) + H_2O [1]
3 Sephadex (G-100 /G-200) + H_2O [1]

Product spectrum
1 Isomaltotriose [1]
2 Isomaltotriose [1]
3 ?

Inhibitor(s)
Iodine [1]; $HgCl_2$ [1]; N-Bromosuccinimide [1]; $CuSO_4$ [1]; $AgNO_3$ [1]

Enzyme Handbook © Springer-Verlag Berlin Heidelberg 1991
Duplication, reproduction and storage in data banks are only
allowed with the prior permission of the publishers

Cofactor(s)/prostethic group(s)

Metal compounds/salts

Turnover number (min^{-1})

Specific activity (U/mg)
    34.3 [1]

K$_m$-value (mM)

pH-optimum
    7.0–7.5 [1]

pH-range
    6.5–8.0 (85% of maximum activity at 6.5 and 8.0) [1]

Temperature optimum (°C)
    37 (assay at) [1]

Temperature range (°C)

---

3 ENZYME STRUCTURE

Molecular weight

Subunits

Glycoprotein/Lipoprotein
    –

---

4 ISOLATION/PREPARATION

Source organism
    Brevibacterium fuscum var. dextranlyticum [1]

Source tissue
    Cell [1]

Localisation in source

Purification
    Brevibacterium fuscum var. dextranlyticum [1]

Crystallization
    –

Cloned
–

Renaturated
–

5 STABILITY

pH
5.6–11.0 (37° C, 12 hours) [1]

Temperature (°C)
50 (pH 7.7, 30 minutes) [1]; 60 (pH 7.5, 30 minutes, less than 30% loss of activity) [1]

Oxidation

Organic solvent

General stability information

Storage

6 CROSSREFERENCES TO STRUCTURE DATABANKS

PIR/MIPS code

Brookhaven code

7 LITERATURE REFERENCES

[1] Sugiura, M., Ito, A., Yamaguchi, T.: Biochim. Biophys. Acta, 350, 61–70 (1974)

Enzyme Handbook © Springer-Verlag Berlin Heidelberg 1991
Duplication, reproduction and storage in data banks are only
allowed with the prior permission of the publishers

1 NOMENCLATURE

EC number
3.2.1.96

Systematic name
Glycopeptide-D-mannosyl-N^4-(N-acetyl-D-glucosaminyl)$_2$-asparagine
1,4-N-acetyl-beta-glucosaminohydrolase

Recommended name
Mannosyl-glycoprotein endo-beta-N-acetyl glucosaminidase

Synonymes
Di-N-acetylchitobiosyl beta-N-acetylglucosaminidase
Endo-beta-N-acetylglucosaminidase
Acetylglucosaminidase, endo-.beta.
Endo-.beta.-N-acetylglucosaminidase
Endo-.beta.-acetylglucosaminidase
Endo-.beta.-(1.fwdarw.4)-N-acetylglucosaminidase
Endo-N-acetyl-.beta.-glucosaminidase
Endo-N-acetyl-.beta.-D-glucosaminidase
Endo-beta-N-acetylglucosaminidase H [9, 25 , 31]
Endo-beta-N-acetylglucosaminidase D [19, 24]
Endoglycosidase S [23]
Endo-beta-N-acetylglucosaminidase F [26]
Endo-beta-N-acetylglucosaminidase L [39]
Mannosyl-glycoprotein 1, 4-N-acetamidodeoxy-beta-D-glycohydrolase [44]
More (a group of related enzymes)
Endoglycosidase H (Boehringer)

CAS Reg. No.
37278-88-9

---

2 REACTION AND SPECIFICITY

Catalysed reaction
Endohydrolysis of the di-N-acetylchitobiosyl unit in high-mannose
glycopeptides and glycoproteins containing the
-[Man(GlcNAc)$_2$]Asn-structure. One N-acetyl-D-glucosamine residue
remains attached to the protein, the rest of the oligosaccharide is released
intact

Enzyme Handbook © Springer-Verlag Berlin Heidelberg 1991
Duplication, reproduction and storage in data banks are only
allowed with the prior permission of the publishers

Reaction type
O-Glycosyl bond hydrolysis

Natural substrates
Glycoproteins + H_2O

Substrate spectrum
1 N-Glycans + H_2O (high mannose type N-glycan)
2 Glycoproteins + H_2O (e.g.: ribonuclease B [7, 11, 13, 27], yeast invertase [7, 13], ovalbumin [11, 13, 26, 27, 29, 30], yeast carboxypeptidase [11], Ricinus lectin [11], Taka amylase A [14], IgG [17, 26, 28, 30], Saccharomyces cerevisiae invertase [27], glycoproteins of retrovirus, lymphocytic choriomeningitis virus, Pichinde virus and HLA-A and -B antigens, bromelain, ovomucoid, alpha$_1$-acid glycoprotein, influenza virus hemagglutinin [26], fetuin [26, 38], transferrin [38]) [1–45]
3 Glycopeptides + H_2O [1, 7, 23]
4 Dolichyl pyrophosphoryl oligosaccharides + H_2O [25]
5 Acetyl-Asn(GlcNAc)$_2$(Man)$_6$ + H_2O [29]
6 (Man)$_n$GlcNAcbeta1 --> 4NAcAsn + H_2O [36]
7 (GlcNAc)$_3$ + H_2O [39]
8 (GlcNAc)$_4$ + H_2O [39]
9 Glycoasparagine (asialo-N-acetyl-lactosaminic type) + H_2O [42]
10 More (high-mannose and complex glycoproteins [26], choline in the teichonic acid of cell wall substrate required for catalytic activity [3], C I and C II with different substrate specificity [17], enzyme from Streptococcus griseus and Diplococcus pneumoniae differ in substrate specificity [34], 2 forms with different substrate specificity [36]) [1, 3, 4, 5, 7, 8, 9, 11, 13, 15, 16, 18, 19, 23, 24, 25, 26, 27, 28, 30, 31, 33, 34, 35, 36, 37, 38, 39, 41, 44, 45]

Product spectrum
1 ?
2 Protein (with an attached N-acetyl-D-glucosamine residue) + oligosaccharide
3 Peptide (with an attached N-acetyl-D-glucosamine residue) + oligosaccharide
4 ?
5 Acetyl-AsnGlcNAc + GlcNAc(Man)$_6$ [29]
6 (Man)$_n$GlcNAc + GlcNAcAsn [36]
7 (GLcNAc)$_2$ + GlcNAc [39]
8 (GlcNAc)$_2$ [39]
9 GlcNAc-Asn + oligosaccharides [42]
10 ?

Inhibitor(s)
Mannose (not: endo-beta-N-acetylglucosaminidase H [32]) [19, 32]; Al-
pha-mannosides [19]; Lipoteichonic acid [21]; SDS [25]; Methyl al-
pha-mannoside (not: endo-beta-N-acetylglucosaminidase H) [32]; Yeast
mannan (endo-beta-N-acetylglucosaminidase) [32]; p-Nitrophenyl al-
pha-mannopyranoside [32]; Acetate (not: glucosaminidase 1) [1]; Cations
(inorganic mono- and divalent) [2]; Oligosaccharides (product inhibition)
[4]; $HgCl_2$ [13, 28]; EDTA [17, 33]; p-Chloromercuriphenyl-sulfonate [17, 33];
p-Chloromercuribenzoate [28]

Cofactor(s)/prostethic group(s)

Metal compounds/salts

Turnover number (min^{-1})
More [21]

Specific activity (U/mg)
13.4 [19]; 23.0 [28]; 4.67 [30]; 0.752 [31]; More [1, 3, 7, 8, 13, 14, 44, 17, 18, 21,
27, 33, 37, 39]

K$_m$-value (mM)
0.25 (acetylated glycopeptide) [30]; 0.032 (sidechain-free IgG glycopeptides) [33]; 0.050 ((Man)$_6$(GlcNAc)$_2$Asn) [33]; 0.20
((Man)$_5$(GlcNAc)$_2$Asn-acetyl) [38]; 0.7 (Man(GlcNAc)$_2$Asn-dansyl) [39]; 2.67
(hen ovalbumin) [4]; 0.3 (dansyl-Asn-(GlcNAc)$_4$(Man)$_6$) [28]; 0.30
(dansyl-Asn-(GlcNAc)$_2$(Man)$_6$) [13, 16]; 0.0868 (Taka-amylase A) [14]; 0.20
((Man)$_5$(GlcNAc)$_2$Asn-acetyl) [19]

pH-optimum
5–6 [13, 16, 28]; 4.5 (1) [1]; 3.5 (2 [1]) [1, 12]; 6.5 [3, 19, 30, 43]; 4.0–4.8 [14];
5.5 [18]; 5.0 [4, 21]; 4–6 [8]; 6.5–7.0 [17, 33]; 7.0 (pig) [29]; 4.0–4.5 [39]; 5.9
(F-I) [44]; 5.4 (F-II) [44]

pH-range
2.5–5.0 (2.5: about 70% of maximal activity, 5.0: about 20 % of maximal activity) [1]; 2–9 (about 10% of maximal activity at pH 2 and 9) [13]; 5–9 (less
than 50% of maximal activity at pH 5 and 9) [33]

Temperature optimum (°C)

Temperature range (°C)

Enzyme Handbook © Springer-Verlag Berlin Heidelberg 1991
Duplication, reproduction and storage in data banks are only
allowed with the prior permission of the publishers

3 ENZYME STRUCTURE

Molecular weight

27000 (sedimentation equilibrium analysis, SDS-PAGE, Streptomyces plicatus) [16]
56000 (hen, gel filtration) [18]
280000 (gel filtration, Diplococcus pneumoniae) [19]
150000 (SDS-PAGE, Diplococcus pneumoniae) [19]
27200 (sedimentation equilibrium analysis, Streptomyces griseus) [28]
280000 (gel chromatography, Diplococcus pneumoniae) [38]
49500 (sedimentation equilibrium, Streptomyces plicatus) [39]
52000 (gel filtration, fig, F-I) [44]
175000 (gel filtration, fig, F-II) [44]
64000 (SDS-PAGE, Streptococcus pneumoniae) [3]
32000 (Flavobacterium meningosepticum, SDS-PAGE) [8]
27000 (gel filtration, Flavobacterium sp.) [13]
30000 (SDS-PAGE, Flavobacterium sp.) [13]
31000 (gel filtration, Aspergillus oryzae) [14]

Subunits

Monomer (SDS-PAGE, Streptomyces griseus) [28]
Dimer (2 × 150000, SDS-PAGE, Diplococcus pneumoniae) [38]
Monomer (SDS-PAGE, Flavobacterium sp.) [13]
Dimer (2 × 90000, SDS-PAGE, Bacillus subtilis) [21]

Glycoprotein/Lipoprotein

–

4 ISOLATION/PREPARATION

Source organism

E. coli (recombinant strain) [16]; Human (2 forms: 1, 2 [1]) [1, 10, 43];
Staphylococcus simulans [2]; Streptococcus pneumoniae [3]; Flavobacterium sp. [4, 6, 11, 13]; Mucor hiemalis [5]; Jack bean [7]; Flavobacterium
meningosepticum [8]; Streptomyces plicatus [9, 16, 22, 39]; Streptococcus
griseus [25, 27, 28, 31, 34]; Rat [12, 15, 29, 40, 42, 45]; Hen [18, 35]; Bacillus
subtilis [21]; Pig [29]; Rabbit [41]; Aspergillus oryzae [14]; Clostridium perfringens (2 enzymes: C I / C II) [17]; Diplococcus pneumoniae [19, 24, 30,
34]; Dictyostelium discoideum [23]; Flavobacterium meningosepticum [26];
Clostridium perfringens (C I, C II) [33]; Fig (F-I, F-II [44]) [36, 37, 44]

Source tissue

Vegetative cells [23], Latex [36, 44]; Serum [41]; Brain [43]; Liver [12, 15, 29,
40, 42, 43, 45]; Oviduct [18, 35]; Kidney (not: kidney of sheep, cattle, pig
[10]) [1, 10, 29, 40, 43]; Cell [3, 14]; Meal [7]; Culture fluid [13, 17, 19, 30, 33,
37]; Culture filtrate [11, 16, 28, 39]; Spleen [29, 43]

Localisation in source
Cell-wall (bound) [3, 21]; Periplasmic space [22]; Lysosomes (not [40]) [12, 23]; Cytoplasm [15, 40, 42]

Purification
Human (2 forms: 1, 2) [1]; Streptococcus pneumoniae [3]; Jack bean [7]; Flavobacterium meningosepticum [8]; Streptomyces plicatus [9, 16, 39]; Streptomyces griseus [23, 31]; Flavobacterium sp. [11, 13]; Rat (partial) [15]; Hen [18, 35]; Aspergillus oryzae [14]; Clostridium perfringens [17, 33]; Diplococcus pneumoniae [19, 30]; Bacillus subtilis [21]; Dictyostelium discoideum (partial) [23]; Fig (partial [37]) [37, 44]; Rabbit [41]

Crystallization
–

Cloned
(Streptomyces plicatus enzyme in E. coli) [20]

Renaturated
–

---

5 STABILITY

pH
5–7 [13]; 4.0–6.0 (for at least 4 months) [14]; 4.5–8.5 (37°C, 48 hours) [28]; 3.2–8 (24 hours, F-II, 80–100% of activity) [44]; 5–8 (24 hours, F-I, unstable below 5) [44]

Temperature (°C)
37 (18 hours, unstable [8], 48 hours [28]) [8, 16, 28]; 50 (2 hours, less than 10% loss of activity) [14]; 60 (10 minutes, 90% loss of activity) [14]

Oxidation

Organic solvent

General stability information
Freezing (inactivates); Lyophilization (stable) [8, 28 , 39]; Repeated freezing and thawing (stable) [16, 28, 39]; Bovine serum albumin (stabilizes) [39]

Storage
–20°C (for at least 2 years) [16]; –70°C [18]; 4°C, toluene (for at least 1 year) [18]; –10°C–20°C (inactivation) [18]; 4°C, pH 8.45, dialyzed against 0.01 M potassium phosphate, concentrated by ultrafiltration (for at least 6 months) [28]

Enzyme Handbook © Springer-Verlag Berlin Heidelberg 1991
Duplication, reproduction and storage in data banks are only
allowed with the prior permission of the publishers

6 CROSSREFERENCES TO STRUCTURE DATABANKS

PIR/MIPS code

RBSMHP (H, precursor, Streptomyces plicatus)

Brookhaven code

7 LITERATURE REFERENCES

[1] DeGasperi, R., Li, Y.-T., Li, S.-C.: J. Biol. Chem., 264, 9329–9334 (1989)
[2] Bierbaum, G., Sahl, H.-G.: FEMS Microbiol. Lett., 58, 223–228 (1988)
[3] Garcia, P., Garcia, J.L., Garcia, E., Lopez, R.: Biochem. Biophys. Res. Commun., 158, 251–256 (1989)
[4] Kitabatake, N., Ishida, A., Yamamoto, K., Tochikura, T., Doi, E.: Agric. Biol. Chem., 52, 2511–2516 (1988)
[5] Kadowaki, S., Yamamoto, K., Fujisaki, M., Kumagai, H., Tochikura, T.: Agric. Biol. Chem., 52, 2387–2389 (1988)
[6] Kadowaki, S., Takegawa, K., Yamamoto, K., Kumagai, H., Tochikura, T.: Agric. Biol. Chem., 52, 2105–2106 (1988)
[7] Yet, M.-G., Wold, F.: J. Biol. Chem., 263, 118–122 (1988)
[8] Tarentino, A.L., Plummer, T.H.: Methods Enzymol., 138, 770–778 (1987)
[9] Trimble, R.B., Trumbly, R.J., Maley, F.: Methods Enzymol., 138, 763–770 (1987)
[10] Song, Z., Li, S.-C., Li, Y.-T.: Biochem. J., 248, 145–149 (1987)
[11] Yamamoto, K., Takegawa, K., Fan, J., Kumagai, H., Tochikura, T.: J. Ferment. Technol., 64, 397–403 (1986)
[12] Baussant, T., Strecker, G., Wieruszeski, J.-M., Montreuil, J., Michalski, J.-C.: Eur. J. Biochem., 159, 381–385 (1986)
[13] Yamamoto, K., Kadowaki, S., Takegawa, K., Kumaga, H., Tochikura, T.: Agric. Biol. Chem., 50, 421–429 (1986)
[14] Hitomi, J., Murakami, Y., Saitoh, F., Shigemitsu, N., Yamaguchi, H.: J. Biochem., 98, 527–533 (1985)
[15] Lisman, J.J.W., Van Der Wal, C., Overdijk, B.: Biochem. J., 229, 379–385 (1985)
[16] Tarentino, A.L., Trimble, R.B., Maley, F.: Methods Enzymol., 50 C, 574–580 (1978)
[17] Kobata, A.: Methods Enzymol., 50, 567–574 (1978)
[18] Tarentino, A.L., Maley, F.: Methods Enzymol., 50, 580–584 (1978)
[19] Muramatsu, T.: Methods Enzymol., 50, 555–559 (1978)
[20] Shimatake, H., Rosenberg, M.: Nature, 272, 128–132 (1981)
[21] Rogers, H.J., Taylor, C., Rayter, S., Ward, J.B.: J. Gen. Microbiol., 130, 2395–2402 (1984)
[22] Robbins, P.W., Trimble, R.B., Wirth, D.F., Hering, C., Maley, F., Maley, G.F., Das, R., Gibson, B.W., Royal, N., Bieman, K.: J. Biol. Chem., 259, 7577–7583 (1984)
[23] Freeze, H.H., Etchison, J.R.: Arch. Biochem. Biophys., 232, 414–421 (1984)
[24] Mizuochi, T., Amano, J., Kobata, A.: J. Biochem., 95, 1209–1213 (1984)
[25] Chalifour, R.J., Spiro, R.G.: Arch. Biochem. Biophys., 229, 386–394 (1984)
[26] Elder, J.H., Alexander, S.: Proc. Natl. Acad. Sci. USA, 79, 4540–4544 (1982)
[27] Tarentino, A.L., Plummer, T.H., Maley, F.: J. Biol. Chem., 249, 818–824 (1974)
[28] Tarentino, A.L., Maley, F.: J. Biol. Chem., 249, 811–817 (1974)
[29] Nishigaki, M., Muramatsu, T., Kobata, A.: Biochem. Biophys. Res. Commun., 59, 638–645 (1974)

[30] Koide, N., Muramatsu, T.: J. Biol. Chem., 249, 4897–4904 (1974)
[31] Arakawa, M., Muramatsu, T.: J. Biochem., 76, 307–317 (1974)
[32] Koide, N., Muramatsu, T.: Biochem. Biophys. Res. Commun., 66, 411–416 (1975)
[33] Ito, S., Muramatsu, T., Kobata, A.: Arch. Biochem. Biophys., 171, 78–86 (1975)
[34] Tarentino, A.L., Maley, F.: Biochem. Biophys. Res. Commun., 67, 455–462 (1975)
[35] Tarentino, A.L., Maley, F.: J. Biol. Chem., 251, 6537–6543 (1976)
[36] Chien, S.-F., Weinburg, R., Li, S.-C., Li, Y.-T.: Biochem. Biophys. Res. Commun., 76, 317–323 (1977)
[37] Ogata-Arakawa, M., Muramatsu, T., Kobata, A.: J. Biochem., 82, 611–614 (1977)
[38] Muramatsu, T., Koide, N., Maeyama, Ken-ichi: J. Biochem., 83, 363–370 (1978)
[39] Trimble, R.B., Tarentino, A.L., Evans, G., Maley, F.: J. Biol. Chem., 254, 9708–9713 (1979)
[40] Pierce, R.J., Spik, G., Montreuil, J.: Biochem. J., 180, 673–676 (1979)
[41] Delmotte, F., Kieda, C., Bouchard, M., Monsigny, M. in "Glycoconjugate Res." (Gregory, J.D., Jeanloz, R.W., Ed.) 2, 881–884 (1979)
[42] Pierce, R.J., Spik, G., Montreuil, J.: Biochem. J., 185, 261–264 (1980)
[43] Overdijk, B., Van Der Kroef, W.M.J., Lisman, J.J.W., Pierce, R.J., Montreuil, J., Spik, G.: FEBS Lett., 128, 364–366 (1981)
[44] Li, S.-C., Asakawa, M., Hirabayashi, Y., Li, Y.-T.: Biochim. Biophys. Acta, 660, 278–283 (1981)
[45] Tachibana, Y., Yamashita, K., Kobata, A.: Arch. Biochem. Biophys., 214, 199–210 (1982)

Enzyme Handbook © Springer-Verlag Berlin Heidelberg 1991
Duplication, reproduction and storage in data banks are only
allowed with the prior permission of the publishers

1 NOMENCLATURE

EC number
3.2.1.97

Systematic name
D-Galactosyl-N-acetyl-alpha-D-galactosamine
D-galactosyl-N-acetylgalactosaminohydrolase

Recommended name
Glycopeptide alpha-N-acetylgalactosaminidase

Synonymes
Endo-alpha-N-acetylgalactosaminidase
Endo-alpha-acetylgalactosaminidase

CAS Reg. No.
59793-96-3

---

2 REACTION AND SPECIFICITY

Catalysed reaction
Hydrolysis of terminal D-galactosyl-N-acetyl-alpha-D-galactosaminidic
residues from a variety of glycopeptides and glycoproteins

Reaction type
O-Glycosyl bond hydrolysis

Natural substrates
Glycopeptides and glycoproteins + H_2O (with serine or threonine
O-glycosidic linkages) + H_2O [1–6]

Substrate spectrum
1 Glycopeptides and glycoproteins (with serine or threonine O-glycosidic
   linkages) + H_2O (ir) [1–6]
2 Glycopeptides and glycoproteins (with serine or threonine O-glycosidic
   linkages) + H_2O [2]

Product spectrum
1 Galactosyl-alpha-N-acetyl-D-galactosamine + glycopeptides or
   glycoproteins (with free serine or threonine residues) [1–6]
2 Oligosaccharides + glycopeptides or glycoproteins (with free serine or
   threonine residues) [2]

Enzyme Handbook © Springer-Verlag Berlin Heidelberg 1991
Duplication, reproduction and storage in data banks are only
allowed with the prior permission of the publishers

Inhibitor(s)
Hg^{2+} [1, 5]; p-Chloromercuribenzenesulfonate [3, 6]; EDTA [3, 5, 6]; Mn^{2+}
[3, 5, 6]; Zn^{2+} [3, 6]; Galactose [5]

Cofactor(s)/prostethic group(s)

Metal compounds/salts
Mg^{2+} [3, 6]; Ca^{2+} [3, 6]

Turnover number (min^{-1})

Specific activity (U/mg)
2.18 [1]; 0.0115 [3, 6]; 40 [4]

K_m-value (mM)
1.0–3.7 (asialofetuin) [1, 5]; 3.2 (asialo kappa-casein glycopeptide) [1]; 0.20
(glycopeptide mixture) [3, 6]

pH-optimum
4.5–5.0 (asialofetuin) [1]; 6.0 (glycopeptide mixture) [3, 6]; 7.6 (asialofetuin)
[5]

pH-range
3.0 (not active below, asialofetuin) [1]

Temperature optimum (°C)
40–45 (asiolofetuin) [1]

Temperature range (°C)

---

3 ENZYME STRUCTURE

Molecular weight
160000 (gel filtration, Alcaligenes sp.) [1, 5]
190000 (gel electrophoresis, Diplococcus pneumoniae) [4]
160000 (gel filtration, Diplococcus pneumoniae) [5]

Subunits
Monomer (Alcaligenes sp., gel electrophoresis) [1]

Glycoprotein/Lipoprotein
–

---

4 ISOLATION/PREPARATION

Source organism
Alcaligenes sp. [1]; Diplococcus pneumoniae [3–6]

Source tissue

Localisation in source

Purification
Alcaligenes sp. [1]; Diplococcus pneumoniae [3–6]

Crystallization
–

Cloned
–

Renaturated
–

---

5 STABILITY

pH
4.5–6.5 [1]

Temperature (°C)
55 (not stable above) [1]

Oxidation

Organic solvent

General stability information

Storage
2 months, –20°C [4]; 6 months, 4°C [5]

---

6 CROSSREFERENCES TO STRUCTURE DATABANKS

PIR/MIPS code

Brookhaven code

Enzyme Handbook © Springer-Verlag Berlin Heidelberg 1991
Duplication, reproduction and storage in data banks are only
allowed with the prior permission of the publishers

7 LITERATURE REFERENCES

[1] Fan, J.Q., Kadowaki, S., Yamamoto, K., Kumagai, H., Tochikura, T.: Agric. Biol. Chem., 52, 1715–1723 (1988)
[2] Iwase, H., Ishii, I., Ishihara, K., Tanaka, Y., Omura, S., Hotta, K.: Biochem. Biophys. Res. Commun., 151, 422–428 (1988)
[3] Kobata, A., Takasaki, S.: Methods Enzymol., 50, 560–567 (1978)
[4] Glasgow, L.R, Paulson, J.C., Hill, R.L.: J. Biol. Chem., 252, 8615–8623 (1977)
[5] Umemoto, J., Bhavanandan, V.P., Davidson, E.A.: J. Biol. Chem., 252, 8609–8614 (1977)
[6] Endo, Y., Kobata, A.: J. Biochem., 80, 1–8 (1976)

1 NOMENCLATURE

EC number
3.2.1.98

Systematic name
1, 4-Alpha-D-glucan maltohexaohydrolase

Recommended name
Glucan 1, 4-alpha-maltohexaosidase

Synonymes
Exo-maltohexao-hydrolase
Maltohexaohydrolase, exo-
Exomaltohexaohydrolase
Exo-maltohexaose hydrolase

CAS Reg. No.
72561-12-7

---

2 REACTION AND SPECIFICITY

Catalysed reaction
Hydrolysis of 1, 4-alpha-O-glucosidic linkages in amylaceous polysaccharides so as to remove successive maltohexaose residues from the
non-reducing chain ends

Reaction type
O-Glycosyl bond hydrolysis

Natural substrates
Polysaccharides (amylaceous) + H_2O

Substrate spectrum
1 Starch + H_2O [1, 4]
2 Amylose (short-chain) + H_2O [3, 4]
3 Maltooligosaccharides (6 glucose units and shorter) + H_2O [3]
4 Glycogen + H_2O [4]
5 More (overview) [3]

Enzyme Handbook © Springer-Verlag Berlin Heidelberg 1991
Duplication, reproduction and storage in data banks are only
allowed with the prior permission of the publishers

Product spectrum
1 Maltohexaose (main product) [1]
2 Maltohexaose (main product) [3, 4]
3 Maltooligosaccharides (with less than 6 maltose units) [3]
4 ?
5 More (overview) [3]

Inhibitor(s)
Hg^{2+} [4]; Cu^{2+} [4]; Zn^{2+} [4]; $AgNO_3$ [4]; Iodoacetamide (weak) [1]; p-Chloromercuribenzoate (weak) [1]

Cofactor(s)/prostethic group(s)

Metal compounds/salts
Ca^{2+} (increases activity [4], not: immobilized enzyme [5]) [4]; Sr^{2+} (increases activity) [4]; Mn^{2+} (increases activity of immobilized enzyme) [5]

Turnover number (min^{-1})

Specific activity (U/mg)
124 [1]

K_m-value (mM)
0.46 (short-chain amylose) [3]; 50 (maltotetraose) [3]; 10.02 (maltopentaose) [3]; 7.68 (maltohexaose) [3]

pH-optimum
6.8 [4]

pH-range
3.0–10.0 (3.0: more than 70% of activity maximum, 10.0: more than 50% of activity maximum) [4]; 5.5–8.0 (more than 80% of maximum activity at 5.5 and 8.0) [4]

Temperature optimum (°C)
50 [4]; 52 (immobilized enzyme) [5]

Temperature range (°C)
20–80 [4]

3 ENZYME STRUCTURE

Molecular weight
54000 (gel filtration, Aerobacter aerogenes) [4]
73000 (SDS-PAGE, H-I-1, Bacillus sp.) [1]
59000 (SDS-PAGE, H-I-2, Bacillus sp.) [1]
80000 (SDS-PAGE, H-II, Bacillus sp.) [1]

Subunits

Glycoprotein/Lipoprotein
–

---

4 ISOLATION/PREPARATION

Source organism
Bacillus sp. H-167 (3 enzymes: H-I-1, H-I-2, H-II) [1]; Aerobacter aerogenes
(Klebsiella pneumoniae) [2, 3, 4, 5]

Source tissue
Culture medium [1]

Localisation in source
Extracellular [2, 3]

Purification
Bacillus sp. H-167 [1]; Aerobacter aerogenes [4]

Crystallization
–

Cloned
–

Renaturated
–

---

5 STABILITY

pH
6.0–9.0 (20% loss of activity, 60 minutes, 40° C) [4]; 4.0–11.0 (immobilized
enzyme) [5]

Temperature (°C)
40 (extremely unstable at) [4]; 60 (all activity lost after 15 minutes) [4]; More
(thermostability of immobilized enzyme increases about 10°C compared to
native enzyme) [5]

Oxidation

Organic solvent

General stability information
Substrate (protects) [4]

Storage

Enzyme Handbook © Springer-Verlag Berlin Heidelberg 1991
Duplication, reproduction and storage in data banks are only
allowed with the prior permission of the publishers

6 CROSSREFERENCES TO STRUCTURE DATABANKS

PIR/MIPS code

Brookhaven code

7 LITERATURE REFERENCES

[1] Hayashi, T., Akiba, T., Horikoshi, K.: Agric. Biol. Chem., 52, 443–448 (1988)

[2] Nakakuki, T., Azuma, K., Kainuma, K.: Carbohydr. Res., 128, 297–310 (1984)

[3] Monma, M., Nakakuki, T., Kainuma, K.: Agric. Biol. Chem., 47, 1769–1774 (1983)

[4] Kainuma, K., Wako, K., Kobayashi, S., Nogami, A., Suzuki, S.: Biochim. Biophys. Acta, 410, 333–346 (1975)

[5] Nakakuki, T., Hayashi, T., Monma, M., Kawashima, K., Kainuma, K.: Biotechnol. Bioeng., 25, 1095–1107 (1983)

1 NOMENCLATURE

EC number
3.2.1.99

Systematic name
1, 5-Alpha-L-arabinan 1, 5-alpha-L-arabinanohydrolase

Recommended name
Arabinan endo-1, 5-alpha-L-arabinosidase

Synonymes
Endo-1, 5-alpha-L-arabinanase
Arabinase, endo-1, 5-.alpha.-L-
Endo-alpha-1, 5-arabanase [5]
Endo-arabanase [5]

CAS Reg. No.
75432-96-1

---

2 REACTION AND SPECIFICITY

Catalysed reaction
Endohydrolysis of 1, 5-alpha-L-arabinofuranosidic linkages in
1,5-arabinans (also acts on beet-arabinan, slowly)

Reaction type
O-Glycosyl bond hydrolysis

Natural substrates
Arabinans + H_2O

Substrate spectrum
1. 1, 5-Alpha-L-arabinan + H_2O [1, 3, 4]
2. Arabinan + H_2O (beet and apple juice ultrafiltration retentate arabinan [1, 5]) [1, 3, 4, 5]
3. Arabinose trisaccharide + H_2O (weak) [1]
4. Arabinose tetrasaccharide + H_2O (weak) [1]
5. Arabinose pentasaccharide + H_2O [1]
6. Galacton + H_2O (potato galacton, weak) [1]
7. Cell walls + H_2O (apple cell wall [5], sycamore cell wall [3]) [3, 5]
8. More (potato-disc macerating activity [4], inactive towards p-nitrophenyl-alpha-L-arabinoside [1], not: p-nitrophenyl al-pha-L-arabinofuranoside [4], phenyl alpha-L-arabinofuranoside, p-nitrophenyl beta-D-galactopyranoside, arabinoxylan, gum arabic [3]) [1, 3, 4]

Enzyme Handbook © Springer-Verlag Berlin Heidelberg 1991
Duplication, reproduction and storage in data banks are only
allowed with the prior permission of the publishers

Product spectrum
1 Arabinose disaccharide + arabinose [1, 3]
2 Disaccharides + trisaccharides [1]
3 ?
4 ?
5 ?
6 ?
7 Arabinose (34%) + galactose (10%) [5]
8 ?

Inhibitor(s)
Hg^{2+} [3, 4]; Fe^{2+} [4]; Fe^{3+} [4]; More (L-arabonic-gamma-lactone, D- and L-galactonic-gamma-lactones) [4]

Cofactor(s)/prostethic group(s)

Metal compounds/salts

Turnover number (min^{-1})

Specific activity (U/mg)
1.2 (best substrate: 1, 5-alpha-L-arabinan) [1]; 46. 5 [5]; 48.8 [3]; More [4]

K_m-value (mM)
26 (apple juice ultrafiltration retentate arabinan) [1]; 9.3 (apple juice ultrafiltration retentate arabinan, linearised) [1]; 4.2 (1, 5-alpha-L-arabinan) [1]

pH-optimum
5.0 [1]; 6.0 [3, 4, 6]

pH-range
3–8 [4]; 4–9 [3, 6]

Temperature optimum (°C)
50 [1]

Temperature range (°C)

3 ENZYME STRUCTURE

Molecular weight
35000 (SDS-PAGE, Aspergillus niger) [1]
32000 (Bacillus subtilis) [3, 6]
33000 (SDS-PAGE, Bacillus subtilis) [4]

Subunits

Glycoprotein/Lipoprotein
Glycoprotein (Aspergillus niger) [1]

---

4 ISOLATION/PREPARATION

Source organism
Aspergillus niger [1]; Bacillus subtilis [3, 4, 5, 6]; Clostridium felsineum var. sikokianum [2]

Source tissue
Culture medium [3, 4, 6]

Localisation in source
Extracellular [6]

Purification
Aspergillus niger [1]; Bacillus subtilis [3, 4, 5, 6]

Crystallization
−

Cloned
−

Renaturated
−

---

5 STABILITY

pH

Temperature (°C)

Oxidation

Organic solvent

General stability information

Storage

---

6 CROSSREFERENCES TO STRUCTURE DATABANKS

PIR/MIPS code

Brookhaven code

Enzyme Handbook © Springer-Verlag Berlin Heidelberg 1991
Duplication, reproduction and storage in data banks are only
allowed with the prior permission of the publishers

7 LITERATURE REFERENCES

[1] Rombouts, F.M., Voragen, A.G.J., Searle-van Leeuwen, M.F., Geraeds, C.C.J.M., Schols & Pilnik, H.A.: Carbohydr. Polym., 9, 25–47 (1988)
[2] Kaji, A., Anabuki, Y., Taki, H., Oyama, Y., Okada, T.: Kagawa Daigaku Nogakubu Gakujutsu Hokoku, 15, 40–44 (1963)
[3] Kaji, A., Saheki, T.: Biochim. Biophys. Acta, 410, 354–360 (1975)
[4] Yoshihara, O., Kaji, A.: Agric. Biol. Chem., 47, 1935–1940 (1983)
[5] Voragen, A.-G.-J., Geerst, F., Pilnik, W. in "Util. Enzymes Technol. Aliment., Symp. Int." (Dupuy P., Ed.), 497–502 (1982)
[6] Weinstein, L., Albersheim, P.: Plant Physiol., 63, 425–432 (1979)

1 NOMENCLATURE

EC number
3.2.1.100

Systematic name
1, 4-Beta-D-mannan mannobiohydrolase

Recommended name
Mannan 1, 4-beta-mannobiosidase

Synonymes
Mannobiohydrolase, exo-1, 4-.beta.-
Exo-.beta.-mannanase
Exo-1, 4-beta-mannobiohydrolase
Exo-beta-Mannanase

CAS Reg. No.
81811-49-6

---

2 REACTION AND SPECIFICITY

Catalysed reaction
Hydrolysis of 1, 4-beta-D-mannosidic linkages in 1, 4-beta-D-mannans, so as to remove successive mannobiose residues from the non-reducing chain ends

Reaction type
O-Glycosyl bond hydrolysis
More (also catalyzes transglycosylation) [1]

Natural substrates
1, 4-Beta-D-mannans + H_2O
Galactomannans + H_2O (degradation) [2]

Substrate spectrum
1 Mannotriose + H_2O [1]
2 Mannotetraose + H_2O [1]
3 Mannopentaose + H_2O [1]
4 Codium + H_2O [1]
5 Coffee mannans + H_2O [1]
6 Mannobiosylmannitol + H_2O [1]
7 Mannotetraosylmannitol + H_2O [1]
8 More (not: mannobiose, p-nitrophenyl-beta-D-mannoside, konjac glucomannan, guar gum galactomannan)

Enzyme Handbook © Springer-Verlag Berlin Heidelberg 1991
Duplication, reproduction and storage in data banks are only
allowed with the prior permission of the publishers

Product spectrum

1 ?
2 ?
3 ?
4 Mannobiose (only) [1]
5 Mannobiose (only) [1]
6 Mannobiose + mannitol [1]
7 Mannobiose + mannosyl mannitol [1]
8 ?

Inhibitor(s)

Ag^+ [1]; Hg^{2+} [1]; Cu^{2+} [1]; Pb^{2+} [1]; Zn^{2+} [1]; Fe^{3+} [1]; EDTA [1];
p-Substituted mercuribenzoate [1]

Cofactor(s)/prostethic group(s)

Metal compounds/salts

Turnover number (min^{-1})

Specific activity (U/mg)

86.5 [1]

K_m-value (mM)

0.51 (mannobiose) [1]; 0.24 (mannotetraose) [1]; 0.13 (mannopentaose) [1]

pH-optimum

6.0 [1]

pH-range

Temperature optimum (°C)

37 (assay at) [1]

Temperature range (°C)

3 ENZYME STRUCTURE

Molecular weight

64000 (gel filtration, Aeromonas sp.) [1]

Subunits

Glycoprotein/Lipoprotein

—

2

4 ISOLATION/PREPARATION

Source organism
Aeromonas sp. F-25 [1]; Guar [2]

Source tissue
Culture fluid [1]; Endosperm [2]; Cotyledons [2]

Localisation in source
Extracellular [1]

Purification
Aeromonas sp. F-25 [1]

Crystallization
–

Cloned
–

Renaturated
–

5 STABILITY

pH
5.0–8.5 (20°C, 20 hours) [1]

Temperature (°C)
45 (15 minutes, stable below) [1]; 55 (15 minutes, complete loss of activity) [1]

Oxidation

Organic solvent

General stability information

Storage

6 CROSSREFERENCES TO STRUCTURE DATABANKS

PIR/MIPS code

Brookhaven code

7 LITERATURE REFERENCES

[1] Araki, T., Kitamikado, M.: J. Biochem., 91, 1181–1186 (1982)
[2] McCleary, B.V.: Phytochemistry, 22, 649–658 (1983)

Enzyme Handbook © Springer-Verlag Berlin Heidelberg 1991
Duplication, reproduction and storage in data banks are only
allowed with the prior permission of the publishers

1 NOMENCLATURE

EC number
3.2.1.101

Systematic name
1, 6-Beta-D-mannan mannanohydrolase

Recommended name
Mannan endo-1, 6-beta-mannosidase

Synonymes
Endo-alpha-1 -- > 6-D-mannanase [1]
Endo-1, 6-beta-mannanase

CAS Reg. No.

2 REACTION AND SPECIFICITY

Catalysed reaction
Random hydrolysis of 1, 6-beta-D-mannosidic linkages in unbranched
1,6-mannans (enzyme hydrolyzes 1, 6-alpha-D- and not
1,6-beta-D-mannosidic linkages) [1]

Reaction type
O-Glycosyl bond hydrolysis

Natural substrates
Beta-D-1, 6-mannans + H_2O

Substrate spectrum
1 Beta-D-1, 6-mannan (unbranched, with more than 80% alpha 1-- >
  6-D-mannosidic linkages) [1]
2 Mannooligosaccharides + H_2O (alpha$_1$ -- > 6-linked) [1]
3 More (enzyme hydrolyzes 1, 6-alpha-D- and not 1, 6-beta-D-mannosidic
  linkages [1], smallest substrate: alpha-1 -- > 6-mannotriose, not: methyl
  alpha-D-mannoside, p-nitrophenyl alpha-D-mannoside) [1]

Product spectrum
1 Mannose + alpha1 -- > 6-mannooligosaccharides (of various sizes) [1]
2 Mannose + mannobiose + mannotriose [1]
3 ?

Enzyme Handbook © Springer-Verlag Berlin Heidelberg 1991
Duplication, reproduction and storage in data banks are only
allowed with the prior permission of the publishers

Inhibitor(s)

Diisopropyl fluorophosphate [1]; Alpha 1--> 6-mannotriose (reduced) [1]; Cu^{2+} [1]; Fe^{2+} [1]; Mannan (Saccaromyces cerevisiae wilde type) [1]; Alpha 1 --> 6-mannohexaose [1]; p-Chloromercuribenzoate [1]

Cofactor(s)/prostethic group(s)

Metal compounds/salts

Ca^{2+} (required for full activity) [1]

Turnover number (min⁻¹)

440 [1]

Specific activity (U/mg)

1.87 [1]

K_m-value (mM)

4.0 (mannobiose) [1]; 1.0 (mannotriose) [1]; 0.33 (mannopentaose) [1]; 0.072 (mannohexaose) [1]; 0.0044 (alpha 1 --> 6-mannose) [1]

pH-optimum

5.5–7.0 [1]

pH-range

5.0–7.5 [1]

Temperature optimum (°C)

50 (pH 5.5) [1]

Temperature range (°C)

---

3 ENZYME STRUCTURE

Molecular weight

131000 (sedimentation equilibrium study, Bacillus circulans) [1]

Subunits

Monomer (SDS-PAGE, Bacillus circulans) [1]

Glycoprotein/Lipoprotein

–

---

4 ISOLATION/PREPARATION

Source organism

Bacillus circulans (probably) [1]

Source tissue
Culture filtrate [1]

Localisation in source
Extracellular [1]

Purification
Bacillus circulans (probably) [1]

Crystallization
–

Cloned
–

Renaturated
–

---

5 STABILITY

pH

Temperature (°C)
55 (presence of substrate, stable for at least 3 hours) [1]; 60–65 (5 minutes, denatured) [1]

Oxidation

Organic solvent

General stability information
Dialysis against EDTA (loss of activity) [1]

Storage
4°C, 0.1 M citrate phosphate buffer, pH 6.0 or 0.1 M potassium phosphate buffer, pH 6.8, 1 mM $CaCl_2$ (stable for at least 1 months) [1]

---

6 CROSSREFERENCES TO STRUCTURE DATABANKS

PIR/MIPS code

Brookhaven code

---

7 LITERATURE REFERENCES

[1] Nakajima, T., Maitra, S.K., Ballou, C.E.: J. Biol. Chem., 251, 174–181 (1976)

Enzyme Handbook © Springer-Verlag Berlin Heidelberg 1991
Duplication, reproduction and storage in data banks are only
allowed with the prior permission of the publishers

1 NOMENCLATURE

EC number
3.2.1.102

Systematic name
Blood-group-substance 1, 4-beta-D-galactanohydrolase

Recommended name
Blood-group-substance endo-1, 4-beta-galactosidase

Synonymes
Endo-beta-galactosidase

CAS Reg. No.

2 REACTION AND SPECIFICITY

Catalysed reaction
Endohydrolysis of 1, 4-beta-D-galactosidic linkages in blood group A and B substances (hydrolysis of 1,4-beta-D-galactosyl linkage adjacent to a 1,3-alpha-D-galactosyl or N-acetylgalactosaminyl residues and a 1,2-alpha-D-flucosyl residue)

Reaction type
O-Glycosyl bond hydrolysis

Natural substrates
Glycoproteins (1, 4-beta-D-galactosidic linkage) + H_2O

Substrate spectrum
1 Glycoproteins + H_2O (hydrolysis of 1, 4-beta-D-galactosyl linkage adjacent to a 1, 3-alpha-D-galactosyl or N-acetylgalactosaminyl residues and a 1, 2-alpha-D-fucosyl residue)
2 More [3–9]
3 Oligosaccharides + H_2O [1]

Product spectrum
1 Oligosaccharides
2 More [3–9]
3 Hydrolysed oligosaccharides

Inhibitor(s)
$HgCl_2$ [2, 6, 7]; Ag_2SO_4 [2, 6, 7]; p-Chloromercuribenzoate [2, 6, 7]; Cysteine [2, 7]

Enzyme Handbook © Springer-Verlag Berlin Heidelberg 1991
Duplication, reproduction and storage in data banks are only
allowed with the prior permission of the publishers

Cofactor(s)/prostethic group(s)

Metal compounds/salts

Turnover number (min^{-1})

Specific activity (U/mg)
  0.162 [2, 7]; 0.07 [6]

K$_m$-value (mM)
  More [7]

pH-optimum
  6.0 [2, 7]; 4.8–5.2 [6]

pH-range

Temperature optimum (°C)

Temperature range (°C)

---

3 ENZYME STRUCTURE

Molecular weight

Subunits

Glycoprotein/Lipoprotein
  –

---

4 ISOLATION/PREPARATION

Source organism
  Escherichia freundii [1, 4, 8]; Diplococcus pneumoniae [2, 6, 7]; Bacteroides fragilis [9]

Source tissue

Localisation in source

Purification
  Diplococcus pneumoniae [2, 6, 7]

Crystallization
  –

Cloned
  –

Renaturated
  –

5 STABILITY

pH

Temperature (°C)

Oxidation

Organic solvent

General stability information

Storage
Frozen [2, 7]

---

6 CROSSREFERENCES TO STRUCTURE DATABANKS

PIR/MIPS code

Brookhaven code

---

7 LITERATURE REFERENCES

[1] Fukuda, M.N., Matsumura, G.: J. Biol. Chem., 251, 6218–6225 (1976)
[2] Kobata, A., Takasaki, S.: Methods Enzymol., 50, 560–570 (1978)
[3] Mueller, T.J., Li, Y.T., Morrison, M.: J. Biol. Chem., 254, 8103–8106 (1979)
[4] Fukuda, M., Fukuda, M.N., Hakomori, S.I.: J. Biol. Chem., 254, 3700–3703 (1979)
[5] Scudder, P., Hanfland, P., Uemura, K.I, Feizi, T.: J. Biol. Chem., 259, 6586–6592 (1984)
[6] Fukuda, M.: Biochemistry, 24, 2154–2163 (1985)
[7] Takasaki, S., Kobata, A.: J. Biol. Chem., 251, 3603–3609 (1976)
[8] Fukuda, M.N., Fukuda, M., Hakomori, S.I.: J. Biol. Chem., 254, 5458–5465 (1979)
[9] Scudder, P., Lawson, A.M., Hounsell, E.F., Carruthers, R., Childs, R.A., Feizi, T.: Eur. J. Biochem., 168, 585–593 (1987)

1 NOMENCLATURE

EC number
3.2.1.103

Systematic name
Keratan-sulfate 1, 4-beta-D-galactanohydrolase

Recommended name
Keratan-sulfate endo-1, 4-beta-galactosidase

Synonymes
Endo-beta-galactosidase
Galactosidase, keratosulfate endo-. beta.-
Keratan sulfate endogalactosidase
Keratanase

CAS Reg. No.
55072-01-0

---

2 REACTION AND SPECIFICITY

Catalysed reaction
Endohydrolysis of 1, 4-beta-D-galactosidic linkages in keratan sulfate
(hydrolyses the 1,4-beta-D-galactosyl linkages adjacent to a 1,3-beta-D-N-acetylglucosaminyl residue, also acts on some non-sulfated oligo-saccharides, but only acts on blood group substances when the 1,2-linked fucosyl residues have been removed)

Reaction type
O-Glycosyl bond hydrolysis

Natural substrates
Keratan sulfate + H_2O

Substrate spectrum
1 Keratan sulfate (1, 4-beta-D-galactosyl linkage adjacent to a 1,3-beta-D-N-acetylglucosaminyl residue) + H_2O
2 Non sulfated polysaccharides (only acts on blood group substances when the 1, 2-linked fucosyl residues have been removed) + H_2O
3 More [12, 13, 18]
4 Sulfated glycoproteins + H_2O [4]
5 Glycospingolipids + H_2O [5, 7, 10, 14]
6 Polyglycosylceramids + H_2O [10]
7 Glycosaminglycan + H_2O [10]
8 Oligosaccharides + H_2O [4, 6, 7, 13, 15]

Enzyme Handbook © Springer-Verlag Berlin Heidelberg 1991
Duplication, reproduction and storage in data banks are only
allowed with the prior permission of the publishers

Product spectrum
1 Beta-D-2-acetamido-2-deoxy-6-O-sulfoglucosyl-(1, 3)-D-galactose [1]
2 ?
3 More [3, 7, 12–14, 18]
4 ?
5 ?
6 ?
7 ?
8 ?

Inhibitor(s)
Monosulfate disaccharide [1]; Na^+ [1]; Ca^{2+} [1], Ba^{2+} [1]; Mn^{2+} [1]; Zn^{2+} [1, 10, 11]; Fe^{3+} [11], Hg^{2+} [8–10]; Ag^+ [8–10]; Cu^{2+} [8–11], Fe^{2+} [11]; p-Chloromercuribenzoate [8–11]; Oligosaccharides [11]

Cofactor(s)/prostethic group(s)

Metal compounds/salts
K^+ (activates) [1]

Turnover number (min^{-1})

Specific activity (U/mg)
0.209 [1]; 0.004 [4, 10]; 0.5 [3]; 44 [6, 8]; 0.11 [7]; 156 [11]; 3.11 [15]

K_m-value (mM)
8.3 (polysaccharides) [1]; 0.35 (keratan sulfate) [6]; 3.91 (lacto-N-tetraose) [6]; 0.3 (paragloboside) [6]

pH-optimum
7.2–7.4 [3]; 5.5–5.8 [7, 10]; 6.0 [8, 9]; 5.7 [11]

pH-range
5.0–7.5 [1]; 6.5–8.0 [3]

Temperature optimum (°C)
37 [1, 3]; 55 [11]

Temperature range (°C)
30–40 [1, 3]

3 ENZYME STRUCTURE

Molecular weight
33000–35000 (Escherichia freundii, gel filtration, SDS-gel electophoresis) [6]
28000–30000 (Escherichia freundii, Flavobacterium keratolyticus, Bacteroides fragilis, gel filtration, gel electrophoresis) [7–11]

Subunits
  Monomer (Escherichia freundii, SDS-gel electrophoresis) [6]

Glycoprotein/Lipoprotein
  –

---

4 ISOLATION/PREPARATION

Source organism
  Pseudomonas sp. IFO 13309 [1–3, 14, 17]; Actinobacillus spec. IFO 133310 [2]; Escherichia freundii [4–7, 10, 14, 15–17]; Flavobacterium keratolyticus [8, 9, 14]; Cocobacillus sp. [9]; Bacteriodes fragilis [11, 13, 16]

Source tissue

Localisation in source

Purification
  Pseudomonas sp. [1, 3]; Escherichia freundii [6, 7, 15], Flavobacterium keratolyticus [8, 9]; Bacteroides fragilis [11, 13]

Crystallization
  –

Cloned
  –

Renaturated
  –

---

5 STABILITY

pH
  5–10 [8, 9]; 4.5–5.5 [10]; 3.5–9–0 [11]

Temperature (°C)
  37 (unstable) [8–11]

Oxidation

Organic solvent

General stability information
  Labile [1]; Dialyzed (unstable) [6]; Ca^{2+} (stabilizes) [8]; Bovine serum albumin (stabilizes) [11]

Storage
  –20°C [1, 3, 8, 9, 11]; –4°C (2–3 months) [11]

Enzyme Handbook © Springer-Verlag Berlin Heidelberg 1991
Duplication, reproduction and storage in data banks are only
allowed with the prior permission of the publishers

6 CROSSREFERENCES TO STRUCTURE DATABANKS

PIR/MIPS code

Brookhaven code

7 LITERATURE REFERENCES

[1] Horton, D.S.P.Q., Michelacci, Y.M.: Eur. J. Biochem., 161, 139–147 (1986)
[2] Nakazawa, K., Suzuki, N., Suzuki, S.: J. Biol. Chem., 250, 905–911 (1975)
[3] Nakazawa, K., Suzuki, S.: J. Biol. Chem., 250, 912–917 (1975)
[4] Fukuda, M., Matsumara, G.: Biochem. Biophys. Res. Commun., 64, 465–471 (1975)
[5] Fukuda, M., Watanabe, K., Hakomori, S.I.: J. Biol. Chem., 253, 6814–6819 (1978)
[6] Fukuda, M.: J. Biol. Chem., 256, 3900–3905 (1981)
[7] Nakagawa, H., Yamada, T., Chien, J.L., Gardas, A., Kitamikado, M., Li, D.C., Li, Y.T.: J. Biol. Chem., 255, 5955–5959 (1980)
[8] Kitamikado, M., Ito, M., Li, Y.T.: J. Biol. Chem., 256, 3906–3909 (1981)
[9] Kitamikado, M., Ito, M., Li, Y.T.: Methods Enzymol., 83, 619–625 (1982)
[10] Li, Y.T., Nakagawa, H., Kitamikado, M., Li, S.C.: Methods Enzymol., 83, 610–619 (1982)
[11] Scudder, P., Uemura, K.C., Dolby, J., Fukuda, M.N., Feizi, T.: Biochem. J., 213, 484–494 (1983)
[12] Hounsell, E.F., Feeny, J., Scudder, P., Tang, P.W., Feizi, T.: Eur. J. Biochem., 157, 375–384 (1986)
[13] Scudder, P., Tang, P.W., Hounsell, E.F., Lawson, A.M., Mehemet, H., Feizi, T.: Eur. J. Biochem., 157, 365–373 (1986)
[14] Ito, M., Hirabayashi, Y., Yamagata, T.: J. Biochem., 100, 773–780 (1986)
[15] Fukuda, M.N., Matsumura, G.: J. Biol. Chem., 251, 6218–6225 (1976)
[16] Scudder, P., Hanfland, P., Uemura, K.I., Feizi, T.: J. Biol. Chem., 259, 6586–6592 (1984)
[17] Fukuda, M.: Biochemistry, 24, 2154–2163 (1985)
[18] Scudder, P., Lawson, A.M., Hounsell, E.F., Carruthers, R.A., Childs, R.A., Feizi, T.: Eur. J. Biochem., 168, 585–593 (1987)

1 NOMENCLATURE

EC number
3.2.1.104

Systematic name
Cholesteryl-beta-D-glucoside glucohydrolase

Recommended name
Steryl-beta-glucosidase

Synonymes

CAS Reg. No.

2 REACTION AND SPECIFICITY

Catalysed reaction
Cholesteryl-beta-D-glucoside + H_2O →
→ cholesterol + D-glucose

Reaction type
O-Glycosyl bond hydrolysis

Natural substrates
Cholesteryl-beta-D-glucoside + H_2O

Substrate spectrum
1 Cholesteryl-beta-D-glucoside + H_2O
2 Sitosteryl-beta-D-glucoside + H_2O
3 More (does not act on some related sterols such as coprostanol)

Product spectrum
1 Cholesterol + D-glucose
2 Sitosterol + D-glucose
3 ?

Inhibitor(s)
Gluconolactone [1]; Amygdalin [1]; Androstenolone glucoside [1]; Sitostyrol glucoside [1]

Cofactor(s)/prostethic group(s)

Metal compounds/salts

Enzyme Handbook © Springer-Verlag Berlin Heidelberg 1991
Duplication, reproduction and storage in data banks are only
allowed with the prior permission of the publishers

Turnover number (min^{-1})

Specific activity (U/mg)
  More [1]

K$_m$-value (mM)

pH-optimum
  5.2–5.6 [1]

pH-range

Temperature optimum (°C)
  30 [1]

Temperature range (°C)

---

3 ENZYME STRUCTURE

Molecular weight
  65000 (Sinapis alba, gel filtration) [1]

Subunits

Glycoprotein/Lipoprotein
  –

---

4 ISOLATION/PREPARATION

Source organism
  Sinapis alba [1]

Source tissue
  Seedlings [1]

Localisation in source
  Cell membrane structures (1000–15000 g) [1]

Purification
  Sinapis alba seedlings [1]

Crystallization
  –

Cloned
  –

Renaturated
  –

5 STABILITY

pH

Temperature (°C)

Oxidation

Organic solvent

General stability information

Storage

6 CROSSREFERENCES TO STRUCTURE DATABANKS

PIR/MIPS code

Brookhaven code

7 LITERATURE REFERENCES

[1] Kalinowska, M., Wojciechowski, Z.A.: Phytochemistry, 17, 1533–1537 (1978)

Enzyme Handbook © Springer-Verlag Berlin Heidelberg 1991
Duplication, reproduction and storage in data banks are only
allowed with the prior permission of the publishers

1 NOMENCLATURE

EC number
3.2.1.105

Systematic name
Strictosidine beta-D-glucohydrolase

Recommended name
Strictosidine beta-glucosidase

Synonymes

CAS Reg. No.

2 REACTION AND SPECIFICITY

Catalysed reaction
Strictosidine + H_2O →
→ strictosidine aglycone + D-glucose

Reaction type
O-Glycosyl bond hydrolysis

Natural substrates
Strictosidine (precursor of indole alkaloids) + H_2O

Substrate spectrum
1 Strictosidine (precursor of indole alkaloids) + H_2O
2 More (does not act on a number of closely related glycosides)

Product spectrum
1 Strictosidine aglycone + D-glucose
2 ?

Inhibitor(s)
Gluconolacton [1]

Cofactor(s)/prostethic group(s)

Metal compounds/salts

Turnover number (min^{-1})

Specific activity (U/mg)
0.318 [1]

Enzyme Handbook © Springer-Verlag Berlin Heidelberg 1991
Duplication, reproduction and storage in data banks are only
allowed with the prior permission of the publishers

K_m-value (mM)
  0.1 (strictosidine, isoenzyme II) [1], 0.2 (strictosidine, isoenzyme I) [1]

pH-optimum
  6.0–6.4 [1]

pH-range
  5.5–8.0 [1]

Temperature optimum (°C)
  30 [1]

Temperature range (°C)

---

3 ENZYME STRUCTURE

Molecular weight
  230000–450000 (Catharanthus roseus, gel filtration) [1]

Subunits

Glycoprotein/Lipoprotein
  –

---

4 ISOLATION/PREPARATION

Source organism
  Catharanthus roseus [1]

Source tissue

Localisation in source

Purification

Crystallization
  –

Cloned
  –

Renaturated
  –

---

5 STABILITY

pH

Temperature (°C)

Oxidation

Organic solvent

General stability information

Storage

---

6 CROSSREFERENCES TO STRUCTURE DATABANKS

PIR/MIPS code

Brookhaven code

---

7 LITERATURE REFERENCES

[1] Hemscheidt, T., Zenk, M.H.: FEBS Lett., 110, 187–191 (1988)

Enzyme Handbook © Springer-Verlag Berlin Heidelberg 1991
Duplication, reproduction and storage in data banks are only
allowed with the prior permission of the publishers

1 NOMENCLATURE

EC number
3.2.1.106

Systematic name
Mannosyl-oligosaccharide glucohydrolase

Recommended name
Mannosyl-oligosaccharide glucosidase

Synonymes
$Glc_3Man_9NAc_2$ oligosaccharide glucosidase [2]
Glucosidase, mannosyloligosaccharide
Trimming glucosidase I

CAS Reg. No.
78413-07-7

---

2 REACTION AND SPECIFICITY

Catalysed reaction
Exohydrolysis of the non-reducing terminal glucose residue in the mannosyl-oligosaccharide $Glc_3Man_9GlcNAc_2$

Reaction type
O-Glycosyl bond hydrolysis

Natural substrates
$(Glucose)_3$-$(mannose)_9$-(N-acetylglucosamine)$_2$ + H_2O [1]
More (first step of processing of oligosaccharides after transfer from dolichyl pyrophosphate [2], involved in the formation of high-mannose and complex glycoproteins)

Substrate spectrum
1 $(Glucose)_3$-$(mannose)_9$-(N-acetylglucosamine)$_2$ [1–3]
2 $(Glucose)_2$-$(mannose)_9$-(N-acetylglucosamine)$_2$ (slight [2], not [1]) [2]
3 More (also acts, more slowly, on the corresponding glycolipids and glycopeptides, not: $(glucose)_1$-$(mannose)_9$-(N-acetylglucosamine)$_2$ [2], similar enzyme(s) from calf [4] and rat [5, 6] hydrolyze: $(glucose)_1$-$(mannose)_9$-(N-acetylglucosamine)$_1$, $(glucose)_2$-$(mannose)_9$-(N-acetylglucosamine)$_1$ and $(glucose)_3$-$(mannose)_9$-(N-acetylglucosamine)$_1$) [2, 4–6]

Enzyme Handbook © Springer-Verlag Berlin Heidelberg 1991
Duplication, reproduction and storage in data banks are only
allowed with the prior permission of the publishers

Product spectrum
1 Glucose + $Glc_2Man_9NAc_2$
2 Glucose + $Glc_2Man_9NAc_2$
3 ?

Inhibitor(s)
1-Deoxynojirimycin [1, 3]; Kojibiose [1]; N-Methyl-1-deoxynojirimycin [1];
N-Methyl-1-deoxynojirimycin [1]; N-5-Carboxypentyl-1-deoxynojirimycin [1];
p-Aminophenyl-beta-thioglucoside [2]; Octyl-beta-glucoside [2];
Tris/maleate [2]

Cofactor(s)/prostethic group(s)

Metal compounds/salts
More (metal ions not required) [1, 2]

Turnover number (min^{-1})

Specific activity (U/mg)
More [1, 3]

K_m-value (mM)

pH-optimum
6.7 [1]; 6.8 [2]; 6.2 [3]

pH-range
6.2–7.2 (half maximal activities at) [1]; 6.0–8.5 (6.0: 35 % of maximal activity,
8.5: 10% of maximal activity, 5.8: no activity) [2]

Temperature optimum (°C)
37 (assay at) [1]

Temperature range (°C)

3 ENZYME STRUCTURE

Molecular weight
95000 (SDS-PAGE, Saccharomyces cerevisiae) [1]
85000 (SDS-PAGE, calf) [3]
320000–350000 (gel chromatography, calf) [3]

Subunits
Tetramer (4 × 85000, SDS-PAGE, gel chromatography, calf) [3]

Glycoprotein/Lipoprotein
Glycoprotein (with high-mannose oligosaccharides) [1]

4 ISOLATION/PREPARATION

Source organism
Saccharomyces cerevisiae [1, 2]; Calf [3]; More (similar enzyme(s) from calf [4] and rat [5, 6] hydrolyze: (glucose)$_1$-(mannose)$_9$-(N-acetylglucosamine)$_1$, (glucose)$_2$-(mannose)$_9$-(N-acetylglucosamine)$_1$ and (glucose)$_3$-(mannose)$_9$-(N-acetylglucosamine)$_1$) [4–6]

Source tissue
Liver [3]; Cell [2]

Localisation in source
Microsomes [3]; More (distributed between particulate and supernatant fractions) [2]

Purification
Saccharomyces cerevisiae (partial [2]) [1, 2]; Calf [3]

Crystallization
–

Cloned
–

Renaturated
–

---

5 STABILITY

pH

Temperature (°C)

Oxidation

Organic solvent

General stability information
Bovine serum albumin (enhances stability during assay) [2]; Purification (quite stable at all stages) [2]; Lyophilization (enzyme obtained after step 3 of purification, stable to lyophilization without dialysis) [2]

Storage
4°C, presence of detergent ($t_{1/2}$: 20 days) [3]

Enzyme Handbook © Springer-Verlag Berlin Heidelberg 1991
Duplication, reproduction and storage in data banks are only
allowed with the prior permission of the publishers

6 CROSSREFERENCES TO STRUCTURE DATABANKS

PIR/MIPS code

Brookhaven code

7 LITERATURE REFERENCES

[1] Bause E., Erkens, R., Schweden, J., Jaenicke, L.: FEBS Lett., 206, 208–212 (1986)
[2] Kilker, R.D., Saunier, B., Tkacz, J.S., Herscovics, A. : J. Biol. Chem., 256, 5299–5303 (1981)
[3] Hettkamp, H., Legler, G., Bause, E.: Eur. J. Biochem., 142, 85–90 (1984)
[4] Michael, J.M., Kornfeld, S.: Arch. Biochem. Biophys., 199, 249–258 (1980)
[5] Grinna, L.S., Robbins, P.W.: J. Biol. Chem., 254, 8814–8818 (1979)
[6] Grinna, L.S., Robbins, P.W.: J. Biol. Chem., 255, 2255–2258 (1980)

1 NOMENCLATURE

EC number
3.2.1.107

Systematic name
Protein-alpha-D-glycosyl-1, 2-beta-D-galactosyl-L-hydroxylysine
glucohydrolase

Recommended name
Protein-glucosylgalactosylhydroxylysine glucosidase

Synonymes
2-O-Alpha-D-glucopyranosyl-5-O-beta-D-galactopyranosylhydroxy-L-lysine
glucohydrolase [3]

CAS Reg. No.

2 REACTION AND SPECIFICITY

Catalysed reaction
Protein alpha-D-glucosyl-1, 2-beta-D-galactosyl-L-hydroxylysine + H_2O →
→ D-glucose + protein beta-D-galactosyl-L-hydroxylysine

Reaction type
O-Glycosyl bond hydrolysis

Natural substrates
Protein alpha-D-glucosyl-1, 2-beta-D-galactosyl-L-hydroxylysine + H_2O
(degradation of hydroxylysine-linked disaccharides derived from collagen)
[1, 2]

Substrate spectrum
1 Protein alpha-D-glucosyl-1, 2-beta-D-galactosyl-L-hydroxylysine
   + H_2O [1]
2 2-O-Alpha-D-glucopyranosyl-O-beta-D-galactopyranosylhydroxylysine
   (N-acetylated) + H_2O [1–3]
3 2-O-Alpha-D-glucopyranosyl-O-beta-D-galactopyranosylhydroxylysine
   + H_2O [1–3]
4 Basement membrane + H_2O [1, 3]
5 More (requires free, positively charged epsilon-amino group of
   hydroxylysine) [1]

Enzyme Handbook © Springer-Verlag Berlin Heidelberg 1991
Duplication, reproduction and storage in data banks are only
allowed with the prior permission of the publishers

Product spectrum

1 D-Glucose + protein beta-D-galactosyl-L-hydroxylysine [1]
2 Glucose + galactosyl-hydroxylysine (N-acetylated)
3 Glucose + galactosyl-hydroxylysine
4 ?
5 ?

Inhibitor(s)

D-Glucono-1,5-lactone [3]; Glucosamine [1, 2]; Mannosamine [1, 2]; Cu^{2+} [1, 2]; Zn^{2+} [1, 2]; p-Chloromercuribenzoate [2, 3]; NaCl [3]

Cofactor(s)/prostethic group(s)

Metal compounds/salts

More (no divalent cation required) [3]

Turnover number (min^{-1})

Specific activity (U/mg)

18.7 [1]; 8.3 [2]

K_m-value (mM)

2.2 (2-O-alpha-D-glucopyranosyl-O-beta-D-galactopyranosylhydroxylysine) [1]; 5.9 (2-O-alpha-D-glucopyranosyl-O-beta-D-galactopyranosyl-hydroxylysine) [2]; 5.7 (glucogalactosylhydroxylysine) [3]

pH-optimum

5.3 [1]; 5.8 [2]; 4.4–4.7 [3]

pH-range

3.5–8 [1]; 3.0–7.5 (low activity at 3.0 and 7.5) [3]; 5.8–7.0 (5.8: optimum, 7.0: 50% of maximal activity) [2]

Temperature optimum (°C)

37 (assay at) [1, 2]

Temperature range (°C)

3 ENZYME STRUCTURE

Molecular weight

65000 (gel filtration, chicken) [1]
75000 (SDS-PAGE, chicken) [1]

Subunits

Glycoprotein/Lipoprotein

–

---

4 ISOLATION/PREPARATION

Source organism
Chicken [1]; Rat [2, 3]

Source tissue
Embryo homogenate [1]; Kidney (cortex) [3]; Spleen [2]; Lung [2]

Localisation in source

Purification
Chicken [1]; Rat [2, 3]

Crystallization

–

Cloned

–

Renaturated

–

---

5 STABILITY

pH

Temperature (°C)

Oxidation

Organic solvent

General stability information
Bovine serum albumin (stabilizes purified enzyme) [3]

Storage

Enzyme Handbook © Springer-Verlag Berlin Heidelberg 1991
Duplication, reproduction and storage in data banks are only
allowed with the prior permission of the publishers

6 CROSSREFERENCES TO STRUCTURE DATABANKS

PIR/MIPS code

Brookhaven code

7 LITERATURE REFERENCES

[1] Hamazaki, H., Hotta, K.: J. Biol. Chem., 254, 9682–9687 (1979)
[2] Hamazaki, H., Hotta, K.: Eur. J. Biochem., 111, 587–591 (1980)
[3] Sternberg, M., Spiro, R.G.: J. Biol. Chem., 254, 10329–10336 (1979)

1 NOMENCLATURE

EC number
3.2.1.108

Systematic name
Lactose galactohydrolase

Recommended name
Lactase

Synonymes
Lactase-phlorizin hydrolase [2]
Lactase/phlorizin hydrolase [3]
More (enzyme from intestinal mucosa is isolated as a complex which also catalyses the reaction of EC 3.2.1.62 (see EC 3.2.1.23))

CAS Reg. No.

2 REACTION AND SPECIFICITY

Catalysed reaction
Lactose + H_2O →
→ D-glucose + D-galactose

Reaction type
O-Glycosyl bond hydrolysis

Natural substrates
Lactose
Cellulose (possibly lactose/phlorizin hydrolase plays a role in final hydrolysis of cellulose in those species where cellulose is primarily attached by microorganisms) [3]

Substrate spectrum
1 Lactose + H_2O [2, 3, 4]
2 Phlorizin + H_2O(2.5% the rate of lactose hydrolysis) [2–4]
3 Cellobiose + H_2O (reduced rate) [2, 4]
4 Hetero-beta-glycosides + H_2O (synthetic hetero-beta-glucosides, reduced rate) [2]
5 Cellotriose + H_2O [3]
6 Cellotetraose + H_2O [3]
7 Cellulose + H_2O (low activity) [3]
8 p-Nitrophenyl-beta-galactoside + H_2O [4]
9 o-Nitrophenyl-beta-galactoside + H_2O [4]
10 p-Nitrophenol-beta-glucoside + H_2O [4]
11 More [3]

Enzyme Handbook © Springer-Verlag Berlin Heidelberg 1991
Duplication, reproduction and storage in data banks are only
allowed with the prior permission of the publishers

Product spectrum
1 D-Glucose + D-galactose
2 ?
3 Glucose + ?
4 ?
5 ?
6 ?
7 ?
8 p-Nitrophenol + galactose
9 o-Nitrophenyl + galactose
10 p-Nitrophenol + glucose
11 ?

Inhibitor(s)
Phlorizin (only lactase activity) [2, 3, 4]; Brain cerebrosides [2]; Tris [3, 4]

Cofactor(s)/prostethic group(s)

Metal compounds/salts

Turnover number (min^{-1})

Specific activity (U/mg)
22 [1]; 16.1 [2]; 18 [3]; 32 [4]

K_m-value (mM)
4.4 (cellobiose) [4]; 21 (lactose) [1]; 21.8 (lactose) [2]; 21 (lactose) [4]; 0.44
(phlorizin) [4]; 2.78 (cellobiose) [2]; 0.4 (phlorizin) [2]; 5.1 (lactose) [3]; More
(K_m for phlorizin too low to be determined exactly, less than 0.026) [3]

pH-optimum
5.0–5.5 [1]; 5.6–6.2 [3]; 5.6 [2]; 5.8–6.0 [4]

pH-range
4.8–7.0 (little activity below pH 4.8) [1]; 3.2–10 (little activity below pH 3.2, 10:
50% of maximal activity) [3]; 4.2–9.0 (little activity outside the range) [4]

Temperature optimum (°C)
37 (assay at) [1, 3]

Temperature range (°C)

3 ENZYME STRUCTURE

Molecular weight
320000 (gel filtration, pig [3], human [4], lactase-phlorizin hydrolase,
amphiphilic form) [3, 4]
280000 (gel filtration, pig [3], human [4], lactase-phlorizin hydrolase,
hydrophilic form) [3]

Subunits
Dimer (2 × 160000, SDS-PAGE, pig [3], human [4]) [3, 4]

Glycoprotein/Lipoprotein
More (one or both of the subunits carries a hydrophobic, detergent-bearing
segment) [3, 4]

---

4 ISOLATION/PREPARATION

Source organism
Rat [1]; Human [1, 3, 4]; Monkey (lactase-phlorizin hydrolase complex) [2];
Pig [3]

Source tissue
Small intestine [1–4]

Localisation in source

Purification
Rat [1]; Monkey (lactase-phlorizin hydrolase complex) [2]; Pig [3]; Human
[4]

Crystallization
–

Cloned
–

Renaturated
–

---

5 STABILITY

pH
5.5 (30% loss of activity after 0.5 minutes, 55% loss of activity after 120
minutes) [1]

Enzyme Handbook © Springer-Verlag Berlin Heidelberg 1991
Duplication, reproduction and storage in data banks are only
allowed with the prior permission of the publishers

Temperature (°C)
46 (neutral pH, inactivated) [1]; 37 (neutral pH, inactivated) [1]; 49 (60%
loss of activity after 45 minutes [3], 60% loss of activity after 10 minutes [4])
[3, 4]

Oxidation

Organic solvent

General stability information
Repeated thawing and freezing (loss of activity) [4]

Storage
−20°C, several months [1, 4]; 4°C, several weeks [4]

---

6 CROSSREFERENCES TO STRUCTURE DATABANKS

PIR/MIPS code

Brookhaven code

---

7 LITERATURE REFERENCES

[1] Schlegel-Haueter, S., Hore, P., Kerry, K.R., Semenza, G.: Biochim. Biophys. Acta, 258, 506–519 (1972)
[2] Ramswamy, S., Radhakrishnan, A.N.: Biochim. Biophys. Acta, 403, 446–455 (1975)
[3] Skovbjerg, H., Noren, O., Sjöström, H., Danielsen, E. M., Enevoldsen, B.S.: Biochim. Biophys. Acta, 707, 89–97 (1982)
[4] Skovbjerg, H., Sjöström, H., Noren, O.: Eur. J. Biochem., 114, 653–661 (1981)

1 NOMENCLATURE

EC number
3.2.1.109

Systematic name
Galactosylaminoglycan glycanohydrolase

Recommended name
Endogalactosaminidase

Synonymes

CAS Reg. No.

---

2 REACTION AND SPECIFICITY

Catalysed reaction
Endohydrolysis of galactosaminidic linkages in poly(D-galactosamine)

Reaction type
O-Glycosyl bond hydrolysis

Natural substrates
Galactosaminoglycan

Substrate spectrum
1 Oligogalactosaminoglycan + H_2O [1]
2 Galactosaminoglycan (high molecular weight, obtained from Neurospora) + H_2O [1]
3 More (sporeling detaching activity, not: N-acetyl-oligogalaactosaminoglycan and chitosan) [1]

Product spectrum
1 Oligogalactosaminoglycan (cleaved) [1]
2 Galactosaminoglycan (cleaved) [1]
3 ?

Inhibitor(s)
EDTA [1]; Cu^{2+} [1]

Cofactor(s)/prostethic group(s)

Metal compounds/salts

Enzyme Handbook © Springer-Verlag Berlin Heidelberg 1991
Duplication, reproduction and storage in data banks are only
allowed with the prior permission of the publishers

Turnover number (min^{-1})

Specific activity (U/mg)

K$_m$-value (mM)

pH-optimum
    5.4 (sporeling-detaching activity) [1]

pH-range
    3–8 (very low activity at pH 3 and 8) [1]

Temperature optimum (°C)
    35 (assay at) [1]

Temperature range (°C)

---

3 ENZYME STRUCTURE

Molecular weight

Subunits

Glycoprotein/Lipoprotein
    –

---

4 ISOLATION/PREPARATION

Source organism
    Streptomyces griseus [1]

Source tissue
    Culture filtrate [1]

Localisation in source

Purification
    Streptomyces griseus [1]

Crystallization
    –

Cloned
    –

Renaturated
    –

5 STABILITY

pH
2 (20°C, inactivated within a few minutes) [1]; 7 (quite stable at neutrality)
[1]

Temperature (°C)
100 (half-life: 3 minutes) [1]; 20 (pH 2, inactivated within a few minutes) [1];
More (heat stability increases at less acid pH) [1]

Oxidation

Organic solvent

General stability information
Repeated freezing and thawing (no loss of activity) [1]

Storage
–20°C, 8 months, distilled water, repeated freezing and thawing [1]

---

6 CROSSREFERENCES TO STRUCTURE DATABANKS

PIR/MIPS code

Brookhaven code

---

7 LITERATURE REFERENCES

[1] Reissig, J.L., Lai, H.-H., Glasgow, J.E.: Can. J. Biochem., 53, 1237–1249 (1975)

Enzyme Handbook © Springer-Verlag Berlin Heidelberg 1991
Duplication, reproduction and storage in data banks are only
allowed with the prior permission of the publishers

1 NOMENCLATURE

EC number
3.2.1.110

Systematic name
D-Galactosyl-3-(N-acetyl-beta-D-galactosaminyl)-L-serine
mucinaminohydrolase

Recommended name
Mucinaminylserine mucinaminidase

Synonymes
Endo-alpha-N-acetylgalactosaminidase [1]
Endo-alpha-N-acetyl-D-galactosaminidase [2, 3]
Endo-alpha-N-acetyl galactosaminidase [4]

CAS Reg. No.

---

2 REACTION AND SPECIFICITY

Catalysed reaction
D-Galactosyl-3-(N-acetyl-beta-D-galactosaminyl)-L-serine + H_2O →
→ D-galactosyl-3-N-acetyl-beta-D-galactosamine + L-serine

Reaction type
O-Glycosyl bond hydrolysis

Natural substrates
D-Galactosyl-3-(N-acetyl-beta-D-galactosaminyl)-L-serine + H_2O
D-Galactosyl-3-(N-acetyl-beta-D-galactosaminyl)-L-threonine + H_2O

Substrate spectrum
1 D-Galactosyl-3-(N-acetyl-beta-D-galactosaminyl)-L-serine + H_2O [1, 3]
2 D-Galactosyl-3-(N-acetyl-beta-D-galactosaminyl)-L-threonine + H_2O
  [1, 3]
3 p-Nitrophenyl-2-acetamido-2-deoxy-3-O-beta-D-galactopyranosyl-alpha-
  D-galactopyranoside (synthetic) + H_2O [2]
4 o-Nitrophenyl-2-acetamido-2-deoxy-3-O-beta-D-galactopyranosyl-alpha-
  D-galactopyranosides (synthetic) + H_2O [2]
5 More (not:
  methyl-2-acetamido-2-deoxy-3-O-beta-D-galactopyranosyl-alpha-D-
  galactopyranoside [2]) [2, 3, 4]

Enzyme Handbook © Springer-Verlag Berlin Heidelberg 1991
Duplication, reproduction and storage in data banks are only
allowed with the prior permission of the publishers

Product spectrum

1 D-Galactosyl-3-N-acetyl-beta-D-galactosamine + L-serine
2 D-Galactosyl-3-N-acetyl-beta-D-galactosamine + threonine
3 ?
4 ?
5 ?

Inhibitor(s)

EDTA [1, 3]; p-Chloromercuribenzenesulfonate [1]; Mn^{2+} [1, 3]; Hg^{2+} [3];
Zn^{2+} [1];
p-Nitrophenyl-2-acetamido-2-deoxy-3-O-beta-D-galactopyranosyl-beta-D-
galactopyranoside [2]

Cofactor(s)/prostethic group(s)

Metal compounds/salts

Mn^{2+} (probably includes a tightly bound magnesium or calcium ion) [1];
Ca^{2+} (probably includes tightly bound magnesium or calcium ion) [1]

Turnover number (min⁻¹)

Specific activity (U/mg)

0.001824 [1]

K_m-value (mM)

1.0 (asialofetuin glycopeptide fraction C) [3]; 0.26 (human erythrocyte membrane glycoprotein) [1]; 0.25
(p-nitrophenyl-2-acetamido-2-deoxy-3-O-beta-galactopyranosyl-alpha-
galactopyranoside) [2]; 0.23
(o-nitrophenyl-2-acetamido-2-deoxy-3-O-beta-galactopyranoside) [2]

pH-optimum

6.0 [1]; 5.5–7.0 [2]; 7.6 (asiafetuin glycopeptide fraction C) [3]

pH-range

5.5–7.0 (half-maximal activity at) [1]; 5.5–8.5 (5.5: 25% of maximal activity,
8.5: 55% of maximal activity) [3]

Temperature optimum (°C)

37 (assay at) [1]

Temperature range (°C)

3 ENZYME STRUCTURE

Molecular weight

160000 (gel filtration, Diplococcus pneumoniae) [3]

Subunits

Glycoprotein/Lipoprotein

—

---

4 ISOLATION/PREPARATION

Source organism
Diplococcus pneumoniae [1–4]

Source tissue
Culture medium [1]

Localisation in source

Purification
Diplococcus pneumoniae (partial [4]) [1, 3, 4]

Crystallization

—

Cloned

—

Renaturated

—

---

5 STABILITY

pH

Temperature (°C)

Oxidation

Organic solvent

General stability information

Storage
–20°C, 3 months (no loss of activity) [3]

---

6 CROSSREFERENCES TO STRUCTURE DATABANKS

PIR/MIPS code

Brookhaven code

Enzyme Handbook © Springer-Verlag Berlin Heidelberg 1991
Duplication, reproduction and storage in data banks are only
allowed with the prior permission of the publishers

7 LITERATURE REFERENCES

[1] Endo, Y., Kobata, A.: J. Biochem., 80, 1–8 (1976)
[2] Umemoto, J., Matta, K.L., Barlow, J.J., Bhavanandan, V. P.: Anal. Biochem., 91, 186–193 (1978)
[3] Umemoto, J., Bhavanandan, V.P., Davidson, E.A.: J. Biol. Chem., 252, 8609–8614 (1977)
[4] Bhavanandan, V.P., Umemoto, J., Davidson, E.A.: Biochem. Biophys. Res. Commun., 70, 738–745 (1976)

1 NOMENCLATURE

EC number
3.2.2.1

Systematic name
N-D-Ribosyl-purine ribohydrolase

Recommended name
Purine nucleosidase

Synonymes
Nucleosidase
Purine.beta.-ribosidase
Purine nucleoside hydrolase
Purine ribonucleosidase [6]
Ribonucleoside hydrolase
Nucleoside hydrolase
N-Ribosyl purine ribohydrolase [7, 8]
Nucleosidase g [17]

CAS Reg. No.
9025-44-9

---

2 REACTION AND SPECIFICITY

Catalysed reaction
An N-D-ribosylpurine + H_2O →
→ a purine + D-ribose

Reaction type
N-Glycosyl bond hydrolysis

Natural substrates
Purine nucleosides + H_2O (ureide metabolism [1, 2], purine salvage [21], purine metabolism [4, 5]) [1, 2, 4, 5, 21]

Enzyme Handbook © Springer-Verlag Berlin Heidelberg 1991
Duplication, reproduction and storage in data banks are only
allowed with the prior permission of the publishers

Substrate spectrum
1 Purine nucleosides + H_2O (ir [18, 19], xanthosine [1, 4, 18], inosine [1, 4, 6, 8, 17, 18, 19, 20], guanosine [1, 6, 17, 19, 20, 21], adenosine [1, 4, 17, 18, 19, 20, 21]) [1, 4, 6, 8, 17–21]
2 Pyrimidine nucleosides (uridine [1, 4, 15], thymidine [1], cytidine [1, 19], pyrimidine nucleosides not hydrolyzed [6]) [1, 4, 15, 19]
3 NAD^+ + H_2O [8]
4 5'-AMP + H_2O [8, 17]
5 $NADP^+$ + H_2O [8]
6 Deamino-NAD^+ + H_2O [8]
7 More (only ribofuranosides hydrolysed [18], not: adenosine [6], xanthosine [6, 17], pyrimidine nucleosides [6], purine 2'-deoxyribonucleosides deoxyribosides [6], uridine [18, 20], cytidine [4, 18, 20], thymidine [20]) [4, 6, 8 , 15, 17, 18, 19, 20]

Product spectrum
1 Purine + D-ribose (ir [18, 19])
2 Pyrimidine + D-ribose
3 Adenine + nicotinamide ribose diphosphate ribose [8]
4 Adenine + ribose monophosphate
5 Adenine + nicotinamide ribose diphosphate ribose phosphate [8]
6 Adenine + deamino nicotinamide diphosphate ribose
7 ?

Inhibitor(s)
EDTA [20]; Iodoacetate [20]; NaCN [20]; 5'-Chloro-adenosine [17]; 7-Beta-adenosine [17]; Adenosine monophosphate [17]; $HgCl_2$ [8, 20]; Uric acid riboside [19]; Hypoxanthine [15]; Pyrazolo (3, 4-d)pyrimidine analogue (of adenosine) [6]; Inosine [1, 15]; Cytidine [15]; Adenosine [1, 6, 15, 17, 18]; Guanosine [1, 15, 18]; Uridine [1]; Xanthosine (not [17]) [1, 6, 19]; 2'-Deoxyadenosine [6]; Methylxanthine [16]; Theophylline [16]; Cytidine [15]; Potassium borate [15]; p-Hydroxymercuribenzoate [15]; SDS [15]; Inorganic phosphate [13]; More [1, 15]

Cofactor(s)/prostethic group(s)
More (activity increased by addition of 2-mercaptoethanol or reduced glutathione) [20]

Metal compounds/salts
$MgCl_2$ (increases activity) [20]; More (no effect of divalent cations [8], no requirement for Mg^{2+} [18]) [8, 18]

Turnover number (min⁻¹)

Specific activity (U/mg)
1.5 [18]; 0.432 [1]; 55.0 [8]; More [6, 20]

K$_m$-value (mM)
0.80 (xanthosine) [1]; 0.83 (inosine) [1]; 2.5 (uridine) [15]; 3.0 (NAD$^+$) [8]; 2.9
(5'-AMP) [8]; 1.6 (inosine) [8]; 0.25 [13]; 4.1 (inosine) [20]; 7.5 (guanosine)
[20]; 9 (adenosine) [20]; 0.3 (inosine) [4]

pH-optimum
7–8 [4]; 3–4.0 [20]; 8.5 (uridine) [15]; 6.5 (inosine, uridine) [19]; 6.7
(guanosine) [6]; 4.0–4.5 (NAD) [8]; 7.5–8.5 (broad) [1]

pH-range
5–9.5 [15]; 5–8 [19]; 5–6 (no activity above 6) [20]

Temperature optimum (°C)
75 [8]; 55 [15]; 60 [20]

Temperature range (°C)
37–75 (higher activity at 37°C than at 75°C) [9]; 20–60 [15]; 20–70 (low activity at 20°C and 70°C) [20]

---

3 ENZYME STRUCTURE

Molecular weight
160000 (gel filtration, Vigna unguiculata) [1]
205000 (Leishmania donovani, gel filtration) [6]

Subunits
Oligomer (x × 30600, SDS-PAGE, Vigna unguiculata) [1]

Glycoprotein/Lipoprotein
–

---

4 ISOLATION/PREPARATION

Source organism
Vigna unguiculata [1]; Glycine max [2, 4]; Trypanosomatids [3]; Leishmania
donovani [6]; Pisum sativum [4]; Trypanosoma cruzi [21]; Neurospora
crassa [7, 13]; Aspergillus niger [8, 20]; Bacillus cereus [9, 12, 14]; Bacillus
subtilis [10, 11]; Crithidia fasciculata [15, 16]; Aspergillus foetidus [17]; Lactobacillus delbrueckii [19]; Yeast [18]; Spirochetes [22]

Source tissue
Nodules (N$_2$-fixing) [1, 2]; Epimastigotes [6]; Mycelium [8]; Cell culture [9];
Spores [12, 14]

Localisation in source
More (particulate fraction) [11]; Cell wall [10, 14]

Enzyme Handbook © Springer-Verlag Berlin Heidelberg 1991
Duplication, reproduction and storage in data banks are only
allowed with the prior permission of the publishers

Purification

Vigna unguiculata [1]; Leishmania donovani [6]; Aspergillus niger [8, 20]; Crithidia fasciculata [15]; Aspergillus foetidus [17]; Yeast [18]; Lactobacillus delbrueckii [19]

Crystallization

–

Cloned

–

Renaturated

–

5 STABILITY

pH

3.0–9.0 (90 minutes, 37°C) [8]

Temperature (°C)

60 (20 minutes, 90% loss of activity [15], 60 minutes complete loss of activity [20], 30 minutes, 30% loss of activity [20]) [15, 20]; 50 (20 minutes) [15]

Oxidation

Organic solvent

Octyl alcohol (resistant to) [12]

General stability information

Dialysis (loss of activity even against alkaline buffers) [19]; Lyophilization (stable [15], loss of activity even against alkaline buffers [19]) [15, 19]

Storage

Frozen, 6 months, 0.1 M phosphate buffer, pH 7.5 or 0.005 M pyrophosphate buffer, pH 8.6 [19]; –20°C, 30 days (95% of activity retained, purified enzyme) [6]; 4°C or frozen, for months [15]; 4°C, 0.1 M citrate buffer, pH 4.0, more than 3 months [20]; More [19, 20]

6 CROSSREFERENCES TO STRUCTURE DATABANKS

PIR/MIPS code

Brookhaven code

7 LITERATURE REFERENCES

[1] Atkins, C.A., Storer, P.J., Shelp, B.J.: J. Plant Physiol., 134, 447–452 (1989)
[2] Larsen, K., Jochimsen, B.U.: Plant Physiol., 85, 452–456 (1987)
[3] Hassan, H.F., Coombs, G.H.: Comp. Biochem. Physiol., 84B, 217–223 (1986)
[4] Christensen, T.M.I.E., Jochimsen, B.U.: Plant Physiol., 72, 56–59 (1985)
[5] Hassan, H.F., Coombs, G.H.: Exp. Parasitol., 59, 139–150 (1985)
[6] Koszalka, G.W., Krenitsky, T.A.: J. Biol. Chem., 254, 8185–8193 (1979)
[7] Mattoo, A.K., Dandekar, A.M., Trivedi, J.P., Majmudar, G.H., Patel, D.M.: Z. Allg. Mikrobiol., 19, 253–260 (1979)
[8] Kuwahara, M., Fujii, T.: Can. J. Biochem., 56, 345–348 (1978)
[9] Agrawal, P.K., Narayan, R., Gollakota, K.G.: Biochem. Biophys. Res. Commun., 63, 562–570 (1975)
[10] Halvorson, H., Church, B.: Bacteriol. Rev., 21, 112–131 (1957)
[11] Lawrence, N.L.: J. Bacteriol., 70, 577–582 (1955)
[12] Agrawal, P.K., Narayan, R., Gollakota, K.G.: Biochem. Biophys. Res. Commun., 60, 111–117 (1974)
[13] Mattoo, A.K., Shah, Z.M.: Z. Allg. Mikrobiol., 14, 581–591 (1974)
[14] Srivastava, O.P., Fitz-James, P.C.: Can. J. Microbiol., 27, 408–416 (1981)
[15] Dewey, V.C., Kidder, G.W.: Arch. Biochem. Biophys., 157, 380–387 (1973)
[16] Nolan, L.L., Kidder, G.W.: Biochem. Biophys. Res. Commun., 91, 253–262 (1979)
[17] Reese, E.T., Maguire, A.H.: J. Bacteriol., 96, 1696–1699 (1968)
[18] Heppel, L.A., Hilmoe, R.J.: J. Biol. Chem., 198, 683–694 (1952)
[19] Takagi, Y., Horecker, B.L.: J. Biol. Chem., 225, 77–86 (1956)
[20] Hassan, M.M., El-Zainy, T.A., Alla, A.M.: Egypt. J. Chem., 22, 189–196 (1979)
[21] Gutteridge, W.E., Davies, M.J.: FEBS Lett., 127, 211–214 (1981)
[22] Canale-Parola, E., Kidder, G.W.: J. Bacteriol., 152, 1105–1110 (1982)

Enzyme Handbook © Springer-Verlag Berlin Heidelberg 1991
Duplication, reproduction and storage in data banks are only
allowed with the prior permission of the publishers

1 NOMENCLATURE

EC number
3.2.2.2

Systematic name
Inosine ribohydrolase

Recommended name
Inosine nucleosidase

Synonymes
Inosinase
Nucleosidase, inosine
Inosine-guanosine nucleosidase [3]

CAS Reg. No.
9030-95-9

---

2 REACTION AND SPECIFICITY

Catalysed reaction
Inosine + H_2O →
→ hypoxanthine + D-ribose

Reaction type
N-Glycosyl bond hydrolysis

Natural substrates
Inosine + H_2O (purine catabolism [2], nucleoside recycling) [3]

Substrate spectrum
1 Inosine + H_2O [1–5]
2 Xanthosine + H_2O [2, 4]
3 Purine ribosides (nebularine) + H_2O [2]
4 6-Mercaptopurine riboside + H_2O [2, 3]
5 8-Azainosine + H_2O [2]
6 Adenosine + H_2O [2]
7 Guanosine + H_2O [2]

Product spectrum
1. Hypoxanthine + D-ribose
2. Xanthine + D-ribose [2]
3. Purine + D-ribose [2]
4. 6-Mercaptopurine + D-ribose [2]
5. 8-Azahypoxanthine + D-ribose [2]
6. Adenine + ribose [2]
7. Guanine + ribose [2]

Inhibitor(s)
ATP [1]; Adenine [1, 4]; Hypoxanthine [1, 4]; 6-Mercaptopurine riboside [3]; Nucleotides [4]; ADP [4]; ATP [4]; UTP [4]

Cofactor(s)/prostethic group(s)

Metal compounds/salts

Turnover number (min^{-1})

Specific activity (U/mg)
2.33 [4]; More [2]

K_m-value (mM)
0.065 (inosine) [2]; 0.0025 (inosine) [3]; 0.65 (inosine, pH 7.1) [4]; 0.85 (inosine, pH 9, 0) [4]; 1.2 (xanthosine, pH 7.1) [4]

pH-optimum
8 [2]; 7–9 (inosine) [4]; 7 (xanthosine) [4]

pH-range
5.5–9.75 (5.5: about 15% of maximal activity, 9.75: about 45% of maximal activity) [2]; 6–10 (inosine, 6 / 10: about 30% of maximal activity) [4]; 6–8.5 (xanthosine, 6: about 75% of maximal activity, 8.5: about 25% of maximal activity) [4]

Temperature optimum (°C)
25 (assay at) [2]; 37 (assay at) [4]

Temperature range (°C)

3 ENZYME STRUCTURE

Molecular weight
62000 (gel filtration, Lupinus luteus) [2]

Subunits

Glycoprotein/Lipoprotein

–

4 ISOLATION/PREPARATION

Source organism
Azotobacter vinelandii [1, 4]; Lupinus luteus [2]; Helianthus tuberosus [3]; E. coli [5]

Source tissue
Seed [2]; Seedlings [2]; Cell [4]

Localisation in source

Purification
Lupinus luteus (partial) [2]; Azotobacter vinelandii [4]

Crystallization

–

Cloned

–

Renaturated

–

5 STABILITY

pH

Temperature (°C)

Oxidation

Organic solvent

General stability information
Freezing and thawing (tolerated) [2]

Storage

Enzyme Handbook © Springer-Verlag Berlin Heidelberg 1991
Duplication, reproduction and storage in data banks are only
allowed with the prior permission of the publishers

6 CROSSREFERENCES TO STRUCTURE DATABANKS

PIR/MIPS code

Brookhaven code

7 LITERATURE REFERENCES

[1] Yoshino, M., Tsukada, T.: Int. J. Biochem., 20, 971–975 (1988)
[2] Guranowski, A.: Plant Physiol., 70, 344–349 (1982)
[3] Le Floc'h, F., Lafleuriel, J.: Phytochemistry, 20, 2127–2129 (1981)
[4] Yoshino, M., Tsukada, T., Tsushima, K.: Arch. Microbiol., 119, 59–64 (1978)
[5] Koch, A.L.: J. Biol. Chem., 223, 535–549 (1956)

1 NOMENCLATURE

EC number
3.2.2.3

Systematic name
Uridine ribohydrolase

Recommended name
Uridine nucleosidase

Synonymes
Nucleosidase, uridine
Uridine hydrolase [9]

CAS Reg. No.
9025-47-2

---

2 REACTION AND SPECIFICITY

Catalysed reaction
Uridine + H_2O →
→ uracil + D-ribose

Reaction type
N-Glycosyl bond hydrolysis

Natural substrates
Uridine + H_2O

Substrate spectrum
1 Uridine + H_2O [1–9]
2 5'-Methyluridine + H_2O [2, 6]
3 More (only active against uridine, narrow specificity) [1, 9]

Product spectrum
1 Uracil + D-ribose
2 Uracil + 5'-methylribose
3 ?

Enzyme Handbook © Springer-Verlag Berlin Heidelberg 1991
Duplication, reproduction and storage in data banks are only
allowed with the prior permission of the publishers

Inhibitor(s)
o-Phenanthroline [2]; EDTA [2, 3]; Ribosylthymidine [2, 4, 6];
Glucose-6-phosphate [2, 3, 4, 6]; Chelating agents [3]; Ribose [2–4, 6];
$CuCl_2$ [1–3]; $FeCl_2$ [1]; $MnCl_2$ [2, 3]; NiCl [2, 3]; NaCN [2, 3];
Ribulose-5-phosphate [2, 3]; Ribose-5-phosphate [2, 3]; $ZnCl_2$ [2, 3]; $CdCl_2$
[2, 3]; $CoCl_2$ [2, 3]; CMP [7]; GMP [7]; XMP [7]; GTP [7]; CTP [7]; IMP [7];
UMP [7]; p-Chloromercuribenzoate [9]; N-Ethylmaleimide [9]; Zn^{2+} [9];
Co^{2+} [9]; Ag^+ [9]; Hg^{2+} [9]; Cu^{2+} (+ ascorbic acid) [9]; More (inactivation
with proteinase A) [10]

Cofactor(s)/prostethic group(s)

Metal compounds/salts
Cu^{2+} (metal content: 1g-atom Cu^{2+} per mole enzyme [2, 3]); More (no metal ions required) [9]

Turnover number (min^{-1})

Specific activity (U/mg)
0.77 [1]; 1.56 [2, 4]

K_m-value (mM)
0.1 (uridine) [1]; 0.86 (uridine) [2, , 4, 6]; 1.66 (5-methyluridine) [2, 4, 6]; 1.0
(uridine) [9]; More [7, 9]

pH-optimum
7.5 [1]; 7.1 [2]; 7.4 [9]; 7.0–7.2 [4, 6]; 6.8–7.0 [7]

pH-range
6.0–9.0 (6.0: 39% of maximum activity, 9.0: 10% of maximum activity) [1];
6.0–8.0 (6.0, 8.0: about 35% of maximum activity) [9]

Temperature optimum (°C)
60 [1]; 45 [9]

Temperature range (°C)
30–60 (linear increase of activity between 30°C and 50°C, sharp drop of activity above 60°C) [1]; 30–60 (30°C : about 60% of maximal activity, 60°C: about 30% of maximal activity) [9]

3 ENZYME STRUCTURE

Molecular weight
55000 (gel filtration, Pisum sativum) [8]
30000 (gel filtration, yeast) [7]
117000 (barley, malt, gel filtration) [1]
32500 (gel filtration, yeast) [4]

Subunits
    Dimer (2 × 58000, SDS-PAGE, barley /malt [1], 2 × 17000, SDS-PAGE, yeast
    [4]) [1, 4]

Glycoprotein/Lipoprotein
    –

---

4 ISOLATION/PREPARATION

Source organism
    Barley (malt) [1]; Sacchyromyces cerevisiae [2–7, 10]; Pisum sativum [8];
    Phaseolus radiatus [9]

Source tissue
    Seeds [8]; Seedlings [9]

Localisation in source

Purification
    Barley (malt) [1]; Saccharomyces cerevisiae [2, 5, 6, 7]; Phaseolus radiatus
    [9]

Crystallization
    –

Cloned
    –

Renaturated
    –

---

5 STABILITY

pH
    5–9 (4°C, loss of activity below pH 5.0) [1]

Temperature (°C)
    55 (10 minutes) [7]; 58 (10 minutes, 40% loss of activity) [7]; 50 (15 minutes,
    absence of substrate) [1]; 60 (15 minutes, absence of substrate, 50% loss
    of activity) [1]; 65 (10 minutes, 100% loss of activity) [7]

Oxidation

Organic solvent

General stability information
    Dialysis (retained 95% of original activity after dialysis against 0.01 M
    Tris-HCl buffer, pH 7.5, 24 hours) [9]

Enzyme Handbook © Springer-Verlag Berlin Heidelberg 1991
Duplication, reproduction and storage in data banks are only
allowed with the prior permission of the publishers

Storage

4°C, 100 mM phosphate buffer, pH 7.2, 15 days [2, 6]; –20°C, several months [2]

---

6 CROSSREFERENCES TO STRUCTURE DATABANKS

PIR/MIPS code

Brookhaven code

---

7 LITERATURE REFERENCES

[1] Lee, W.J.: J. Am. Soc. Brew. Chem., 45, 131–135 (1987)

[2] Magni, G.: Methods Enzymol., 51, 290–296 (1978) (Review)

[3] Magni, G., Natalini, P., Ruggieri, S., Vita, A.: Biochem. Biophys. Res. Commun., 69, 724–730 (1976)

[4] Vita, A., Natalini, P., Ipata, P.L., Magni, G.: Boll. Soc. Ital. Biol. Sper., 50, 1077–1083 (1974)

[5] Corradetti, E., Natalini, P., Ipata, P.L., Magni, G.: Boll. Soc. Ital. Biol. Sper., 50, 1070–1076 (1974)

[6] Magni, G., Fioretti, E., Ipata, P.L., Natalini, P.: J. Biol. Chem., 250, 9–13 (1975)

[7] Raggi-Ranieri, M., Ipata, P.L.: Ital. J. Biochem., 20, 27–43 (1971)

[8] Murray, M.G., Ross, C.: Phytochemistry, 10, 2645–2648 (1971)

[9] Achar, B.S., Vaidyanathan, C.S.: Arch. Biochem. Biophys., 119, 356–362 (1967)

[10] Magni, G., Natalini, P., Santarelli, I., Ruggieri, S., Vita, A.: DHEW Publ. (NIH) (U.S.) NIH-79–1591, Ltd Proteolysis Microorg., 87–96 (1979)

1 NOMENCLATURE

EC number
3.2.2.4

Systematic name
AMP phosphoribohydrolase

Recommended name
AMP nucleosidase

Synonymes
Nucleosidase, adenylate
Adenylate nucleosidase
Adenosine monophosphate nucleosidase

CAS Reg. No.
9025-45-0

---

2 REACTION AND SPECIFICITY

Catalysed reaction
AMP + H_2O →
→ adenine + D-ribose 5-phosphate (mechanism [5, 6, 7, 8, 17])

Reaction type
N-Glycosyl bond hydrolysis

Natural substrates
AMP + H_2O (adenylate metabolism [10, 11], enzyme is responsible for adenylate energy charge and adenylate pool size [12], conversion of adenine nucleotides to inosine or guanine nucleotides [13], major enzyme of AMP catabolism in E. coli [15], regulates degradation and interconversion of purine nucleotides in Azotobacter vinelandii [27]) [10, 11, 12, 13, 15, 27]

Substrate spectrum
1 AMP + H_2O [1–34]
2 2-Amino AMP + H_2O (at a low rate) [22]
3 8-Aza AMP + H_2O (at a low rate) [22]
4 2'-Deoxy-5'-AMP + H_2O (at a low rate) [22]
5 3'-Deoxy-5'-AMP + H_2O (at a low rate) [22]
6 NMN + H_2O (at a low rate) [22]
7 More [25]

Enzyme Handbook © Springer-Verlag Berlin Heidelberg 1991
Duplication, reproduction and storage in data banks are only
allowed with the prior permission of the publishers

Product spectrum
1 Adenine + D-ribose 5-phosphate
2 ?
3 ?
4 ?
5 ?
6 ?
7 ?

Inhibitor(s)
IMP (poor inhibitor [22]) [22, 25, 27]; GMP (poor inhibitor [22]) [22, 27]; UMP (poor inhibitor) [22]; CMP (poor inhibitor) [22]; 2'-AMP (poor inhibitor) [22]; 3'-AMP (poor inhibitor) [22]; Adenosine (poor inhibitor) [22]; Inosine (poor inhibitor) [22]; Tubercidin 5'-phosphate [22, 30]; 8-BrAMP [22]; 8-Azido AMP [22]; 4-Aminopyrazolo(3, 4-d) pyrimidine-1-ribonucleotide [22]; 8-[[[(2, 2, 5, 5-Tetramethyl-1-oxy-3-pyrrolidinyl) car-bamoyl]methyl]thio]AMP [22, 23]; Mn^{2+} [25]; Orthophosphate (protection by substrate and allosteric activator [29]) [14, 15, 25, 27, 29, 31, 34]; Ar-senate [25]; Flavodoxin [26]; Flavinmononucleotide [26]; Citrate [27]; NH_3 [27]; AMP (substrate inhibition at high concentration [31]) [30, 31]; Formycin 5'-monophosphate [2, 4, 14, 15, 22]; Adenine [7]; N^6-Methyladenosine [16]; Transition metals (high concentration, free metal cation) [18]

Cofactor(s)/prostethic group(s)
ATP ($MgATP^{2-}$: alosteric activator [6, 7, 8, 14, 15, 16, 17], absolute require-ment for ATP and Mg^{2+} [27], $MnATP^{2-}$: allosteric activator [6]) [6, 7, 8, 14, 15, 16, 17]

Metal compounds/salts
Mg^{2+} ($MgATP^{2-}$ complex: allosteric activator [16, 17, 19, 21, 22, 30, 31], MgATP activates [6, 7, 8, 14, 15, 16], completely inactive in absence of Mg^{2+} [25], absolte requirement for ATP and Mg^{2+} [27], $MgATP^{2-}$: most effective activator, ATP^{4-} or magnesium pyrophosphate less effective [31]) [6, 7, 8, 14, 15, 16, 17, 19, 21, 22, 25, 27, 30, 31]; Mn^{2+} ($MnATP^{2-}$: allosteric acivator [17], replaces Mg^{2+} [25]) [17, 25]; Rb^+ (activates) [19]; Cs^+ (activates) [19]; K^+ (activates) [19]; Na^+ (activates) [19]; Li^+ (activates) [19]; Ca^{2+} (replaces Mg^{2+}) [25]; Co^{2+} (replaces Mg^{2+}) [25]; More (all alkaline earth metal-ATP complexes are essential activators of the enzyme, free alkaline earth activate in allosteric manner [18], transition metal-ATP complexes also activate [18], Rb^+, Cs^+: most effective monovalent cation activators, followed by K^+, Na^+, NH_4^+, Li^+ [19]) [18, 19]

Turnover number (min^{-1})

Specific activity (U/mg)
0.3 (mutant enzyme) [14]; 2.0 [21]; 17 (E. coli) [24]; 34 (Azotobacter vinelandii) [24, 30]; More [27]

K_m-value (mM)
0.160 (AMP, mutant enzyme) [14]; 0.120 ($MgATP^{2-}$) [21, 22]; 0.084 (8-aza AMP) [22]; 0.250 (2-amino AMP) [22]; 0.240 (3'-deoxy-5'-AMP) [22]; 0.074 (dAMP) [22]

pH-optimum
7.8 [7, 25]

pH-range
6–9 (6: about 20% of activity maximum, 9: about 50% of activity maximum) [27]

Temperature optimum (°C)
37 (assay at) [27]; More (thermodynamics) [7]

Temperature range (°C)
More (thermodynamics) [7]

---

3 ENZYME STRUCTURE

Molecular weight
320000 (Azotobacter vinelandii, gel filtration, mutant enzyme [14], equilibrium sedimention [24]) [14, 24]
280000 (E. coli, gel filtration) [15]
360000 (gel filtration, Azotobacter vinelandii) [21]
104000 (sedimentation equilibrium analysis, Azotobacter vinelandii) [32]
370000 (gel filtration, sedimentation velocity experiments, Azotobacter vinelandii) [33]
325000 (Archibald approach to equilibrium method, Azotobacter vinelandii) [33]
More [1, 4]

Subunits
Octamer (active form is an octamer, Azotobacter vinelandii) [32]
Tetramer (4 × 26000, gel filtration and sedimentation equilibrium analysis in guanidine hydrochloride containing dithiothreitol, Azotobacter vinelandii) [32]
Oligomer (asymetric subunit of Azotobacter vinelandii enzyme, 9 or 12 subunits each 54000) [1]

Enzyme Handbook © Springer-Verlag Berlin Heidelberg 1991
Duplication, reproduction and storage in data banks are only allowed with the prior permission of the publishers

Hexamer (Azotobacter vinelandii, mutant enzyme, 6 × 54000, SDS-PAGE) [14]
Trimer (asymetric subunit of E. coli enzyme, 3 × 53800) [1]
Oligomer (denaturing gel electrophoresis, E. coli, x × 52000) [15]
Hexamer (SDS-PAGE, A. vinelandii, 6 × 54000) [24]
More (protomer: 60000, dimer: 120000, tetramer: 240000, dimer and tetramer are interconvertible, association into tetramer in presence of AMP, ATP, GMP, IMP, ITP, divalent or trivalent anions [28]) [4, 28, 33]

Glycoprotein/Lipoprotein
–

---

4 ISOLATION/PREPARATION

Source organism
E. coli [1, 4, 9, 15, 16, 24]; Azotobacter vinelandii (mutant enzyme [14]) [1, 6, 7, 8, 9, 11, 12, 13, 14, 17, 18, 19, 20, 21, 22, 24, 25, 26, 27, 28, 30, 31, 32, 33, 34]; Azotobacter beijerinckii [10]

Source tissue

Localisation in source

Purification
Azotobacter vinelandii (mutant enzyme [14]) [14, 21, 24, 25, 27, 33, 34]; E. coli [24]

Crystallization
(E. coli [1, 4], Azotobacter vinelandii [1, 24, 33]) [1, 4 , 24, 33]

Cloned
[3]

Renaturated
–

---

5 STABILITY

pH

Temperature (°C)
60 (pH 6.5, stable) [25]; 70 (pH 6.5, rapidly destroyed) [25]

Oxidation

Organic solvent

General stability information

Divalent or trivalent anions (e.g. SO_4^{2-}, succinate or citrate stabilize) [27, 34]; K_2HPO_4 (stablize) [3]; Inorganic salts (stabilize native enzyme) [21]; Substrate (stabilizes) [21]; $MgATP^{2-}$ (stabilizes) [21]; More [33]

Storage

−70°C, dry ice-ethanol mixture (stable for at least 1 year) [24, 33]; 4°C, ammonium sulfate solution (stable for at least 1 year) [24, 33]; −10°C, 2 months (no loss of activity) [25]; Frozen (50% loss of activity overnight) [34]; 4°C or at rooom temperature, K_2SO_4 (stable) [34]

---

6 CROSSREFERENCES TO STRUCTURE DATABANKS

PIR/MIPS code
A33364 (Escherichia coli)

Brookhaven code

---

7 LITERATURE REFERENCES

[1] Giranda, V.L., Berman, H.M., Schramm, V.L.: J. Biol. Chem., 264, 15674–15680 (1989)
[2] Giranda, V.L., Berman, H.M., Schramm, V.L.: Biochemistry, 27, 5813–5818 (1988)
[3] Leung, H.B., Schramm, V.L.: J. Biol. Chem., 259, 6972–6978 (1984)
[4] Giranda, V.L., Berman, H.M., Schramm, V.L.: J. Biol. Chem., 261, 15307–15309 (1986)
[5] Parkin, D.W., Schramm, V.L.: Biochemistry, 26, 913–920 (1987)
[6] Mentch, F., Parkin, D.W., Schramm, V.L.: Biochemistry, 26, 921–930 (1987)
[7] DeWolf, W.E., Emig, F.A., Schramm, V.L.: Biochemistry, 25, 4132–4140 (1986)
[8] Skoog, M.T.: J. Biol. Chem., 261, 4451–4459 (1986)
[9] Parkin, D.W., Schramm, V.L.: J. Biol. Chem., 259, 9418–9425 (1984)
[10] Marriott, I.D., Dawes, E.A., Rowley, B.I.: J. Gen. Microbiol., 125, 375–382 (1981)
[11] Yoshino, M., Tsukada, T., Murakami, K., Tushima, K.: Arch. Microbiol., 128, 222–227 (1980)
[12] Schramm, V.L., Leung, H.: J. Biol. Chem., 248, 8313–8315 (1973)
[13] Yoshino, M., Ogasawara, N.: J. Biochem., 72, 223–233 (1972)
[14] Leung, H.B., Schramm, V.L.: J. Biol. Chem., 256, 12823–12829 (1981)
[15] Leung, H.B., Schramm, V.L.: J. Biol. Chem., 255, 10867–10874 (1980)
[16] DeWolf, W.E., Markham, G.D., Schramm, V.L.: J. Biol. Chem., 255, 8210–8215 (1980)
[17] Schramm, V.L., Reed, G.H.: J. Biol. Chem., 255, 5795–5801 (1980)
[18] Murakami, K., Yoshino, M.: Biochim. Biophys. Acta, 613, 153–159 (1980)
[19] Yoshino, M., Murakami, K., Tsushima, K.: Biochim. Biophys. Acta, 570, 118–123 (1979)

[20] Yoshino, M., Murakami, K., Tsushima, K.: Experientia, 35, 578–579 (1979)
[21] Schramm, V.L., Hochstein, L.I.: Biochemistry, 10, 3411–3417 (1971)
[22] DeWolf, W.E., Fullin, F.A., Schramm, V.L.: J. Biol. Chem., 254, 10868–10875 (1979)
[23] DeWolf, W.E., Schramm, V.L.: J. Biol. Chem., 254, 6215–6217 (1979)
[24] Schramm, V.L., Leung, H.B.: Methods Enzymol., 51, 263–271 (1978) (Review)
[25] Hurwitz, J., Heppel, L.A., Horecker, B.L.: J. Biol. Chem., 226, 525–540 (1957)
[26] Yoshino, M., Murakami, K., Tsushima, K.: J. Biochem., 80, 839–843 (1976)
[27] Yoshino, M.: J. Biochem., 68, 321–329 (1970)
[28] Ogasawara, N., Yoshino, M., Asai, J.: J. Biochem., 68, 331–340 (1970)
[29] Schramm, V.L., Fullin, F.A.: J. Biol. Chem., 253, 2161–2167 (1978)
[30] Schramm, V.L.: J. Biol. Chem., 251, 3417–3424 (1976)
[31] Schramm, V.L.: J. Biol. Chem., 249, 1729–1736 (1974)
[32] Yoshino, M., Takagi, T.: J. Biochem., 74, 1151–1156 (1973)
[33] Schramm, V.L., Hochstein, L.I.: Biochemistry, 11, 2777–2783 (1972)
[34] Yoshino, M., Ogasawara, N., Suzuki, N., Kotake, Y.: Biochim. Biophys. Acta, 146, 620–622 (1967)